Organ Transplantation
2nd Edition

Frank P. Stuart, M.D.
Feinberg School of Medicine
Northwestern University
Northwestern Memorial Hospital
Chicago, Illinois, U.S.A.

Michael M. Abecassis, M.D., M.B.A.
Feinberg School of Medicine
Northwestern University
Northwestern Memorial Hospital
Chicago, Illinois, U.S.A.

Dixon B. Kaufman, M.D., Ph.D.
Feinberg School of Medicine
Northwestern University
Northwestern Memorial Hospital
Chicago, Illinois, U.S.A.

LANDES
BIOSCIENCE

GEORGETOWN,
TEXAS U.S.A.

Organ Transplantation
2nd Edition
VADEMECUM
LANDES BIOSCIENCE
Georgetown, Texas, U.S.A.

Copyright ©2003 Landes Bioscience

Printed in the U.S.A.

Please address all inquiries to the Publisher:
Landes Bioscience, 810 S. Church Street, Georgetown, Texas, U.S.A. 78626
Phone: 512/ 863 7762; FAX: 512/ 863 0081

ISBN: 1-57059-675-1

While the authors, editors, sponsor and publisher believe that drug selection and dosage and the specifications and usage of equipment and devices, as set forth in this book, are in accord with current recommendations and practice at the time of publication, they make no warranty, expressed or implied, with respect to material described in this book. In view of the ongoing research, equipment development, changes in governmental regulations and the rapid accumulation of information relating to the biomedical sciences, the reader is urged to carefully review and evaluate the information provided herein.

Library of Congress Cataloging-in-Publication

CIP applied for but not recieved at time of publication.

Contents

Editors

Frank P. Stuart, M.D.
James Roscoe Miller Professor of Surgery
Feinberg School of Medicine
Northwestern University
Chief, Division of Organ Transplantation
Northwestern Memorial Hospital
Chicago, Illinois, U.S.A.
Chapters 2, 3, 19, 20, Appendix I

Michael M. Abecassis, M.D., M.B.A.
Associate Professor of Surgery
Feinberg School of Medicine
Northwestern University
Director, Liver Transplantation
Northwestern Memorial Hospital
Chicago, Illinois, U.S.A.
Chapters 3, 9, 19, 20

Dixon B. Kaufman, M.D., Ph.D.
Associate Professor
and Vice Chair of Surgery
Feinberg School of Medicine
Northwestern University
Chicago, Illinois, U.S.A.
Chapters 6, 7, 8, 19, 20

Contributors

P. Stephen Almond, M. D.
Assistant Professor of Surgery
Feinberg School of Medicine
Northwestern University
Children's Memorial Hospital
Chicago, Illinois, U.S.A.
Chapter 13 (Kidney)

Estella M. Alonso, M.D.
Associate Professor, Pediatrics
Feinberg School of Medicine
Northwestern University
Director of Hepatology and Liver
    Transplantation
Children's Memorial Hospital
Chicago, Illinois, U.S.A.
Chapter 13 (Liver)

Jarold A. Anderson
Chief Executive Officer
Regional Organ Bank of Illinois
Chicago, Illinois, U.S.A.
Chapter 4

Vincent T. Armenti, M.D., Ph.D.
Department of Surgery
Thomas Jefferson University Hospital
Philadelphia, Pennsylvania, U.S.A.
Chapter 21

Carl Lewis Backer, M.D.
A.C. Buehler Professor of Surgery
Surgical Director, Heart Transplantation
Children's Memorial Hospital
Feinberg School of Medicine
Professor of Surgery
Northwestern University
Division of Pediatric Cardiovascular-
   Thoracic Surgery
Chicago, Illinois, U.S.A.
Chapter 13 (Heart)

Giacomo P. Basadonna, M.D., Ph.D.
Professor of Surgery
University of Massachusetts School
   of Medicine
Chief of Transplantation
Worcester, Massachusetts, U.S.A.
Essay 9

Thomas D. Batiuk, M.D.
Assistant Professor of Medicine
Division of Nephrology
Indiana University
Indianapolis, Indiana, U.S.A.
Chapter 1

Andres T. Blei, M.D.
Professor of Medicine
Division of Gastroenterology
   and Hepatology
Professor of Surgery
Division of Transplantation
Feinberg School of Medicine
Northwestern University
Northwestern Memorial Hospital
Chicago, Illinois, U.S.A.
Chapter 9

Matthew Blum, M.D.
Assistant Professor of Surgery
Division of Cardiothoracic Surgery
Northwestern Memorial Hospital
Chicago, Illinois, U.S.A.
Chapter 12

Robert E. Brannigan, M.D.
Assistant Professor of Surgery
Department of Urology
Feinberg School of Medicine
Northwestern University
Chicago, Illinois, U.S.A.
Essay 4

Ronald W. Busuttil, M.D., Ph.D.
Professor and Chief of Liver
   Transplantation
University of California in Los Angeles
   Medical Center
Los Angeles, California, U.S.A.
Essay 6

Patricia M. Campbell, M.B., Ch.B.
Assistant Professor of Medicine
Division of Nephrology
University of Alberta
Edmonton, Alberta, Canada
Chapter 1

Cristine J. Cooper, Ph.D.
Department of Immunology
University of Washington
Seattle, Washington, U.S.A.
Essay 2

Lisa A. Coscia, R.N., BSN
Department of Surgery
Thomas Jefferson University Hospital
Philadelphia, Pennsylvania, U.S.A.
Chapter 21

Anthony M. D'Alessandro, M.D.
Professor of Surgery
Division of Organ Transplantation
University of Wisconsin
   Medical School
Madison, Wisconsin, U.S.A.
Chapter 5

John M. Davison, M.D.
Department of Surgery
Thomas Jefferson University Hospital
Philadelphia, Pennsylvania, U.S.A.
Chapter 21

Alberto de Hoyos, M.D.
Chief Resident
Cardiothoracic Surgery
Northwestern Hospital
Chicago, Illinois, U.S.A.
Chapter 12

Andre DeWolf, M.D.
Professor of Clinical Anesthesiology
Feinberg School of Medicine
Northwestern University
Chief, Transplant Anesthesiology
Northwestern Memorial Hospital
Chicago, Illinois, U.S.A.
Chapter 14

Pamela J. Fink, M.D.
Department of Immunology
Universityof Washington
Seattle, Washington, U.S.A.
Essay 2

Steven Flamm, M.D.
Assistant Professor of Medicine
Section of Gastroenterology
    and Hepatology
Assistant Professor of Surgery
Division of Transplantation
Feinberg School of Medicine
Northwestern University
Northwestern Memorial Hospital
Chicago, Illinois, U.S.A.
Chapter 9

John E. Franklin, M.D., M.Sc.
Associate Professor of Psychiatry
Associate Dean
Feinberg School of Medicine
Northwestern University
Northwestern Memorial Hospital
Chicago, Illinois, U.S.A.
Chapter 15

Amy L. Friedman, M.D.
Assistant Professor of Surgery
Chief, Liver Transplantation Services
Yale University School of Medicine
New Haven, Connecticut, U.S.A.
Essay 9

Jonathan P. Fryer, M.D.
Assistant Professor of Surgery
Feinberg School of Medicine
Northwestern University
Director, Small Bowel
    Transplantation Program
Northwestern Memorial Hospital
Chicago, Illinois, U.S.A.
Chapters 9, 10, Essay 12

David A. Fullerton, M.D.
Professor of Surgery
Feinberg School of Medicine
Northwestern University
Chief, Division of Cardiothoracic Surgery
Northwestern Memorial Hospital
Chicago, Illinois, U.S.A.
Chapters 11, 13 (Lung)

William J. Gaughan, M.D.
Department of Surgery
Thomas Jefferson University Hospital
Philadelphia, Pennsylvania, U.S.A.
Chapter 21

Nelson Goes, M.D.
Research Fellow
Harvard University
Boston, Massachusetts, U.S.A.
Chapter 1

Luis Graca, Ph.D.
Sir William Dunn School of Pathology
University of Oxford
Oxford, U.K.
Essay 1

Stephen D. Haid
Independent Consultant
Haid Enterprises
McKinney, Texas, U.S.A.
Chapter 4

Philip F. Halloran, M.D.
Director, Division of Nephrology
    and Immunology
Department of Medicine
University of Alberta
Edmonton, Alberta, Canada
Chapter 1

Peter S. Heeger, M.D.
Department of Immunology
and the Urologic Institute
Cleveland Clinic Foundation
Cleveland, Ohio, U.S.A.
Essay 3

Bernhard J. Hering, M.D.
Feinberg School of Medicine
Northwestern University
Northwestern Memorial Hospital
Chicago, Illinois, U.S.A.
Chapter 8

Keith Horvath, M.D.
Assistant Professor of Surgery
Division of Cardiothoracic Surgery
Feinberg School of Medicine
Northwestern University
Chicago, Illinois, U.S.A.
Chapter 11

Abhinav Humar, M.D.
Department of Surgery
University of Minnesota
Minneapolis, Minnesota, U.S.A.
Essay 8

Peter S. Hurst, D.D.S.
Clinical Professor
Feinberg School of Medicine
Northwestern University
Chairman, Department of Dentistry
Director, Hospital Dental Center
Northwestern Memorial Hospital
Chicago, Illinois, U.S.A.
Essay 10

Yoogoo Kang, M.D.
Chairman, Merryl and Sam Israel
   Professor
Department of Anesthesiology
Tulane University School of Medicine
New Orleans, Louisiana, U.S.A.
Chapter 14

Bruce Kaplan, M.D.
Professor of Medicine
   and Pharmacology
Central Florida Kidney Foundation
   Eminent Scholar Chair
Medical Director Kidney
   and Pancreas Transplant
University of Florida Gainesville
Gainesville, Florida, U.S.A.
Chapter 18

Simon Kimm
Medical Student
Feinberg School of Medicine
Northwestern University
Chicago, Illinois, U.S.A.
Chapters 6-8

James Kisthard
Organ Recovery Specialist
Life Center Northwest
Seattle, Washington, U.S.A.
Chapter 4

Alan Koffron, M.D.
Feinberg School of Medicine
Northwestern University
Northwestern Memorial Hospital
Chicago, Illinois, U.S.A.
Chapter 9

Susan M. Lerner, M.D.
Liver Transplant Fellow
University of California at Los Angeles
   Medical Center
Los Angeles, California, U.S.A.
Essay 6

Joseph R. Leventhal, M.D., Ph.D.
Assistant Professor of Surgery
Feinberg School of Medicine
Northwestern University
Division of Organ Transplantation
Northwestern Memorial Hospital
Chicago, Illinois, U.S.A.
Chapter 17, Essays 11 and 12

Marc I. Lorber, M.D.
Professor of Surgery
Chief, Organ Transplant
   and Immunology
Yale University School of Medicine
New Haven, Connecticut, U.S.A.
Essay 9

James F. Markmann, M.D., Ph.D.
Associate Professor of Surgery
University of Pennsylvania
Multi-Organ Transplantation Program
Hospital of the University
   of Pennsylvania
Philadelphia, Pennsylvania, U.S.A.
Essay 6

Arthur J. Matas, M.D.
Professor of Surgery
Director of Renal Transplantation
University of Minnesota
Minneapolis, Minnesota, U.S.A.
Essays 7, 8

Constantine Mavroudis, M.D.
Willis J. Potts Professor
Surgeon-in-Chief
Children's Memorial Hospital
Professor of Surgery
Northwestern University
Feinberg School of Medicine
Chicago, Illinois, U.S.A.
Chapter 13 (Heart)

Herwig-Ulf Meier-Kriesche, M.D.
Associate Professor of Medicine
Clinical Director, Renal Transplantation
University of Florida
Gainesville, Florida, U.S.A.
Chapter 18

Michael J. Moritz, M.D.
Department of Surgery
Thomas Jefferson University Hospital
Philadelphia, Pennsylvania, U.S.A.
Chapter 21

Robert Nadler, M.D.
Assistant Professor of Urology
Head, Section of Endurology
  and Stone Disease
Feinberg School of Medicine
Northwestern University
Chicago, Illinois, U.S.A.
Essay 4

Thomas E. Nevins, M.D.
University of Minnesota
Minneapolis, Minnesota, U.S.A.
Essays 7

Elfriede Pahl, M.D.
Associate Professor of Pediatrics
Northwestern University Medical School
Medical Director, Cardiac Transplantation
Division of Cardiology
Children's Memorial Hospital
Chicago, Illinois, U.S.A.
Chapter 13 (Heart)

Roslyn M. Paine, A.M., L.S.W.
Transplant Social Worker
Northwestern Memorial Hospital
Chicago, Illinois, U.S.A.
Chapter 15

Israel Penn, M.D. (deceased)
Professor of Surgery
University of Cincinnati
Cincinnati, Ohio, U.S.A.
Essay 5

Paige Porrett, M.D.
University of Pennsylvania
Multi-Organ Transplantation Program
Hospital of the University
  of Pennsylvania
Philadelphia, Pennsylvania, U.S.A.
Essay 6

John S. Radomski, M.D.
Department of Surgery
Thomas Jefferson University Hospital
Philadelphia, Pennsylvania, U.S.A.
Chapter 21

William A. Schlueter, M.D.
Assistant Professor of Medicine
Feinberg School of Medicine
Northwestern University
Director, Outpatient Dialysis
Chicago, Illinois, U.S.A.
Chapter 17

Laurance Sherman, M.D., J.D.
Professor Emeritus
Department of Pathology
Feinberg School of Medicine
Northwestern University
Chicago, Illinois, U.S.A.
Chapter 14

James H. Southard, Ph.D.
Professor of Surgery
University of Wisconsin Medical School
Madison, Wisconsin, U.S.A.
Chapter 5

Valentina Stosor, M.D.
Assistant Professor of Medicine
Division of Infectious Diseases
Feinberg School of Medicine
Northwestern University
Chicago, Illinois, U.S.A.
Chapter 16

Riccardo A. Superina, M.D.
Professor of Surgery
Feinberg School of Medicine
Northwestern University
Director of Transplant Surgery
Children's Memorial Hospital
Chicago, Illinois, U.S.A.
Chapter 13 (Liver)

Hermann Waldmann, Ph.D.
Sir William Dunn School of Pathology
University of Oxford
Oxford, U.K.
Essay 1

Gary A. Wilson, M.D.
Department of Surgery
Thomas Jefferson University Hospital
Philadelphia, Pennsylvania, U.S.A.
Chapter 21

Preface

Organ transplantation is increasingly complex and at the same time increasingly effective. The lengthening waiting list for cadaver organs now exceeds the supply several-fold. Despite its high profile, not more than 25,000 organs are transplanted each year in the United States with a population of 270 million. Most practicing physicians encounter only a few transplant recipients during a year of practice. This volume was written as a quick, but comprehensive, reference for medical students, residents, fellows, nurses, and practicing physicians who interface intermittently with recipients and transplant teams. It contains twenty-one chapters and twelve essays; together they present the standard of practice and also controversial issues such as the ethical dilemma of long waiting lists, noncompliance with long-term immunosuppression, the relationship between acute and chronic rejection, the living organ donor, the older cadaver donor, laparoscopic nephrectomy, retransplantation, organ banks and the national transplant network's criteria for allocating organs to potential recipients, and the promise of xenotransplantation. Appendix I includes detailed information about immunosuppressive drugs.

We thank colleagues who have so generously shared their wisdom and insights in this volume and we solicit comments from the reader about improving content and presentation of the material.

Frank P. Stuart, M.D.

Michael M. Abecassis, M.D., M.B.A.

Dixon B. Kaufman, M.D., Ph.D.

ACKNOWLEDGEMENT

The editors recognize and would like to thank Carolyn E. Johnson and Sue Benning for their essential skill and advice in executing the second edition of *Organ Transplantation*.

Immunologic Concepts

Philip F. Halloran, Thomas D. Batiuk, Nelson Goes
and Patricia M. Campbell

INTRODUCTION

The dream of replacing a diseased human organ with one from a dead person is ancient: legend states that Saints Cosmas and Damien in the fourth century A.D. miraculously transplanted a leg from a dead man. Such a creature would be a chimera, named after the "mingled monster" of Homer's Iliad. The scientific study of the biology of transplanting tissue dates to the first years of this century, when Little and Tyzzer[1] defined the Laws of Transplantation, paraphrased as: "isografts succeed; allografts are rejected." The clinical practice of transplantation is governed by these laws. This chapter introduces the immunologic events in transplantation, and in particular the molecular basis of these events, to be supplemented by reviews.[2-14] Table 1.1 summarizes our approach, and Table 1.2 presents some useful terms. A recurrent theme is the "allo" relationship, which describes the relationship between two members of the same species who are not genetically identical. Thus we can describe alloantigens, allografts, and alloantibody.

THE FATE OF ALLOGRAFTS

Allografts are usually rejected in one of three patterns: acute rejection; accelerated and hyperacute rejection; and chronic rejection.

ACUTE REJECTION

Around 5-7 days the tissue begins to manifest signs of two processes: inflammation and specific cell injury. The inflammation is manifested by infiltration with mononuclear cells, accompanied by edema and reduced blood flow; specific destruction of parenchymal and endothelial cells by infiltrating lymphocytes, coupled with decreased perfusion, cause a rapid loss of function. Destruction of blood vessels frequently leads to late infarction of some or all of the tissue.

ACCELERATED AND HYPERACUTE REJECTION

If certain organs, particularly kidneys, are transplanted into a recipient who has high levels of preformed antibodies against donor alloantigens of the graft endothelium, particularly HLA class I (see below) or ABO blood group antigens, *hyperacute rejection* follows. The antibodies on the endothelium fix complement, which attracts polymorphs, and destroy the endothelium within hours or even minutes. Hyperacute rejection is usually prevented by "crossmatching", i.e., testing the recipient's serum for complement-dependent antibodies against donor lymphocytes.[15]

Organ Transplantation, 2nd edition, edited by Frank P. Stuart, Michael M. Abecassis and Dixon B. Kaufman. ©2003 Landes Bioscience.

Newly synthesized class II molecules in the ER cannot bind peptide because a portion of the *invariant chain* occupies the peptide groove.[39] Invariant chain guides the class II from the ER through the Golgi apparatus to an acidic compartment of endosomes.[40-42] Proteins taken into the cell by endocytosis enter the acidic endosome and are broken down by proteases. Invariant chain protecting the class II groove is also degraded in the endosome,[43] freeing the groove to bind peptide. Peptides 13-25 amino acids in length occupy the grooves of class II molecules.[43] Class II molecules may "select" peptides by protecting fragments of larger proteins from degradation.[44] A larger peptide bound in the class II groove could hang out the ends and the exposed ends may be "trimmed".[45] After peptide binding, class II is stable and is transported to the cell surface.

Endocytic vesicles from the cell surface sample the external environment and also receive self membrane-bound molecules. Thus DR1 molecules often contain peptides from self MHC class I and II.[43,46]

In B cells antigen binds to the B cell receptor and is internalized into the endosome. Such antigenic proteins are broken down into peptides, bound by class II, and exported to the cell surface to permit T cells to help the B cell to make an antibody response (see below). In addition, the endosome may receive cytosol-derived peptides transported via chaperones of the heat shock protein 70 (hsp70) family.[47,48] This enables class II to present some endogenously derived peptides.[34]

PROTEASOMES AND PEPTIDE TRANSPORTERS

To permit cytosolic peptides to be displayed by class I molecules, proteins from the cytosol must be broken down to short peptides, and the peptides must have access to class I grooves in the ER. This requires mechanisms to degrade proteins and to transport the peptides into the ER. Peptides are generated by *proteasomes*, large cytoplasmic complexes containing protease activities. Genes for two proteasome components are located in the class II region, although their function is to assist class I products.[49-51] The proteasome genes are termed LMP2 and LMP7 (large multifunctional protease genes). They are polymorphic subunits of the proteasome complex which lyses cytoplasmic proteins.[52-54] The transporters (called TAPs or *transporters associated with antigen processing*) are TAP1 and TAP2.[51,55-64] The transporters are located in the membrane of the ER. Polymorphisms occur in the TAP genes but the importance of these is unknown.

Thus cytosolic proteins are digested into peptides by proteosomes, access the ER via transporters, and engage the groove. The LMP and TAP genes, like the class I heavy chain and the class II genes, are upregulated by IFN-γ.[49]

ANTIGEN RECOGNITION MOLECULES

A specialized Ig domain—the variable or V domain—is found at the N terminal of Ig light (L) and heavy (H) chains, all TCR chains, and CD4, CD8, and ICAM-1 molecules. In V domains, loop 3 between strand C and strand D forms two more β strands C' and C", and joins β strands C, F, and G to form a five-stranded β-sheet (C", C', C, F, G) (Fig. 1.1).

In antigen recognition receptors (TCR, immunoglobulins), the V domains are highly variable or "hypervariable" to permit specific recognition of many different antigens. The variability is confined to loops 2, 3, and 6 (Fig. 1.1). These loops form the "complementarity determining regions" or CDRs: loop 2 forms CDR1, loop 3 CDR2 and loop 6 CDR3. *The CDRs form the combining sites in antibodies and T cell receptors that recognize specific antigens.* The six CDRs determine the antigenic specificity.

IMMUNOGLOBULIN, B-CELL RECEPTORS AND ANTIBODY

An antibody molecule is formed by two L chains and two H chains. Each L or H chain has a variable region, V_H or V_L, which is a single V domain, and a constant region, C_H or C_L. The L chain constant region is one Ig domain. The C_H region consists of three or four Ig domains. The V regions of the L and H chains pair to form the antigen binding site: the three CDRs of V_H, plus the three CDRs of V_L. H chains are of five types, designated by the Greek letter for the Ig class in which they are found: α, IgA; γ, IgG; μ, IgM; δ, IgD; and ϵ, IgE. In transplantation the most relevant Ig classes are IgM and IgG.

B lymphocytes and their progeny, plasma cells, make immunoglobulin. Immunoglobulin can serve as the antigen receptor or can be released into the circulation. Each clone of B cells expresses only one type of L chain (lambda or kappa) with one type of V_L region. It can make only V_H, but can associate this with different C_H and thus switch the class of Ig which it is making. Switching C_H while retaining the same V_H and the same L chain is called *Ig class switching*. Since the clone makes only one V_L and one V_H, it can make only one antigen specificity.

The B cell antigen receptor on naive B cells, i.e., never exposed to antigen, is monomeric IgM. Some B cells also have IgD receptors. After antigen exposure they undergo class switching and express IgG, IgA, or IgE on their membranes. Stimulation of the cell results in massive production of soluble antibody.

T-CELL RECEPTOR (TCR)

The TCR (Fig. 1.4) is a dimer of nonidentical α and β chains. There is a second TCR, which is a dimer of γ and δ chains, but most allorecognition can be attributed to $\alpha\beta$ receptors. Each TCR α or β chain resembles an Ig light chain, having V and C regions, with the addition of a membrane anchor and intracytoplasmic region. The TCR V region is believed to be similar to the Ig V region.[65] The V domain is hypervariable in loops 2, 3, and 6, forming CDR1, CDR2, and CDR3 in each V region of the dimer. The V_α and V_β regions dimerize face to face with their CDR3s adjacent in the center and their CDR1 and 2 on the outsides. Despite the fact that the TCR structure is not solved, inferential evidence confirms this model.[66] The $\gamma\delta$ receptor may be similar.

HOW THE TCR ENGAGES MHC

It is likely that all six CDRs of the TCR engage the upper surface of the MHC.[67,68] The outer regions of the TCR (CDRs 1 and 2) engage the α-helices of the MHC, and the central region (the CDR3s) engages the peptide. One model is that the TCR-α chain CDRs engage the α-helix of the α1 domain of class I or class II.[66] The

1

genes, but the demands on MHC products are very different from those on TCRs and antibodies. MHC products are antigen presenting structures which must exist in many forms in the human population but few forms in any one individual. Thus MHC genes encode the proteins without random generation of diversity, but with enormous numbers of alleles in the population.

Antigen receptors (Ig and TCR) generate diversity randomly from a high number of genes encoding V regions of the L and H chains. These Ig genes rearrange in B cell precursors to randomly generate great diversity in selected sites, the CDRs. Thus, unlike the MHC alleles, the TCR and antibody genes combine germ line diversity with massive randomly generated somatic diversity to give each person an enormous repertoire of V region specificities by which antibody or TCRs can engage antigen. The potential repertoires of Ig and TCR chains is estimated at 10^6 to 10^9 specificities each.

Each Ag recognition structure involves combining two different chains (heavy chain with light chains in the Ig molecule, and α with β, or γ with δ in the TCR). The potential diversity created by combining such diverse molecules increases beyond 10^{10} for antibody and beyond 10^{15-18} for TCR $\alpha\beta$ and TCR $\gamma\delta$.

In the case of MHC genes, the polymorphism is mainly confined to the bases encoding the amino acids lining the groove. In the TCR and Ig genes, the diversity is mostly confined to the regions encoding CDRs.

WHAT IS ALLORECOGNITION?

When T cells of a recipient encounter allogeneic MHC, in the context of appropriate additional signals, stimulation of some of the recipient T-cell clones occurs. How allorecognition occurs in vivo is not clear. Small numbers of amino acid differences in the donor MHC can lead to strong responses. This could be because (1) they alter groove shape and thus determine peptide occupation of the groove; (2) they change the shape of the upper surface of the native molecule and change the interaction with the TCR; or (3) they make MHC peptides antigenic.

The donor MHC differences can be presented by either a direct or indirect pathway of presentation. "Direct" refers to recipient T cells recognizing donor MHC molecules on donor antigen presenting cells. Direct recognition could reflect recognition of α-helix differences affecting the contact sites for TCRs on the α-helices;[80] or differences in the peptides in the groove.[81,82]

"Indirect" presentation of donor MHC requires recipient antigen presenting cells with peptides of donor MHC molecules in their grooves. Recent evidence has emphasized the importance of the indirect pathway, particularly since immunity and tolerance can be induced by peptide alone.[83,84]

THE POTENTIAL IMPORTANCE OF PEPTIDES OF DONOR MHC ANTIGENS

Peptides from MHC class I proteins are prominent among peptides occupying the class I groove[85-87] and peptides of class I and II and invariant chain are prominent in the class II groove of DR1.[43,45,46,88] This has given rise to the possibility that a major component of -- across an MHC difference is due to recognition of MHC peptides in the donor (direct) or host (indirect) MHC grooves.

Indirect presentation of allogeneic donor MHC peptides in self MHC class II grooves (and possibly in class I grooves) by host antigen presenting cell (APC) must involve recognition of differences in amino acid sequences. Indirect presentation is a distinct possibility for triggering CD4 T cells, and could generate "help" for both T cell and antibody responses as well as inflammation akin to "delayed type hypersensitivity". However, graft injury by cytotoxic T cells must involve direct recognition.

T-CELL RECOGNITION AND TRIGGERING[8,10,11]

Synopsis: Engagement of the TCR and CD4 or CD8 activates protein tyrosine kinases (TKs) associated with the intracytoplasmic portions of the receptor. TKs trigger second messengers and initiate several signalling pathways which eventually alter proteins which regulate the transcription of genes for cytokines and cytokine receptors. This locks the T cell into activation. Signals provided by additional membrane receptors such as CD28 also play a key role ("second signals").

THE NATURE OF TCR TRIGGERING

The binding of sufficient TCRs to MHC molecules is a necessary condition for T cell activation by antigen. The signal requires the CD3 complex, which includes γ, δ, ϵ, and the long ζ chains.[89] How does TCR binding to MHC alter CD3? This problem is generally explained by one of two mechanisms:

1. Conformational change: engagement of the V regions alters remote parts of the TCR, which in turn alters the CD3 complex; or
2. Crosslinking: the TCR complexes are brought together by engaging antigen and activate one another. Dimerization of class II molecules may serve to bring TCRs together to aid triggering. This would imply that class II recognition may proceed through a complex of two TCRs and two CD4s.

Crosslinking is a common mechanism of triggering of receptors in general. Class I may be able to form multimers,[90] and the dimeric nature of CD8 and of class II suggest that crosslinking could occur.[79] Nevertheless, TCR-mediated T cell activation in vivo may reflect molecular changes triggered by the assembly of the TCR-CD3-CD4 or CD8 complex, in which the CD4 or CD8 molecules play key roles, particularly if the affinity of the TCR for the MHC is low.[91]

The CD3 complex is the transducer which tells the interior of the cell that the TCR has engaged MHC. The ζ chains interact directly with the tyrosine kinases. Meanwhile CD4 (or CD8) engage the MHC, and assembly of the complex brings a series of tyrosine kinases together.

THE KEY ROLE OF TYROSINE KINASES

The CD3-TCR complex is associated with at least three TKs: ZAP, $p59^{fyn}$, and $p50^{csk}$. CD4 and CD8 are associated with another protein tyrosine kinase, $p56^{lck}$. TKs phosphorylate the tyrosine residues in the CD3 molecule ζ chain, in key transduction molecules, and in one another. The functions of $p56^{lck}$ and $p59^{fyn}$ have been shown in knockout mice to be nonoverlapping: $p59^{fyn}$ knockouts have defective TCR signalling and $p56^{lck}$ have defective thymocyte development.[92,93] The tyrosine kinase p50[csk 94] may be particularly involved in negative signalling (tolerizing) the T cell when the TCR is signalled. TKs activate many signalling pathways, including:

1. Ras: the tyrosine kinases can activate the Ras pathway through recently discovered intermediate proteins such as Shc[95] and GRB2;[96,97] activation of Ras then triggers a cascade which can activate enzymes such as mitogen activated protein kinase (MAP kinase) and eventually impact on cell division.

2. PLC-γ1, which lyses the membrane phospholipid phosphatidyl inositol bisphosphate to yield IP$_3$ and diacylglycerol (DAG). DAG activates protein kinase C (PKC) and inositol trisphosphate (IP$_3$) binds to receptors on the ER to release stored calcium and raise intracytosolic Ca^{2+} levels. The high Ca^{2+} is then sustained by increased calcium entry through channels in the plasma membrane to maintain high cytosolic Ca^{2+} concentrations.[98]

3. Phosphatidyl inositol-3-kinase and several others.

Each of these pathways has multiple consequences leading to expression of many genes, blast transformation, mitosis, and expression of effector functions. The calcium-dependent pathway is critical for T-cell activation and important in transplantation. High intracellular calcium activates calcium-regulated enzymes, particularly the enzyme calcineurin (CN). This is a calcium- and calmodulin-dependent serine phosphatase. It activates transcription factors for some key cytokines, particular members of the nuclear "factor of activated T cells" or NF•AT family. CN is the target for some of the most important immunosuppressive agents, cyclosporine and tacrolimus (FK506).

Within minutes, mRNA is transcribed from the "immediate" genes, which do not require new protein synthesis. Some of these are transcription factors. The newly synthesized transcription factors, plus the newly activated factors, now activate a second set of genes. The mRNAs and products for IL-2, IFN-γ, and other cytokines and certain cytokine receptors then appear.

Costimulation ("Signal 2")

When the naive T cell encounters alloantigen, it requires other signals before proceeding with activation,[99] in keeping with the classic two-signal model of lymphocyte activation.[100] Signal 1 is the allogeneic MHC antigen, which must be at a high density to trigger a primary T-cell response. High antigen expression may be one reason why antigen presenting cells (dendritic cells and macrophages) are required. "Signal 2" is the nonantigen signal provided by antigen presenting cells.

(A classic belief in immunology is that when T cells engage antigen without appropriate second signals, anergy results. This renders the identity of the second signals crucial for transplantation and immunosuppression. If we could block them, we might induce anergy.)

"Signal 2" may involve certain adhesion molecules of the Ig superfamily, notably B7-1 and B7-2 (also called B70) on the APC, engaging CD28 on the T cells.[101-105] CD28 activates systems in the T cell which synergize with the signals from the T-cell receptor. CD28 amplifies and prolongs signal 1, increasing IL-2 transcription and prolonging the half life of IL-2 mRNA.[106] In CD28 knockout mice, T-cell triggering can still occur, indicating that other systems can compensate.[107] Other signals from the antigen presenting cell, which could contribute to signalling, include other *adhesion molecule* ligand receptor pairs on the APC and T cell respectively (ICAM-1-LFA-1 and LFA-3-CD2), and cytokines such as IL-1 and IL-6 produced by the antigen presenting cell.

Stimulation of the primary T-cell response may require all of these, in a "conversation" between T cells and APCs initiated by high density of the allogeneic class II molecules on the APCs in the context of cytokines and adhesion molecules. The signals from the triggered CD4 T cells then activate the APCs to increase the signals to the T cell in a cascade of reciprocal activation.

One of the key sites for regulating signal 2 may be a expression of CD40 ligand. CD45 is a tyrosine phosphatase on the surface of all marrow-derived cells whose function may be to keep the key tyrosine in tyrosine kinases (*lck* and *fyn*) dephosphorylated and ready to participate in triggering.[108]

DETAILS OF SIGNAL TRANSDUCTION AND T-CELL ACTIVATION: CONTROL OF CYTOKINE EXPRESSION

PLC-γ1, activated by tyrosine phosphorylation, lyses membrane phosphatidyl inositol bisphosphate (PIP$_2$), releasing DAG and IP$_3$. DAG activates PKC which is also activated through other pathways, including calcium flux. PKC activation leads to the transcription of several genes which encode transcription factors such as *fos* and *jun* which form the complex called AP-1, composed of the Jun and Fos proteins.[109]

IP$_3$ binds to receptors on the endoplasmic reticulum which release calcium into the cytosol. The high cytosolic calcium is then sustained by changes in membrane transport.[110] The high calcium activates calcium-dependent enzymes, one of which is CN. CN activates cytosolic factors called NF-AT, which is free to translocate from the cytosol to the nucleus.[111,112] When cytoplasmic and nuclear factors assemble to form the full NF-AT complex, transcription of IL-2 mRNA begins. While the NF-AT sites account for the majority of inducible IL-2 expression, it is likely that the NF-κB site[113] and the octamer site are also critical. The characteristic behavior of the IL-2 gene requires the interaction of multiple transcription factors binding to these sites.

Similar events occur with other cytokine genes, although less is known about them. The result is a wave of transcription of cytokine mRNAs. Note that this is the "second wave" of protein synthesis, the first being the nuclear factors which control the cytokine promoters. In this sense the cytokines are "early", not "immediate" genes.[113]

Naive CD4 T cells make predominantly IL-2 in their first encounter with antigen, whereas previously stimulated or memory T cells make other cytokines. IL-2 engages its receptor, and other cytokines engage through their receptors, giving waves of receptor triggering and signal transduction. The cell becomes committed to activation, differentiation, mitosis, and clonal expansion. Effector functions emerge such as cytotoxicity in CD8 cells. Eventually the molecules associated with

1

memory and recirculation, such as the "very late antigens" or VLA molecules,[114] appear.

CYTOKINES AND THEIR RECEPTORS[9]

The term "cytokine" includes the interleukins, interferons, and colony stimulating factors of the hematopoietic and host defense system. They are protein mediators which signal cells through specific membrane receptors. Cytokines and their receptors are related in structure and function to protein hormones and their receptors. Cytokines have certain characteristics:

1. Short half-life: cytokine mRNAs and cytokines themselves have short half-lives to permit fine regulation.
2. Relatively small size: the typical cytokine gene is about 4-5 kb in length, with about four exons. Numerous AT sequences at its 3' end confer a short half life on the mRNA.[115] The protein is typically a polypeptide chain of about 10-20 kD, often glycosylated and/or multimerized to a higher molecular weight.
3. α-helical structure: many cytokines are folded into a bundle of four to six α-helices, sometimes with very short β strands. Exceptions include TNF-α, a sandwich or "jelly roll" of antiparallel β strands, and TGF-β, which has both α-helices and β-pleated sheets.[116]
4. Multimer formation is common: IFN-γ and TGF-β are dimers, and TNF-α is a trimer.[117]
5. Cytokines are generally not stored but are synthesized and secreted when needed. They are not usually expressed as membrane proteins, but some have membrane-bound variants, e.g., TNF-α[118] and IL-1.
6. The main control of cytokine production is transcriptional, although post-transcriptional control is known, e.g., for TNF-α.[119]

Cytokines often act in concert with other cytokines: interactions (synergy, competition, and antagonism) are common. Cytokines are pleiotropic (i.e., have many effects) and redundant (i.e., have overlapping effects). Cytokines commonly induce other cytokines in a cascade. Self-amplifying circuits are common to facilitate rapid potent responses. The potency of the cytokine response is impressive as is well known to the clinician who observes the cytokine release syndrome after OKT3 treatment (see below).

Some cytokines are produced in normal tissues at low levels and affect growth, development, and homeostasis, e.g., the maturation of T and B lymphocytes. But their most characteristic effects are in inflammation and host response to injury or infection.

"Knockout mice" are providing important insights into the roles of cytokines and their receptors.[120-123] In knockout mice, both copies of the target gene have been mutated to prevent expression. Such strategies may underestimate the importance of the deleted structure because the deletion forces the embryo to use other cytokines to develop, thereby maximizing apparent redundancy. Moreover, the laboratory mouse, protected from many of the usual pathogens of its species, tolerates immune defects which would be more serious in the natural environment.

Surprises arise in knockouts: for example the IL-2 and IL-10 knockouts, as well as some TCR knockouts, get inflammatory bowel disease for unknown reasons.[124,125]

Cytokine receptors are typically multimers of different transmembrane proteins, one or more which have an external ligand-binding domain, and an intracytoplasmic signalling domain. One or more chains may bind the cytokine with high affinity, but the multimer is required for internalization and/or signalling. Cytokine receptors are classified into families on the basis of their external, ligand binding domain.[126-128]

1. The hemopoietins, e.g., IL-2R β chain, use a 200 kd external domain with four conserved cysteines and one tryptophan residue at the N terminal, and aromatic residues (Trp-Ser-X-Trp-Ser) at the C terminal. A few receptors in this group have typical Ig domains in their extracellular regions.
2. The interferon and IL-10 receptors, e.g., IFN-γR, have two external domains distantly related to Ig domains, with characteristic conserved cysteines.
3. The TNF receptor and its relatives have an external domain with cysteine-rich repeats.[128]
4. IL-8 and its relatives have a "seven pass" membrane receptor associated with G proteins, similar to many endocrine receptors.

Unlike the cytokines, which are often predominantly α-helical, the external ligand binding domain of a hematopoietin or interferon receptor is often two β-pleated sheets. A second chain of the receptor may or may not actually engage the cytokine: in the IL-2 receptor it does, but in the IFN-γ receptor the binding site is formed by the single receptor protein with the second receptor component presumably playing other roles. Binding of the cytokine to the external domain of the receptor may alter the cytoplasmic domain, triggering second messengers usually through a kinase, usually a protein tyrosine kinase or less commonly a serine/threonine kinase. The signalling systems are similar to those already described: PTKs activate PLC-γ, PI-3 kinase and other second messengers with downstream activation of serine-threonine kinases, e.g., PKC and release of intracellular calcium.

The final effect is often on transcription factors, but other events are common, such as direct effects on membrane receptors or cytoplasmic effector mechanisms.

Signal transduction, through the IFN-γ receptor,[129] is a useful example of a cytokine system which we can watch in operation in transplant rejection. IFN-γ engages the IFN-γR and activates two tyrosine kinases, JAK1 and JAK2, which phosphorylate a factor called STAT 91. This induces transcription of selected genes by moving to the nucleus and engaging specific sites in their promoters. We will expand on some features of the IFN-γ response later as an example of cytokine signal transduction. The TNF receptor acts through a sphingomyelin pathway to induce NF-κB to be released from its cytoplasmic binding protein (IκB) to enter the nucleus and bind to specific DNA regulatory sites.[118]

From the above, the passage of signals from hematopoietin and interferon receptors to the interior of the cell involves the regulation of tyrosine phosphorylation.

1

The cytoplasmic regions of many membrane receptors for protein hormones have intrinsic tyrosine kinase activity, but cytokine receptors are associated with separate tyrosine kinases (like JAKs which associated with the IFNL-γR). Engagement of the receptor by its ligand activates the tyrosine kinase activity, which results in phosphorylation of one or more key tyrosine residues in the cytoplasmic region of the receptor. This phosphorylated tyrosine can then be recognized by other proteins via specific regions in those proteins called "src homology-2" or SH2 domains.[130,131] The sequence "ligand-receptor-tyrosine kinase activation-tyrosine phosphorylation—recognition and binding of a second messenger via its SH2 domain activation of second messenger by tyrosine phosphorylation—is probably a common pattern for linking membrane receptors to second messengers like STAT proteins.

Several cytokine receptors, including IL-2R, apparently utilize a signal transduction pathway which involves the activation and phosphorylation of an enzyme called the "Target of Rapamycin" or TOR. TOR in turn activates p70 S6 kinase.[132,133] The role of TOR was discovered because the immunosuppressive drug rapamycin acts at this point. The role of TOR kinase is probably crucial in the initiation of cell division by cytokines. TOR acts to increase the translation of existing mRNAs for proteins which control the cell cycle.

SPECIFIC IMMUNE RESPONSES OF T CELLS AND B CELLS

Synopsis: Specific lymphocyte activation leads to cell cycling (clonal expansion), T cell/B cell-antigen presenting cell interactions, altered cell traffic, and altered expression of many genes in the transplanted organ and elsewhere in the host. The lymphocyte population changes. Many lymphocyte activation events may actually occur within the graft, as opposed to the lymphoid organs.[114] Three lines of lymphocyte differentiation lead to effector mechanisms, which require massive clonal expansion to become quantitatively important:

1. *The delayed type hypersensitivity response, principally engineered by cytokines from CD4 T cells;*
2. *The B-cell antibody response, dependent on CD4 T cell help;*
3. *The cytotoxic T-cell response by CD8 cells.*

Activated CD4 cells influence other cells through two mechanisms: the production of cytokines, which interact through their receptors to signal the target cell, and direct interaction through their TCRs and adhesion and signalling molecules. Direct interactions must involve the same MHC plus peptide for which the CD4 T cell is primed. In direct interactions, the release of cytokines is directional, focused on the target by the TCR and the adhesion molecules.[134,135] Activated CD4 cells help CD8 cells to become cytotoxic and B cells to make antibody and activate macrophages and endothelial cells to mediate delayed type hypersensitivity.

DIVISION OF LABOR AMONG CD4 CELLS: "TH1" AND "TH2" CYTOKINES[8]

The primary function of CD4 T cells is to produce cytokines, which they do more efficiently than CD8 cells. Naive CD4 T cells produce primarily IL-2, with increasing amounts of assorted other cytokines. With prolonged stimulation, e.g., in cloning experiments in vitro and under certain conditions in vivo,[136] CD4 T cells cease production of some cytokines and increase production of others in characteristic patterns: a "TH1" pattern or "TH2" pattern. The TH1 pattern of cytokine production is IFN-γ, lymphotoxin, and IL-2. The TH2 pattern consists of IL-4, IL-5, IL-6, and IL-10 (Table 1.4). These are called the TH1 and TH2 cytokines respectively. CD4 T cells can be found which produce exclusively TH1 or TH2 cytokines and are called TH1 and TH2 T cells or subsets. But typically the intermediate forms are much more frequent.

No cytokine or surface antigen constitutes an exclusive marker for any CD4 phenotype: for example, IL-10 can be produced by CD4$^+$ T cells of the TH2 characteristics but also by many TH1 cells in human and by many non T cells.[136] IFN-γ is a "TH1 cytokine", but most of it is made by other T cells (CD8 T cells and CD4 T cells not fitting the TH1 definition) or by NK cells. We prefer to reserve the term TH1 and TH2 "subsets" for circumstances where we know that discreet populations, rather than a continuum, can be shown to exist. Under most circumstances TH1 and TH2 cytokines are *not* made by CD4 T cells which fulfill the criteria for TH1 and TH2 subsets.

TH2 cytokines are more important in helping B cells. In vitro the activation of resting B cells to proliferate and differentiate requires cytokines, particularly of the TH2 type, e.g., IL-4, IL-5.[137-139] In addition, IL-10 enhances in vitro viability of B cells and upregulates MHC class II expression on resting small dense B cells from mouse spleens.[136,139]

TH1 cytokines can enhance or suppress B cell responses, according to the relative amounts of IL-2 and IFN-γ produced.[140,141] IL-2 in large amounts enhances differentiation, proliferation, and Ig production. IFN-γ in low concentrations enhances certain antibody responses but in high amounts suppresses both proliferation and Ig secretion and can be cytotoxic to activated B cells. TH2 cytokines favor IgE and IgG1 responses (through IL-4),[120] whereas TH1 cytokines in mice induce IgG2a.[140]

Table 1.4. Some cytokine phenotypes of mouse CD4 T-cell clones

Cytokine Phenotype	THO	TH1	TH2
GM-CSF	++	++	+
TNF-α	++	++	+
IL-3	++	++	++
TH2 cytokines:			
IL-4	++	–	++
IL-5	++	–	++
IL-6	++	–	++
IL-10	++	–	++
TH1 cytokines:			
IL-2	++	++	–
IFN-γ	++	++	–
Lymphotoxin	++	++	–

1

The generation of CD8⁺ cytotoxic T cells is enhanced by both TH1 and TH2 cytokines. IL-2, IFN-γ, IL-4, and IL-5 all enhance the generation of CTLs although IL-2 is most effective.[142-145]

TH1 cytokines are important mediators of delayed type hypersensitivity (DTH)[146] *(see below)*. TH1 and TH2 cytokines cross-regulate. TH1 and TH2 cytokines tend to be mutually inhibitory[136,147] regardless of what cell is producing them. IFN-γ inhibits the proliferation of TH2 clones and IL-10 suppresses both cytokine production and proliferation of TH1 clones. IL-10 inhibits IFN-γ production by TH1 clones by 90% and inhibits production of TH1 cytokines by CTL clones and LGL. IL-10 acts at the level of antigen presenting cells and their relatives such as skin Langerhans cells.[148]

T cells from IL-2 knockout mice have disturbed cytokine production[121] in vitro, overproducing IL-4, IL-6, and IL-10. In vivo, such mice have increased serum levels of IgG1 and IgE due to increased IL-4 production. For IL-10 knockouts the results are less clearcut.

REVISING THE CONCEPT OF "SUPPRESSOR T CELLS"

Mixing nonresponding and responding populations of T cells can shut off the responders. This "contagious" unresponsiveness used to be attributed to a special class of "suppressor T cells", typically carrying CD8 markers, but it is now clear that antigen-specific T cells with unique suppressor function cannot be isolated. Suppression is a cell population phenomenon, not attributable to unique specialized cell class. All cytokines have both positive and negative effects, and the cells which produce them cannot be assigned a uniquely positive or negative function except in relationship to one specific set of circumstances and one specific target system. For example, TH2 cytokines such as IL-4 can suppress DTH responses but help IgE responses. Thus T cells producing IL-4 can have many simultaneous functions, positive and negative, depending on where the IL-4 is received. So suppression by IL-4 is really a characteristic of the cell that receives IL-4, not the cell that produces it. Similarly, cells producing TGF-β are negative regulators in some types of inflammation and positive in others.[149-155]

Thus negative regulation or suppression can often be explained without postulating the existence of specialized suppressor cells. In clinical transplantation, where negative regulation is vital to success, the agenda has shifted from suppressor T cells to a detailed analysis of the role of particular molecules in negative regulation of graft injury and inflammation, such as TH2 cytokines and TGF-β.

THE ALLOANTIBODY RESPONSE AGAINST MHC ANTIGENS

B-cell activation normally takes place in germinal centers (GC) of draining lymph nodes or spleen, but may occur in the graft infiltrate of a transplant. The surface Ig of the B cell, sIg, engages the polymorphic regions of the donor MHC, particularly the α-helices, in the native, nondenatured, unprocessed form. The MHC antigen is probably shed from donor cells. This leads to B cell triggering and internalization of the antigen. The mechanisms of signalling through sIg involves a receptor complex on the B cell similar to the CD3 complex on T cells.[156]

The result is that a signal for B cell triggering is delivered, activating intracellular pathways which include the calcium-dependent pathway.

To recruit antigen-specific T cell help, B cells must present peptides of allogeneic MHC antigen in the groove of its class II antigens. To accomplish this, the allogeneic MHC antigen bound by sIg is endocytosed, through proteins around the receptor termed α and β,[156] and presented as peptides in the class II groove of the B cell. Host B cells thus present peptides of donor MHC to host CD4 T cells. Antigen presentation by a B cell is crucial for the T-cell response.[157] T cell-B cell interactions are weak unless the T cell recognizes its cognate antigen on the B cell. The CD4 T cells may initially be sensitized by antigen on host or donor dendritic cells[158] because antigen specific B cells are uncommon in the early stages of the response before they are triggered and undergo clonal expansion. CD4 T cells, B cells, and DCs presumably interact in multi-cell complexes.

T cell-B cell engagement involves a variety of adhesion and signalling interactions, including CD4 with class II, CD40 ligand with CD40, LFA1 with ICAM-1, and CD2 with LFA3, CD5 with CD72, etc. Several cytokines are also transmitted from the CD4 cell to the B cell and B-cell signalling molecules such as CD40 help to trigger the T cell. The expression of adhesion molecules, cytokines, and cytokine receptors increases. The signals to the B cell from the T-helper determine whether the B cell will progress towards antibody production and memory, or toward anergy/apoptosis (programmed suicide).[159] Apoptosis is regulated by the gene *bcl-2* in lymphocytes:[160] mice with knockouts of *bcl-2* gene have spontaneous suicide of their lymphoid tissues and lymphocytes.[161]

If the signals are correct, the B cell undergoes massive clonal expansion and differentiation. Ig, initially expressed on the B cell membrane, can now be released in large quantities as circulating antibody against MHC and other alloantigens.

WHAT SITES ON THE MHC DOES ALLOANTIBODY RECOGNIZE?

Alloantibody recognizes the "nonself" sites in the α-helix and the ends of the β-pleated sheets that are due to the effect of polymorphic amino acids. The most abundant and important Ig class produced is IgG, which has two antigen binding sites. Each IgG can engage only one site in the MHC molecule. The other binding site of the alloantibody can engage the same region of another MHC molecule. The alloantibody usually binds to the side of an MHC sheet and helix domain or to the top of one α-helix, not across the groove like the TCR. However, one IgG molecule will not fix complement efficiently: an adjacent IgG molecule is needed. The best way of assuring that such IgG complexes will be assembled is to have multiple clones responding to different sites in the mismatched molecule. This is usually the case with clinically important anti-MHC responses: they are polyclonal and react with several sites on the MHC molecule.

Does the peptide in the MHC groove influence antibody binding to the MHC? Perhaps, because the peptide may alter the shape of the domain, as well as possibly directly contacting the antibody in a few cases. Alloantibodies specific for the MHC allele *plus a specific peptide* are known[162] and would escape detection in our usual antibody screening programs. Alloantibody which required a specific peptide would

1

usually react with too few MHC molecules to be quantitatively important. It is conceivable that alloantibody recognizing abundant tissue specific peptides in MHC alleles could act as tissue specific alloantibodies in rejection, e.g., anti-endothelial antibodies. This could help to resolve the old problem of tissue specific alloantibodies such as anti-endothelial antibodies.[163]

CYTOTOXIC CD8 RESPONSE

Whether the naive CD8 cell requires an APC for its primary stimulation is less well established than for the CD4 cell. The presence of the CD4 cytokines and possibly direct contact from CD4 have been suggested to be necessary for the CD8 cell to be triggered. However, CD8 cells can also be directly triggered without CD4 cells at times, as shown in CD4 deficient or class II deficient mice.

With time and clonal expansion, the CD8 cell acquires the ability to be cytotoxic for target cells. Cytotoxicity is direct lysis of target cells in suspension with the targets undergoing programmed cell death (apoptosis). Functional cytolytic ability correlates with the expression of serine esterases (granzymes) and perforins.[164] Although both are sequestered in cytoplasmic granules, perforins and granzymes are regulated differently. Another mechanism of target cell lysis is the interaction of a TNF-like molecule on the T cell (Fas ligand) with a TNF-receptor-like molecule on the target (Fas). Cytolytic ability also requires adhesion molecule interaction between the cytotoxic T cell and the target cell.

THE POSSIBLE ROLE OF NATURAL KILLER (NK) CELLS IN ALLORECOGNITION

NK cells can lyse cells with little or no class I, apparently being inhibited by expression of class I. This may reflect recognition of the class I groove by an NK receptor.[165-167] Little is known about such receptors, and the role of NK cells in transplantation is uncertain.

ORGANIZATION OF INFLAMMATION[13]

The inflammation in the graft is analogous in some respects to the delayed type hypersensitivity reaction (DTH), exemplified by the classic skin reaction to tuberculin. DTH is an in vivo phenomenon with no single in vitro correlate. It is manifest histologically as a heterogeneous nonspecific inflammation with edema, fibrin accumulation, T-cell infiltration (both specific and nonspecific), B cells, numerous macrophages, and lesser numbers of other leukocytes, and endothelial changes. The key events in DTH are cytokine production (especially TNF-α and β, IFN-γ and IL-1), altered expression and function of adhesion molecules, and nonspecific activation of many bone marrow-derived cells, particularly macrophages. Although usually ascribed to CD4$^+$ T cells, DTH reactions mediated by CD8 T cells have been described. The result is graft inflammation.

THE ADHESION MOLECULES

These sets of molecules, which are involved in all levels of the immune response and inflammation, are classified into three groups: the Ig superfamily; the integrins; and the selectins.

Adhesion Molecules of the Ig Superfamily

The principal members are ICAM-1, ICAM-2, VCAM, CD2, CD58, CD28, CTLA4, B7-1, B7-2. These tend to be involved in signalling as well as adhesion. Their expression is increased by pro-inflammatory TH1 cytokines. Ig superfamily members generally interact with other Ig superfamily members or with integrins.

ICAM-1 is a chain of five Ig domains with a membrane anchor and an intracytoplasmic region. Its N terminal domain binds the integrin LFA1. The N terminal V domain of ICAM-1 uses the CDR2-like loop to interact with LFA-1.[168] The interactions of CD2-LFA3; CD28-B7, as well as Ig domain interactions with integrins, may follow these principles. Detailed modelling of the interactions involving the Ig superfamily will permit the design of better monoclonal antibodies or other antagonists.

Integrins

Integrins are heterodimers of an α chain and a β chain. The integrins are classified on the basis of the β chain they employ as β_1, β_2, or β_3 integrins. Each β chain can potentially be combined with many different α chains. β_1 integrins are important markers of memory and recirculation in T cells (the VLA group). β_2 integrins are important in leukocyte adherence reactions (LFA-1, Mac-1).[169] Both β_1 and β_2 integrins are activation-dependent with low avidity in the unactivated state, but high avidity following T-cell activation. Integrins are also associated with diapedesis and intracellular signalling.

Selectins

The selectins are large molecules with three characteristics: lectins (sugar residues with the ability to bind to sugars on other molecules), epidermal growth factor-like motifs, and short consensus repeats (2-9). Each also possesses intracytoplasmic domains. The name selectin helps us to remember these features: S (short consensus repeats), e (epidermal growth factor-motif), and lectins.

There are three members, named for the cells that express them. E (endothelial)-selectin, is induced by IL-1 and TNF. Its ligand is L(leukocyte)-selectin, which is important for both endothelial binding during inflammation and as a recirculation receptor. L-selectin also binds to P(platelet)-selectin, which is stored in granules of platelets and endothelial cells and is released in response to clotting cascade products.

Selectin interactions are weak under flow conditions and serve as first step adhesion receptors. By slowing leukocyte passage, they expose the leukocytes to the local environment and other endothelial surface molecules. Selectins are involved in all types of tissue injury and may be important mediators of reperfusion injury in transplanted organs. Antibody against P-selectin has been used to ameliorate reperfusion injury of lungs, presumably by inhibiting the interaction of neutrophils with injured endothelium.[170]

The Roles of Adhesion Reactions

The leukocyte interacts with endothelium through interactions between the selectins. The result is loose binding permitting the leukocyte to roll along the

1

Table 1.5. Characteristics of chemokines

	α Subfamily	β Subfamily
Chromosome location	4	17
Structure	C - X - C	C - C
Subfamily members	IL-8, Gro-α, β-thromboglobulin	MIP-1α, MIP-1β, RANTES, MCR-1, MCAF
Target cells	Neutrophils	Monocytes, T cells

CHANGES IN THE TARGET TISSUE

In an inflamed tissue, the expression of many surface molecules increases, usually because cytokines increase the transcription of the genes. The example of IFN-γ triggering the transcription of the MHC class I and II genes in vivo is the best known, but many adhesion molecules can also be induced on endothelium, inflammatory cells, and parenchymal cells. We shall outline the features of the induction of class I and II MHC molecules by IFN-γ.

THE TRANSCRIPTIONAL REGULATION OF MHC GENES[184,185]

The level of MHC expression determines the immunogenicity of tissues and their sensitivity to immune injury, and some increased MHC expression is invariably seen in acute T cell-mediated rejection. The expression of MHC proteins in tissues is primarily regulated by transcriptional control.

MHC genes behave as "housekeeping genes" (as opposed to tissue specific genes), which are either expressed or expressible in most tissues, but to varying degrees. This implies that the chromatin structure of their regulatory regions is available for transcription factors in many tissues.

In the normal mammal, constitutive class I expression is widespread but highly variable between cell types. A component of IFN-γ-induced expression is common even in normal hosts.[186] Constitutive class II expression is confined to B cells. The class II expression in dendritic cells and some macrophages in normal individuals may reflect low levels of cytokines such as IFN-γ, IL-4, and GM-CSF. Class II expression found in some normal epithelia probably also reflects cytokine induction.

MHC PROMOTERS

The level of MHC expression is closely related to the steady state mRNA levels and probably reflects the activity of the promoter in regulating transcription. The class I and II promoters are highly conserved.

The characteristic DNA sequence in the class I promoter is a class I regulatory element (CRE) at about -160 to -200 bp from the start site of transcription, overlapping an interferon consensus sequence (ICS) at about -140 to -160 bp. The CRE is a series of overlapping palindromic sequences which are sites of binding of transcriptional regulatory proteins and are necessary for the tissue specific basal and induced expression of class I. Protein binding to the CRE seems to correlate with constitutive class I expression.[184,185,187] The ICS probably binds proteins which are regulated by IFN-α/β, and IFN-γ and acts by increasing transcription in concert with the CRE. TNF-α also acts on the CRE, probably through NF-κB proteins.

The class II promoter contains a conserved region at about -60 to -100 bp which contains sequences termed the X box, Y box, and a spacer between them. The X and Y boxes are occupied by proteins in the basal state and probably are the elements giving class II genes their characteristic patterns of regulation in the basal and cytokine induced state, but other elements participate.[188] The key regulation of class II genes is the class II transactivator, or CIITA.

NORMAL AND INDUCED IFN-γ PRODUCTION AND REGULATION

IFN-γ is produced by T cells (CD4, CD8) and NK cells. IFN-γ production is an important event in rejection, with both adverse and favorable effects. IFN-γ-mediated MHC induction is probably necessary but not sufficient for rejection, and IFN-γ can induce accelerated rejection.[189] IFN-γ is produced by the specifically triggered T cells and is also capable of triggering its own release, probably from NK cells with the appearance of large granular lymphocytes (LGL).[190] Thus the LGLs may serve as an amplifier to increase the release of IFN-γ.

IFN-γ RECEPTOR TRIGGERING

Two IFN-γ receptors bind the IFN-γ homodimer, each engaging the N terminal of one unit and the C terminal of the other.[129] Receptor crosslinking leads to membrane-to-cytoplasm signal transduction via mechanisms involving the large intracytoplasmic domain of the receptor. The mechanism involves a protein kinase: the receptor becomes phosphorylated,[191-193] and one tyrosine in the intracytoplasmic portion of the receptor has been shown to be essential to the biological activity of the receptor.[194] The receptor has additional subunits, encoded on chromosome 21 and chromosome 16 in the human. Tyrosine kinases (JAK1 and JAK2) then phosphorylate the cytoplasmic form of a transcription factor, interferon stimulated gene factor 3, in particular, the p91 component, now called STAT-1. This then moves into the nucleus to activate transcription of genes with IFN-γ activated sites. Some of these induced mRNAs encode products which are transcription factors.

The details of the pathway from the IFN-γ receptor to MHC promoters remain to be elucidated; it is unclear why MHC expression tends to be induced later than some other genes, e.g., 24-48 hours after IFN-γ administration. It is likely that MHC induction requires the synthesis of IFN-induced transcription factors such as IRF-1. In the case of class I induction, the signal transduction pathway used by IFN-γ seems to require some of the same steps as are used by IFN-α/β.[195] These proteins probably affect the ICS. In the case of class II, the new protein induced by IFN-γ is CIITA.[195a]

TARGET INJURY

CANDIDATE MECHANISMS OF SPECIFIC DONOR CELL INJURY IN REJECTION

The hallmark of acute T-cell mediated rejection is injury to the endothelial and parenchymal cells, initially reversible, but eventually becoming irreversible and proceeding to infarction. Inflammation is probably necessary but not sufficient for rejection injury. The parenchymal injury is usually conceptualized as apoptosis of individual parenchymal cells triggered by cytotoxic T cells. Many cytokines such as TNF-α are expressed in rejecting or inflamed grafts,[178] but no single cytokine has been shown to mediate rejection injury. Understanding of what constitutes rejection injury should begin with the pathology, not with immunologic theory.

THE PATHOLOGY OF ACUTE REJECTION

International collaborations have classified the histologic lesions which correlate with rejection.[196,197] Classifications are all based on the concept that donor cell injury, not the inflammatory infiltrate or interstitial edema, defines rejection. Thus tubulitis in kidney transplants, myocyte necrosis in heart transplants, injury to the biliary epithelium of liver transplants, and injury to the epithelium of small airways in lung transplants, constitute rejection. In general, areas of high MHC class I and II expression, either basal or inducible, are important targets of acute rejection.

Tubulitis in renal transplants refers to invasion by lymphocytes which cross the basement membrane and attack the basolateral membrane of the epithelial cells, where MHC products are expressed (Fig. 1.6). Bile ductule invasion, damage to small airway epithelium, and myocyte necrosis probably involve analogous mechanisms. The lymphocytes are believed to be T cells expressing cytotoxic molecules, but more details on the cells in these lesions are needed.

The endothelium of small arteries and arterioles in all types of grafts is damaged in the lesion known as intimal arteritis or endothelialitis. (Such lesions are often missed in biopsies: for example, endomyocardial biopsies of rejecting heart transplants are relatively poor at sampling arteries.) Lymphocytes adhere to the endothelium, infiltrate beneath it and lift up the endothelial cells. The result is increased resistance, perhaps due to loss of endothelial regulation of vasomotion, increased coagulation, and eventual loss of perfusion and downstream ischemia.

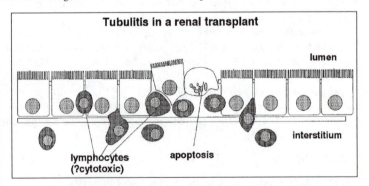

Fig. 1.6. Acute tubulitis. Lymphocytes infiltrate through the basement membrane and recognize alloantigens expressed on the MHC of graft epithelial cells and mediate cell death via apoptosis or cell lysis.

The candidate mediators of specific cell injury include cytokines, Fas and granule contents (serine esterases and perforins), both concentrated on the target cell by receptor directed exocytosis, and in some cases cytotoxic alloantibody. Serine esterases are expressed in the infiltrate of rejecting grafts.[198] At least some of the injured graft cells probably die by apoptosis. Numerous cytokines are found in the infiltrate of rejecting grafts or in the serum, but the roles of these mediators are not established. Some may cause injury, but some may reflect the response to injured tissue. Both CD4 and CD8 T cells are present in rejection and neither has an exclusive role.[199]

There are nonspecific as well as alloantigen-specific lymphocytes in the cellular infiltrate. Macrophages are abundant within rejecting grafts and may play a role in the immune injury. Macrophages make a wealth of cytokines, growth factors, eicosanoids, enzymes, procoagulant activities, NO, etc, and may contribute to the parenchymal and endothelial cell injury and dysfunction in vascularized grafts. But the majority of early injury is probably due to specific T cells.

THE ROLE OF ANTIBODY IN ACUTE REJECTION

Alloantibody can play a major role in acute transplant rejection, especially in the increasing population of recipients sensitized to MHC antigens. EC are important targets for alloantibody. The sequence of events in antibody-mediated rejection seems to involve endothelial dysfunction and injury, via complement and neutrophils, followed by vasospasm, ischemic injury, fibrin and/or platelet deposition, and infarction or hemorrhage.

Hyperacute rejection is predominantly a problem in renal transplantation, mediated by preformed antibodies against HLA class I molecules or by antibodies such as ABO blood group antigens.[15] A population of antibodies against poorly defined endothelial antigens of arteries ("anti-endothelial antibodies") also mediates hyperacute and accelerated rejection.[200] Anti-class II antibodies rarely mediate hyperacute rejection. A positive B-cell crossmatch is frequently due to antibodies which are not class II specific. For example, antibodies against B cells are often autoantibodies. Low levels of anti-class I can also produce a positive B-cell crossmatch with a negative T cell crossmatch because B cells are relatively rich in class I. Thus a positive B-cell crossmatch may have several explanations.

Successful immunosuppressive strategies usually suppress primary alloantibody as well as T cell responses, probably by suspending help from CD4 T cells, but do little to preformed antibody and may have difficulty suppressing secondary antibody responses.

Anti-class I-mediated rejection of kidney transplants can be recognized clinically.[201,202] Typically a transplant into a presensitized patient with a negative crossmatch functions initially, then suddenly loses function after 1-7 days. The kidney may rapidly develop acute tubular necrosis secondary to severe decrease of perfusion. The pathology shows evidence of endothelial injury in the microcirculation, rather than tubulitis or endothelialitis. Neutrophils may be present. The demonstration of antibody against donor class I can aid the diagnosis. OKT3 can sometimes suppress this rejection by abrogating T-cell help. Anti-class I-mediated acute rejection of the heart may also occur.

1

HOST AND GRAFT ADAPTATION

Synopsis: Despite immunosuppression, transplantation could not be successful if adaptive changes favoring prolonged graft survival did not occur in both the host and the graft. The adaptive changes in the graft may reflect the loss of the donor bone marrow-derived cells, with the loss of "signal 2". The antigen specific adaptive changes in the host are dependent on the continuous presence of the antigens of the graft — and on immunosuppressive therapy, in many or most patients. The host probably develops a state of partial peripheral tolerance.

HOST ADAPTATION

The encounter of the immune system with antigen can result in a positive response, a negative response ("tolerance"), or no response ("neglect"), depending on the circumstances in which the antigen is presented. Tolerance is defined as a state of antigen-specific unresponsiveness induced by exposure to antigen, typically under conditions of immaturity, injury, or drug therapy. The ability to induce tolerance is vital for self and nonself discrimination and to randomly generate potentially autoreactive cells. T cell tolerance is classified by location: central versus peripheral.

CENTRAL TOLERANCE

The principal central mechanism of tolerance in the thymus during T-cell ontogeny is clonal deletion by apoptosis.[203] Intrathymic injections of antigen can induce tolerance in rats, but these approaches have not yet been successful in primates.[84,204] Central tolerance is believed to have little role in transplantation although microchimerism with donor cells could play a role centrally. In general, chimerism can induce tolerance only in significant levels, and microchimerism does not correlate with true tolerance.

PERIPHERAL TOLERANCE

Successful transplantation involves a degree of peripheral tolerance. Studies of transgenic mice expressing foreign MHC antigens in peripheral tissues have recently been particularly helpful for understanding peripheral tolerance (reviewed in ref. 206). These and other models suggest several possible mechanisms:

First, in some models, *clonal expansion then clonal deletion* occurs, causing peripheral tolerance. This is particularly true for responses to "superantigens", which delete previously expanded clones as an outcome of powerful immune responses probably by programmed cell death.[207] Lack of co-stimulatory signals (IL-1, adhesion molecules) may promote peripheral clonal deletion. Overall, however, peripheral clonal deletion is not a prominent mechanism.

Clonal anergy, i.e., paralysis without deletion, is demonstrable in some circumstances. In some MHC class I transgenic mice, tolerance is the result of anergy and is dependent on the continuous presence of Ag and the lack of IL-2.[208] Exogenous IL-2 reverses the state of anergy. Some patients receiving long-term immunosuppression with functioning allografts simulate this state. In some models of MHC class II transgenic mice, T cells exhibit low reactivity against class II in vitro, with no in vivo pathology,[209] a form of "neglect". IFN-γ may abrogate some tolerant states.[210] Cytokines of the TH2 type, e.g., IL-10, may suppress IL-2 and

IFN-γ expression,[211,212] but it is difficult to imagine long-term high levels of cytokine production as a mechanism for maintaining tolerance. Clinical immunosuppressive treatment, particularly with cyclosporine and steroid, may also act in this way, selectively reducing IL-2 and IFN-γ production.

A variety of other mechanisms could be important:

1. Down-regulation of TCRs and co-receptors (CD4 or CD8);[213]
2. "Veto cells" (these are T cells which inactivate T cells which try to recognize them);[214]
3. Antigen-specific T cells actively maintaining unresponsiveness, especially CD4 T cells.[215] CD4 T cells producing TH2 cytokines could decrease the production of TH1 cytokines from other lymphocytes in a "contagious" fashion;
4. Anti-idiotypes. Idiotypes are antigen combining sites, either of TCRs or antibodies, and anti-idiotypes are antibodies which are directed against them. The extensive literature on anti-idiotype antibodies and idiotype-specific regulatory T cells has not led to examples of negative regulation unequivocally mediated by an idiotype/anti-idiotype interaction. There is evidence for the role of anti-idiotypes in turning off anti-HLA antibody responses.[216]

Adaptive Changes in the Graft

With time, if the graft survives, the inflammation subsides, and the induced expression of adhesion molecules and MHC antigens in the graft returns toward normal. There is a progressive loss of the donor antigen presenting cells, replaced by the recipient cells. Thus both direct antigen presentation by donor cells and indirect presentation by host cells subside. Injuries, including ischemic and reperfusion injury, rejection, and viral infection, can promote inflammatory changes and sustain the immune process.[217] The changes of inflammation and those of tissue repair in response to injury overlap. Immunologic and nonimmunologic injury can both therefore lead to a common pathway of chronic inflammation which manifests itself in sub-acute or chronic rejection. Thus injury may sustain the host APC and antigen expression burden of the graft, sustaining immunologic activity and preventing the stabilization of the host graft relationship.

Microchimerism

The transfer of tissue from a donor to a recipient transfers some bone marrow-derived cells, some of which are stem cells. The donor bone marrow-derived cells can persist and establish bone marrow *microchimerism*, i.e., permanent persistence of small numbers of bone marrow derived stem cells of donor type, presumably due to establishment of a few donor stem cells. This would link transplantation-induced peripheral tolerance with classic neonatal tolerance in mice,[218] which is probably a chimeric state. Microchimerism after blood transfusion may explain the well known blood transfusion effect and why matching of HLA antigens between the blood donor and recipient helps to establish the hyporesponsive state. Some long-term transplant recipients have evidence of microchimerism,[219] even decades after the transplant. Persistent donor cells could be the result or the cause of host hyporesponsiveness. Microchimerism in long-term survivors could lead to central tolerance and clonal deletion by colonization of the host thymus by donor stem cells.

Implantation of allogeneic tissues in the thymus before allografting is an experimental strategy for inducing some central tolerance in rodents.[220] It remains a challenge to demonstrate that this technique works in large animals and man.

CHRONIC REJECTION

This is a process whereby a successful graft begins to develop a slow deterioration in function, usually with nonspecific features which do not easily make for diagnosis. Each organ has unique features, but certain themes recur, including:

1. Thickening of the intima of arteries and arterioles due to smooth muscle cell invasion and proliferation;
2. A degree of parenchymal atrophy and interstitial fibrosis which may or may not represent ischemia.

The organ specific features are:

1. Heart: severe diffuse concentric coronary artery disease extending into small vessels.
2. Kidney: some cases have proteinuria and a variable glomerular lesion termed "transplant glomerulopathy". Hypertension is frequent.
3. Lung: obliterative bronchiolitis dominates the picture with marked narrowing of the respiratory bronchiole.
4. Liver: destruction of the bile ductules (vanishing bile ducts) may be the dominant lesion.

Chronic rejection often follows acute rejection, and some observers believe that acute rejection, incompletely reversed, is the harbinger of chronic rejection. Despite its nonspecific features, chronic rejection may result from a specific immune response. Earlier beliefs that alloantibody causes chronic rejection have now been tempered by the realization that relatively few cases have evidence of donor-specific antibody. The immune mechanisms appear to be additive with other factors related to the age, acute injury, hypertension, etc. The final common pathway may have elements in common with other chronic diseases or aging.[221]

REFERENCES

1. Little CC, Tyzzer EE. Further experimental studies on the inheritance of susceptibility to a transplantable tumor carcinoma (J.W.A.) of the Japanese Waltzing Mouse. J Med Res 1916; 33:393-453.
2. Strom TB. Molecular immunology and immunopharmacology of allograft rejection. Kidney Int 1992; 42(38):182-187.
3. Krensky AM, Weiss A, Crabtree G et al. T-lymphocyte-antigen interactions in transplant rejection. N Engl J Med 1990; 322:510-517.
4. Rosenberg AS, Singer A. Cellular basis of skin allograft rejection: an in vivo model of immune-mediated tissue destruction. Annu Rev Immunol 1992; 10:333-358.
5. Halloran PF, Broski AP, Batiuk TD et al. The molecular immunology of acute rejection: an overview. Transplant Immunology 1993; 1:3-27.
6. Weissman IL. Developmental switches in the immune system. Cell 1994; 76:207-218.

7. Nossal GJV. Negative selection of lymphocytes. Cell 1994; 76:229-239.

8. Paul WE, Seder RA. Lymphocyte responses and cytokines. Cell 1994; 76:241-251.

9. Kishimoto T, Taga T, Akira S. Cytokine signal transduction. Cell 1994; 76:253-262.

10. Weiss A, Littman DR. Signal transduction by lymphocyte antigen receptors. Cell 1994; 76:263-274.

11. Janeway CA Jr, Bottomly K. Signals and signs for lymphocyte responses. Cell 1994; 76:275-285.

12. Germain RN. MHC-dependent antigen processing and peptide presentation: providing ligands for T lymphocyte activation. Cell 1994; 76:287-299.

13. Springer TA. Traffic signals for lymphocyte recirculation and leukocyte emigration: the multistep paradigm. Cell 1994; 76:301-314.

14. Sprent J. T and B memory cells. Cell 1994; 76:315-322.

15. Patel R, Terasaki PI. Significance of the positive crossmatch test in kidney transplantation. N Engl J Med 1969; 280:735-780.

16. Paul LC, Fellstrom B. Chronic vascular rejection of the heart and the kidney—have rational treatment options emerged. Transplantation Overview. Transplantation 1992; 53:1169-1179.

17. Williams AF. A year in the life of the immunoglobulin superfamily. Immunol Today 1987; 8:298-303.

18. Campbell RD, Trowsdale J. Map of the human MHC. Immunol Today 1993; 14:349-352.

19. Bodmer JG, Marsh SGE, Albert ED et al. Nomenclature for factors of the HLA system, 1991. Eur J Immunogen 1992; 19:327-344.

20. Ruppert J, Sidney J, Celis E et al. Prominent role of secondary anchor residues in peptide binding to HLA-A2.1 molecules. Cell 1993; 74:929-937.

21. Abastado JP, Casrouge A, Kourilsky P. Differential role of conserved and polymorphic residues of the binding groove of MHC class I molecules in the selection of peptides. J Immunol 1993; 151:1569-1575.

22. Sette A, Sidney J, Oseroff C et al. HLA DR4w4-binding motifs illustrate the biochemical basis of degeneracy and specificity in peptide-DR interactions. J Immunol 1993; 151:3163-3170.

23. Bjorkman PJ, Saper MA, Samraoui B et al. Structure of the human class I histocompatibility antigen, HLA-A2. Nature 1987; 329:506-512.

24. Bjorkman PJ, Saper MA, Samraoui B et al. The foreign antigen binding site and T cell recognition regions of class I histocompatibility antigens. Nature 1987; 329:512-518.

25. Brown JH, Jardetzky TS, Gorga JC et al. Three-dimensional structure of the human class II histocompatibility antigen HLA-DR1. Nature 1993; 364:33-39.

26. Connolly JM, Hansen TH, Ingold AL et al. Recognition by CD8 on cytotoxic T lymphocytes is ablated by several substitutions in the class I a3 domain: CD8 and the T-cell receptor recognize the same class I molecule. Proc Natl Acad Sci USA 1990; 87:2137-2141.

27. Potter TA, Rajan TV, Dick RF II et al. Substitution at residue 227 of H-2 class I molecules abrogates recognition by CD8-dependent, but not CD8-independent, cytotoxic T lymphocytes. Nature 1989; 337:73-75.

28. Konig R, Huang LY, Germain RN. MHC class II interactions with CD4 mediated by a region analogous to the MHC class I binding site for CD8. Nature 1992; 356:796-798.

29. Belch MP, Madrigal JA, Hildebrand WH et al. Unusual HLA-B alleles in two tribes of Brazilian Indians. Nature 1992; 357:326-329.

30. Watkins DI, McAdam SN, Liu X et al. New recombinant HLA-B alleles in a tribe of South American Indians indicate rapid evolution of MHC class I loci. Nature 1992; 357:329-333.
31. Howard J. Fast forward in the MHC. Nature 1992; 357:284-285.
32. Klein J, Satta Y, O'hUigin C. The molecular descent of the major histocompatibility complex. Annu Rev Immunol 1993; 11:269-295.
33. Brodsky FM, Guagliardi LE. The cell biology of antigen processing and presentation. Annu Rev Immunol 1991; 9:707-744.
34. Malnati MS, Marti M, La Vaute T et al. Processing pathways for presentation of cytosolic antigen to MHC class II-restricted T cells. Nature 1992; 357:702-704.
35. Peters PJ, Neefjes JJ, Oorschot V et al. Segregation of MHC class II molecules from MHC class I molecules in the Golgi complex for transport to lysosomal compartments. Nature 1991; 349:669-676.
36. Williams DB, Barber BH, Flavell RA et al. Role of β_2-microblogulin in the intracellular transport and surface expression of murine class I histocompatibility molecules. J Immunol 1989; 142:2796-2806.
37. Degen E, Williams DB. Participation of a novel 88-kD protein in the biogenesis of murine class I histocompatibility molecules. J Cell Biol 1991; 112:1099-1115.
38. Degen E, Cohen-Doyle MF, Williams DB. Efficient dissociation of the p88 chaperone from major histocompatibility complex class I molecules requires both β_2-microglobulin and peptide. J Exp Med 1992; 175:1653-1661.
39. Teyton L, O'Sullivan D, Dickson PW et al. Invariant chain distinguishes between the exogenous and endogenous antigen presentation pathways. Nature 1990; 348:39-44.
40. Neefjes JJ, Stollorz V, Peters PJ et al. The biosynthetic pathway of MHC class II but not class I molecules intersects the endocytic route. Cell 1990; 61:171-183.
41. Guagliardi LE, Koppelman B, Blum JS et al. Co-localization of molecules nvolved in antigen processing and presentation in an early endocytic compartment. Nature 1990; 343:133-139.
42. Lotteau V, Teyton L, Peleraux A et al. Intracellular transport of class II MHC molecules directed by invariant chain. Nature 1990; 348:600-605.
43. Chicz RM, Urban RG, Lane WS et al. Predominant naturally processed peptides bound to HLA-DR1 are derived from MHC-related molecules and are heterogeneous in size. Nature 1992; 358:764-768.
44. Donermeyer DL, Allen PM. Binding to Ia protects an immunogenic peptide from proteolytic degradation. J Immunol 1989; 142:1063-1068.
45. Rudensky AY, Preston-Hurlburt P, Al-Ramadi BK et al. Truncation variants of peptides isolated from MHC class II molecules suggest sequence motifs. Nature 1992; 359:429-431.
46. Chicz RM, Urban RG, Gorga JC et al. Specificity and promiscuity among naturally processed peptides bound to HLA-DR alleles. J Exp Med 1993; 178:27-47.
47. Sargent CA, Bunham I, Trowsdale J et al. Human major histocompatibility complex contains genes for the major heat shock protein HSP70. Proc Natl Acad Sci USA 1989; 86:1968-1972.
48. Vanbuskirk A, Crump BL, Margoliash E et al. A peptide binding protein having a role in antigen presentation is a member of the HSP70 heat shock family. J Exp Med 1989; 170:1799-1809.
49. Driscoll J, Brown MG, Finley D et al. MHC-linked LMP gene products specifically alter peptidase activities of the proteasome. Nature 1993; 365:262-264.
50. Powis SH, Tonks S, Mockridge I et al. Alleles and haplotypes of the MHC-encoded ABC transporters TAP1 and TAP2. Immunogenetics 1993; 37:373-380.

51. Kelly A, Powis SH, Kerr LA et al. Assembly and function of the two ABC transporter proteins encoded in the human major histocompatibility complex. Nature 1992; 355:641-644.

52. Glynne R, Kerr LA, Mockridge I et al. The major histocompatibility complex-encoded proteasome component LMP7: alternative first exons and post-translational processing. Eur J Immunol 1993; 23:860-866.

53. Kelly A, Powis SH, Glynne R et al. Second proteasome-related gene in the human MHC class II region. Nature 1991; 353:667-668.

54. Glynne R, Powis SH, Beck S et al. A proteasome-related gene between the two ABC transporter loci in the class II region of the human MHC. Nature 1991; 353:357-360.

55. Spies T, Cerundolo V, Colonna M et al. Presentation of viral antigen by MHC class I molecules is dependent on a putative peptide transporter heterodimer. Nature 1992; 355:644-646.

56. Powis SJ, Townsend ARM, Deverson EV et al. Restoration of antigen presentation to the mutant cell line RMA-S by an MHC-linked transporter. Nature 1991; 354:528-531.

57. Attaya M, Jameson S, Martinez CK et al. Ham-2 corrects the class I antigen-processing defect in RMA-S cells. Nature 1992; 355:647-649.

58. Spies T, De Mars R. Restored expression of major histocompatibility class I molecules by gene transfer of a putative peptide transporter. Nature 1991; 351:323-324.

59. Powis SH, Mockridge I, Kelly A et al. Polymorphism in a second ABC transporter gene located within the class II region of the human major histocompatibility complex. Proc Natl Acad Sci USA 1992; 89:1463-1467.

60. Powis SJ, Deverson EV, Coadwell WJ et al. Effect of polymorphism of an MHC-linked transporter on the peptides assembled in a class I molecule. Nature 1992; 357:211-215.

61. Monaco JJ. Structure and function of genes in the MHC class II region. Curr Opin Immunol 1993; 5:17-20.

62. Goldberg AL, Rock KL. Proteolysis, proteasomes and antigen presentation. Review article. Nature 1992; 357:375-379.

63. Parham P. Antigen processing: transporters of delight. Nature 1990; 348:674-675.

64. Monaco JJ, McDevitt HO. Identification of a fourth class of proteins linked to the murine major histocompatibility complex. Proc Natl Acad Sci USA 1982; 79:3001-3005.

65. Davis MM, Bjorkman PJ. T-cell antigen receptor genes and T-cell recognition. Nature 1988; 334:395-402.

66. Hong SC, Chelouche A, Lin RH et al. An MHC interaction site maps to the amino-terminal half of the T-cell receptor a chain variable domain. Cell 1992; 69:999-1009.

67. Kasibhatla S, Nalefski EA, Rao A. Simultaneous involvement of all six predicted antigen binding loops of the T cell receptor in recognition of the MHC/antigenic peptide complex. J Immunol 1993; 151:3140-3151.

68. White J, Pullen A, Choi K et al. Antigen recognition properties of mutant Vβ+ T cell receptors are consistant with an immunoglobulin-like structure for the receptor. J Exp Med 1993; 177:119-125.

69. Weber S, Traunecker A, Oliveri F et al. Specific low-affinity recognition of major histocompatibility complex plus peptide by soluble T-cell receptor. Nature 1992; 356:793-796.

70. Ryu SE, Kwong PD, Truneh A et al. Crystal structure of an HIV-binding recombinant fragment of human CD4. Nature 1990; 348:419-426.

71. Brady RL, Dodson EJ, Dodson GG et al. Crystal structure of domains 3 and 4 of rat CD4: relation to the NH_2-terminal domains. Science 1993; 260:979-983.

72. Fleury S, Lamarre D, Meloche S et al. Mutational analaysis of the interaction between CD4 and class II MHC: class II antigens contact CD4 on a surface opposite the gp 120-binding site. Cell 1991; 66:1037-1049.

73. Cammarota G, Scheirle A, Takacs B et al. Identification of a CD4 binding site on the β_2 domain of HLA-DR molecules. Nature 1992; 356:799-801.

74. Vignali DAA, Moreno J, Schiller D et al. Species-specific binding of CD4 to the β_2 domain of major histocompatibility complex class II molecules. J Exp Med 1992; 175:925-932.

75. Leahy DJ, Axel R, Hendrickson WA. Crystal structure of a soluble form of the human T cell coreceptor CD8 at 2.6 Å resolution. Cell 1992; 68:1145-1162.

76. Sanders SK, Fox RO, Kavathas P. Mutations in CD8 that affect interactions with HLA class I and monoclonal anti-CD8 antibodies. J Exp Med 1991; 174:371-379.

77. Wheeler CJ, von Hoegen P, Parnes JR. An immunological role for the CD8 β-chain. Nature 1992; 357:247-249.

78. Newberg MH, Ridge JP, Vining DR et al. Species specificity in the interaction of CD8 with the a3 domain of MHC class I molecules. J Immunol 1992; 149:136-142.

80. Grandea AG,III, Bevan MJ. A conservative mutation in a class I MHC molecule outside the peptide binding groove stimulates responses to self peptides. J Immunol 1993; 151:3981-3987.

81. Jorgensen JL, Esser U, Fazekas de St.Groth B et al. Mapping T-cell receptor-peptide contacts by variant peptide immunization of single-chain transgenics. Nature 1992; 355:224-230.

82. Udaka K, Tsomides TJ, Eisen HN. A naturally occurring peptide recognized by alloreactive $CD8^+$ cytotoxic T lymphocytes in association with a class I MHC protein. Cell 1992; 69:989-998.

83. Sayegh MH, Khoury SJ, Hancock WW et al. Induction of immunity and oral tolerance with polymorphic class II major histocompatibility complex allopeptides in the rat. Proc Natl Acad Sci USA 1992; 89:7762-7766.

84. Sayegh MH, Perico N, Imberti O et al. Thymic recognition of class II major histocompatibility complex allopeptides induces donor-specific unresponsiveness to renal allografts. Transplantation 1993; 56:461-465.

85. Hunt DF, Henderson RA, Shabanowitz J et al. Characterization of peptides bound to the class I MHC molecule HLA-A2.1 by mass spectrometry. Science 1992; 255:1261-1263.

86. van Bleek GM, Nathenson SG. The structure of the antigen-binding groove of major histocompatibility complex class I molecules determines specific selection of self-peptides. Proc Natl Acad Sci USA 1991; 88:11032-11036.

87. Madden DR, Gorga JC, Strominger JL et al. The structure of HLA-B27 reveals nonamer self-peptides bound in an extended conformation. Nature 1991; 353:321-325.

88. Chen BP, Madrigal A, Parham P. Cytotoxic T-cell recognition of an endogenous class I HLA peptide presented by a class II HLA molecule. J Exp Med 1990; 172:779-788.

89. Lerner A, D'Adamio L, Diener AC et al. CD3 zeta/nu/theta locus is colinear with and transcribed antisense to the gene encoding the transcription factor Oct-1. J Immunol 1993; 151:3152-3162.

90. Krishna S, Benaroch P, Pillai S. Tetrameric cell-surface MHC class I molecules. Nature 1992; 357:164-167.

91. Williams AF, Beyers AD. At grips with interactions. Nature 1992; 356:746-747.

92. Appleby MW, Gross JA, Cooke MP et al. Defective T cell receptor signaling in mice lacking the thymic isoform of p59^{fyn}. Cell 1992; 70:751-763.

93. Penninger J, Kishihara K, Molina T et al. Requirement for tyrosine kinase p56^{lck} for thymic development of transgenic αδ T cells. Science 1993; 260:358-361.

94. Chow LML, Fournel M, Davidson D et al. Negative regulation of T-cell receptor signalling by tyrosine protein kinase ^{csk}. Nature 1993; 365:156-160.

95. Ravichandran KS, Lee KK, Songyang Z et al. Interaction of Shc with the zeta chain of the T-cell receptor upon T-cell activation. Science 1993; 262:902-905.

96. Rozakis-Adcock M, McGlade J, Mbamalu G et al. Association of the Shc and Grb2/Sem5 SH2-containing proteins is implicated in activation of the Ras pathway by tyrosine kinases. Nature 1992; 360:689-692.

97. Egan SE, Giddings BW, Brooks MW et al. Association of Sos Ras exchange protein with Grb2 is implicated in tyrosine kinase signal transduction and transformation. Nature 1993; 363:45-51.

98. Berridge MJ. Inositol trisphosphate and calcium signalling. Nature 1993; 361:315-325.

99. Schwartz RH. Co-stimulation of T lymphocytes: The role of CD28, CTLA-4, and B7/BB1 in interleukin-2 production and immunotherapy. Cell 1992; 71:1065-1068.

100. Lu Y, Granelli-Piperno A, Bjorndahl JM et al. CD28-induced T-cell activation. Evidence for a protein-tyrosine kinase signal transduction pathway. J Immunol 1992; 149:24-29.

101. Cohen J. New protein steals the show as 'costimulator' of T cells. Science 1993; 262:844-845.

102. Azuma M, Ito D, Yagita H et al. B70 antigen is a second ligand for CTLA-4 and CD28. Nature 1993; 366:76-79.

103. Freeman GJ, Gribben JG, Boussiotis VA et al. Cloning of B7-2: A CTLA-4 counter-receptor that costimulates human T-cell proliferation. Science 1993; 262:909-911.

104. Freeman GJ, Borriello F, Hodes RJ et al. Uncovering of functional alternative CTLA-4 counter-receptor in B7-deficient mice. Science 1993; 262:907-909.

105. Hathcock KS, Laszlo G, Dickler HB et al. Identification of an alternative CTLA-4 ligand costimulatory for T-cell activation. Science 1993; 262:905-907.

106. June CH, Ledbetter JA, Gillespie MA et al. T-cell proliferation involving the CD28 pathway is associated with cyclosporine-resistant interleukin 2 gene expression. Mol Cell Biol 1987; 7:4472-4481.

107. Shahinian A, Pfeffer K, Lee KP et al. Differential T-cell costimulatory requirements in CD28-deficient mice. Science 1993; 261:609-612.

108. Donovan JA, Koretzky GA. CD45 and the immune response. J Am Soc Nephrol 1993; 4:976-985.

109. Jain J, McCaffrey PG, Valge-Archer VE et al. Nuclear factor of activated T cells contains Fos and Jun. Nature 1992; 356:801-804.

110. Putney JW Jr., Bird GSTJ. The signal for capacitative calcium entry. Cell 1993; 75:199-201.

111. Ullman DS, Northrop JP, Verweij CL et al. Transmission of signals from the T-lymphocyte antigen receptor to the genes responsible for cell proliferation and immune function: the missing link. Annu Rev Immunol 1990; 8:421-452.

112. Jain J, McCaffrey PG, Miner Z et al. The T-cell transcription factor NFAT$_p$ is a substrate for calcineurin and interacts with Fos and Jun. Nature 1993; 365:352-355.

113. Arima N, Kuziel WA, Gardina TA et al. IL-2-induced signal transduction involves the activation of nuclear NF-κB expression. J Immunol 1992; 149:83-91.

114. Elices MJ, Osborn L, Takada Y et al. VCAM-1 on activated endothelium interacts with the leukocyte integrin VLA-4 at a site distinct from the VLA-4/fibronectin binding site. Cell 1990; 60:577-584.

115. Cosman D. Control of messenger RNA stability. Immunol Today 1987; 8:16-17.

116. Schlunegger MP, Grütter MG. An unusual feature revealed by the crystal structure at 2.2 Å resolution of human transforming growth factor-β2. Nature 1992; 358:430-434.

117. Jones EY, Stuart DI, Walker NPC. Structure of tumor necrosis factor. Nature 1989; 338:225-228.

118. Fiers W. Tumor necrosis factor. Characterization at the molecular, cellular and in vivo level. FEBS Lett 1991; 285:199-212.

119. Spriggs DR, Deutsch S, Kufe DW. Genomic structure, induction, and production of TNF-α. Immunol Ser 1992; 56:3-34.

120. Kühn R, Rajewsky K, Müller W. Generation and analysis of Interleukin-4 deficient mice. Science 1991; 254:707-710.

121. Schorle H, Holtschke T, Hünig T et al. Development and function of T cells in mice rendered interleukin-2 deficient by gene targeting. Nature 1991; 352:621-624.

122. Paul WE. Poking holes in the network. Nature 1992; 357:16-17.

123. Rothe J, Lesslauer W, Lötscher H et al. Mice lacking the tumor necrosis factor receptor 1 are resistant to TNF-mediated toxicity but highly susceptible to infection by Listeria monocytogenes. Nature 1993; 364:798-802.

124. Goverman J, Woods A, Larson L et al. Transgenic mice that express a myelin basic protein-specific T-cell receptor develop spontaneous autoimmunity. Cell 1993; 72:551-560.

125. Strober W, Ehrhardt RO. Chronic intestinal inflammation: an unexpected outcome in cytokine or T-cell receptor mutant mice. Cell 1993; 75:203-205.

126. Bazan JF. Haemopoietic receptors and helical cytokines. Immunol Today 1990; 11:350-354.

127. Taga T, Kishimoto T. Cytokine receptors and signal transduction. FASEB J 1993; 7:3387-3396.

128. Loetscher H, Brockhaus M, Dembic Z et al. Two distinct tumor necrosis factor receptors - members of a new cytokine receptor gene family. Oxf Surv Eukaryot Genes 1991; 78:119-142.

129. Farrar MA, Schreiber RD. The molecular cell biology of interferon-γ and its receptor. Annu Rev Immunol 1993; 11:571-611.

130. Waksman G, Kominos D, Robertson SC et al. Crystal structure of the phosphotyrosine recognition domain SH2 of V-src complexed with tyrosine-phosphorylated peptides. Nature 1992; 358:646-653.

131. Petsko GA. Fishing in Src-infested waters. Nature 1992; 358:625-626.

132. Kuo CJ, Chung J, Fiorentino DF et al. Rapamycin selectively inhibits interleukin-2 activation of p70 S6 kinase. Nature 1992; 358:70-73.

133. Chung J, Kuo CJ, Crabtree GR et al. Rapamycin-FKBP specifically blocks growth-dependent activation of and signaling by the 70 kD S6 protein kinases. Cell 1992; 69:1227-1236.

134. Mosmann TR. Directional release of lymphokines from T cells. Immunol Today 1988; 10:306-307.

135. Mosmann TR, Coffman RL. Th1 and Th2 cells: different patterns of lymphokine secretion lead to different functional properties. Annu Rev Immunol 1989; 10:145-173.

136. Mosmann TR, Moore KW. The role of IL-10 in crossregulation of TH1 and TH2 responses. Immunol Today 1991; 12:A49-A53.

137. Boom WH, Liano D, Abbas AK. Heterogeneity of helper/inducer T lymphocytes. II. Effects of interleukin 4- and interleukin 2-producing T-cell clones on resting B lymphocytes. J Exp Med 1988; 167:1352-1363.

138. Rasmussen R, Takatsu K, Harada N et al. T cell-dependent hapten-specific and polyclonal B cell responses require release of interleukin-5. J Immunol 1988; 140:705-712.

139. Go NF, Castle BE, Barrett R et al. Interleukin-10, a novel B cell stimulatory factor: unresponsiveness of X chromosome-linked immunodeficiency B cells. J Exp Med 1990; 172:1625-1631.

140. Mosmann TR, Coffman RL. Heterogeneity of cytokine secretion patterns and functions of helper T cells. Adv Immunol 1989; 46:111-147.

141. Erard F, Corthesy P, Nabholtz M et al. Interleukin-2 is both necessary and sufficient for the growth and differentiation of lectin-stimulated cytolytic T lymphocyte precursors. J Immunol 1985; 134:1644-1652.

142. Simon MM, Landolfo S, Diamantstein T et al. Antigen- and lectin-sensitized murine cytolytic T lymphocyte-precursors require both interleukin-2 and endogenously produced immune (gamma) interferon for their growth and differentiation into effector cells. Curr Top Microbiol Immunol 1986; 126:173-185.

143. Pfeifer JD, McKenzie DT, Swain SL et al. B cell stimulatory factor 1 (interleukin 4) is sufficient for the proliferation and differentiation of lectin-stimulated cytolytic T lymphocyte precursors. J Exp Med 1987; 166:1464-1470.

144. Takatsu K, Kikuchi YH, Takahashi T et al. Interleukin-5, a T-cell derived B-cell differentiation factor also induces cytotoxic T lymphocytes. Proc Natl Acad Sci USA 1987; 84:4234-4238.

145. Cher DJ, Mosmann TR. Two types of murine helper T-cell clone: 2. Delayed-type hypersensitivity is mediated by TH1 clones. J Immunol 1987; 138:3688-3694.

146. Greenbaum LA, Horowitz JB, Woods A et al. Autocrine growth of CD4+ T cells. Differential effects of IL-1 on helper and inflammatory T cells. J Immunol 1988; 140:1555-1560.

147. Killar L, MacDonald G, West J et al. Cloned, Ia-restricted T cells that do not produce interleukin-4 (IL-4)/B cell stimulatory factor I (BSF-I) fail to help antigen-specific B cells. J Immunol 1987; 138:1674-1679.

148. Enk AH, Angeloni VL, Udey MC et al. Inhibition of Langerhans cell antigen-presenting function by IL-10. A role for IL-10 in induction of tolerance. J Immunol 1993; 151:2390-2398.

149. Shull MM, Ormsby I, Kier AB et al. Targeted disruption of the mouse transforming growth factor-β1 gene results in multifocal inflammatory disease. Nature 1992; 359:693-699.

150. Wahl SM, Allen JB, Costa GL et al. Reversal of acute and chronic synovial inflammation by anti-transforming growth factor β. J Exp Med 1993; 177:225-230.

151. Chen RH, Ebner R, Derynck R. Inactivation of the type II receptor reveals two receptor pathways for the diverse TGF-β activities. Science 1993; 260:1335-1338.

152. Koff A, Ohtsuki M, Polyak K et al. Negative regulation of G1 in mammalian cells: Inhibition of cyclin E-dependent kinase by TGF-β. Science 1993; 260:536-539.

153. Roberts AB, Sporn MB. Physiological actions and clinical applications of transforming growth factor-β (TGF-β). Growth Factors 1993; 8:1-9.

154. Santambrogio L, Hochwald GM, Saxena B et al. Studies on the mechanisms by which transforming growth factor-β (TGF-β) protects against allergic encephalomyelitis. J Immunol 1993; 151:1116-1127.

155. Bogdan C, Paik J, Vodovotz Y et al. Contrasting mechanism for suppression of macrophage cytokine release by transforming growth factor-β and IL-10. J Biol Chem 1992; 267:23301-23308.

156. Patel KJ, Neuberger MS. Antigen presentation by the B-cell antigen receptor is driven by the α/β sheath and occurs independently of its cytoplasmic tyrosines. Cell 1993; 74:939-946.

157. Croft M, Swain SL. B-cell response to fresh and effector T helper cells. Role of cognate T-B interaction and the cytokines IL-2, IL-4, and IL-6. J Immunol 1991; 146:4055-4064.

158. O'Rourke AM, Mescher MF. Cytotoxic T-lymphocyte activation involves a cascade of signalling and adhesion events. Nature 1992; 358:253-255.

159. Williams GT, Smith CA. Molecular regulation of apoptosis: genetic controls on cell death. Cell 1993; 74:777-779.

160. Bissonnette RP, Echiverri F, Mahboubi A et al. Apoptotic cell death induced by c-myc is inhibited by bcl-2. Nature 1992; 359:552-554.

161. Nakayama KI, Nakayama K, Negishi I et al. Disappearance of the lymphoid system in Bcl-2 homozygous mutant chimeric mice. Science 1993; 261:1584-1588.

162. Sherman LA, Chattopadhyay S, Biggs JA et al. Alloantibodies can discriminate class I major histocompatibility complex molecules associated with various endogenous peptides. Proc Natl Acad Sci USA 1993; 90:6949-6951.

163. Pober JS, Cotran RS. Immunologic interactions of T lymphocytes with vascular endothelium. Adv Immunol 1991; 50:261-302.

164. Podack ER, Hengartner H, Lichtenheld MG. A central role of perforin in cytolysis? Annu Rev Immunol 1991; 9:129-157.

165. Karlhofer FM, Ribaudo RK, Yokoyama WM. MHC class I alloantigen specificity of Ly-49+ IL-2-activated natural killer cells. Nature 1992; 358:66-70.

166. Storkus WJ, Alexander J, Payne JA et al. Reversal of natural killing susceptibility in target cells expressing transsected class I HLA genes. Proc Natl Acad Sci USA 1989; 86:2361-2364.

167. Storkus WJ, Salter RD, Alexander J et al. Class I-induced resistance to natural killing: identification of nonpermissive residues in HLA-A2. Proc Natl Acad Sci USA 1991; 88:5989-5992.

168. Berendt AR, McDowall A, Craig AG et al. The binding site on ICAM-1 for plasmodium falciparum-infected erythrocytes overlaps, but is distinct from, the LFA-1-binding site. Cell 1992; 68:71-81.

169. Albelda SM, Buck CA. Integrins and other cell adhesion molecules. FASEB J 1990; 4:2868-2880.

170. Winn RK, Liggitt D, Vedder NB et al. Anti-P-selectin monoclonal antibody attenuates reperfusion injury to the rabbit ear. J Clin Invest 1993; 92:2042-2047.

171. Abe Y, Sekiya S, Yamasita T et al. Vascular hyperpermeability induced by tumor necrosis factor and its augmentation by IL-1 and IFN-γ is inhibited by selective depletion of neutrophils with a monoclonal antibody. J Immunol 1990; 145:2902-2907.

172. Gerritsen ME, Bloor CM. Endothelial cell gene expression in response to injury. FASEB J 1993; 7:523-533.

173. Remuzzi G, Benigni A. Endothelin in the control of cardiovascular and renal function. Lancet 1993; 342:589-593.

174. Langrehr JM, Hoffman RA, Lancaster JR Jr et al. Nitric oxide—a new endogenous immunomodulator. Transplantation 1993; 55:1205-1212.

175. Bogdan C, Vodovotz Y, Paik J et al. Traces of bacterial lipopolysaccharide suppress IFN-γ-induced nitric oxide synthase gene expression in primary mouse macrophages. J Immunol 1993; 151:301-309.

176. Albina JE, Abate JA, Henry WL Jr. Nitric oxide production is required for murine resident peritoneal macrophages to suppress mitogen-stimulated T–cell proliferation. Role of IFN-γ in the induction of the nitric oxide-synthesizing pathway. J Immunol 1991; 147:144-148.

177. Mantovani A, Bussolino F, Dejana E. Cytokine regulation of endothelial cell function. FASEB J 1992; 6:2591-2599.

178. Morel D, Normand E, Lemoine C et al. Tumor necrosis factor alpha in human kidney transplant rejection—analysis by in situ hybridization. Transplantation 1993; 55:773-777.

179. Oppenheim JJ, Zachariae COC, Mukaida N et al. Properties of the novel proinflammatory supergene "intercrine" cytokine family. Annu Rev Immunol 1991; 9:617-648.

180. Blanchet O, Bourge JF, Zinszner H et al. DNA binding of regulatory factors interacting with MHC-class-1 gene enhancer correlates with MHC-class-1 transcription level in class-1-defective cell lines. Int J Cancer 1991; 6:138-145.

181. Kasama T, Strieter RM, Standiford TJ et al. Expression and regulation of human neutrophil-derived macrophage inflammatory protein 1α. J Exp Med 1993; 178:63-72.

182. Taub DD, Conlon K, Lloyd AR et al. Preferential migration of activated CD4+ and CD8+ T cells in response to MIP-1a and MIP-1β. Science 1993; 260:355-358.

183. O'Hehir RE, Yssel H, Verma S et al. Clonal analysis of differential lymphokine production in peptide and superantigen induced T-cell anergy. Int Immunol 1991; 3:819-826.

184. Glimcher LH, Kara CJ. Sequences and factors: A guide to MHC class-II transcription. Annu Rev Immunol 1992; 10:13-49.

185. Halloran PF, Madrenas J. Regulation of MHC transcription. Transplantation Overview. Transplantation 1990; 50:725-738.

186. Hardy KJ, Sawada T. Human γ interferon strongly upregulates its own gene expression in peripheral blood lymphocytes. J Exp Med 1989; 170:1021-1026.

187. Burke PA, Hirschfeld S, Shirayoshi Y et al. Developmental and tissue-specific expression of nuclear proteins that bind the regulatory element of the major histocompatibility complex class I gene. J Exp Med 1989; 169:1309-1321.

188. Sloan JH, Hasegawa SL, Boss JM. Single base pair substitutions within the HLA-DRA gene promoter separate the functions of the X1 and X2 boxes. J Immunol 1992; 148:2591-2599.

189. La Rosa FG, Talmage DW. Major histocompatibility complex antigen expression on parenchymal cells of thyroid allografts is not by itself sufficient to induce rejection. Transplantation 1990; 49:605-609.

190. Halloran PF, Autenried P, Ramassar V et al. Local T cell responses induce widespread MHC expression. Evidence that IFN-γ induces its own expression in remote sites. J Immunol 1992; 148:3837-3846.

191. Larner AC, David M, Feldman GM et al. Tyrosine phosphorylation of DNA binding proteins by multiple cytokines. Science 1993; 261:1730-1733.

192. Ruff-Jamison S, Chen K, Cohen S. Induction by EGF and interferon-γ of tyrosine phosphorylated DNA binding proteins in mouse liver nuclei. Science 1993; 261:1733-1736.

193. Sadowski HB, Shuai K, Darnell JE Jr et al. A common nuclear signal transduction pathway activated by growth factor and cytokine receptors. Science 1993; 261:1739-1744.

194. Shuai K, Stark GR, Kerr IM et al. A single phosphotyrosine residue of Stat91 required for gene activation by interferon-γ. Science 1993; 261:1744-1746.

195. Müller M, Briscoe J, Laxton C et al. The protein tyrosine kinase JAK1 complements defects in interferon-α/β and -γ signal transduction. Nature 1993; 366:129-135.

195a. Reith W, Muhlethalet-Mottet A, Krzysztof M et al. The molecular basis of MHC class II deficiency and transcriptional control of MHC class II gene expression. Microbes and Infection 1999; I: 839-846.

196. Solez K, Alexlsen RA, Benediktsson H et al. International standardization of criteria for the histologic diagnosis of renal allograft rejection: The Banff working classification of kidney transplant pathology. Kidney Int 1993; 44:411-422.

197. Billingham ME, Cary NRB, Hammond ME et al. A working formulation for the standardization of nomenclature in the diagnosis of heart and lung rejection: Heart rejection study group. J Heart Transplant 1990; 9:587-593.

198. Mueller C, Gershenfeld HK, Lobe CG et al. A high proportion of T lymphocytes that infiltrate H-2-incompatible heart allografts in vivo express genes encoding cytotoxic cell-specific serine proteases, but do not express the MEL-14-defined lymph node homing receptor. J Exp Med 1988; 167:1124-1136.

199. Hall BM. Cells mediating allograft rejection. Transplantation 1991; 51:1141-1151.

200. Cerilli J, Brasile L, Galouzis T et al. The vascular endothelial cell antigen system. Transplantation 1985; 49:286-289.

201. Halloran PF, Wadgymar A, Ritchie S et al. The significance of the anti-class I antibody response. I. Clinical and pathologic features of anti-class I mediated rejection. Transplantation 1990; 49:85-91.

202. Halloran PF, Schlaut J, Solez K et al. The significance of the anti-class I response II. Clinical and pathologic features of renal transplants with anti-class I-like antibody. Transplantation 1992; 53:550-555.

203. Von Boehmer H. Developmental biology of T cells in T cell-receptor transgenic mice. Annu Rev Immunol 1990; 8:531-556.

204. Remuzzi G, Rossini M, Imberti O et al. Kidney graft survival in rats without immunosuppressants after intrathymic glomerular transplantation. Lancet 1991; 337:750-753.

206. Rocha B, Von Boehmer H. Peripheral selection of the T-cell repertoire. Science 1991; 251:1225-1228.

207. Kawabe Y, Ochi A. Programmed cell death and extrathymic reduction of V$\beta8^+$ CD4$^+$ T cells in mice tolerant to *Staphylococcus aureus* enterotoxin B. Nature 1991; 349:245-248.

208. Morahan G, Allison J, Miller JF. Tolerance of class I histocompatibility antigens expressed extrathymically. Nature 1989; 339:622-624.

209. Lo D, Burkley LC, Flavell RA et al. Tolerance in transgenic mice expressing class II major histocompatibility complex on pancreatic acinar cells. J Exp Med 1989; 170:87-104.

210. Sarvetnick N, Shizuru J, Liggitt D et al. Loss of pancreatic islet tolerance induced by β-cell expression of interferon-γ. Nature 1990; 346:844-847.

211. Takeuchi T, Lowry RP, Konieczny B. Heart allografts in murine systems. The differential activation of Th2-like effector cells in peripheral tolerance. Transplantation 1992; 53:1281-1294.

212. Howard M, O'Garra A. Biological properties of interleukin 10. Immunol Today 1992; 13:198-200.

213. Schönrich G, Kalinke U, Momburg F et al. Downregulation of T-cell receptors on self-reactive T cells as a novel mechanism for extrathymic tolerance induction. Cell 1991; 65:293-304.

214. Fink PJ, Shimonkevitz RP, Bevan MJ. Veto cells. Annu Rev Immunol 1988; 6:115-137.

215. Hall BM, Gurley KE, Dorsch SE. Specific unresponsiveness in rats with prolonged cardiac allograft survival after treatment with cyclosporine. Mediation of specific suppression by T helper/inducer cells. J Exp Med 1985; 162:1683-1694.

216. Hardy MA, Suciu-Foca N, Reed E et al. Immunodulation of kidney and heart transplants by anti-idiotypic antibodies. Ann Surg 1991; 214:522-530.

217. Shoskes D, Parfrey NA, Halloran PF. Increased major histocompatibility complex antigen expression in unilateral ischemic acute tubular necrosis in the mouse. Transplantation 1990; 49:201-207.

218. Billingham RE, Brent L, Medawar PB. Actively acquired tolerance of foreign cells. Nature 1953; 172:603-606.

219. Starzl TE, Demetris AJ, Trucco M et al. Chimerism after liver transplantation for type IV glycogen storage disease and type 1 Gaucher's disease. N Engl J Med 1993; 328:745-749.

220. Remuzzi G, Perico N. Induction of unresponsiveness via intrathymic inoculation. Lancet 1991; 338:450.

221. Halloran PF, Melk A, Barth C. Rethinking chronic allograft nephropathy: The concept of accelerated senenence. J Am Soc Nephrology 1999; 10: 167-18.

Overview of Living and Deceased Organ Donors, Immunosuppression and Outcomes

Frank P. Stuart

ORIGINS OF IMMUNOSUPPRESSION AND SOLID ORGAN TRANSPLANTATION

Out of turmoil, destruction, and death in World War II came research that led to the artificial kidney and recognition that allograft destruction was an immunological event. Peter Medawar described the accelerated second set rejection of mouse skin grafts while trying to understand why skin allografts always failed on severely burned pilots of the British Royal Air Force. Meanwhile, in Rotterdam, Willem Kolff developed a primitive artificial kidney to treat renal failure that followed severe crush injury during bombing of that city. In the early post-war years, Boston's Peter Bent Brigham Hospital assembled a team, which included Kolff that began human kidney transplantation and refined the artificial kidney. The first kidney transplant at the Brigham, in 1947, came from a cadaver donor and was revascularized by the recipient's antecubital vessels. The graft failed after three days, but it functioned long enough to clear the recipient's uremic coma and permitted recovery from reversible injury to the native kidneys. Next, the team developed the surgical procedure for implanting kidneys in the iliac fossa of dogs and performed it on 15 human recipients between 1949 and 1951. Immunosuppression had not yet arrived and the grafts failed. However, the iliac fossa operation worked well and was extended to kidney transplantation between identical twins in 1954. From 1955 until 1962, cadaver kidneys were transplanted in Paris and Boston with immunosuppression from whole body irradiation and adrenocortical steroids; there were a few hard-earned, short-term successes. The era of pharmacologic immunosuppression began in 1962 with a mother-to-son kidney transplant for which the recipient was treated with both steroids and the antiproliferative drug 6-mercaptopurine. The kidney functioned more than 20 years.

By 2003 three hundred thousand Americans with renal failure were maintained on chronic dialysis; seventeen thousand received kidney transplants in 2002 (8000 from deceased donors and 9000 from living donors). Graft and recipient survival beyond the critical first year exceeds 90 percent in many centers for kidney, pancreas, liver, heart, lung and intestine. At least 100,000 individuals wait for organ transplants in the United States.

Organ Transplantation, 2nd edition, edited by Frank P. Stuart, Michael M. Abecassis and Dixon B. Kaufman. ©2003 Landes Bioscience.

Table 2.1. *Immunosuppressive agents*

Generic Name	Brand Name	Company	Date of FDA Approval
Prednisone	Deltasone	Upjohn	<1960
Methylprednisolone	Solu-Medrol	Upjohn	<1960
Mercaptopurine	Purinethol	Glaxo Wellcome	<1960
Azathioprine	Imuran	Glaxo Wellcome	1969
Antithymocyte Globulin (equine)	ATGAM	Upjohn	1972*
Cyclosporine-A	Sandimmune	Novartis	1983
Muromonab-CD3 (OKT-3)	Orthoclone OKT-3	Ortho Biotech/ Johnson & Johnson	1986
Cyclosporine Microemulsion	Neoral	Novartis	1994
Tacrolimus	Prograf	Fujisawa	1995
Mycophenolate Mofetil	CellCept	Roche	1996
Daclizumab	Zenapax	Roche	1997
Basiliximab	Simulect	Novartis	1998
Antithymocyte Globulin (Rabbit)	Thymoglobulin	SangStat	1999
Sirolimus	Rapamune	Wyeth-Ayerst	1999
Rituximab	Rituxan	IDEC/Genetech	2001
Alemtuzumab	Campath	Berlex	2001

Immunosuppressive agents approved for use by United States Food and Drug Administration (FDA) since 1960. Detailed description of each agent appears in the appendix to this volume.
*No longer available

2

CADAVERIC DONORS

The Association of Organ Procurement Organizations (AOPO) estimates that potential cadaveric organ donors have stabilized at 11,000 to 14,000 yearly. A few (less than 5 percent) are non-heart-beating donors, but in the vast majority cardiovascular circulation remains intact until organs are removed. Despite broad public awareness of the need for and benefit from organ transplants, less than half of the immediate relatives of potential cadaveric donors give their consent to remove vital organs. Consequently, the number of donors is stalled between 6,000 and 7,000 each year. Moreover, the quality of organs has diminished as age of the typical cadaveric donor increases and the cause of death shifts from head trauma to stroke in 44 percent of donors. Transplant centers and potential recipients have turned increasingly to organs from "extended criteria cadaveric donors". Altogether in 2001 roughly 6,100 cadaveric donors provided 7773 kidney, 800 simultaneous pancreas/kidney, 446 pancreas, 25 islet, 48 intestine, 4,800 liver, 992 lung and 1973 heart transplants, but nearly 100,000 potential recipients remained on the waiting lists at the end of the year.

The gap between waiting list and available cadaveric organs widens every year. Few believe that the number of cadaveric organ transplants will increase more than one or two percent yearly during the next decade. Perhaps educational programs in primary and secondary school will lead the next generation to broader support of cadaveric organ donation based on altruism, enlightened self interest, or some form of compensation to the donor's estate.

The supply of cadaveric donors is just as inadequate in many European countries with presumed consent laws as it is in English speaking countries, where consent to remove organs must be obtained from the closest relative. In many countries where transplantation is not a high priority, individuals who are moribund and considered not salvageable are not admitted to intensive care units. Temporary ventilatory support is withheld and cardio circulatory collapse follows. In other potential donors ventilatory support may be initiated, but inadequate attention is paid to maintaining normal physiologic function in organ systems after brain death has been diagnosed. The striking exceptions are Spain, Austria and Belgium, all of which have presumed consent laws. In all three countries national government recognizes maximum retrieval of transplantable organs as a national priority. Spain in particular has engaged participation of anesthesiologists in most of its hospitals to lead identification of potential donors, implementation of protocols to improve cardiopulmonary function, approaching potential donor families and continuing donor medical management until organs are removed.[1,2] Spain's rate of cadaveric organ donation approaches 40 per million population which is twice the rate in the United States.

The message from Europe is that presumed consent alone will not yield high donor rates. But, presumed consent coupled with national determination and systems to identify and stabilize potential donors in every trauma center and emergency room will maximize the opportunity for cadaveric organ transplantation. However, most English speaking countries are unlikely to implement presumed consent laws. Presumed consent for removal of transplantable organs springs from the Napoleonic legal code that gave final authority for autopsy of dead bodies to the state rather than the family. As organ transplantation evolved, removal of transplantable organs was viewed as the first steps of an autopsy and thus covered under presumed consent. But under English common law it was the next of kin who was responsible for appropriate disposition of the relative's body and for granting permission for autopsy. Thus, maximizing the potential for cadaveric organ transplantation in most English speaking countries will require informed and willing public motivated by altruism or something else.

That half of the relatives of potential organ donors in the United States refuse consent implies that the public lacks information and understanding of organ donation or lacks long-term self-interest and sufficient altruism. For the past two decades clergy have supported organ donation in their sermons, staff from organ procurement organizations have spoken at schools and to many organizations, and television has dramatized the miracle of transplantation; so the public is rather well informed. Self-interest should also serve to motivate, because no one knows when failure of one of his own vital organs might call for transplantation; but self interest may not be immediate enough to motivate consent. Consequently, many believe that financial incentives passed through the potential donor's estate, perhaps as a funeral benefit or other consideration, are needed to increase motivation.

Debate intensified in 2002 over financial and other incentives to stimulate consent. Boards of Directors of the American Medical Association, UNOS, and the American Society of Transplant Surgeons voted to study and test changes to U.S. law that would permit compensating cadaveric donor's estates and burial costs. The American College of Surgeons and the Board of the National Kidney Foundation expressed vigorous opposition for even the testing of offering financial incentives to donate. A non financial incentive discussed at the December 2002, Congress on Ethics in Organ Transplantation (Munich) would reward family members of a cadaveric organ donor with preferred status on the transplant waiting list; that should appeal to enlightened self interest. Moreover, it is consistent with current UNOS policy that gives waiting list priority to living kidney donors who may need a kidney transplant in the future. The Munich Congress passed resolutions that allocation policies should aim at giving equal concern and respect to all potential recipients; equity and justice in organ allocation are as important as seeking maximum utility; and all societies should make every effort to maximize cadaver organ donation. Even with these resolutions the Congress recognized that living donor kidney transplantation should be encouraged and adopted as widely as possible.[3,4]

UNOS (United Network for Organ Sharing) is a not-for-profit corporation that has operated the nation's Organ Procurement and, Transplant Network (OPTN) since 1985 after the United States Congress passed the National Organ Transplant Act (NOTA). UNOS operates the OPTN under contract with the United States Department of Health and Human Services and ultimately reports to the Secretary of the Department. UNOS and its representative working committees achieved broad consensus with respect to the set of rules that governed organ transplantation until 1998-2000. But disagreement over rules for allocating scarce cadaveric

2

livers eventually prevented consensus, and in 2000 the Secretary appointed a committee of forty members to serve as an Advisory Committee on Transplantation. Satisfactory rules for sharing livers fell into place without resorting to the new committee, but the Advisory Committee convened in November 2002 at the Secretary's request to consider issues related to well-being of living donors, shortage of cadaveric donors and equal access to organ transplantation. The Advisory Committee made nine recommendations designed to increase cadaveric organ donation, two recommendations to encourage equal access to minority populations and seven recommendations with respect to living donors.[5] The seven that concern living donors will be presented later in this review, but the eleven recommendations on cadaveric donation and equal access are these:

1. That legislative strategies be adopted that will encourage medical examiners and coroners not to withhold life-saving organs and tissues from qualified organ procurement organizations.

2. That the Secretary of HHS, in concert with the Secretary of Education, should recommend to states that organ and tissue donation be included in core curriculum standards for public education as well as in the curricula of professional schools, including schools of education, schools of medicine, schools of nursing, schools of law, schools of public health, schools of social work and pharmacy schools.

3. That in order to ensure best practices, organ procurement organizations and the OPTN be encouraged to develop, evaluate, and support the implementation of improved management protocols of potential donors.

4. That in order to ensure best practices at hospitals and organ procurement organizations (OPO), the following measure should be added to the CMS (Center for Medicare/Medicaid Services). Conditions of Participation: Each hospital with more than 100 beds should identify an advocate for organ and tissue donation from within the hospital clinical staff.

5. That in order to ensure best practice at hospitals and OPOs, the following measure should be added to the CMS Conditions of Participation: Each hospital should establish, in conjunction with its OPO, policies and procedures to manage and maximize organ retrieval from donors without a heartbeat.

6. That the following measure be added to the CMS Conditions of Participation: Hospitals shall notify OPO prior to the withdrawal of life support to a patient, so as to determine that patient's potential for organ donation. If it is determined that the patient is a potential donor, the OPO shall reimburse the hospital for appropriate costs related to maintaining that patient as a potential donor.

7. That the regulatory framework provided by CMS for transplant and OPO certification should be based on principles of continuous quality improvement. Subsequent failure to meet performance standards established under such principles should trigger quality improvement processes under the supervision of the Health Resources and Services Administration (HRSA).

8. That all hospitals, particularly those with more than 100 beds, be strongly encouraged by CMS and the Agency for Health Care Research and Quality

Table 2.2. Evolution of immunosuppressive regimens used to treat recipients of cadaver kidney transplants, 1960-2003

Year	Prednisone	Azathioprine	Anti Thymus Globulin (equine)	Anti Thymus Globulin (rabbit)	Cyclosporine	Muromonab-CD3	Cyclosporine Micro Emulsion	Tacrolimus	Micophenolate Mofetil	Daclizumab	Basiliximab	Sirolimus	Alemtuzumab	Rejection Episode	Graft Survival	Nephrotoxicity	Hyper-cholesterolemia	Diabetes
1965	X	X												80%	40%		X	X
1975	X	X	X											60%	65%		X	X
1983	X	X	X		X									45%	80-90%	X	X	X
1983	X	X			X									45%	80-90%	X	X	X
1986	X	X			X	X								45%	80-90%	X	X	X
1995	X	X					X							40%	80-90%	X	X	X
1995	X	X		X			X							35%	80-90%	X	X	X
1995	X	X				X	X							35%	80-90%	X	X	X
1995	X	X						X						35%	80-90%	X	X	X
1995	X	X	X					X						35%	80-90%	X	X	X
1995	X	X				X		X		*				35%	80-90%	X	X	X
1996	X						X		X					25%	80-90%	X	X	X
1996								X	X									

*Reduced side effects with lower dosages in current use

Table 2.2, cont'd. Evolution of immunosuppressive regimens used to treat recipients of cadaver kidney transplants, 1960-2003

Year	Prednisone	Azathioprine	Anti Thymus Globulin (equine)	Anti Thymus Globulin (rabbit)	Cyclosporine	Muromonab-CD3	Cyclosporine Micro Emulsion	Tacrolimus	Micophenolate Mofetil	Daclizumab	Basiliximab	Sirolimus	Alemtuzumab	Rejection Episode	Graft Survival	Nephrotoxicity	Hyper-cholesterolemia	Diabetes
														Cadaver Kidney Transplants Incidence by End of First Post-Transplant Year				
1996	X	X					X		X					35%	80-90%	X	X	X
1997	X	X					X			X				30%	80-90%	X	X	X
1997	X							X	X	X				45%	85-90%		X	X
1998	X	X					X				X			30%	85-90%	X	X	X
1998	X							X	X		X			15%	90-95%	X	X	X
1998								X	X		X			15%	85-90%	X		X
1999	X	X					X					X		40%	85-90%		X	X
1999	X	X						X	X			X		10-15%	90-95%	X	X	X
1999	X			X										10-15%	90-95%	★	★	★
2000								X	X		X			10-15%	95%	★	★	★
2001								X			X			10-15%	95%	★	★	★
2003								X	X		X		X	<10%	>95%	★	★	★

(AHCRQ) to implement policies such that the failure to identify a potential organ donor and/or refer such a potential donor to the OPO in a timely manner be considered a serious medical error. Such events should be investigated and reviewed by the hospitals in a manner similar to that for other major adverse healthcare events.

9. That the Joint Commission on Accreditation of Healthcare Organizations (JCAHO) strengthen its accreditation provisions regarding organ donation, including consideration of treating as a sentinel event the failure of hospitals to identify a potential donor and/or refer a donor to the relevant OPO in a timely manner. Similar review should be considered by the National Committee on quality Assurance (NCQA).

10. That specific methods be employed to increase the education and awareness of patients at dialysis centers as to transplant options available to them.

11. That research be conducted into the causes of existing disparities in organ transplant rates and outcomes with the goal of eliminating those disparities.

LIVING DONORS

Modern kidney transplantation began in 1954 between identical twins and was extended to other living and cadaveric donors as control of rejection evolved. The living donor's remaining kidney compensates quickly and long-term health and life span are not different from matched controls.[6] In response to 4-5 year waiting times and less invasive laparoscopic techniques for donor nephrectomy, the number of living donor kidney transplants increased dramatically from 1800 in 1988 to 4432 in 1999 and 6445 in 2001. Midway through 2002 living donor transplants overtook cadaveric kidney transplants and approached 9000 for the year. If living donor kidney transplantation continues to grow at its present pace, the waiting lists will probably shrink and cadaveric kidneys will become more available to individuals with no possible living donor.

Although the intestine was transplanted only 48 times in 2001, approximately ten of the transplants came from living donors. Living donors will almost certainly be an important source of intestinal transplants to infants, small children and adults as they find their place in management of intestinal failure. Of the remaining four transplantable organs, living donation occurs rarely for pancreas and lung and never for the heart.

Liver, the second most commonly transplanted organ after the kidney (4944 in 2001) will also depend increasingly on living donors. Adult to child living donor liver transplantation was initiated in 1987 when the left lateral segments were transplanted from mother to infant child. The mother's liver mass returned to normal within a few weeks and the donated segments grew in the daughter, who is now a healthy 15-year-old.[7] In the last six years liver donation from healthy adults was extended to left lobe (for small adolescents and adults) and right lobe for larger adult recipients. Two donors are known to have died among the first 500 adult to adult living donor transplants in the United States. Because many individuals with end-stage liver disease die for lack of cadaveric liver transplants, adult to adult living donor transplantation will almost certainly continue. But risk of donation must be minimized, and both donor and recipient and their families must understand the early and long-term risks. Toward that end the National Institutes of Health recently initiated a seven-year study of living donors in ten centers that perform adult living donor liver transplantation.[8] In a further effort to protect living donors from pressure to donate or other kinds of exploitation, the Advisory Committee on Transplantation of the U.S. Department of Health and Human Services recommended a detailed set of standards to the Secretary concerning living donors. It emphasized that all living donors must undergo a thorough consent process, be provided with a donor advocate, and be enrolled in a national registry to monitor the donor's long-term health.[5] The committee's seven recommendations are:

1. That the following ethical principles and informed consent standards be implemented for all living donors. Ethical principles of consent to being a live organ donor should include the view that the person who gives consent to becoming a live organ donor must be competent (possessing decision making capacity); willing to donate; free from coercion; medically and psychosocially suitable; fully informed of the risks and benefits as a donor; and fully informed of the risks, benefits, and alternative treatments available to the recipient.

2. That each institution that performs living donor transplantation provide an independent donor advocate physician to ensure that the informed consent standards and ethical principles are applied to the practice of all live organ donor transplantation.

3. That a database of health outcomes for all live donors be established and funded through and under the auspices of HHS.

4. That serious consideration be given to the establishment of a separate resource center for living donors and their families.

5. That the present preference in OPTN allocation policy given to prior living organ donors who subsequently need a kidney be extended so that any living organ donor would be given preference as a candidate for any organ transplant, should one become needed.

6. That the requirements for HLA typing of liver transplant recipients and/or living liver donors should be deleted.

7. That a process be established that would verify the qualifications of a center to perform living donor liver or lung transplantation.

During the past two years governments, medical professional societies and the public throughout the world have turned much more attention to living organ donation. Attitudes and practices differ widely. Europe, Australia, North American and South America permit altruistic living donation but prohibit commerce and profit. On other continents, a black market is tolerated in some countries, and one country, Iran, has regulated living kidney donation such that the donor is compensated and the waiting list for transplantation has disappeared without permitting commercialism or a black market.[4]

In the United States bills were introduced during the last Congress to provide limited financial support to living donors and to maintain registries to monitor the donor's long-term well being; none was enacted into law. Three new bills were introduced as the Congress convened in January 2003. Senate Bill 178 would close loopholes and inconsistencies in Medicare laws as they affect transplant recipients. Known as the Comprehensive Immunosuppressive Drug Coverage for Transplant Patients Act of 2003, it would also extend drug coverage to Medicare beneficiaries for as long as it is needed. The Senate's Living Donor Protection Act (S.186) would assure that living organ donors are not denied insurance nor are they subject to discriminatory premiums because of their living donor status. The Organ Donation Improvement Act of 2003 (H.R. 399) would award grants to help cover expenses of people who volunteer to or become living organ donors. The grants would be permitted to pay for travel and subsistence costs and specified incidental costs incurred by living donors if the recipient's annual income did not exceed $35,000.00; presumably a recipient with income over $35,000.00 would be expected to help the living donor with his expenses.

Great Britain's National Health Service is considering lifting the ban on living donor compensation. Donation incurs considerable out of pocket cost, which in many instances might serve as a financial disincentive to otherwise altruistic donation. Tests, hospitalization, medical evaluation and care before and after the donor operation are covered as part of the recipient's cost. But time lost from work during evaluation and during recovery after donation is not compensated; nor is travel to and from or housing for donor and family members at the medical center. In a study of 22 donors after right hepatectomy for living related liver transplant in Germany, nine experienced adverse financial effects. On average, donors resumed work after nine weeks and felt fully recovered after thirteen weeks. The decision to donate was easy or not difficult for 21 and 20 of them would "do it again." Whether or not living donors are rewarded, it seems entirely appropriate that the financial penalty be mitigated at least partially.[9]

Black markets disadvantage both donor and recipient, because the living donors are usually inadequately evaluated. Illness in the donor compromises both the donor's recovery and survival of the donated organ in the recipient. A survey of 305 individuals in India who sold a kidney at least six years earlier showed that the average family income declined by one-third after nephrectomy .[10,11] Three-fourths of the donors were still in debt six years later (96% sold their kidneys to pay off debts). Eighty six percent reported deterioration in their health and 79 percent would not do it again. Clearly sale of organs does not serve as an escape route from poverty. In the unregulated black market where organ brokers roam city slums looking for sellers, the after effects of selling an organ may make escape even more difficult. Unregulated sale of organs is not a win-win proposition.

The live organ donor consensus group of the American Society of Transplant Surgeons reaffirmed in 2000 the position of the Transplantation Society (an international organization) that "organs and tissues should be given without commercial consideration or commercial profit."[12] The groups' position draws heavily on Christian belief that selling organs deprives the donation of its ethical quality. In contrast, contemporary Jewish law and ethics agree that donation motivated by altruism, rather than monetary gain and greed, is a most pious act, but deny that altruism is the only ethical basis for donation. Jewish tradition holds that the religious and ethical value of a good deed is not diminished by lack of "proper motivation."[13] However, Jewish tradition does not open the door for an organ trade that exploits the poor. Rather, the ethical status of non-altruistic sale of kidneys is inextricably connected to solving a series of pragmatic problems such as creating a system that ensures that potential donors are properly informed, evaluated, cared for and not exploited. Without such arrangements, ethical non-altruistic kidney donation is but a theoretical possibility.

Delegates at the recent International Congress on Ethics in Organ Transplantation, sponsored by the German Academy of Transplantation in December 2002, in Munich struggled for three days before endorsing eight resolutions on living organ donors and financial incentives:

1. Living donor kidney transplantation should be adopted as widely as possible,
2. Non-directed living kidney donation is ethically acceptable and should be permitted,
3. Kidneys derived from non-directed donation should be allocated using the standard cadaver allocation criteria,
4. The suitability of living related and unrelated organ donors should be assessed by the same criteria,
5. Living organ donors and their families must be adequately insured against the risk of death and disability caused by the act of donation,
6. There should be no financial disincentives to living donors. All donors should be legally entitled to reimbursement of those expenses incurred solely by the act of donation,
7. Appropriate compensation for pain, discomfort and inconvenience suffered by living donors is morally acceptable and may be adopted in a regulated fashion,
8. Individual countries will need to study alternative, locally relevant models, considered ethical in their societies, which would increase the number of transplants, protect and respect the donor, and reduce the likelihood of rampant, unregulated commerce.[4]

Iran, perhaps more than any other country, has already implemented a system of safeguards that allows altruistic and financial motivation to coexist without exploitation of either the donor or the recipient. Central features of the program were presented at the Munich Congress. Iran's program is open only to its own citizens. There are no foreign donors and no foreign recipients. No one can come to Iran to buy a kidney. There is no black market, no commercialism, no middleman or company to sell kidneys. No patient can buy a kidney; it is the government that offers the reward to all living donors, related or unrelated, and the donor is free to refuse the reward. Rich and poor are transplanted equally. By the end of 2002 there was no kidney waiting list in Iran and 1200 kidneys were transplanted yearly. Most of the living donors have been men in their 30's whose initial interest was the reward, but who clearly valued saving someone's life too.

IMMUNOSUPPRESSION

Clinical immunosuppression has always been a problem of balancing prevention or control of rejection with loss of protection against an array of infectious agents and mutant cells. Immunosuppression is intense immediately after transplantation and then diminishes during the first year as the allograft loses some of its immunogenicity and the recipient's immune system begins to adapt to the graft in various ways. That is, homeostasis appears but almost never to the point that immunosuppression can be discontinued without risking onset of rejection. Some recipients experience acute rejection despite fairly vigorous immunosuppression. Most acute rejection occurs within the first few months and nearly all can be reversed by intensifying immunosuppression. Few recipients experience chronic rejection in the absence of a preceding acute rejection episode or noncompliance with long-term maintenance immunosuppression. Thus, transplant centers continually search for regimens that prevent that first acute rejection episode, and they regularly warn recipients about the danger of noncompliance.

Immunosuppressive regimens have evolved steadily from broad attack on nearly all rapidly dividing cells in the body to combinations of drugs and biological agents that interfere specifically with those parts of the immune system most responsible for rejection. Review of the time-line for approval of immunosuppressive agents by the United States Food and Drug Administration during the past 30 years provides a framework for understanding the slow but steady conquest of acute rejection and the application of organ transplantation from the kidney to all vital abdominal and thoracic organs (Table #1). Information taken from the manufacturer's package inset for each of the drugs listed in Table #1 is presented as an appendix to the chapters and essays in this volume.

Clinical regimens for immunosuppression changed with the introduction of each new drug or therapeutic antibody. Typical regimens are shown in Table #2 for the years since the early 1960s. The approximate incidences for acute rejection and graft survival after one year are recorded for each regimen. Prior to 1960 prednisone was used to control various inflammatory processes and as a substitute for hydrocortisone. Mercaptopurine was an antimetabolite used to treat myelocytic leukemia. The observation in 1959 that mercaptopurine prolonged skin graft survival in rabbits led to its use in human transplant recipients in 1962.

Burroughs Wellcome subsequently introduced azathioprine which was a less toxic modification of mercaptopurine; azathioprine, also known as Imuran, was included in almost all immunosuppressive regimens from 1969 until 1996 when mycophenolate mofetil (CellCept) began to displace it. CellCept, like Imuran, is an inhibitor of cell division and nucleotide metabolism, but unlike Imuran, it affects lymphocytes primarily and largely spares most other rapidly dividing cell populations such as bone marrow and gut epithelium. CellCept inhibits the enzyme inosine monophosphate dehydrogenase, which is crucial to de novo synthesis of guanosine monophosphate. Many rapidly dividing populations of cells have a shunt pathway that bypasses the need for de novo synthesis, but activated T and B lymphocytes have an absolute requirement for de novo synthesis of guanosine monophosphate in order to accomplish clonal expansion. The net effect is immunosuppression with relatively few side-effects. Both Imuran and CellCept can depress bone marrow cell lines and intestinal epithelial cells, but CellCept achieves much more immunosuppression before reaching toxic side-effects such as anemia,

Table 2.3.   *Immunosuppressive Regimen for Kidney Transplant Recipients, Northwestern Memorial Hospital at Northwestern University*

Day	Alemtuzumab[1]	Tacrolimus[2]	MMF	Methyl Prednisone
0 (operative)	30 mg I.V. infusion over 2 hours during surgery			500 mg at start of surgery
1		2 mg bid	750-1000 mg bid	250 mg
2		2 mg bid	750-1000 mg bid	125 mg
3		2 mg bid	750-1000 mg bid	Discontinue
4		2 mg bid	750-1000 mg bid	

[1] Alemtuzumab omitted for HLA identical donor recipient pair; [2]Maintain trough concentration at 5-10 ng/ml.

leukopenia and diarrhea, and the side-effects are easily controlled by reducing the dose of CellCept.

Equine antithymocyte globulin, introduced as ATGAM by Upjohn in 1972, built on the observation reported in 1960 that rabbit antirat lymphocyte serum prolonged skin graft survival in rats. Antilymphocyte and antithymocyte globulins are extremely potent agents that deplete the host's lymphoid tissue. They are usually given intravenously on a daily basis for as long as 14 days, and are effective as induction treatment to delay (prevent) acute cellular rejection and also to reverse it. Because polyclonal cross species antisera against human lymphocytes/thymocytes are easy to prepare in rabbits, goats, and horses, many transplant centers provided their own local product. Their side-effects are limited to fever, leukopenia, thrombocytopenia, and anemia, which respond to prednisone or dose reduction. In 1999 SangStat introduced a polyclonal antithymocyte globulin that had been used effectively in Europe for nearly 20 years; it is prepared in rabbits and marketed as Thymoglobulin.

In 1986 OrthoBiotech (Johnson & Johnson) introduced the first therapeutic monoclonal antibody (orthoclone OKT-3). It was directed specifically against the T lymphocyte CD3 receptor for alloantigen presented by transplanted organs. Like the polyclonal antibodies, it delayed or prevented rejection and also reversed most acute cellular rejection episodes. Because it could be infused into a peripheral vein in less than a minute it was much simpler than ATGAM to administer and quickly became the preferred treatment of prednisone-resistant acute cellular and mild to moderate vascular rejection; it was also less expensive. However, polyclonal antilymphocyte sera are more effective than the monoclonal antibody if the transplant biopsy shows aggressive infiltrates of both T and B lymphocyte lines (i.e., plasma cells). Polyclonal antisera include antibodies against surface antigens shared by both T- and B-lymphocytes regardless of which cell type is used to immunize the animal that produces the antiserum. Until the early 1990s, transplant centers in the Untied States were equally divided between those that used and those that did not use antilymphocyte preparations (ATGAM, OKT-3, or a product prepared locally) as part of initial induction of immunosuppression.

Cyclosporine A (Sandimmune) introduced by Novartis (Sandoz) in 1983 dramatically reduced the incidence of acute-rejection episodes, increased kidney graft survival at one year and permitted rapid development of liver, heart and lung transplantation. Cyclosporine allowed less dependence on prednisone. Cyclosporine is a calcineurin inhibitor that preferentially suppresses activation of T lymphocytes by inhibiting production of the lymphokine interleukin-2. In 1994 Novartis introduced a microemulsion of cyclosporine (marketed as Neoral) which was absorbed more readily in the upper intestine. Side-effects of cyclosporine include fine tremor of the hand, hirsutism, gingival hyperplasia, increased appetite and hyperlipidemia (both cholesterol and triglycerides) which can generally be managed by lowering the dose. But, its main problematic side-effect was constriction of preglomerular arterioles which causes hypertension; if arteriolar constriction is sustained, the arteriole undergoes hyaline degeneration, narrowing, and the ischemic glomerulus becomes sclerotic. Although cyclosporine increased the number of transplants surviving beyond one and two years, nephrotoxicity took its toll not only on transplanted kidneys but also on healthy native kidneys of heart, lung and liver recipients. Despite fewer early rejection episodes in recipients of cadaver kidneys treated with cyclosporine, the fraction of transplanted kidneys that survive beyond five years was not higher than for earlier regimens that did not contain cyclosporine. Cyclosporine nephrotoxicity is such a serious side-effect that 10% of heart allograft recipients lost their native kidneys and were on maintenance dialysis within ten years. Many transplant centers kept maintenance doses and 12-hour blood trough levels lower than the manufacturer's recommendation; they accepted a few more acute rejection episodes rather than risk loss of kidney grafts to drug-induced glomerular ischemia and sclerosis.

Prograf/tacrolimus/FK 506 was introduced by Fujisawa in 1995. Like cyclosporine it is a calcineurin inhibitor, but its side-effects are different and it appears to induce less nephrotoxicity for equivalent immunosuppression. Like cyclosporine it can cause tremor but does not cause hyperlipidemia, hirsutism, gingival hyperplasia or increased appetite. Prograf can induce diabetes but usually not with doses of 3 mg twice daily or less. Prograf increases the bioavailability (AUC-Area Under the Curve) of CellCept when both drugs are taken concurrently.

Monoclonal antibodies against the IL-2a receptor (CD25) were introduced in 1997 by Roche (Zenapax/Daclizumab) and by Novartis in 1998 (Simulect/ Basiliximab). Both are intravenous preparations with no side-effects and both reduced acute rejection episodes significantly during the first six post-transplant months. The Roche preparation is infused just before transplantation and then every other week for four more doses. The Novartis preparation is infused only twice, once just before transplantation and again four days later. Because a course of Simulect requires only two infusions and costs less, it is preferred by many transplant centers. Most transplant centers that used antilymphocyte antibodies (either polyclonal or monoclonal) as part of early post-transplant immunosuppression soon switched to one of the anti IL-2 alpha monoclonals.

Rapamune (rapamycin/sirolimus) is the most recent antirejection drug and was released by the U.S. Food and Drug Administration in 1999. Like cyclosporine and tacrolimus it is a fungal product, but it does not inhibit calcineurin and is not nephrotoxic. It inhibits lymphocyte effects driven by certain cytokines, particularly IL-2. Its major side-effects is thrombocytopenia and hyperlipidemia, which can be controlled by reducing the dose or adding lipid-lowering agents. Rapamycin may also afford protection against chronic rejection by blocking proliferation of vascular endothelial and smooth muscle cells.

From the introduction of mercaptopurine/azathioprine in 1962 until cyclosporine was approved by the FDA 21 years later in 1983, organ transplantation was limited to the kidney. Bone marrow suppression markedly limited the amount of immunosuppression that could be achieved with azathioprine. Relatively high doses of prednisone provided most of the immunosuppressive effect. Even when a fourteen-day course of antilymphocyte serum became the third component of the regimen, steroids still carried most of the load. Bacterial sepsis and slow wound healing were such severe problems that liver, lung and heart transplantations were limited to a few hardy pioneering centers. The main contribution of cyclosporine after 1983 was that it reduced rejection, increased early graft survival, and reduced dependence on steroids; bacterial sepsis decreased; wounds healed more quickly; and successful transplantation of liver, heart and lung was quickly achieved in many centers.

But for kidney transplantation, cyclosporine was a mixed blessing. Despite fewer rejection episodes and increased graft survival beyond a year, the number of grafts surviving beyond the fifth year was scarcely more than before cyclosporine. Many kidney grafts eventually succumbed to ischemia and fibrosis caused by constriction of preglomerular arterioles. Cyclosporine had raised early transplant outcomes to a new plateau where they remained another 12-13 years until the appearance of two new drugs (Prograf and CellCept) and two monoclonal antibodies directed against the IL-2 receptor between 1995 and 1998. The net effect of these four agents has been less nephrotoxicity and much less dependency on prednisone. Many believe that Prograf is less nephrotoxic than cyclosporine, and CellCept is not nephrotoxic at all. Optimism has developed in just the last two or three years that some of the most bothersome side-effects of immunosuppression may be on the way out: nephrotoxicity, hyperlipidemia, accelerated vascular disease, osteonecrosis, osteoporosis and other steroid-related problems, and posttransplant diabetes. By 1998 many centers were discontinuing prednisone 6 and 12 months after transplantation. Most kidney recipients tolerated weaning from oral prednisone, but rejection was triggered in 15-20 percent of the weaned recipients. In mid-1998 Northwestern Memorial Hospital's transplant center removed oral prednisone completely from the immunosuppressive regimen used for kidney recipients. The regimen included basiliximab (day 0,3), methylprednisolone for 3 days, maintenance tacrolimus and mycophenolate mofetil, but no prednisone. Incidence of first year rejection episodes remained 10-15 percent and graft survival exceeded 95 percent.

Rapamycin/sirolimus (Wyeth-Ayerst Rapamune) was approved in 1999. Sirolimus, cyclosporine microemulsion, azathioprine and steroids limited acute renal allograft rejection to 10 percent with 90 percent first year graft survival. Regimens that combine anti CD-25 induction, early methylprednisolone and tacrolimus with either sirolimus or mycophenolate mofetil, with or without maintenance prednisone limit first year rejection episodes to 10 per cent and permit 95 percent graft survival.

The two most recent agents with immunosuppressive properties, which entered clinical practice in 2001, are monoclonal antibodies against lymphocyte cell surface determinants CD20 and CD52. Rituximab targets CD20, which is found on the surface of both normal B lymphocytes and 90 percent of B cell non-Hodgkin's lymphomas. It produces profound depletion of B lymphocytes in peripheral blood and elsewhere. The primary indication for its use is to treat B cell lymphomas, but it has also been used to reverse antibody mediated rejection in a heart transplant and to decrease production of HLA antibodies in a pre-sensitized potential kidney recipient prior to successful transplantation.[14,15] The second antilymphocytic monoclonal is alemtuzumab (Campath), which binds to CD52 found on all B and T lymphocytes, a majority of monocytes, macrophages and NK cells, and a subpopulation of granulocytes. It induces profound depletion of its targeted cells through antibody dependent lysis and is indicated in treatment of non-Hodgkin's lymphoma and chronic lymphocytic leukemia. Alemtuzumab has also been used in the early "induction" phase of several immunosuppressive regimens for kidney transplantation. A single intraoperative infusion of 30 mg induces complete absence of lymphocyte cells from the peripheral blood for 7-10 days with gradual recovery after six months. Alemtuzumab is a murine (rat) anti CD52 developed and humanized at the University of Cambridge in Herman Waldmann's Laboratory.[16] Because one or two infusions induce profound lymphopenia for several weeks, Calne treated 31 cadaveric kidney recipients with it in an attempt to induce "prope" (almost) tolerance at Cambridge. Each recipient received a 20 mg intravenous infusion of alemtuzumab on the operative and first postoperative days. Maintenance immunosuppression consisted of a single drug, cyclosporine, in moderate doses. Twenty-nine grafts had good postoperative function beyond the first year. Six rejection episodes during the first year were reversed and one recipient died.[17] Knechtle and Kirk report similar experience with alemtuzumab and rapamycin as the only maintenance drug in kidney recipients.[18,19] The transplant center at Northwestern Memorial Hospital has treated more than 200 kidney recipients with a prednisone free protocol that begins with a single 30 mg intraoperative infusion of alemtuzumab followed by oral maintenance with tacrolimus 1-2 mg twice daily and mycophenolate mofetil 750-1000 twice daily. Corticosteroids were restricted to the operating room (500 mg) and the first two postoperative days (250 mg, 125 mg). Details of the regimen are shown in Table 3. First year patient survival, graft survival and incidence of acute rejection are 99%, 97% and 8%. Mean serum creatinine at one year is 1.3 mg/dl. Incidence of infection of all kinds is less than 5 percent.[20,21] The alemtuzumab, tacrolimus based prednisone free protocol has been extended to simultaneous kidney and pancreas

2

recipients at Northwestern with two modifications: corticosteroids are administered for 4-5 days and sirolimus is used in place of mycophenolate mofetil; outcomes are similar to transplantation of the kidney alone.[22] As an anti lymphocyte induction immunosuppressive agent, alemtuzumab is at least equivalent in its lymphopenic effect to polyclonal antilymphocyte globulin or muramonab-CD-3 but less expensive, because a single infusion is sufficient. It facilitates prednisone free maintenance but does not eliminate the need for other maintenance drugs in relatively low doses.

Vincenti recently reviewed several new classes of immunosuppressive drugs that are under investigation but not yet approved by the FDA.[23] The major targets of new agents are cell-surface molecules important in immune cell interactions (especially the costimulatory pathway), signaling pathways that activate T cells, T cell proliferation and trafficking, and recruitment of immune cells responsible for rejection. The most promising include a humanized OKT-3, humanized anti-CD11a (anti-LFA1), humanized anti B7.1/B7.2, a second generation CTLA4Ig, LEA29y, an anti CD45RB, FK778, (a leflunomide analog), FTY720 and several antagonists to chemokine receptors (CCR1, CXCR3 and CCR5). The FDA is expected to approve modifications or variations on several classes of drugs that are already approved but is unlikely to approve any new class of immunosuppressive drug during the next few years.

To summarize evolution of transplant immunosuppression, seven new classes of drugs have been introduced since 1970, and all of them act primarily or exclusively to deplete lymphocytes or inhibit their function.

Class	Function	Name	Date
Polyclonal antibody	Lymphocyte depletion	Antihymocyte globulin	1972,1998
Monoclonal antibody	Lymphocyte depletion	Alemtuzumab anti CD52 Rituximab anti CD20	2001 2001
Monoclonal antibody	Binds T Lymphocyte receptor for antigen	Muromonab-CD-3	1986
Monoclonal antibody	Binds T Lymphocyte receptor for Interleukin-2	Daclizumab anti CD25 Basiliximab anti CD25	1997 1998
Calcineurin inhibitor	Inhibits IL-2 synthesis by T Lymphocytes	Cyclosporine Tacrolimus	1983 1995
Inosine Mono-phosphate dehydrogenase inhibitor	Blocks purine synthesis and proliferation of T and B lymphocytes	Mycophenolate mofetil	1995
Inhibitor of Regulatory kinase activation (target of Rapamycin)	Blocks Interleukin activation and proliferation of T lymphocytes	Sirolimus	1999

Four of the seven classes are biologic agents, antibodies, directed against lymphocytes; the other three classes are chemical reagents, medium sized molecules of less than 1000 (kd) in molecular weight.

2

All four of the antibody classes are administered intravenously as short courses ranging from a single dose to intermittent dosing for as long as eight weeks. Two of the classes (polyclonal antithymocyte globulin and monoclonal antibodies against lymphocyte cell surface differentiation markers) act by destroying and depleting a wide array of T and B lymphocytes, monocytes, macrophages, and NK cells through cell lysis. The two remaining classes of biologic agents are also monoclonal antibodies; muromonab-CD3 blocks the T-lymphocyte receptor for antigen while daclizumab and basiliximab block the T lymphocyte receptor for IL-2. These last two classes of antibody interfere with T lymphocyte response to antigens introduced by the transplanted organ.

Immunosuppressive regimens throughout the past four decades have been divided into two camps: those that use only chemical reagents and those that combine chemical reagents with antibodies. Both camps included corticosteroids among the chemical reagents. Between 1964 and 1986 polyclonal anti-human lymphocyte globulin was the only antibody class available and chemical reagents consisted only of corticosteroids and azathioprine until 1983. The combination of antibody with steroids and azathioprine was superior to steroids and azathioprine alone. But after cyclosporine joined steroids and azathioprine in 1983 the two camps parted ways. Advocates of chemical reagents alone claimed graft and patient survival equal to that of drugs plus antibody. Advocates of adding antibody to drugs claimed that the combination permitted delayed onset of first rejection, lower early doses of cyclosporine with less nephrotoxicity and better renal function at no cost with respect to incidence of rejection episodes and graft survival. Moreover it simplified management of delayed graft function which affected up to one-third of cadaveric kidney transplants.

The two camps, drugs versus drugs plus antibody, still disagree. The drug camp has new drugs (tacrolimus, MMF and sirolimus) and the antibody camp has new antibodies. Both camps have learned to manage with less calcineurin inhibitor (so as to reduce nephrotoxicity) and both camps have largely replaced azathioprine with MMF, but corticosteroids, a relic from the 1960s that lacks any specificity for lymphocytes and is the most devastating of all immunosuppressive drugs with respect to long-term crippling side effects, persist as part of the regimen in both camps throughout the world.

STEROID FREE IMMUNOSUPPRESSION

Many liver recipients can be weaned from maintenance corticosteroids in the first or second post-transplant year with little risk of inducing rejection. But for other organs, acute rejection follows in 15 percent of recipients weaned from prednisone.[24] Are the recipients who are refractory to weaning dependent on prednisone? Or do they simply need increased doses of the other drugs in the regimen to make the transition from steroids successfully? Perhaps high doses of maintenance steroids in the first post transplant months/years inhibit an active facilitative response that would otherwise permit homeostasis without maintenance steroids.

Rather than weaning from maintenance steroids, it may be simpler and more effective to start with a potent regimen that combines antibody and drugs but either restricts post transplant corticosteroids to a few days or avoids them altogether. In 2000, 2001 and 2002 more than ten centers for kidney, kidney/pancreas and islet transplantation reported excellent outcomes with up to five-year follow-up with regimens that excluded steroids completely or discontinued them within the first post-transplant week.[17-22,25-33] Most of the regimens combine drugs with an antibody. Incidence of acute rejection episodes during the first year was 10-15 percent or less. Graft function up to four years was equal to or better than steroid regimens. An earlier 1997 report that compared three cyclosporine-based regimens in kidney recipients showed increased vertebral bone density during the first 18 post-transplant months for cyclosporine alone but decreased vertebral density for cyclosporine plus steroids or cyclosporine plus both steroids and azathioprine.[34] The recently reported steroid free regimens probably will also increase bone density and spare recipients from many of the crippling side effects of osteoporosis.

The Edmonton steroid free regimen for pancreatic islets is a breakthrough. Steroids are especially toxic to islets. Investigators at the University of Alberta combined antibody (an IL-2 receptor antagonist or modified OKT3) with tacrolimus and sirolimus. Recipients usually underwent two islet infusions in the course of a year. Twenty-four recipients who have had both infusions have one year insulin independence of 87.5 percent and two year independence of 70 percent.[25,26]

We have avoided maintenance steroids at Northwestern Memorial Hospital in more than 500 kidney transplant recipients with two different antibody induction regimens since mid 1998. All recipients were treated with tacrolimus and either MMF or sirolimus. Each also received intravenous methylprednisolone daily for three days only (500 mg, 250 mg, 125 mg, stop). From mid 1998 until September 2001 the induction antibody basiliximab was given in the operating room and on the third post-transplant day. After September 2001, intraoperative antibody treatment consisted of a single 30 mg I.V. infusion of alemtuzumab. First year patient and graft survival in both groups are 99% and 97% respectively. Acute rejection episodes occurred in 9 percent of recipients in both groups. The main difference between the two induction antibodies in these prednisone free protocols is that the onset of first rejection episode was earlier with basiliximab (7.5 days) than with alemtuzumab (107 days). In general, post transplant management is simpler and cost is less if first rejection episodes can be delayed until 30 days or more after transplantation. Avoidance of maintenance prednisone imposed no penalty upon graft or patient survival, level of renal function or freedom from rejection episodes.[20,21]

OUTCOMES

All transplant centers are required to report patient and graft survival (and much more) to UNOS. Data collected by UNOS are transferred to the Scientific Registry of Transplant Recipients for analysis and preparation of reports that have begun to appear on the Internet every six months in July and January.[35] Each organ specific report covers a 30- month cohort of recipients followed for at least one year after transplantation. The reports list one-year patient and graft survival for the entire United States and individually for each of the centers. The reports also indicate the expected outcomes for each center; these are based on analysis of multiple risk factors in the cohort. Finally the report includes a statistical P-value to indicate whether outcomes were higher, lower or not different from expected.

A brief summary of the January 2003 report for kidney transplants is representative of all solid organ outcomes. Graft and patient survival are increasing nationwide, but surprisingly large variation exists among centers. Nationwide first year graft and patient survival are 90.86% and 95.44%. Among nearly 250 centers fifteen had higher graft survival than expected ($P<0.05$) and sixteen had lower than expected ($P<0.05$). Patient survival was higher than expected in fourteen centers and lower in thirteen. First year graft survival ranged from 97 percent in higher centers to 80 percent in lower centers. The higher centers were three to four percentage points above their expected outcome and the lower were six to eight points below expected. First year patient survival ranged from 86 percent to 98 percent. The higher centers were two points above expected and lower centers were five to six points below expected.

Outcomes for liver, simultaneous kidney/pancreas, and heart transplants exhibit the same wide range with respect to graft and patient survival one-year after transplantation. In 109 liver transplant centers average graft survival was 80.69%. Expected graft survival among the 109 centers ranged from 75% to 85%. Actual graft survival ranged from 67% to 93%. Five centers had higher than expected and six centers had lower than expected graft survival ($P<.05$). Recipient survival was 86.27% for all 109 centers; expected survival ranged from 81% to 89% and actual recipient survival ranged from 70% to 94% among the centers. Four exceeded expected survival and six were below ($P<.05$).

For 123 centers that perform simultaneous kidney and pancreas transplants, the average survival for kidney, pancreas and patient respectively were 91.90%, 84.96% and 94.86%. Individual centers ranged from 74% to 97% for kidney survival, 60% to 96% for pancreas and 80% to 98% for patient survival. None of the 123 centers had higher than expected kidney or patient survival and two had higher than expected pancreas graft survival. Five had lower than expected survival for kidney, four for pancreas and two for patient.

Among 130 heart transplant centers, graft and patient survival averaged 84.50% and 84.81%. Expected survival ranged from 78% to 89% for graft and 78% to 89% for patient. Actual survival ranged from 50% to 95% for grafts and 60% to 98% for recipients. Five centers exceeded and nine centers were lower than expected for graft survival. Four centers exceeded and eight were lower than expected for patient survival.

The internet outcomes reports for all of these organ transplants suggest that centers with higher outcomes have mastered technical and recipient selection issues and have learned how to assemble the current large array of immunosuppressive drugs into effective regimens, while centers with lower outcomes have not. Now that semi-annual updates on outcomes at all centers are available for the public and each transplant center to review, sophisticated consumers will know which centers to avoid. Centers with lower outcomes than expected will also know whom to call for advice.

REFERENCES

1.   Valero R. Donor management: One step forward. Am J Transplantation 2002;2(8):693-694.
2.   Vilardell J. Cadaveric organ procurement optimization. Transplantation 2002;74:54(4) Suppl.
3.   Conference Proceedings; Pabst Science. Publishers, www.pabst-publishers.de.
4.   Warren J. Commerce in organs. Transplant News 2003;13:1-3,No1.
5.   Warren J. Advisory panel calls on HHS Secretary. Transplant News 2003;13:1-3 No. 3.
6.   Ramcharan T, Matas AJ. Long-term (20-37 years) follow-up of living kidney donors. Am J Transplant 2002;2:959-964.
7.   Personal communication with Peter Whitington, MD, Chicago Children's Memorial Hospital.
8.   Adult-to-adult living donor liver transplant cohort study (A2ALL), National Institute of Diabetes and Digestive and Kidney Diseases, website:www.nih-2all.org.
9.   Karliova M, Malago M, Valentin-Gamazo C, et al. Living-related liver transplantation from the view of the donor: A 1-year follow-up survey. Transplantation 2002;73:1799-1804.
10. Goyal M, Mehta RL, Schneiderman LJ, et al. Economic and health consequences of selling a kidney in India. JAMA 2002;288:1589-1593.
11. Rothman DJ, Ethical and social consequences of selling a kidney. JAMA 2002;288:1640-1644.
12. Abecassis M, Adams M, Adams P, et al. For the live organ donor consensus groups. Consensus statement on the live organ donor. JAMA 2002;284(22):2919-2926.
13. Grazi RV, Wolowelsky JB. Non-altruistic kidney donations in contemporary Jewish Law and Ethics. Transplantation 2003;75:250-252.
14. Aranda JM, Scornik JC, Normann SJ, et al. Anti-CD20 monoclonal antibody (rituximab) therapy for acute cardiac humoral rejection: A case report. Transplantation 2002;73:907-910.
15. Leventhal JR. Northwestern University Medical School, personal communication.
16. Waldmann H, Hale G, Cobbold S. Appropriate targets for monoclonal antibodies in the induction of transplantation tolerance. Phil.Trans.R.Soc.Lond.D 2001;356:659-663.
17. Calne R, Moffatt SD, Friend PJ, et al. Campath 1H allows low dose cyclosporine monotherapy in thirty-one cadaver renal allograft recipients. Transplantation 1999;68:1613-1616.
18. Knechtle SJ, Pirch JD, Becker BN, et al. A pilot study of Campath 1H induction plus rapamycin mono-therapy in renal transplantation. Transplantation 2002;74(4):32.
19. Kirk AD, Hale DA, Hoffmann SC, et al. Results from a human tolerance trial using Campath 1H with or without infiliximab. Transplantation 2002;74(4):33.
20. Stuart FP, Leventhal JR, Kaufman DB, et al. Alemtuzumab (Campath IH) facilitates prednisone-free immunosuppression in kidney transplant recipients. Transplantation 2002;74(4):121.
21. Leventhal JR, Gallon LG, Kaufman DB, et al. Alemtuzumab (Campath 1H) facilitates prednisone-free immunosuppression in kidney transplant recipients. Am J Transplant, American Transplant Congress Abstracts, May 2003.

2

22. Kaufman DB, Leventhal JR, Gallon LG. Pancreas transplantation in the prednisone-free Era. Am J Transplant, American Transplant Congress Abstracts, May 2003.

23. Vicenti F. What's in the pipeline? New immunosuppressive drugs in transplantation. Am J Transplant 2002;2:898-903.

24. Hricik DE. Steroid-free immunosuppression in kidney transplantation: An Editorial Review. Am J Transplant 2002;2:19-24.

25. Shapiro AM, Lakey JR, Ryan EA, et al. Islet transplantation in seven patients with type I diabetes mellitus using a glucocorticoid free immunosuppressive regimen. N Engl J Med 2000;343:230.

26. Shapiro AM, Ryan ER, Paty B, et al. Human islet transplantation can correct diabetes. Transplantation 2002;74:119. (suppl vol 4).

27. Matas AJ, Ramcharan T, Paraskevas S, et al. Rapid discontinuation of steroids in living donor kidney transplantation: A Pilot Study. Am J Transplant 2001;1:278-283.

28. Kaufman DB, Leventhal JR, Koffron AJ, et al. A prospective study of rapid corticosteroid elimination in simultaneous pancreas-kidney transplantation. Transplantation 2002;73(2):169-177.

29. Boots JMM, Christiaans MHL, Van Duijnhoven EM, et al. Early steroid withdrawal in renal transplantation with tacrolimus dual therapy: A pilot study. Transplantation 2002;74:1703-1709.

30. Cole E, Landsberg D, Russell D, et al. A pilot study of steroid free immunosuppression in the prevention of acute rejection in renal allograft recipients. Transplantation 2001;72(5):845.

31. Leventhal JR, Kaufman DB, Gallon LG. Four-year single center experience with prednisone-free immunosuppression in 432 kidney transplant recipients. Am J Transplant, American Transplant Congress Abstracts, May 2003.

32. Kaufman DB, Leventhal JR, Fryer JP, Abecassis MI, Stuart FP. Kidney transplantation without prednisone. Transplantation 2000;69:S133.

33. Birkeland SA. Steroid-free immunosuppression in renal transplantation: A long-term follow-up of 100 consecutive patients. Transplantation 2001;71:1089.

34. Aroldi A, Tarantino A, Montagnino B, et al. Effect of three immunosuppressive regimens on vertebral bone density in renal transplant recipients. Transplantation 1997;63:380-386.

35. http://www.ustransplant.org. Scientific Registry of Transplant Recipients.

Organ Allocation in the United States

Frank P. Stuart and Michael Abecassis

3

THE NATIONAL ORGAN TRANSPLANT ACT (NOTA)

In 1984, the National Organ Transplant Act was passed by Congress to address the need for better coordination and distribution of scarce organs. The Act established a national task force to study transplantation issues and to create a National Organ Procurement and Transplantation Network (OPTN). The OPTN was started in 1986 and a Scientific Registry of Transplant Recipients (SRTR), a data gathering and tracking service on transplants, began operation in late 1987. Both were funded by the Health Resources and Services Administration (HRSA), an agency of the U.S. Department of Health and Human Services (DHHS), through contracts awarded to the United Network for Organ Sharing (UNOS) in Richmond, Virginia. UNOS now serves as the umbrella organization for national organ procurement, transplantation, and statistical information. The primary function of the OPTN is to maintain a national computerized list of patients waiting for organ transplants. All hospital transplant centers, organ procurement organizations, and tissue typing laboratories are required to meet the requirements for voting membership in the OPTN. Its purpose is to ensure equitable access to organs for critically ill and medically qualified patients and to guarantee that scarce organs are procured and used safely and efficiently.

ORGAN PROCUREMENT ORGANIZATIONS (OPOS)

Organ Procurement Organizations (OPOs) coordinate activities relating to organ procurement in a designated service area. There are 63 OPOs (often referred to as organ banks) throughout the United States. Their service areas do not overlap. Some include parts of a state, and others include one or more states. The Health Care Finance Administration (HCFA) of the department of Health and Human Care Finance Administration (HCFA) of the Department of Health and Human Services designates and regulates OPOs and sets the criteria by which their performance is judged. OPOs evaluate potential donors, discuss donation with family members, and arrange for the surgical removal of donated organs. OPOs are also responsible for preserving organs and arranging for their distribution according to national organ sharing policies established by the OPTN.

THE DIVISION OF TRANSPLANTATION (DOT)

Within HRSA, the Division of Transplantation (DOT), in the Office of Special Programs, administers the OPTN and the SRTR. Other DOT activities include providing technical assistance to the 63 OPOs, working with public and private

Organ Transplantation, 2nd edition, edited by Frank P. Stuart, Michael M. Abecassis and Dixon B. Kaufman. ©2003 Landes Bioscience.

organizations to promote donation, serving as a national resource to professional associations, health providers, health insurers, state health departments, and the media about donation and transplantation, and managing the contract with the National Marrow Donor Program to administer the National Bone Marrow Registry for Unrelated Donors.

UNOS HISTORY

In the mid-1960s, an important development occurred that had a major effect on organ transplantation. It was determined that by transplanting a cadaveric donor kidney into a recipient that matched genetically, graft survival could be increased. As a result of this development, several transplant centers began to share kidneys as a means of extending kidney survival. Preliminary results of shipping kidneys between centers were successful. With this experience, the Kidney Disease and Control (KDC) Agency of the Public Health Service awarded seven contracts to transplant centers throughout the United States. The purpose of the contracts was to prove the feasibility of procuring kidneys in one place and preserving, matching, and transporting them in a viable condition for transplantation.

The Southeastern Regional Organ Procurement Program (SEROPP) was awarded one of these contracts on June 27, 1969. SEROPP originally had a membership of eight transplant programs in four states and the District of Columbia. It implemented a computerized on-line kidney matching system in December 1969.

In 1975, responding to the increase in activity, the South-Eastern Organ Procurement Foundation (SEOPF) was incorporated with 18 members in a six-state area.

Responding to requests from non-SEOPF transplant centers to utilize the computer system for registering potential recipients and sharing kidneys, the United Network for Organ Sharing (UNOS) was established in January 1977. UNOS was designed to utilize the benefits of a computerized system for matching kidneys nationally. The ultimate objective was to better utilize procured kidneys while improving outcome. UNOS granted access to the computer registry and matching program to any transplant program within the United States. The registry in the late 1970s included not only kidney recipients, but those awaiting other organs as well.

By 1982, UNOS was becoming more of a national sharing network, and because of the complexity of sharing kidneys over a large portion of the country, SEOPF and UNOS created "The Kidney Center." The Kidney Center was staffed 24 hours a day with personnel who could run the computer and locate recipients for kidneys and other organs, arrange kidney transportation, maintain and update registry files for those who requested it, and attempt to locate organs through the UNOS/STAT system for patients who were critically ill. Recipients listed on the computer were assigned "status" codes to reflect urgency of need. When a match was found, the kidney was offered to the recipient center and transplanted there with arrangements made by SEOPF. The transportation of other organs (hearts, livers) remained the responsibility of the donor center since the recipient center sent its own team of surgeons to retrieve the non-renal organ. In 1984, the

Kidney Center became known as the "Organ Center" to reflect its activity with other organs.

In anticipation of changes occurring both in the field of transplantation and in the legislative arena, UNOS was incorporated as a private, non-profit voluntary membership organization in 1984. This action was recommended by two committees, working separately, to determine if UNOS should incorporate to meet the changing demands of the transplant field. UNOS was classified for federal tax purposes as a medical, scientific, and educational organization. The primary mission of the organization was to operate the computerized national recipient registry for patients in need of transplantation and to coordinate the placement of organs procured in the United States through the Organ Center. UNOS was the only organization of its kind offering services to the entire nation. Transplant programs, organ procurement organizations, and histocompatibility laboratories joined UNOS to participate in the efficient and effective distribution of organs for transplantation.

The goals of UNOS, as outlined in the Articles of Incorporation, were to:
- establish a national Organ Procurement and Transplantation Network under the Public Health Services Act;
- improve the effectiveness of the nation's renal and extra-renal organ procurement, distribution, and transplantation systems by increasing the availability of, and access to, donor organs for patients with end-stage organ failure;
- develop, implement, and maintain quality assurance activities; and
- systematically gather and analyze data and regularly publish the results of the national experience in organ procurement and preservation, tissue typing, and clinical organ transplantation.

The UNOS Board of Directors, composed of one representative from each member institution, governed the organization. UNOS and SEOPF remained closely intertwined, sharing office space, computer hardware, and personnel.

In 1986, UNOS sought and was awarded the federal contract to establish and operate the national Organ Procurement and Transplantation Network. With the awarding of the contract, UNOS changed its operation to accommodate the mandates of the law. In making the changes, UNOS sought input from the transplant community and its Board of Directors. UNOS also seriously considered the recommendations of the Task Force on Organ Transplantation. During the first year of operation as the national OPTN, UNOS enrolled new members and elected a new Board of Directors to conform with OPTN contract requirements. While the original Board of Directors consisted of a representative of each member, the new board included representatives of groups of members. As mandated by contract, the board was composed of 15 transplant surgeons and physicians and 16 non-physicians. Non-physicians were representatives of the following UNOS member categories: Independent Organ Procurement Agencies (two representatives), transplant coordinators (two representatives), Tissue Typing Laboratories (two representatives), Voluntary Health Organizations and Public Members (ten representatives). Public members represented the fields of ethics, law, religion, behavioral, and social sciences and included patients, patient advocates, and non-transplant physicians. Surgeons and physicians represented each of the ten UNOS geographic regions (one each), and in addition included a President, Immediate Past President, Vice President, Treasurer, and Secretary (total = 15). UNOS later provided for a heart transplant representative to be elected to the Board of Directors, bringing the total number of board members to 32. In addition to enrolling members and creating a governing body, UNOS established an administrative organization with an executive director and assistant executive director and internal departments including: Technical Services and Computer Operations, Professional Education, Communications, Travel, Finance, Membership and Personnel. Later changes in the administrative organization included the addition of a Research and Policy Department with more specific responsibilities for supporting the scientific and policy-making functions of the OPTN.

For administrative purposes, UNOS divided the country into eight geographic regions. Due to size discrepancies and organ sharing concerns, several of the regions were altered to create a ninth, tenth, and eleventh region by the fall of 1989.(Fig. 1) Each region was assigned a UNOS staff administrator to assist in coordinating regional activities and to provide input to the UNOS committees and Board of Directors.

Also in the first year of operation, UNOS created 11 permanent standing committees: Communications, Education, Ethics, Finance, Foreign Relations, Transportation, Membership and Professional Standards, Heart Transplantation, Organ Procurement and Distribution, Histocompatibility and Scientific Advisory. An ad hoc committee on Patient Affairs was later made a permanent standing commit-

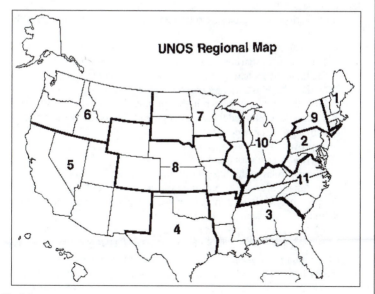

Fig. 3.1.

tee. Ad hoc committees presently include Donations, Multiple Listing, Organ Procurement Organizations, and Pediatrics. Currently, committee members are recommended by the regional councilor and are selected to provide broad and experienced input into all committee activities. The president makes the appointment. Committees receive input from regional subcommittees, from the transplant community, and from the public. Each individual member of the OPTN is represented in all deliberations by the transplant organization or institution for whom he or she works.

UNOS established by-laws, membership criteria, and operating policies during its first year of operation. It also established a mechanism for public input into the policy-making process. Additionally, UNOS established a Scientific Registry under a two-year contract with the federal government. The Scientific Registry contains pre- and post-transplant data on all solid organ recipients in the United States as well as data on all donor referrals and actual organ donors.

Membership in the national OPTN was mandated in the Omnibus Budget Reconciliation Act of 1986. This legislation required that transplant centers be members of the OPTN and abide by its rules and policies or stop transplanting organs. If a transplant center elected not to join but to continued transplanting, that center would no longer be eligible to participate in the federal Medicare/Medicaid programs. As a result, membership in UNOS is no longer voluntary, and therefore policies set by UNOS govern all organ transplantation programs in the United States.

One of the provisions of NOTA in 1984 was that the Secretary of HHS would eventually submit a set of OPTN operating rules to the Federal Register; at that time, rules of the OPTN would acquire the force of law. The purpose of any set of rules would be as follows:

- The effectiveness of cadaveric organ procurement and distribution is improved;
- Access to an optimal organ transplant is improved and increased;
  - The system for sharing renal and extra-renal organs is improved so as to:
  - Facilitate the matching of renal and extra-renal donor organs with potential recipients based on criteria established for each organ;
  - Provide a system by which highly immunologically pre-sensitized patients will be afforded the broadest possible opportunity to be matched with an acceptable donor;
  - Improve transplant outcome; and
  - Decrease organ wastage.
- Quality control is assured by collection, analysis, and publication of data on organ donation, procurement, and transplantation; and
- The professional skills of those involved in organ procurement and transplantation is maintained and improved.

In 1984, UNOS was incorporated as a legal entity, and in 1986, SEOPF gave it its computer matching system. The foundation also gave UNOS the 24-Alert voice-activated computerized matching system for non-renal vascular organs, developed for the North American Transplant Coordinators Organization in Pittsburgh. UNOS received a contract from the federal government effective October 1, 1986, to put in place an organ procurement and transplant network. This network was to develop a national policy to assure equitable organ allocation. A point system developed through the University of Pittsburgh and later published in the New England Journal of Medicine was offered to UNOS by Dr. Thomas Starzl to be used nationally for allocating kidneys, livers, and thoracic organs. The UNOS Board of Directors adopted Dr. Starzl's point system in May 1987 and implemented it on October 1, 1987, the date that the Organ Procurement and Transplant Network (OPTN) became operational.

In June 1988, the Board of Directors approved an allocation system for hearts and heart-lung combinations. This new system was not based on points, but instead allocated organs first locally, then to recipients within a 500-mile radius of the donor hospital, followed by recipients within a 500-1,000 mile concentric circle, and then finally, to all recipients beyond a 100-mile radius. These organs were allocated first to Status I patients (those patients who were critically ill and in urgent need of a transplant), and secondly, to Status II patients (all other potential heart recipients). (See UNOS Policy 3.7.) That policy went into effect January 4, 1989. At the February 1989 Board Meeting, the Board of Directors approved a modification of the point system for renal allocation that put a higher emphasis on antigen matching while maintaining a major emphasis on the length of time potential recipients had been waiting. Additionally, the new match process only considered the percent reactive antibodies of the recipient if the level exceeded 80 percent reactive antibodies and a preliminary negative crossmatch was available (see Policy 3.5). A simple pancreas allocation policy was developed in 1989 (Policy 3.6.10). Also in 1989, a slight modification was made to the liver allocation policy found in section 3.6 of the policies. The current organ allocation policy for each organ follows this chapter.

THE ORGAN CENTER

In 1982, the UNOS Organ Center was developed by SEOPF through a grant from the American Kidney Fund. The Organ Center was established to assist organ procurement coordinators with organ placement according to established protocols and to arrange transportation for those organs to the recipient center. The Organ Center is staffed 24 hours a day with trained personnel to help assure that organs are allocated, shipped, and delivered in a timely and appropriate fashion so that more patients can be transplanted with suitable organs. The Organ Center maintains the minimum acceptance criteria that each United States center uses for sharing organs. This information is updated periodically to be of the most benefit.

The Organ Center is used by most of the nation's transplant centers and organ procurement organizations for sharing kidneys. Organ Center personnel receive the information from the donor center, access the computer for matches, and telephone potential recipient transplant centers and organ procurement organizations until they find a transplant center willing to accept the organ. Once they have identified the center, Organ Center personnel hook up a three-way telephone conversation between the donor center, the Organ Center, and the recipient transplant center. This assures a clear understanding of the expectations by all parties. Next, the Organ Center arranges transportation from the donor center.

ALLOCATION OF ABDOMINAL ORGANS

KIDNEY ALLOCATION

Kidneys are allocated on a local, regional, and national basis with the exception of mandatory sharing of six antigen matched kidneys. The allocation of cadaveric kidneys is made at the local level according to a point system. Patients on the local waiting list are offered kidneys in descending sequence with the patient with the highest number of points receiving the highest priority. A local area is defined by either the individual transplant center recipient list or a shared list of recipients within a defined procurement area which can be no larger than the OPO and service area designated by HCFA. The point system includes blood group, time of waiting, quality of antigen match, panel reactive antibody, and pediatric status. Medical urgency is not considered for kidney or pancreas allocation. A pay back system to the OPO of origin exists for six antigen match shared kidneys.

PANCREAS/KIDNEY ALLOCATION

Combined kidney/pancreas transplants are typically allocated according to the kidney allocation policies.

LIVER ALLOCATION

Organs are offered on a local, regional, and national basis. A point system similarly exists which includes blood group, time waiting, and degree of medical urgency. For every potential liver recipient, the acceptable donor size is determined and used as preliminary stratification.

Upon approval of the OPTN Board of Directors, a transplant center or an OPO may assign to each of the point systems' criteria, points other than the number of points set forth by OPTN policy. In 2000 UNOS adopted the Model for End-stage Liver Disease (MELD) system for predicting the prognosis of patients with end-stage liver disease. The score relies on three laboratory parameters, bilirubin, prothrombin time (INR) and creatinine. A modification of the MELD system known as PELD has been adopted for allocating cadaveric livers to children. Both systems have been modified to take into account the patient with hepatocellular carcinoma, which can spread before bilirubin, prothrombin time and creatinine rise. Most agree that the new system is an improvement over the previous one, which depended heavily on waiting time, subjective prediction that death was likely within 7 days, and hospitalization in an intensive care unit.

INCREASING DISPARITY BETWEEN ORGAN SUPPLY AND WAITING LISTS

The past ten years have witnessed remarkable progress. Outcomes are vastly improved; the number of transplant centers has increased so that they are available throughout the country (over 100 centers each for heart, liver, and pancreas, 250 centers for kidney, 25 centers for lung, and 15 centers for intestine); large numbers of transplant physicians and surgeons have been trained; the waiting list for all organ approaches 90,000 individuals; yet the number of cadaver donors has stalled at approximately 6000 each year. Many die each year while waiting for an organ transplant.

Because supply and demand are so far out of balance, consensus on fairness and utility in allocation alogorithms is increasingly more difficult to achieve. Vocal articulate advocates of particular points of view have lobbied Congress and HHS for changes in the allocation rules. Dramatic stories in newspapers, magazines, and television have become commonplace.

RECENT DEVELOPMENTS

In 2000 the Secretary appointed a committee of 40 members to serve as an Advisory Committee on Transplantation. The Advisory Committee convened in November 2002 at the Secretary's request to consider issues related to well-being of living donors, shortage of cadaveric donors and equal access to organ transplantation. The Advisory Committee made nine recommendations designed to increase cadaveric organ donation, two recommendations to encourage equal access to minority populations and seven recommendations with respect to living donors. All eighteen of the Advisory Committee's recommendations are presented in Chapter 2 of this edition as part of a general review of long waiting lists and the increasing importance of living organ donors. Transplant related initiatives of the 2003 U.S. Congress and Senate are also part of that review.

Organ Procurement Organizations

Stephen D. Haid, James A. Kisthard and Jarold A. Anderson

INTRODUCTION

Organ procurement organizations (OPOs) are entities that play an integral role in the organ transplantation process through the provision of all activities related to organ donation. This includes education of the general public, education of medical professionals in hospitals, assisting hospitals with the development of written policies and procedures, obtaining family consent, medical evaluation of potential donors, the surgical removal of organs, organ preservation, organ distribution, and follow-up with participants of the recovery process. The OPO is also responsible for required reporting to the national Organ Procurement and Transplantation Network (OPTN), the Center for Medicare and Medicaid Services (CMS) and, in some cases, state health organizations.

Initially, OPOs were formed within academic transplant hospitals, typically in the department of surgery, to support the hospital's kidney transplant program. In the late 1960s and early 1970s, a few of these entities formed their own governing structures and separated from the transplant hospitals to form independent, not-for-profit corporations. As more OPOs began operating as separate corporations, several bonded together to form a trade association, the Association of Independent Organ Procurement Agencies (AIOPA). This organization has evolved to include independent and hospital-based OPOs and is now the Association of Organ Procurement Organizations (AOPO).

Through the 1970s and midway through the 1980s, OPOs were essentially unregulated. Typically organ allocation occurred only locally and was at the direction of the transplant program(s). Organ sharing beyond the local programs was driven by expediency and, to some extent, medical priority. Early efforts to allocate organs via a structured system were coordinated through various organizations including individual OPOs, the Southeastern Organ Procurement Foundation (SEOPF) and the North American Transplant Coordinators Organization (NATCO). Authorized by the National Organ Transplant Act of 1984, the United Network for Organ Sharing (UNOS) currently holds a federal contract to be the OPTN. In recent years, UNOS has established and now oversees the national organ sharing system. All OPOs are required to be members of the OPTN.

During the development of the National Organ Transplant Act and shortly after its passage, there was a substantial shift from hospital-based OPOs to independent OPOs. Another remarkable effect of the legislation was a striking reduction in the number of OPOs. Currently, OPOs are regulated in terms of governance, function and performance. As reporting requirements have increased, so have

Organ Transplantation, 2nd edition, edited by Frank P. Stuart, Michael M. Abecassis and Dixon B. Kaufman. ©2003 Landes Bioscience.

performance expectations. Most OPOs now expend significant resources promoting organ donation initiatives through public, professional and legislative avenues.

LEGISLATION AND REGULATION

Although organ procurement rates experienced moderate, but steady, growth during the 1970s and early 1980s, the growth rate of the transplant waiting list was much larger. The gap between supply and demand caused patient groups to insist on a fair system of organ allocation that would provide equitable access to organs on a national level. Legislators responding to their constituents rushed to introduce bills to deal with this issue. Among the numerous legislators participating in this effort were Senators Ted Kennedy, Orin Hatch and Dan Quayle, and Representatives Dan Marriott, Edward Madigan and Henry Waxman. However, a bill introduced in 1983 by a Tennessee Democrat, Congressman Al Gore, ultimately changed the face of history with respect to organ procurement and transplantation. After several days of hearings, Congressman Gore drafted legislation in October 1983 titled the National Organ Transplant Act.[1] The bill underwent numerous revisions until it was passed into law on October 19, 1984.

The National Organ Transplant Act was an amendment to the Public Health Service Act and it was a landmark statute for the transplant world. Most other federal legislation that followed has been tied to this important law. The law was divided into four parts—Titles I-IV. Title I established a task force charged with examining issues related to human organ procurement and transplantation, making an assessment of immunosuppressive medications used in transplantation, and presenting a report to the Secretary of the Department of Health and Human Services (DHHS). The task force held its first meeting in February 1985 and submitted its final report in April 1986. The task force outlined 60 recommendations in its 232-page published report.[2] Table 4.1 lists several of the recommendations that directly affected OPOs.

Title II of the act dealt with organ procurement activities. Section 371 defined OPO qualifications including non-profit status, service area size, board composition and functional capabilities. Regulations regarding OPO qualifications have been revised several times, and current regulations will be addressed later in this chapter. Section 372 established the OPTN. The law provided initial funding for

---

Table 4.1 Task force recommendations affecting OPOs

◊ The enactment of uniform state laws for the determination of death
◊ The enactment of legislation requiring implementation of policies on organ donation and required request
◊ The development of minimum performance standards for OPOs
◊ Public education on organ donation targeted to minority populations
◊ Incorporation of organ procurement and transplantation into the curriculum of nursing and medical schools
◊ Certification of organ procurement specialists
◊ Certification of not more than one OPO in any one service area
◊ OPO governance similar to that described for the OPTN
◊ A single national system for organ sharing

establishment and operation of the OPTN and set forth its qualifications, functions and board composition. Section 373 established a scientific registry to be awarded either by grant or contract. This registry was to include information on transplant outcomes. It was intended to allow patients and professionals to evaluate the scientific and clinical status of organ transplantation on an on-going basis. It has subsequently become the primary source of information for transplant patients to evaluate organ-specific outcomes at individual transplant centers. Section 375 of Title II established the Office of Organ Transplantation. This office was to coordinate organ procurement activities under Title XVIII of the Social Security Act (Medicare), conduct public education about organ donation, provide technical assistance to OPOs, and provide an annual report to Congress on the status of organ donation. The Office of Organ Transplantation was later made a permanent part of the federal government when it was designated as a division

---

Table 4.2. Summary of qualification requirements for OPO designation

---

◊   Must qualify as a nonprofit entity
◊   Must have accounting procedures sufficient to maintain fiscal stability and to obtain payments from transplant centers for organs provided
◊   Must have an agreement with the Secretary of DHHS for Medicare reimbursement
◊   Must have an appropriately defined service area
◊   Must have a director and sufficient staff to be effective in recovering organs from the OPO's service area
◊   Must have a Board of Directors with authority to recommend donation policy and which meets composition requirements defined in these regulations
◊   Must have a documented working relationship to identify potential organ donors with at least 75% of the hospitals that have organ recovery capabilities and which participate in the Medicare and Medicaid programs
◊   Must have a systematic approach to identifying potential donors and acquiring all usable organs from those potential donors
◊   Must arrange for tissue typing
◊   Must have a system for allocating organs equitably in compliance with OPTN rules and with CDC *Guidelines for Preventing Transmission of Human Immunodeficiency Virus Through Transplantation of Human Tissue and Organs*
◊   Must arrange for transportation of donated organs to transplant centers
◊   Must coordinate its activities with area transplant centers
◊   Must have cooperative arrangements with tissue banks
◊   Must maintain data which demonstrates compliance with performance standards
◊   Must maintain data and records in a format which could be easily transferred to a successor OPO to facilitate uninterrupted service
◊   Must have procedures to assure confidentiality of patient records
◊   Must conduct professional education
◊   Must ensure that donor screening is performed by an appropriately certified laboratory to comply with OPTN standards and CDC screening guidelines
◊   Must assist hospitals in making routine inquiries about organ donation
◊   Must ensure that donors are tested for HIV markers in compliance with CDC guidelines and OPTN rules
◊   Must provide in a timely manner annual data concerning the population of the OPO's service area, the number of actual donors, and the number of renal and extra-renal organs procured and transplanted

---

under the Health Resources and Services Administration (HRSA). It is now referred to as the Division of Transplantation (DOT) and has taken on the role of overseeing the OPTN contract.

Title III of the National Organ Transplant Act made it unlawful for any person to transfer any human organ for valuable consideration if the transfer affects interstate commerce. The term "valuable consideration" did not include reasonable reimbursement costs associated with the acquisition, preservation and transportation of organs acquired from deceased donors. Title IV dealt with the establishment of a national bone marrow registry.

The Omnibus Budget Reconciliation Act (OBRA) of 1986 defined the requirement that each OPO be certified by Medicare as a qualified OPO.[3] The law further stated that OPOs be re-certified every two years by meeting qualifying criteria and performance standards established by the Secretary of the DHHS. It quickly became evident that numerous OPOs would not qualify under the initial qualifying criteria, especially those criteria related to the size of the OPO's service area and its donor potential. The ability of an OPO to qualify for certification was critical to its very survival. Any OPO not certified by October 1, 1987, would no longer receive payment for Medicare and Medicaid reimbursable expenses. As the deadline for certification drew near, OPOs across the country were merging and consolidating in order to meet the requirements. By the time the first certification process was completed, the number of OPOs had been reduced by approximately 40%.

Regulations related to OPO qualifying criteria and performance standards have been revised several times since 1986 and were last modified in November 2000. CMS still has not provided details about all elements of the new regulations, although one key element is the change in the certification cycle for OPOs from two to four years. The following are several key elements of the regulations used previously.[4]

For an OPO to receive Medicare and Medicaid reimbursement, it must be exclusively designated by CMS to operate in a defined service area. To be the designated OPO for a service area, the OPO must make application to CMS and meet certain requirements including the following:

- The OPO must be certified as a qualified OPO and must be a member of the OPTN.
- The OPO must have a formal agreement with CMS for reimbursement.
- The OPO must have working relationships with hospitals and transplant centers within its service area.
- The OPO must provide cost projections and cost reports to CMS to establish reimbursement rates and must provide data to CMS related to organ recovery activity.
- The OPO must also comply with defined performance standards in order to be redesignated.
- The OPO must provide extensive information regarding its service area, including the size and boundaries, the population, names of the counties, and names of the hospitals with organ recovery capabilities.

A summarized list of designation requirements is shown in Table 4.2. The gov-

---

Table 4.3. Summary of requirements for OPO board composition

◊ Hospital administrators, tissue banks, voluntary health associations and either intensive care or emergency room personnel within the OPO's service area
◊ General public residing in the OPO's service area
◊ A physician or individual with a doctorate degree in the biological sciences who is a specialist in histocompatibility
◊ A physician who is a neurosurgeon or a specialist in neurology
◊ A transplant surgeon from each transplant center affiliated with the OPO

4

erning Boards of OPOs are also subject to composition requirements defined by the regulations. Table 4.3 lists the required member categories for an OPO's Board of Directors. Although an OPO may have more than one board, at least one of the boards must be composed in accordance with the regulations.

Performance standards for OPOs were less stringent prior to January 1, 1996. To meet those standards, each OPO had to demonstrate that it procured from its service area at least 23 kidneys per million population per year and that, of those procured kidneys, at least 19 per million population per year were transplanted.

The current performance standards implemented January 1, 1996, include five performance categories: 1) number of actual donors per million population; 2) number of kidneys recovered per million population; 3) number of extra-renal organs recovered per million population; 4) number of kidneys transplanted per million population; and 5) number of extra-renal organs transplanted per million population. To be redesignated, each OPO must achieve at least 75% of the national mean in four out of the five performance categories per year averaged over the two years prior to redesignation. In theory, all existing OPOs could meet these requirements without any being closed. However, several OPOs have already failed to meet these standards and have been closed. It is anticipated that as the lower-performing OPOs drop out via the redesignation process, the performance mean will continue to rise. On the positive side, a rising mean accomplishes the objective of having mandatory performance standards by raising the overall performance requirements of OPOs. On the other hand, some OPOs will fail and there is no guarantee that there will be an improvement of performance in a given service area with a different OPO. There are many who argue that the current performance standards are inappropriate because they are based solely on population and don't take into account population density, population demographics, trauma referral patterns or other factors that may influence organ donation activity but may be out of the sphere of control of the OPO. The AOPO and others who have criticized the validity of these standards are reviewing alternative standards that may more directly measure OPO performance. It is hoped that once CMS finally publishes the details of its November 2000 regulations, they will address the inadequacies of the January 1, 1996, performance standards.

The regulations also have a direct effect on hospitals. Each donor hospital in the OPO's service area must have an agreement to work with the OPO designated for the service area in which the hospital is located. The hospital may request a waiver to work with a different OPO but must demonstrate that the waiver will

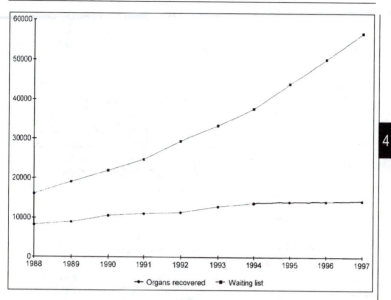

Fig. 4.1. Organ supply vs. demand. Source: United Network for Organ Sharing

improve the rate of organ donation and ensure equitable access to recovered organs.

The regulations also deal with terminations of OPO agreements with CMS. OPOs may terminate voluntarily or involuntarily. If the OPO fails to meet the performance standards described above, CMS may terminate its agreement with the OPO. An OPO's agreement with CMS also may be terminated immediately if CMS determines the OPO is guilty of unsound practices.

A hospital-specific Medicare regulation implemented in August 1998 had a direct impact on OPO operations.[5] All Medicare-certified hospitals must comply with this regulation or risk losing their agreements for Medicare reimbursements. These hospitals must have written agreements with a designated OPO and at least one eye bank and one tissue bank.

They must notify the OPO or the OPO's designated third party of all deaths or imminent deaths in the hospital. It is then the responsibility of the OPO to determine whether or not the individual is medically suitable for organ donation. It is also the responsibility of the OPO or a designated requestor trained by the OPO to discuss organ, tissue and eye donation with the family and obtain the appropriate consent documentation. The regulation also requires hospitals to educate their staffs about organ, tissue and eye donation issues, including identification of donors and maintenance until the recovery can occur.

Previously, hospitals in most states only called the OPO about potential donors. Criteria for such determinations were provided by the OPO. In reality, most OPOs were being notified about a small percentage of the total number of deaths in each hospital. The regulation placed a substantial burden on OPOs and hospitals. A direct impact on the OPO's level of staffing required to handle donor referrals and subsequent donations occurred, and many OPOs found it necessary to expand their telecommunications and information systems capabilities as the level of referral activity increased.

OPO FUNCTIONS

In the late 1960s and early 1970s, OPOs were often located in academic hospitals as programs within the hospital's Department of Surgery or Transplantation. The organ procurement functions of those OPOs were typically limited to assisting in the operating room with kidney recoveries, kidney preservation, and transporting kidneys to neighboring transplant centers if they could not be used locally. At that time, kidneys were preserved almost exclusively by continuous pulsatile preservation. The job of the early procurement coordinators was usually more technical than clinical. The OPO's employees were often technicians, seldom nurses, and almost never business people. It was very common for OPO staff members to have other responsibilities in the hospital or department in which they were employed. Often they were research technicians, dialysis nurses or technicians, heart pump technicians, or operating room nurses or technicians.

Since the mid-1980s, the overwhelming majority of OPOs have become independent of hospitals. The number of hospital-based OPOs still in existence today is very small and often function in a manner similar to independent OPOs. There are, however, a few ways in which hospital-based OPOs differ from independent OPOs. Independent OPOs are self-supporting, nonprofit corporations. Hospital-based OPOs may have segregated finances, but they are still financially tied to a hospital. The hospital-based OPO's finances are reported as part of the hospital's financial reports. The insurance umbrella of the hospital or university typically covers the hospital-based OPO, whereas independent OPOs must obtain their own insurance policies. Employees of hospital-based OPOs are really hospital employees and are subject to the hospital's employment policies. Independent OPOs are companies that have anywhere from a handful to more than 150 employees; many have fewer than 30. Independent OPOs must adhere to state and federal employment laws, and most hire human resource consultants to ensure compliance. A number of OPOs have even hired full-time human resources personnel.

Most independent OPOs rent office space at one or more locations within their service area. A number have purchased their own buildings. Staffing has evolved to the point where many OPOs have departments including procurement, marketing, education, hospital development, human resources, information systems, accounting and others. Registered nurses and degreed nurses dominate the procurement staffs. Directors of OPOs, who had previously been clinical staff promoted from within, are increasingly becoming business or hospital executives hired from outside. Most OPOs have full-time accountants, and many now have full-time information systems specialists. As the focus on public and professional education has increased, most OPOs have hired marketing or education specialists. The annual operating budget of some of the larger OPOs is in the tens of millions of dollars.

Not only do OPOs look like serious corporations, they also act like serious corporations. Concepts such as strategic planning and strict adherence to employment laws that have long been commonplace in corporate America are now common in OPOs. Well-established OPOs have formal in-house training programs, employee handbooks, policies for compliance with environmental and health safety standards, internal performance standards and overall sound business practices.

The primary purpose of OPOs is to coordinate all aspects of organ donation and to maximize the recovery of usable organs for transplantation. This involves many functions beginning with public and professional education, media relations, hospital relations, tissue and eye bank relations, donor evaluation, family counseling and consent, medical management of the donor, and the surgical removal of organs. Additionally, OPO employees are responsible for organ preservation, organ distribution, transportation of organs, follow-up to donor families and medical staff, accounting and reporting, and contributing to industry knowledge. Add to that interpreting organ allocation policies, acting as a liaison between multiple surgical recovery teams and hospital staff, and ensuring compliance with all federal, state, OPTN, OPO, and hospital policies. The staff members of OPOs must juggle numerous medical, ethical, political and regulatory issues simultaneously, and they must do so under intense public, professional and regulatory scrutiny. It is not surprising that the burnout rate for OPO staff, especially clinical coordinators, is extremely high. An important OPO task is to develop employment screening techniques, training programs and retention programs aimed at maintaining adequate staffing experience and staffing levels.

Promoting donation is a key function of OPOs that have recognized that their operations are not driven by organ recovery, but that organ recovery is a result of effective marketing and education. OPOs are motivated by numerous factors to play a leading role in improving the rate of organ donation. Public interest is one of the motivating factors. Although there has been steady, but modest, growth in the number of organs recovered from deceased donors each year, the percent of increase has flattened since 1995. From 1988 to 1994, the number of organs recovered increased an average of 9.1% per year; from 1995 through 2002, the average annual growth was only 1.8%. Even more disturbing is the fact that the transplant waiting list is expanding at a much more rapid rate and shows no signs of slowing. In fact, the average annual increase in the size of the waiting list at year end from 1988 through 2002 was 28.6%.[6] Additionally, OPOs are subject to intense pressure from affiliated transplant programs to provide organs for their patients. The third factor that motivates OPOs to increase the rate of donation is survival. Simply stated, OPOs that fail to meet government-imposed performance standards will be shut down.

In order to impact organ donation rates, OPOs must attempt to modify the attitudes and behaviors of the general public and medical professionals regarding organ donation. According to a survey conducted by the Gallup Organization, while 85% of the public claims to support organ donation, only 28% have signed a license or donor card indicating their intent to donate.[7] Actual consent rates further demonstrate the discrepancy between the stated attitudes of Americans toward donation and their actual behavior. According to the Partnership for Organ Donation, the rate of refusal to consent to donation was 50% in its study group.[8] Assuming that study reflects national behaviors, one out of every two Americans asked to donate refuses. Many OPOs and others are focusing efforts to improve the consent rate through education about organ donation and by encouraging individuals to discuss their wishes to donate with their families. Groups within the general public that have particularly high refusal rates are being studied to determine what factors cause them to refuse to donate. As more is learned about the reasons for refusal, these groups are being targeted for focused education campaigns. Many OPOs have employed full-time staff specifically to coordinate public education campaigns. Interaction with the mass media, which had once been only reactive, is now a primary tool of public education for OPOs. In many OPOs, full-time public relations staff plan media events and work steadfastly to develop relationships with key media representatives in their service areas. This not only facilitates a more proactive approach to media involvement, but also creates a less adversarial environment when difficult news stories arise.

Although public attitude is the most significant determinant of organ donation activity, the attitudes of medical professionals also have a profound impact. A study by the Partnership for Organ Donation revealed that only one third of potential donors in hospitals actually became donors. Twenty-seven percent of the potential donors were either not identified or the family was not asked to donate, while the remaining third refused donation when asked. Although OPOs expend significant resources to develop strong relationships with hospital personnel, there is still a lack of participation among many medical professionals. Some of this can be attributed to personal feelings about donation, some to a lack of clear procedures, and some to a workload that causes them to view donation as a low priority. Virtually all OPOs have marketing or hospital development staff to work closely with hospitals toward an objective of improved participation. In some OPOs, the marketing staff is as large or larger than the clinical staff. These individuals facilitate the donation process by endeavoring to make a seemingly complicated process as simple as possible. They help the hospitals develop written policies and procedures, they provide around-the-clock in-service education programs, they provide role-playing opportunities, and they conduct postrecovery debriefing conferences. Some OPOs even provide debriefing sessions in situations when a referral does not result in a donation. Marketing personnel provide one-on-one support and recognition for hospital employees who participate in the organ donation process. Many OPOs also host annual conferences for nurses and physicians in their service area.

One of the most important activities of the marketing staff is to determine the annual donor potential in every donor hospital in the OPO's service area. This is actually one of the best methods for an OPO to evaluate its own performance, and it is critical to resource planning. If an OPO can identify which hospitals have the highest donor potential, it can focus more of its resources toward those hospitals. Further, if the OPO can determine which hospitals are falling short of their potential, it can reallocate its resources to improve performance in those hospitals. Knowing this information creates an opening for OPOs to give direct feedback to hospital administrators about the level of donor potential versus actual recoveries for any given period. The mechanisms for determining donor potential vary, but most involve some sort of retrospective review of medical records. Each OPO utilizes the methodology and criteria for donor suitability that best meet local needs. While this may be useful on a local level, the lack of consistency makes it impossible to determine donor potential at the national level. Beginning in 1997, the AOPO conducted a pilot project designed to develop a methodology for estimating organ donor potential. A secondary motive of this project was to provide the data necessary to develop better national performance standards based on true donor potential rather than the current standards that are based on population. There have been many estimates of the national donor potential calculated by numerous methods involving extrapolation. The AOPO Death Record Review study recently projected the national donor potential at 11,000 to 14,000 potential donors per year.

Relationships with eye and tissue banks can have a direct impact on the performance of an OPO. By association, medical professionals and members of the general public often assume that the OPO, the eye bank and the tissue bank are a single entity. While in some cases this may be true, it most frequently is not. It is important for these three entities to coordinate education programs, donor referrals and recoveries to provide the smoothest possible procurement service to donor hospitals. Any complications in the process that can be attributed to poor communications between OPOs, eye banks and tissue banks can create a risk that hospital participants or public attitudes will be compromised. It is the responsibility of OPOs to take the lead in coordinating the activities of the three entities since it is mandated that all hospital deaths must be reported to the OPO.

THE ORGAN PROCUREMENT PROCESS

Virtually all OPOs recover multiple organs from donors, whenever possible. Additionally, some OPOs also recover eyes and tissues. It is the responsibility of the OPO to initially evaluate potential donors for medical suitability. This requires the clinical coordinators to have extensive medical knowledge about the physiology and function of multiple organ and tissue systems. The coordinators must be skilled at reviewing, sometimes voluminous, medical records for pertinent information that may provide insight about organ function or that may identify contraindications to donation. Coordinators also must be resourceful in determining past medical history and high-risk behaviors. This information is obtained through previous hospital admissions, as well as discussions with nurses, attending and family physicians, and friends and family members. Obviously, the coordinators must exercise the utmost sensitivity when discussing these issues with family members and friends.

Clinical coordinators and other staff receive specific training in counseling grieving family members about the organ donation process. Some OPOs also train designated requestors, who typically are hospital employees. This is especially important in situations where the hospital is located a significant distance from the OPO. It is imperative that the families of potential donors are approached regarding the option of donation with compassion and sensitivity, recognizing the sudden loss of their loved one. As noted previously in this chapter, OPOs are required either to speak directly to the family about consent for donation or they must be involved by training a designated requestor. Many past studies have shown that personnel from OPOs are more effective than nurses, physicians or hospital clergy in securing consent. However, recent studies have demonstrated that the consent rate is even higher when the discussion is conducted jointly by a member of the OPO along with a member of the hospital staff. It is important that the OPO coordinator provide complete information to the family about the donation process, including the timeframe for completion. The coordinator also must provide updates to the family if there are unexpected delays. It is imperative that the family not feel pressured or harassed. Their decision about donation should be respected even if their answer is no. Coordinators must be cognizant of the critical balance between the desire to get consent for a given donation and the possible ill will that could result from alienating a potential donor's family.

Once a donor has been identified and the family has consented to donation, it is critical that appropriate medical management is provided to ensure that the organs are functioning optimally at the time of recovery. Before the OPO can begin its involvement in medical management, death must be declared and documented appropriately in the donor's medical record. The first step in the donor management process is to determine the current status of organ function. This is accomplished by physical examination, review of past and current medical records, obtaining necessary laboratory tests and other diagnostic tests or consultations. A detailed discussion of medical management of deceased organ donors is beyond the scope of this chapter, but several key objectives are described as follows. Perfusion and oxygenation are the two main goals of donor management. Maintenance of normal blood pressure, fluid electrolytes and blood oxygen levels are the key ingredients in accomplishing those objectives. In the process of managing the donor to achieve optimal function of one organ, coordinators must be careful not to compromise the function of another organ. For example, it is desirable to maintain a brisk diuresis in the kidneys up to the moment of surgical recovery. However, overhydrating a donor may cause excess fluid in the lungs and may compromise pulmonary function. It is important for the medical management of the donor to be performed in a manner consistent with the optimum function of all transplantable organs.

The coordinators also are responsible for ordering laboratory tests to determine the presence of any transmissible diseases such as HIV, hepatitis or other systemic infections. All OPOs have Medical Directors or physicians designated to oversee and assist as necessary in the screening and medical management of donors. Their level of involvement in a given case depends on the complexity of the case and the experience level of the coordinator. Additionally, physicians from each of the receiving transplant teams may request specific tests or management parameters. It is the role of the coordinator of the host OPO to coordinate the numerous variations that often occur with different recovery teams and be as responsive as possible to the needs of each. Of course, special requests must not interfere with sound donor management and should not be permitted to compromise one organ to the benefit of another.

After the evaluation and management of the donor is in progress, the coordinator must place the organs. The federal government, through the contracted OPTN, regulates organ allocation. Coordinators must register each donor with the OPTN, and allocation is determined through computer matching by the OPTN. Once a transplant center has accepted an organ, it is the responsibility of the coordinator from the host OPO to communicate with the coordinator from the receiving center to coordinate and schedule the surgical recovery. The host coordinator often assists in obtaining local transportation for a team flying in from a distant location. The coordinator also should determine and assist with any special needs of that team. In some situations, one team may be removing all organs. More typically, several teams are involved in each surgical procedure. The coordinator of the host OPO is responsible for coordinating the arrival of each team and discussing the order of the surgical procedure with members of the surgical teams.

Typically, each recovery team provides its own preservation fluids and supplies. The coordinators are responsible for preparing preservation solutions and making them available to the surgical recovery team at the appropriate time during the procedure. The length of warm and cold ischemic periods are important to the anticipated function of transplanted organs, and the coordinators are responsible for documenting when these periods begin and end. Other times, such as the incision time and the time of drug administration also are documented during the procedure. Although most organs are preserved by static cold storage, some OPOs preserve kidneys by continuous pulsatile perfusion. This requires special knowledge and technical skills including surgical skills and the operation of perfusion equipment. Each preservation method has its advantages and disadvantages, but both work well provided that the length of preservation is kept within acceptable parameters.

The coordinator for each recovery team is responsible for appropriate packaging and labeling of organs, tissue typing materials, and any specimens that will accompany their organ(s). This must be done in strict compliance with OPTN policies to prevent errors and provide consistency. The receiving transplant center or laboratory may refuse organs or tissue samples that are not appropriately labeled. The OPO must have established relationships with histocompatibility laboratories for tissue typing and crossmatching. Ideally, blood or tissue samples are delivered to the histocompatibility laboratory prior to the start of the surgical recovery. However, in distant locations, it may not be practical or cost effective to arrange for prerecovery tissue typing. Lymph nodes, blood and other tissue samples are collected during the surgical recovery and the coordinator is responsible for arranging for those samples to be delivered to the histocompatibility laboratory.

Organs are transported in several ways. Sometimes they accompany the recovery team back to the transplant hospital; sometimes they are shipped by commercial or charter aircraft unaccompanied. In the case of pulsatile perfusion, the coordinator or perfusion technician attends to the machine whether the kidney is used locally or at a distant center.

Organ donation only occurs through the good will of the general public and the participation of the medical community. Successful OPOs do an effective job of following-up with members of the donor's family, medical professionals and others who were involved in a given case. An organ recovery is an enormous event involving from 20 to 100 individuals. This includes all the immediate family members, nurses and physicians in the emergency room, ICU and operating room, hospital administrators, hospital security, transportation personnel and a host of others. Prompt feedback to each of these individuals is generally very meaningful to them. Letting them know how much they are appreciated and how much their efforts contributed to saving and improving the lives of others can motivate them to participate in the future. At the very least, it helps them to feel appreciated for participating in a process that can be very stressful and emotional. Typically the feedback is in the form of a letter, but it can also come in other ways. Appropriately timed phone calls to the family are sometimes helpful and provide an opening for the family members to ask any unanswered questions they may have regarding the process. De-briefing meetings with medical professionals have been shown to be an effective way to allow staff members to ask questions or simply vent their feelings. It is also the responsibility of the OPO to follow-up on blood cultures or any other laboratory tests that were not completed prior to the recovery and report the results to receiving transplant centers.

Organ transplantation is an effective, but expensive, treatment for end-stage organ failure. Organ acquisition is a significant component of the overall expense, with the procurement-related costs of some organs exceeding $25,000. The OPO bears the initial cost of organ acquisition and is reimbursed by the transplant center that receives the organ. The transplant center then recovers that cost directly from the patient or third party payer. For kidneys, the third party payer is usually Medicare. Since kidneys represent approximately half of the activity for most OPOs, a substantial portion of OPO funding comes from this source. In fact, OPOs are required to file an annual cost report with CMS. If the cost report indicates that the OPO charged more than its actual cost for kidneys, the OPO must pay that amount back to Medicare. Conversely, if the OPO undercharged for those organs, Medicare reimburses the OPO. Where kidney costs and reimbursements are concerned, the OPO must break even with Medicare for reimbursed expenses. Although the cost report is due annually, OPOs may file for an interim adjustment during the year if they can document a substantial loss.

In the case of kidneys, all OPOs charge transplant hospitals a standard acquisition charge. They create a cost center specifically for kidneys and track all expenses attributable to kidney acquisition over the course of a year. This includes direct expenses such as donor hospital charges and transportation, as well as indirect expenses such as professional education, salaries and rent. Direct expenses are relatively simple to identify, but indirect expenses can only be reimbursed by Medicare to the extent those expenses can be tied to kidney acquisition. A portion of salaries and other indirect expenses also are allocated to the acquisition of extra-renal organs. Furthermore, Medicare has strict procedures for determining what expenses can be included in the kidney cost center. Once the OPO has established a financial history, it can accurately project expenses in an annual budget. The standard kidney acquisition fee is then calculated by dividing the projected kidney related expenses by the number of projected kidney transplants. Since this is only done at the beginning of each fiscal year, variations in actual versus projected costs can easily result.

Revenues that occur as a result of reimbursement for extra-renal organs are similar to reimbursement for kidneys; however, OPOs are not required to break even for extra-renal organs. As nonprofit entities, OPOs are allowed to build and maintain a fund balance, although no revenues in excess of actual cost can be acquired from kidney revenues. For many years, some OPOs operated with little or no cash reserves, which created serious difficulties in times of slow donor activity. As OPOs have become more sophisticated in their financial practices, they have realized that a strong fund balance is essential for the effective operation of their organizations. Many OPOs include financial planning as part of their annual strategic planning process. Although presently there is no industry standard, many OPOs are wisely building fund balances equal to several months of operating expenses. Unquestionably, OPOs are under intense public scrutiny to be cost effective. It is important for OPOs to expend their resources wisely and to avoid unnecessary expenses or anything the public might consider extravagant. However, it would be very fiscally irresponsible for an OPO to allow its cash reserves to diminish to a point that routine operations are compromised.

All OPOs must report their financial data to CMS annually. They also are subject to periodic financial audits by CMS. Additionally, most OPOs undergo independent financial audits. They must file a corporate tax return to the Internal Revenue Service. And, even though not required, most OPOs provide detailed information to affiliated transplant centers regarding the determination of their organ acquisition charges. In addition, OPOs have public board members that review their budgets and financial data.

In addition to financial reporting, OPOs also must comply with the data reporting requirements of the OPTN, various offices of the federal government and, in some cases, state government or health associations. OPOs must document and report organ recovery activity, compliance with federal OPO regulations, compliance with health and safety standards, compliance with OPTN membership standards, and compliance with OPTN allocation policies. Occasionally OPOs must respond to inquiries from the Office of the Inspector General and other governmental agencies. In some cases, state laws regarding donation have been enacted that require OPOs to report information to state or local health authorities. Affiliated hospitals certainly expect a high level of reporting from the OPO regarding organ recovery activity, marketing activity and finances.

Organ recovery productivity varies, sometimes dramatically, from one OPO to the next. It is in the public's interest for all OPOs to perform at a high level. High-producing OPOs often demonstrate many innovative practices that have maximized their performance. Conversely, low-performing OPOs often identify unique and sometimes unavoidable areas that detract from their performance. An open exchange of information that contributes to the industry knowledge is healthy and beneficial to all OPOs. This can best be accomplished through active participation in trade associations such as AOPO or professional associations such as NATCO, verbal or poster presentations at national meetings, and the publishing of professional papers. Although most OPOs have developed a high level of technical expertise, no OPO has claimed to discover the secret to maximizing organ donation. There is no single magic formula for improved performance. Rather, this is best achieved through a host of activities that combine to affect the behavior of the public and the medical professionals. The extent to which an OPO can discover and implement effective techniques will ultimately determine its performance.

INITIATIVES TO INCREASE DONATION

Each year the gap between supply and demand for transplantable organs widens. Many initiatives to increase the supply of organs have been attempted, and new ones surface at a steady pace. Some of these are very localized and are undertaken by a single OPO. Some are statewide or regional and may be the work of coalitions between donation-related entities or may be the result of statewide legislation. There are numerous examples of national initiatives by associations, coalitions, congress, private corporations and others. The goal of each is the same, but the approach is usually varied. Some are designed to improve the consent rate, some are intended to motivate the public to donate, some are oriented toward expanding the medical acceptance criteria, and others are focused on improving the caliber of OPOs and their employees. Unfortunately, the overall success of these initiatives has not been dramatic, but combining several different approaches may ultimately yield measurable results.

The first notable initiative was the Uniform Anatomical Gift Act of 1968. This legislation, which has been adopted in some form by all states, described provisions to allow individuals or their immediate family members to legally give consent to allow their organs to be donated at the time of death. This legislation gave rise to the development of donor cards. Various campaigns promoting the use of donor cards have evolved including the placement of an individual's donation status on the driver's license in most states. Additionally, donor cards are available from many other sources. If appropriately executed, a donor card is considered a legal document.

Several other initiatives related to consent issues also have been implemented. The concept of required request was introduced in the 1980s by an ethicist, Arthur Caplan. This is a process whereby donor hospitals must present the option of donation to all potential donors in their hospitals. Required request was first attempted at the state level and eventually became a federal statute. There has been limited success reported with required request. Unfortunately, the definition of potential donor was left to the donor hospital and, in many cases, the option was not presented because the hospital prematurely or erroneously deemed an individual unsuitable for donation.

As described above, this concept was taken one step further with CMS regulations requiring hospitals to report all deaths to OPOs and with OPOs being responsible for determining donor suitability. Failure to comply with these regulations can cause a hospital to lose its Medicare and Medicaid funding. Although these federal regulations were based on a state law in Pennsylvania that reportedly resulted in a 40% increase in donation over a three-year period, the impact on organ donation nationally has been relatively modest. Recently, the concept of First-Person Consent legislation (also referred to as "Donation by Donor Designation") has been adopted by many states. First-Person Consent allows OPOs to recover organs from a person who signed up to be a donor through a registry or a uniform donor card, without the signature of two witnesses or consent from the next-of-kin.

Perhaps the last consent-related initiative is presumed consent. Although not currently practiced in the U.S., the premise of presumed consent is that all individuals are considered organ donors unless there is prior notice of objection to donation by the individual. The rationale for this thinking is that since public attitude polls have demonstrated that most people favor donation, it is safe to assume they are willing to donate unless they give notice to the contrary. This has been tried in other countries with some success, but there has been substantial reluctance to legislate it in the United States. Some states have passed limited presumed consent laws that typically permit donation of eyes or tissues unless there has been a prior notice of objection. In these situations, eyes and tissues are removed without consent from the next-of-kin. There has been limited success in increasing the rate of such donations, but there also have been situations where the family has reacted strongly to the donation that occurred without their consent.

Several local coalitions of OPOs, tissue banks, eye banks, transplant programs, voluntary health associations and other interested individuals have formed. The objective of these groups is to improve local donation rates through education and improved public awareness. There also have been formally organized national coalitions. The Coalition on Donation, formed in the mid 1990s, is a prominent national coalition active today. The objective of this group is to establish one unified national message about organ donation. It has developed donation awareness campaigns that have been widely utilized by OPOs and transplant programs across the country. Other national organizations and all OPOs conduct public education and donor awareness programs. It is difficult to empirically measure the effectiveness of these initiatives, but most agree they are very important and will prove helpful over time.

One of the provisions of the National Organ Transplant Act of 1984 is that buying or selling human organs is prohibited. However, there is an initiative, albeit controversial, to increase public participation in donation designed to induce individuals with financial incentives. These incentives take various forms with the most direct being a cash payment to the immediate next-of-kin of the donor. Others are less direct and include proposed payments for funeral expenses, tax deductions, donations to named charities, life insurance policies, and a plethora of other types of compensation. Proponents argue that everyone benefits from organ transplantation except the donor; therefore the donor's family should be reasonably compensated. They also argue that it is logical to think that more people will be motivated to donate if they are paid than if they are not. Opponents argue that removing altruism will prey on those in lower socioeconomic positions and may actually reduce the donor pool. There is also a concern that family members may be less than forthright about the donor's medical history when tempted with compensation for the donation. Public opinion polls and focus groups have demonstrated a lack of enthusiasm for financial incentives, and some individuals have stated they would not participate for reasons other than altruism. Whether or not financial incentives would increase donation remains to be seen. However the biggest obstacle to financial incentives must be addressed before they can even be tested. One of the provisions of the National Organ Transplant Act of 1984 is that buying or selling human organs is prohibited.

A number of medically oriented initiatives have been attempted to increase the availability of donor organs. For example, in the late 1980s surgeons from Loma Linda University Medical Center began a series of transplants utilizing organs recovered from anencephalic infants. As they and others explored this possible source of donor organs, they encountered a number of obstacles. First, determination and declaration of brain death in anencephalics does not fit traditional guidelines. Second, in 1989 the UNOS Board of Directors endorsed a policy developed by its ethics committee discouraging the use of anencephalic infants as donors. Third, the results of transplants from anencephalic donors were poor when compared to organs recovered from traditional organ donors. The use of anencephalic infants as organ donors has become essentially nonexistent in recent years.

One fairly successful approach to increasing the organ supply has been to broaden the criteria for donor acceptance, but only to the extent that donation can occur without negatively impacting transplant outcomes. As transplantation technology has evolved, transplant physicians have discovered that donor organs that had previously been considered unacceptable are often quite suitable for transplantation. It is not surprising that as donor management and post-transplant care of the recipient have improved, so has the ability to use organs from "expanded donors" a term coined by transplant professionals in the mid 1990s. There are many examples of expanded donors, and undoubtedly the list will continue to grow. Acceptance of organs from older donors, donors with some degree of hypertension, non-heartbeating donors, Hepatitis C positive donors, and other expanded donors all have been used effectively given the appropriate donor/patient circumstances. Some disagreement remains regarding acceptable donor criteria, but this approach has received much interest and has been proven effective in many centers.

Increasing donation by improving the proficiency of procurement personnel and the performance of OPOs has been an ongoing goal of procurement professionals. While it may be difficult to quantify the impact of this approach, its effect can only be positive. After several years of development, the American Board of Transplant Coordinators (ABTC) conducted its first certification exams in 1988.

This voluntary certification is designed to measure competency for transplant clinical coordinators and transplant procurement coordinators. A few years later, the AOPO instituted voluntary accreditation of OPOs. Members of its Accreditation Committee conduct a site visit with each OPO seeking accreditation. They rigorously scrutinize all aspects of the OPO's operations and score them against standards that were developed by the AOPO. Many of the nation's OPOs have been accredited by the AOPO, while others are actively pursuing accreditation.

In 1996, the AOPO completed its first financial benchmark process for its participating members. This was a comprehensive analysis of OPO finances. The objective was to provide to each OPO a comparison of the finances of similar OPOs. National statistics also were made available to participants and presented to members of the association at its annual meting. The concept was to share information that would allow OPOs to determine whether or not they were allocating resources in a manner that would result in high performance. For example, if a low-performing OPO determines that it allocated a substantially lower percentage of its resources to marketing than did higher-performing OPOs, it may adjust that allocation accordingly. Simply stated, this is a process that can help to reveal best financial practices with the hope of improving the overall performance of participating OPOs.

DISCUSSION

Thousand of patients each year receive organs supplied by the nation's OPOs. Without the skill and commitment from the individuals who work in these OPOs, the number of transplant procedures occurring in the United States would be greatly diminished. The ability of OPOs to stimulate participation in organ donation is a key element in meeting the needs of those who are waiting for transplants. This is accomplished through sophisticated marketing and education, innovative practices, contributing to industry knowledge, relationship building, and expert public relations.

Many OPOs have developed a high level of expertise in marketing. Most practice market segmentation and target marketing, and most OPOs expend significant resources in this area. OPOs are constantly trying innovative techniques to improve performance. These range from advancing technology to improving the workplace environment to developing better techniques for stimulating public and professional participation in donation. Sharing information about their successes with these innovative practices is necessary to contribute to the industry knowledge base. This allows others to emulate best practices with an objective of improving overall performance of OPOs throughout the country. As with all successful organizations, OPOs have recognized the importance of effective networking at all levels. They spend a great deal of time building relationships in hospitals, with community leaders, with medical professionals, and among their peers. They also have focused attention on developing relationships with representatives of the news media. This helps to ensure fair reporting when negative news stories about procurement or transplantation arise. Although the transplant community is endeavoring to educate the public about these issues, the public generally is not

4

adequately informed. This situation has added to the mistrust of OPOs and the donation process. Certainly, these problems will require the continued attention of OPOs and the entire transplant community.

Organ procurement is a very complicated process, often involving dozens of people. The primary objective of OPOs is to simplify the process by coordinating the countless tasks and communicating effectively with everyone involved. The extent to which OPOs can accomplish this objective will be paramount to the overall success of transplantation in the United States.

REFERENCES

1.   National Organ Transplant Act of 1984. 42 U.S.C, §201, §273, §2339, §1395, §274 §32, 1984. This landmark was the precursor to most federal transplant legislation that followed.

2.   Task Force on Organ Transplantation. Organ transplantation: Issues and recommendations. DHHS Publication. Washington, DC: Office of Organ Transplantation, 1986. Much of the transplant legislation that followed resulted from the recommendations of this national panel of transplant experts.

3.   Omnibus Budget Reconciliation Act of 1986. 42 U.S.C, §3206, §1395-1396, §273-274, 1986. This legislation first defined OPO qualifications, structure and function.

4.   Conditions for Coverage: Organ Procurement Organizations. 42 CFR, §486, 1996:468-490.

5.   Conditions of Participation for Hospitals. 42 CFR, §482, 1998. This regulation defines current hospital requirements for interacting with OPOs regarding identification and evaluation of potential organ donors. It also defines the mechanism for presenting the option of donation the potential donor's family.

6.   United Network for Organ Sharing. U.S. facts about transplantation. http:\\www.unos.org/3/15/03. This website provides the most current statistics regarding donation and transplantation.

7.   The Gallup Organization, Inc. The American public's attitudes toward organ donation and transplantation. Boston, 1993. This is the most current national public opinion poll regarding organ donation.

8.   Gortmaker SL, Beasley CL, Brigham LE et al. Organ donor potential and performance: size and nature of the organ donor shortfall. Crit Care Med 1996; 24(3):432-439. This article presents data from a multi-regional study regarding the gap between the number of potential organ donors and the number of actual donations. It also presents an estimate of national donor potential.

Procurement and Short-Term Preservation of Cadaveric Organs

Anthony M. D'Alessandro and James H. Southard

ORGAN DONATION

Improvements in immunosuppression, organ preservation, surgical technique, as well as long-term recipient management have led to tremendous success following transplantation. Consequently, more patients than ever before have benefited from transplantation. Unfortunately, the rate of organ donation has not kept pace with the ever-increasing recipient waiting lists. Recent United Network for Organ Sharing (UNOS) statistics reveal that greater than 80,000 patients (Table 5.1) currently await transplantation. For a variety of reasons, some organ procurement organizations (OPOs) have very high organ donation rates while others fall significantly below average. Likewise, consent for organ donation averages approximately 60%, although several OPOs have much higher consent rates. Clearly, much greater emphasis needs to be placed on increasing organ donation. Organizations such as the Coalition on Organ Donation, the American Society of Transplant Surgeons (ASTS), UNOS, and the American Association of Organ Procurement Organizations (AOPO) are leading the way in this effort. Still a critical shortage of organs exists which has resulted in an increase in the use of live donation and an increase in the use of expanded cadaveric donors. Since criteria for the use of organs has expanded significantly, any patient who is declared brain dead or who is being withdrawn from support should be considered as an organ donor.

DETERMINATION OF DEATH

Patients may be declared dead by brain death criteria and by cardiopulmonary criteria. Currently, the majority of organ donors (98%) are declared dead by brain death. The definition of brain death was first examined in a report by the Harvard Medical School in 1968 and guidelines later set for brain death determination in 1981 which led to the "Uniform Determination of Death Act." These criteria are shown in Table 5.2.

Brain death occurs when complete and irreversible loss of brain and brain stem function occurs, which presents clinically as complete apnea, brain stem areflexia, and cerebral unresponsiveness. In order to evaluate a patient clinically for brain death, several preconditions must be met. The patient must be on a ventilator in a coma and have a cause for underlying brain damage. Most cases are caused by trauma, subarachnoid hemorrhage, cerebral abscess or tumor, meningitis, encephalitis, or cerebral hypoxia. Reversible causes of brain stem depression such as

Organ Transplantation, 2nd edition, edited by Frank P. Stuart, Michael M. Abecassis and Dixon B. Kaufman. ©2003 Landes Bioscience.

Table 5.1. National Transplant Waiting List by organ*

Organ	Number of Patients
Kidney	53,813
Liver	16,938
Pancreas	1,396
Kidney-Pancreas	2,412
Intestine	177
Heart	3,814
Heart-Lung	196
Lung	3,839
Overall	80,657

*UNOS data, March 2003

Table 5.2. Criteria for brain death

Prerequisite
   All appropriate diagnostic and therapeutic procedures have been performed and the patient's condition is irreversible.
Criteria (to be present for 30 minutes at least 6 hours after the onset of coma and apnea)
   1. Coma
   2. Apnea (no spontaneous respirations)
   3. Absent cephalic reflexes (pupillary, corneal, oculoauditory, oculovestibular, oculocephalic, cough, pharyngeal, and swallowing)
Confirmatory test
   Absence of cerebral blood flow by radionuclide brain scan

hypothermia and drug intoxication must first be excluded. Trauma patients are often intoxicated with alcohol. Thus, 8 hours should be allowed to pass if alcohol use is suspected before a diagnosis of clinical brain death can be made. Patients in intensive care units may also be under the influence of sedative or paralytic agents.

Clinical testing is relatively straightforward and examines the presence of brain stem reflexes and the presence of total apnea. Five brain stem reflexes should all be absent in order to diagnose brain stem death: pupillary response to light, corneal reflex to touch, vestibulo-ocular reflex using the cold caloric test, the gag reflex, and the apnea test. The apnea test demonstrates the absence of respiratory drive to $PaCO_2$ greater than 50 mmHg. During apnea, the $PaCO_2$ rises by about 2 mmHg/min; thus, if the starting $PaCO_2$ is over 30, the $PaCO_2$ will rise to over 50 mmHg in about 10 minutes. To prevent hypoxia during these 10 minutes, the patient should be preoxygenated prior to the test. Confirmatory studies, although not necessary, include serial electroencephalography and radionuclide scan to assess cerebral perfusion.

Death may also be declared by cardiopulmonary criteria, and in certain instances, particularly when patients are being withdrawn from support, organ donation is possible. This type of donation is referred to as donation after cardiac death (DCD) or non-heart-beating donation. Prior to the Harvard criteria defining brain death in 1968, all organ donors were DCD donors. Although some warm ischemia occurs in these donors, several centers have shown that renal and extrarenal donation is possible. Recently the Institute of Medicine (IOM) reviewed non-heart-beating organ donation, published guidelines, and concluded that NHBDs are a medically and ethically acceptable source of donor organs. Currently, NHBDs comprise 2% of organ donors and this percentage will likely increase since the results of transplantation have been shown to be acceptable.

EVALUATION AND SELECTION OF DONORS

OPOs form a vital link between referring donor hospitals and transplant centers and should be notified as early as possible in order to make the determination of suitability for organ donation.

Obtaining consent for organ donation is of paramount importance in increasing organ donation. A caring sensitive approach by trained individuals that have time to spend with families cannot be overstated. Organ procurement personnel, clergy, and nursing staff play a vital role in this area. Once consent is obtained, a review of the patient's history should focus on the mechanism of death, periods of hypotension or cardiac arrest, need for vasoactive medications, and previous surgery. Likewise, the patient's social history, including alcohol and drug use, should be known. Generalized infectious diseases are ruled out by obtaining human immunodeficiency virus (HIV) antigen, anti-HIV-1, anti-HIV-2, human T-cell lymphotoxic virus (HTLV)-1 and HTLV-2, anti-cytomegalovirus (CMV), anti-hepatitis C virus (HCV), hepatitis B surface antigen (HBSAg) and hepatitis B core antibody. Specific organ function is primarily determined by laboratory data, chest x-ray, electrocardiogram, and echocardiogram.

Since criteria for organ donation are expanding, there are fewer absolute contraindications to organ donation (Table 5.3). Relative contraindications to organ donation have increased since many were previously considered to be absolute contraindications. Table 5.4 should be considered only as a guideline to relative contraindications since many centers have successfully utilized organs from every category listed.

As a general rule, hepatitis C positive donors may be used in hepatitis C positive recipients. Also, as long as hepatic trauma is minimal, aspartate aminotransferase (AST) and alanine aminotransferase (ALT) levels are decreasing, and macrovesicular steatosis is < 60%, the liver may be used. Hepatitis B core antibody positivity is more controversial, but with long-term hepatitis B immune globulin (HBIG) use, transplantation may be indicated depending on the clinical situa-

Table 5.3. Absolute contraindications to cadaveric organ donation

Malignancy outside central nervous system
Prolonged warm ischemia
Long-standing hypertension
Hepatitis B surface antigen
Sepsis
Intravenous drug abuse
Human immunodeficiency virus

Table 5.4. Relative contraindications to organ donation by organ type

Heart/Lung	Liver	Pancreas	Kidney
Age > 50	Age > 60	Age > 55	Age > 60; < 6
High dose inotropes	Hepatic trauma	Amylase elevation	Hypertension
Wall motion abnormalities	AST, ALT elevations	Glucose elevation	Diabetes
Chest trauma	Hepatitis B core antibody	Fatty pancreas	ATN (creatinine ≥ 2.5 mg/dL)
Abnormal CXR	Hepatitis C	Hepatitis C	Hepatitis C
PaO_2 < 350 on FiO_2 1.0	Steatosis	Prolonged warm and cold ischemia	Prolonged warm and cold ischemia
Prolonged cold ischemia	Prolonged warm and cold ischemia		

tion. One of the best indicators of whether or not a liver should be used is the intraoperative assessment of an experienced donor surgeon. This is also true for pancreas donors since glucose levels may be elevated due to exogenously administered glucose and steroids as well as to catecholamine release and insulin resistance from trauma. Likewise, an elevated serum amylase does not always reflect pancreatic trauma and should not in isolation be used to preclude pancreatic organ donation. A history of early renal disease, such as mild hypertension and diabetes, may also be compatible with organ donation. A renal biopsy can be obtained to assess the degree of pathology, if any, prior to transplantation. Likewise, in older donors, if glomerulosclerosis is present, both kidneys may be implanted. In children less than 6 years of age, and depending on size, the kidneys can be implanted separately or en bloc. Although heart and lung donor criteria are somewhat more restrictive, depending on the potential recipient's condition, these criteria can be expanded. Cadaveric heart donors should have a normal chest x-ray, electrocardiogram, isoenzymes, and echocardiogram. Lung donors should not have any chest trauma and should have negative sputum cultures and a $PaO_2 \geq$ 350 torr on an FiO_2 of 1.0. Again, examination of the organs by a skilled heart and lung donor surgeon may be necessary before excluding a potential donor.

Due to the risk of organ dysfunction and failure with increasing cold ischemia time, preservation times should be minimized to avoid exacerbating the current donor shortage. Safe acceptable cold ischemic times vary with each organ and, as a general rule, are as follows: heart/lung 6 hours, liver 12 hours, and pancreas 18 hours. Since delayed renal graft function predicts long-term survival, attempts should be made to limit preservation times. When kidneys are cold stored, they should be transplanted within 18-24 hours, and when machine-perfused within 24-30 hours.

THE EXPANDED DONOR

The expanded donor, previously referred to as the marginal donor, has assumed a much greater role in transplantation due to the critical shortage of organs. Prior to the waiting list reaching its current size, ideal donors were primarily utilized. Ideal donors are young, normotensive, brain-dead donors free of any disease and with minimal warm ischemia times. Table 5.4, which outlines the relative contraindications to transplantation may also be viewed as criteria that define the expanded donor. DCD donors, whether controlled or uncontrolled, should also be included in the expanded donor pool since warm ischemia times are greater and there are higher rates of delayed graft function. Likewise, split liver transplantation, where one donor liver is shared between one adult and one child or between two adults, should also be considered in the expanded donor definition. However, what is important to consider when utilizing expanded donors is the risk of a patient dying on the waiting list versus the risk of dying with transplantation of an organ from an expanded donor. Although graft function may initially be worse and long-term patient and graft survival less than from organs transplanted from ideal donors, the risk of dying has been shown to be less than if the patient continued on the waiting list. As more is learned about the expanded donor, pharmacologic interventions and changes in preservation, such as machine perfusion instead of cold storage, may eventually yield results similar to that obtained from ideal donors.

DONOR RESUSCITATION AND STABILIZATION

Clearly, proficient management of the organ donor before retrieval is of paramount importance. However, what may be equally important is the expeditious removal of organs when a donor's condition is difficult to stabilize. In these instances, the organs should be removed as quickly as possible to avoid the risk of the donor having a cardiac arrest or suffering long periods of hypotension.

The hemodynamic management of the donor is of primary importance and includes maintaining an adequate blood pressure (> 100 mmHg) and urine output (> 100 mL/hr). Once the donor has been declared brain dead, large volumes of fluid and plasma expanders may be necessary to resuscitate the donor to achieve adequate blood pressure and urine output. Hemodynamic monitoring with a central venous catheter (CVP), arterial line, and sometimes a pulmonary artery catheter are usually necessary. Care should be exercised to avoid over-hydration which may cause over-distension of the heart as well as congestion of the lungs and liver which may later affect the function of these organs. Because of the hemodynamic instability caused by severe brain injury due to catecholamine hyperactivity which is followed by hypoactivity, volume alone may not stabilize the donor. Vasopressor support, usually with dopamine, is adequate to stabilize the donor. High-dose dopamine in doses up to 15 μg/kg/min has been shown to be well tolerated. Although vasopressors, such as levarterenol and phenylephrine, should be avoided since they have a greater propensity to cause organ ischemia, they may be necessary to maintain an adequate blood pressure. However, attempts should be made to reduce the dosages by volume resuscitation and the use of dopamine. If these more potent alpha receptor vasopressors are necessary, they should be used with dopamine at renal doses (3-5 µg/kg/min) to mitigate against splanchnic and renal vasoconstriction.

Usually when urine output is low, volume expansion results in increased urine output. However, diuretics, such as furosemide and mannitol which generally should be avoided in organ donors, can be used to increase urine output as long as there is adequate blood pressure and volume expansion (CVP 12). Many times, however, the problem in brain-dead donors is massive urine output caused by the development of diabetes insipidus due to the lack of the antidiuretic hormone, vasopressin. If urine output exceeds 500 mL/hr, a hypotonic diuresis ensues that should be replaced with hypotonic infusions. If polyuria persists despite adequate fluid replacement, vasopressin may be given at a rate of 0.5-2.0 units per hour to slow diuresis to a more manageable level.

Due to the significant hormonal imbalances seen in brain-dead donors, hormonal management may help to stabilize donors. There has been some evidence that administration of intravenous triiodothyronine (T_3) and arginine vasopressin (AVP) may stabilize the brain-dead donor by restoring some of the hormonal imbalances and circulatory instability. Likewise, brain death may cause varying degrees of cortisol depression and steroid replacement therapy with hydrocortisone may be indicated. Additionally, due to the loss of thermoregulatory function with brain death, many organ donors will become hypothermic unless measures are taken to avoid hypothermia and its sequelae. Hypothermia may lead to cardiac arrhythmias, myocardial depression and hypotension leading to poor tissue and organ perfusion. Organ function may also be compromised from decreased oxygen delivery caused by hypothermia. Infusion of warm fluids and external heating devices will help reduce hypothermia and its adverse effects. Another common problem in brain-dead donors is the presence of coagulopathy caused by tissue thromboplastin release. Coagulopathy, although difficult at times to manage, can be treated with administration of packed red blood cells, fresh frozen plasma, and platelets.

Since many OPOs have recently instituted DCD programs, it is important to mention some important differences in donor management. DCD donors are not brain dead due to preservation of brainstem reflexes, but usually have severe neurologic injury from which they will not recover. The decision to withdraw support has been made by the primary physician and family before notification of the OPO. These donors tend to be hemodynamically more stable with fewer vasopressor requirements than brain-dead donors. The withdrawal of support may occur either in the intensive care unit (ICU) setting, where the patient expires and is conveyed to the operating room, or alternatively in the operating room. In either instance, the patient must be pronounced dead by a physician not affiliated with the transplant team. The patients should be fully supported until withdrawal of support is initiated. The administration of vasodilators and anticoagulants at the time of support withdrawal may be given on a case-by-case basis in accordance with IOM guidelines. Likewise, an additional period of 5 minutes must elapse after death is pronounced before initiating organ retrieval. Because of the presence of brainstem reflexes, the family must be informed that if the patient continues to have spontaneous respirations beyond a certain period of time (usually > 1 hr), the patient will be returned to the ward or ICU to expire without organ retrieval. Although organs can be transplanted with up to 1 hour of warm ischemia with good results, warm ischemic times of greater than 1 hour will likely result in less than optimal organ function.

COORDINATION OF MULTIORGAN RETRIEVAL

OPOs serve several vital functions in the organ procurement process including donor referrals, donor family request and consent, and donor management. Additionally, OPOs coordinate the donation process once consent is obtained. Since the majority of organ donations are multiorgan, OPOs must coordinate assessment of each organ system as well as assessing donor history, laboratory values, including ABO type and tissue type, and any noninvasive testing. If an OPO serves one transplant center, coordination is easier since communication is facilitated among the different transplant teams. However, most OPOs serve more than one center and organ placement and team coordination is logistically more challenging. It is not unusual to have several teams present at an organ procurement including teams for the heart, lungs, liver, pancreas, kidneys, and small bowel as well as teams for tissue donation. Communication is extremely important in facilitating organ procurement in such a way that donor hospitals remain committed to organ donation in their communities. Since most of the techniques for organ procurement are fairly standard with minor center variation in techniques, early communication between teams via the OPO will also help to facilitate a smooth recovery. As a general rule, after the donor is brought to the operating room, dissection of the heart and lungs is followed by dissection of the liver and pancreas, small bowel, and kidneys. Removal of organs usually follows the same sequence as the dissection of the specific organs. Alternatively, all intraabdominal organs may be removed en bloc without in situ dissection of the individual organs. This technique is mandatory in organ retrieval from DCDs. Eye, bone, and tissue donation follows removal of all solid organs.

OPOs also serve a vital postrecovery function at donor hospitals by providing feedback on the ultimate placement and transplantation of the organs retrieved. Also, continued community visibility of the OPO and transplant centers through educational programs will help to maintain and increase organ donation so that more patients will ultimately undergo transplantation. Likewise, donor and recipient families, by interfacing with their communities, can have a profound effect on helping to increase awareness and, ultimately, organ donation.

SURGICAL TECHNIQUES OF ORGAN PROCUREMENT

Since most organ procurements involve several organ systems, these combined multiorgan procurements will be described. Once the patient is conveyed to the operating room, prepped and draped, a long incision from the suprasternal notch to the pubis is made. The sternum is split and the cardiac team will open the pericardium, inspect the heart and encircle the superior vena cava, suprahepatic vena cava, and the aorta. The pleural spaces will also be opened and the lungs inspected if being considered for transplantation.

The intraabdominal portion of the organ procurement commences once the heart team has inspected the heart and lungs. It is important to note that as organ procurement has evolved, less dissection has been shown to be advantageous since it reduces vasospasm, warm ischemia, and decreases the length of operation and donor instability. Liver dissection is performed first and usually involves encircling the supraceliac aorta, dividing the common bile duct, gastroduodenal artery, and encircling the portal vein. If the pancreas is being used by a center other than the liver center, dissection of the entire celiac artery to the aorta may be performed with the left gastric and phrenic arteries being ligated and the splenic artery encircled. However, prior to ligating the left gastric artery, the donor surgeon must be sure the left hepatic artery does not arise from the left gastric artery. This arterial anomaly is seen in 15% of cases and is visualized in the gastrohepatic omentum. Another hepatic arterial anomaly is the presence of a right hepatic artery arising from the superior mesenteric artery (SMA). This occurs in approximately 10% of cases and can be palpated posterior to the portal vein and common bile duct. Both hepatic arterial anomalies are compatible with hepatic and pancreatic procurement in all cases. Several techniques of vascular reconstruction are available and usually require the use of donor iliac artery grafts.

A new technique of liver procurement involves in situ donor liver splitting for two recipients. Although some centers perform ex vivo liver splitting, in situ splitting may be associated with less bleeding and fewer biliary complications after transplantation. However, a major disadvantage of in situ liver splitting is the additional 1-2 hours required to perform the procedure.

Pancreas dissection involves a Kocher maneuver to mobilize the duodenum as well as dissection of the posterior pancreas to the level of the inferior mesenteric vein (IMV) which is ligated. The first portion of the duodenum and the small bowel just distal to the ligament of Treitz are stapled and the mesenteric vessels are ligated. If the intestine is being recovered for transplantation, the SMA and superior mesenteric vein (SMV) are dissected but not ligated. Also, since the liver and intestine are both transplanted in some patients with short bowel syndrome, the liver, pancreas, and intestine are recovered en bloc without dissection. The pancreas is usually transplanted with the liver and intestine in order to keep the donor porta hepatis intact.

Renal dissection should be minimal and limited to identification and division of the distal ureters. Dissection of the renal arteries and veins as well as mobilization of the kidney should be done only after the intraabdominal organs are infused with preservation solution. This minimal dissection technique helps to limit renal artery vasospasm and subsequent delayed graft function.

Once preparation of each organ to be retrieved is complete, the patient is given 20,000-30,000 units of heparin followed by cannulation of the distal aorta with a chest tube for eventual administration of preservation solution. Also, just prior to organ retrieval, some teams will administer an α-adrenergic antagonist, such as phentolamine, to prevent vasospasm and to ensure more uniform flushout of the intraabdominal organs. Likewise, the heart/lung team may administer prostacyclin, also a vasodilator, during the procurement. Once the SVC is occluded, the aorta is clamped just proximal to the innominate artery, cardioplegic solution infused, and the caval atrial junction at the level of the diaphragm incised. At the same time, infusion of 1-2 liters of University of Wisconsin (UW) solution is begun via the aortic cannula. The portal vein is then incised, cannulated, and infused with 1 liter of UW solution. Once the heart or heart-lung block is removed, the liver and pancreas are removed followed by removal of the kidneys either en bloc or separately according to the retrieval team preference. Figure 5.1 depicts the appearance of the liver, pancreas, and kidneys after dissection as well as placement of aortic and portal vein cannulas just prior to removal. After removal, the liver and pancreas are flushed with an additional 200-300 cc UW solution via the SMA, celiac artery, and portal vein and stored in sterile plastic bags on ice at 4°C. If the liver and pancreas are being used at different centers, they are separated and stored separately prior to transport.

Fig. 5.1. Cadaver donor multi-organ retrieval. Reprinted with permission requested from: Sollinger HW, Odorico JS, D'Alessandro AM et al. Transplantation. In: Schwartz SI, ed. Principles of Surgery, ed. 7. New York, McGraw-Hill, 1998:361-439.

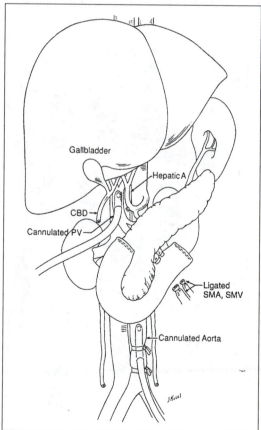

The kidneys, if removed en bloc, are usually separated by dividing the vena cava and aorta longitudinally. This will allow identification of multiple renal arteries from within the aorta without risk of injury. If the kidneys are to be machine perfused instead of cold stored, they may be cannulated en bloc if multiple renal arteries are present or individually if single arteries are present bilaterally. En bloc perfusion requires ligating all lumbar arteries, suturing the proximal aorta, and cannulating the distal aorta. Again, the kidneys are flushed with additional UW solution, placed in sterile plastic bags, and placed on ice at 4°C.

An alternative, rapid en bloc technique of organ retrieval may be used with DCD donors or in donors who have become hemodynamically unstable or who have had cardiac arrest (Fig. 5.2). This technique involves cannulating the femoral artery and vein or the distal aorta and vena cava, clamping the thoracic aorta, and dividing the esophagus, sigmoid colon, and ureters. While flushing the femoral artery or aorta with UW solution, all intraabdominal organs are removed en bloc by dissecting retroperitoneally starting at the level of the diaphragm and ending at the distal aorta and vena cava which are divided. The portal vein is flushed

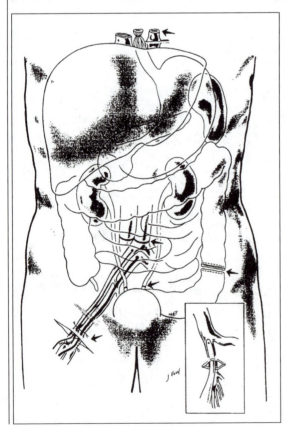

Fig. 5.2. Rapid en bloc retrieval of all intra-abdominal organs. Reprinted with permission requested from: D'Alessandro AM, Hoffmann RM, Knechtle SJ et al. Successful extrarenal transplantation from non-heart-beating donors. Transplantation 1995; 59:977-982.

via the superior mesenteric vein on the back table (inset Fig. 5.2), and the aorta is incised and each orifice flushed with additional UW solution. If the liver and intestine are to be used for transplant, the aorta should not be divided since it may be used as a conduit with both the celiac and SMA attached. The organs may be separated at the donor hospital, or alternatively, stored in plastic bags at 4°C and separated upon return to the transplant center.

SAFE TRANSPORT OF ORGANS

Since organs may be transported from one center to another, uniform packaging and storage is essential to ensure all organs are able to be transplanted upon reaching their destination. All organs must be placed in triple sterile plastic bags as well as a rigid container and placed on ice in a 1-1$^1/_2$" thick polystyrene container. All containers must be labeled, donor paperwork included, and an additional red top tube of blood sent to the receiving center. Depending on the organ and distance to be traveled, transportation may be by ground, commercial flight, or chartered jet. If the organs retrieved are not being sent to other centers, they may be safely stored in triple sterile bags on ice in insulated coolers. The outer container must be moisture resistant and clearly marked with a UNOS donor identification number and a biologic hazard designation label.

SHORT-TERM ORGAN PRESERVATION

INJURY DURING PRESERVATION

Preservation of organs after retrieval is clearly one of the cornerstones of successful transplantation. Although organs vary in their tolerance to cold ischemia, injury to numerous cellular systems begins to occur immediately upon removal. Hypothermia suppresses, to a degree, these changes, but injury during hypothermia still occurs but at a slower rate. Since hypothermic-induced cell swelling is a major source of injury during preservation, most organ preservation solutions are formulated to prevent swelling at cold temperatures. The addition of impermeants such as gluconate, lactobionate, and saccharides such as raffinose, help prevent hypothermic-induced cellular swelling.

Several other phenomena have also been implicated in cell injury during preservation and have been studied extensively. Numerous cellular functions including maintenance of the cellular cytoskeleton requires energy in the form of adenosine triphosphate (ATP). Loss of energy-generating capabilities due to mitochondrial damage or loss of precursors will lead to irreversible cell injury and death upon reperfusion. This concept forms the basis for adding ATP precursors in the form of adenine, adenosine, and ribose to organ preservation solutions. Oxygen-free radical formation after reperfusion has also been implicated in cellular injury during preservation. Suppression of free-radical formation or the addition of free-radical scavengers such as allopurinol, may be beneficial in preservation solutions. Likewise, breakdown of cellular metabolites, such as glycogen and glutathione, may lead to injury and addition of these metabolites may be important in successful organ preservation. Also, activation of catabolic enzymes such as phospholipases and proteases and activation of the arachidonic cascade will lead to cell injury and methods to block their activation may lead to better organ preservation.

CLINICAL ORGAN PRESERVATION

The goals of organ preservation are to maximize organ utilization and maintain excellent organ function while providing safe transport time as well as time for recipient preparation. Currently, the UW solution is primarily used for preservation of intraabdominal organs. Although this solution is used by some centers for heart and lung preservation, many other solutions are also utilized. The components of the UW solution are shown in Table 5.5.

RENAL PRESERVATION

Currently, there are two methods of preserving kidneys for transplantation: static cold storage or continuous machine perfusion. The majority of centers use cold storage due to simplicity but experience a higher rate of delayed graft function (DGF) than with machine perfusion. Although previously not thought to be important, early DGF appears to predict long-term graft survival. Also, with more expanded donors being utilized, including DCD donors, continuous machine perfusion may be beneficial in preserving organ function. The machine perfusion solution is similar to UW cold storage solution except that lactobionate is replaced by gluconate. As a general rule, cold-stored kidneys should be implanted within 18-24 hours and machine-perfused kidneys within 24-30 hours of removal.

PANCREAS PRESERVATION

The UW solution has been used safely to preserve the pancreas on average 16 hours. Although attempts have been made to perfuse the pancreas experimentally, it has not been met with much success. Interestingly, the pancreas clinically appears to tolerate periods of cold ischemia better than the liver.

LIVER PRESERVATION

Liver function and success after transplantation is dependent not only on donor and recipient factors, but also on good preservation. Preservation of the liver not only involves preservation of the parenchyma, but also preservation of the biliary epithelium, as well as the vascular endothelium, particularly the endothelium of the hepatic artery. Prior to the development of UW solution,

Table 5.5. Components of the University of Wisconsin (UW) Solution

Lactobionate (K)	100 mmol/L
KH_2PO_4	25 mmol/L
Glutathione	3 mmol/L
Adenosine	5 mmol/L
$MgSO_4$	5 mmol/L
Allopurinol	1 mmol/L
Raffinose	30 mmol/L
HES	5 g/dL

mOsm/L = 320; pH = 7.4

liver preservation was limited to approximately 6 hours. After the clinical introduction of UW solution, it was believed that extended preservation beyond 12 hours was safe. However, it became apparent that rates of primary nonfunction, biliary complications, and hepatic artery thrombosis increased as preservation time increased. Also, preservation injury may lead to increased rates of rejection via upregulation of MHC class I and II antigens, which in turn may lead to graft loss. Although preservation beyond 12 hours can be achieved, rates of primary nonfunction and initial poor function are increased. For this reason, most transplant centers attempt to limit preservation of the liver to 12 hours or less. In an era of donor shortages, every effort should be made to minimize retransplant rates and this can be achieved by minimizing cold ischemia times. Longer term preservation may only be achieved by machine perfusion which has been shown experimentally to be more successful than cold storage.

HEART AND LUNG PRESERVATION

Although a variety of preservation solutions have been developed for heart and lung preservation, preservation of the intrathoracic organs is still limited to 4-6 hours.

STRATEGIES TO MINIMIZE ISCHEMIC DAMAGE

The use of expanded donors including DCDs where periods of hypotension, hypoxia, and warm ischemia are encountered has provided us with opportunities to examine limits and develop strategies to help minimize damage. Although any period of warm ischemia had previously been thought to be inconsistent with organ donation, most organs will tolerate short periods of warm ischemia. Clinical experience with DCDs indicates that the kidneys, liver, pancreas and the lung will tolerate 30-60 minutes of warm ischemia and will still function adequately after transplantation. Administration of anticoagulation with heparin will help prevent small vessel occlusion and administration of pharmacologic agents, such as phentolamine, will help prevent vasospasm and enhance better flush and preservation of donor organs. Administration of nitric oxide precursors, such as L-arginine and nitroglycerin, either to donors or to preservation solutions has been shown experimentally to mitigate warm ischemic damage. Evidence is mounting supporting continuous machine perfusion of kidneys retrieved from expanded or DCDs. Warm ischemic damage can be limited and perhaps improved during cold preservation by continuously supplying substrates for repair and energy production upon reperfusion. Delayed graft function in machine-perfused kidneys retrieved from DCDs has been shown to be similar to DGF rates in cold-stored kidneys retrieved from ideal donors. Interestingly, brain death itself has been shown to have a detrimental effect on organ function after transplantation. In addition to the marked hormonal imbalances that occur with brain death, organ injury may occur by activating T lymphocytes and the inflammatory response via cytokine release. This response and subsequent organ injury has been shown experimentally to be abrogated by administration of agents that block T-cell costimulation. Since hypothermia-induced cell injury increases with increasing cold ischemia time, preservation times should be minimized, particularly in expanded donors. Likewise, in clinical transplantation, one of the only factors that can be controlled is preservation time and this should be minimized to prevent wastage of organs.

CONCLUSION

Transplantation has become the treatment of choice for patients with end-stage organ failure. Results have improved due to refinements in surgical technique, immunosuppression, preservation, and patient management. Unfortunately, organ donation has not kept pace with the ever-increasing demand for transplantation. Although the techniques described in this chapter on organ procurement and preservation are important, they cannot be applied without the generous gift of organ donation. This is also true for nearly every other advance made in clinical transplantation. Therefore, increasing the number of patients who receive the gift of life through increased organ donation must now be our highest priority.

SELECTED READINGS

1. Institute of Medicine. Non-Heart-Beating Organ Transplantation: Medical and Ethical Issues in Procurement. Washington DC: National Academy Press, 1997.
2. D'Alessandro AM, Hoffmann RM, Knechtle SJ et al. Liver transplantation from controlled non-heart-beating donors. Surgery 2000; 128:579-588.
3. D'Alessandro AM, Odorico JS, Knechtle SJ et al. Simultaneous pancreas-kidney (SPK) transplantation from controlled non-heart-beating donors (NHBDs). Cell Transplant 2000; 9:889-893.
4. Kauffman HM, Bennett LE, McBride MA et al. The expanded donor. Transplant Rev 1997; 11:165-190.
5. The Multi-Organ Donor: Selection and Management. RSD Higgins, JA Sanchez, MI Lorber, JC Baldwin, Eds. Malden, MA, Blackwell Science, 1997.
6. Sollinger HW, Odorico JS, Knechtle SJ et al. Experience with 500 simultaneous pancreas-kidney transplants. Ann Surg 1998; 228:284-296.
8. Problems in General Surgery, Vol. 15: Organ Transplantation. RJ Stratta, Ed. Philadelphia, Lippincott Williams & Wilkins, 1998.
9. D'Alessandro AM, Hoffmann RM, Southard JH. Solution development in organ preservation: the University of Wisconsin perspective. Transplant Rev 1999; 13:1-12.
10. Zaroff JG, Rosengard BR, Armstrong WF et al. Consensus Conference Report-Maximizing use of organs recovered from the cadaver donor: cardiac recommendations. Circulation 2002; 106:836-841.
11. Weber M, Dindo D, Demartines N et al. Kidney transplantation from donors without a heartbeat. N Engl J Med. 2002; 347:248-255.

Kidney Transplantation

Dixon B. Kaufman
Illustrations by Simon Kimm

INTRODUCTION

Kidney transplantation should be strongly considered for all medically suitable patients with chronic and end-stage renal disease. A successful kidney transplant saves lives and greatly enhances quality of life.[1,2] Hundreds of thousands of patients worldwide have received a kidney transplant since the mid-1950's. Currently, in the U.S., there are over 100,000 persons living with a functioning kidney transplant. This number represents only a fraction of the nearly 400,000 persons enrolled in the U.S. end-stage renal disease (ESRD) program. Figure 6.1 illustrates the number of ESRD patients on either dialysis therapy or living with a functional kidney transplant according to age. The median age of a transplant patient is 40 years and that of a patient on dialysis 64 years.

Interestingly, for ESRD patients living in Canada, United Kingdom, Australia, and Sweden the transplantation rates all exceed 50%. The international disparity in renal transplant rates is due, in large part, to the differential degree of access to dialysis therapy and transplantation in the various countries. In 1973, the U.S. Congress enacted Medicare entitlement for the treatment of end-stage renal disease to provide equal access to dialysis and transplantation for all ESRD patients in the Social Security system by decreasing the financial barrier to care. At the time, this was a major step forward in improving the quality of care of the patient with failing kidneys.

Today, access to transplantation is primarily obstructed by the donor organ shortage. Unfortunately, only a minority of patients that could benefit from a kidney transplant ever receive one. Figure 6.2 illustrates, by year, the growing size of the waiting list, the number of kidney transplants performed, and the relatively stagnant number of cadaver organ donors per year.[3] In 2003, approximately 55,000 persons were awaiting kidney transplantation. In 2002, only 14,728 kidney transplants were performed — 8,493 cadaveric and 6,235 living donor transplants.

The annual number of kidney transplants has doubled over the past 18 years. Growth has been largely due to an increase in living donation. At the current pace, the number of living donor kidney transplants will soon exceed cadaveric transplants. Already the number of living donors exceeds cadaveric donors. The growing popularity of living kidney transplantation is due, in part, to the recognition that waiting for a cadaveric kidney is a slow process. Waiting times of 4-5 years are not uncommon. It is also widely appreciated that recipients of living donor

6

Organ Transplantation, 2nd edition, edited by Frank P. Stuart, Michael M. Abecassis and Dixon B. Kaufman. ©2003 Landes Bioscience.

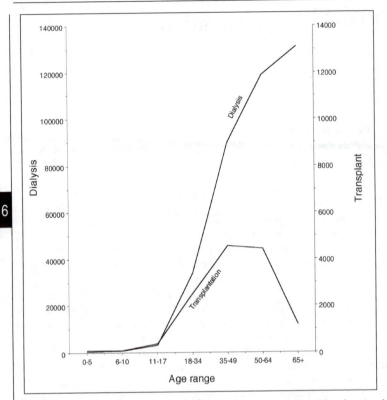

Fig. 6.1. The number of patients on either dialysis therapy or living with a functional kidney transplant enrolled in the U.S. end-stage renal disease program in 2000 according to age.

kidney transplants enjoy outcomes that are superior to those receiving cadaveric transplants. Finally, improvements in the surgical technique using minimally invasive laparoscopic techniques have reduced the reluctance of persons willing to be a living donor.

Cadaveric organs are considered a scarce national resource. The judicious use of cadaveric organs to provide meaningful results for the greatest number of ill patients, without introducing racial bias or inhibiting access, are the underpinning principles of the methodology of cadaveric kidney allocation. Table 6.1 outlines some of the important determinants of the United Network for Organ Sharing (UNOS) cadaver kidney allocation system. The main determinants of kidney allocation include several recipient-specific variables (blood type, degree of sensitization to HLA antigens, pediatric, and donation status), donor variables (HLA matching, expanded criteria status), and accrued waiting time.

It is not a requisite that a patient with renal disease spend time on dialysis to be eligible for a transplant. In fact, outcomes of kidney transplantation are adversely

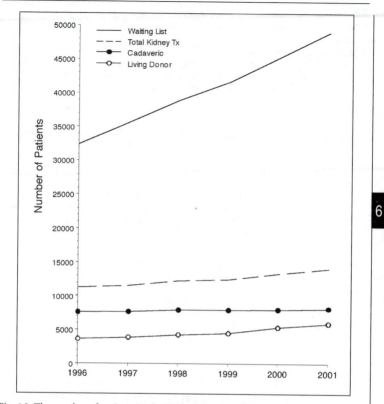

Fig. 6.2. The number of patients in the U.S. waiting for a kidney transplant, receiving a kidney transplant, and the number of cadaver organ donors per year.

affected by prolonged waiting time on dialysis therapy. Therefore, patients early in the course of renal disease need information about, and access to transplantation services. One of the first steps is for the primary care physician or nephrologist to orient the patient to the transplant process. This can occur with that physician or through a referral to the transplant center. Ultimately the evaluation and determination of transplant candidacy occurs at the transplant center.

THE EVALUATION OF CANDIDATES FOR KIDNEY TRANSPLANTATION

Candidates for renal transplantation undergo an extensive evaluation to identify important medical and psychosocial factors that may have an adverse effect on outcome.[4] A thorough evaluation will identify potential pertinent health, social, and financial impediments to a successful transplant that can be solved prior to the procedure. Virtually all transplant programs have a formal committee that meets regularly to discuss the results of evaluation and select suitable candidates for immediate living donor transplantation, or to place on the cadaveric waiting

---

Table 6.1.   UNOS cadaver kidney allocation system

---

1. Blood type O kidneys transplanted only into blood type O recipients
2. Sharing of zero antigen mismatched kidneys
3. Sharing of zero antigen mismatched kidneys to combined K-P candidate
4. Geographic sequence of cadaveric kidney allocation
   A. Local allocation
   B. Regional allocation
   C. National allocation
5. Double kidney allocation: kidneys offered singly unless the donor meets at least two of the following conditions:
   A. Donor age >60 years
   B. Estimated Cr Cl<65 ml/min
   C. Rising serum Cr >2.5 mg%
   D. Adverse donor kidney histology (moderate to severe glomerulosclerosis)
6. Expanded criteria donor kidney allocation. Expanded criteria donors are defined by an "X" in the matrix shown below indicating increased relative risk of graft failure based upon the following factors: age, creatinine, CVA, and hypertension.

Donor Condition

	Age 50 – 59	Age $\geq$ 60
CVA + HTN + Creat > 1.5	X	X
CVA + HTN	X	X
CVA + Creat > 1.5	X	X
HTN + Creat > 1.5	X	X
CVA		X
HTN		X
Creatinine > 1.5		X
None of the above		X

X=Expanded Criteria Donor; CVA=CVA was cause of death; HTN=history of hypertension at any time; Creat > 1.5 = creatinine > 1.5 mg/dl

---

7. The point system for kidney allocation:
   A. Waiting time: (points given when creatinine clearance or calculated GFR £ 20ml/min or initiation of dialysis)
      1 point assigned to the candidate with the longest waiting time; fractions of points assigned to all other patients;
      1 additional point assigned for each full year of waiting time
   B. Quality of match:
      2 points if there are no DR mismatches
      1 point if there is 1 DR mismatch
      0 points if there are 2 DR mismatches
   C. Panel reactive antibody:
      4 points: highly sensitized (PRA $\geq$ 80% recipients with preliminary negative crossmatch)
   D. Pediatric patients:
      4 points: candidates <11 years old
      3 points: candidates 11 to 18 years old
   E. Donation status:
      4 points: transplant candidate who has donated for transplantation within the U.S.
   F. Medical urgency

list. The majority of programs perform the evaluation in the outpatient setting and possess a relatively uniform approach to the diagnosis and treatment of the pertinent medical and psychosocial issues affecting candidacy. The absolute and relative contraindications for listing a patient for kidney transplantation, or for proceeding with transplantation at the final inpatient evaluation, are outlined in Table 6.2.

A. Pre-Existing Morbidities of the Transplant Candidate with Advanced Renal Disease

Patients being evaluated for kidney transplantation have advanced renal disease or renal failure. The scope of the conceivable organ system abnormalities affecting the patients must be appreciated in order to anticipate potential medical problems that may jeopardize the performance of a successful transplant.

Hematologic abnormalities such as anemia and platelet/hemostatic dysfunction are well recognized. The development and extensive use of recombinant human erthyropoietin has dramatically improved treatment of anemia. The use of red blood cell transfusion therapy is now unusual. However, if this has occurred, the patient is at high risk of becoming sensitized by developing anti-HLA Class I cytotoxic antibody. Bleeding disorders are recognized as serious abnormalities in patients with renal insufficiency. Dysfunction of the coagulation cascade, platelet function, and vascular endothelium contribute to the bleeding abnormalities. Augmenting the problem is the use of heparin during hemodialysis, and anti-platelet agents and coumadin to prevent vascular access thrombosis. Conversely, some patients are hypercoagulable, being discovered during work-up for arteriovenous graft thrombotic problems. This has important implications for planning the transplant procedure.

Upper and lower gastrointestinal track abnormalities are very common. In the upper gastrointestinal track there is a high frequency of gastritis and hemorrhage in ESRD patients undergoing hemodialysis. There is also an increase in mortality related to bleeding due, in part, to the bleeding abnormalities described above. Common lower gastrointestinal abnormalities in uremic patients include: diverticulosis, diverticulitis, spontaneous colonic perforation, and prolonged adynamic ileus (pseudo-obstruction). In patients with polycystic kidney disease, the frequency

---

Table 6.2. Contraindications to kidney transplantation

1. Reversible renal disease
2. Recent malignancy
3. Active infection
4. Chronic untreated infection
5. Active glomerulonephritis
6. Advanced forms of major extrarenal complications (coronary artery disease)
7. Life expectancy of less than 1 year
8. Sensitization to donor tissue
9. Noncompliance
10. Active substance abuse
11. Uncontrolled psychiatric disorders

of diverticulosis and diverticulitis is greatly increased over the normal population. The use of aluminum hydroxide, calcium carbonate, analgesic narcotics, and limited fluid intake in the uremic patient may contribute to the development of these colonic disorders.

Hepatic abnormalities as diagnosed by unexpected abnormal liver function tests are often seen in ESRD patients undergoing hemodialysis. Viral hepatitis is the most common etiology in long-term dialysis patients, especially if multiple transfusions of blood products have been required. Most patients are asymptomatic. The prevalence of hepatitis B induced liver dysfunction is approximately 1-2% among dialysis patients. Detection of HB_sAg in patients with abnormal liver function tests is consistent with this diagnosis. The emphasis on PCR methodology to detect hepatitis B virus DNA is indicating that hepatitis B may be more prevalent than was previously appreciated. Hepatitis C is even more prevalent. The second-generation hepatitis C virus antibody test and PCR are redefining the prevalence of HCV positivity in dialysis units. A conservative estimate is that 10-15% of patients are hepatitis C virus positive. Fortunately, hepatitis C virus does not usually lead to cirrhosis of the liver. However, the natural history of progression of liver disease in the immunosuppressed transplant patient is poorly characterized. Other factors, such as drug toxicity and alcohol use, may contribute to liver dysfunction, often present in combination with the viral hepatidities. The physical finding of hepatomegaly does not usually indicate primary liver disease. This physical finding is consistent with chronic passive liver congestion associated with fluid overload or cardiac disease.

The cardiovascular system is profoundly affected in patients with chronic or end-stage renal failure. Uremic patients typically have multiple cardiovascular abnormalities. Patients undergoing hemodialysis have a cardiovascular mortality rate approximately 3 times that of nonuremic patients, and the diabetic, uremic patient has a cardiovascular mortality rate 30 times that of nonuremic patients. The increased mortality is related to atherosclerotic heart disease with myocardial infarction, congestive heart failure, and left ventricular hypertrophy. The increased frequency of coronary artery disease in long-term dialysis patients is multifactorial involving several risk factors: hypertension, hyperlipidemia, and diabetes. Other provocateurs in the development of coronary artery disease include an increase in cardiac output due to AV fistulas/grafts and anemia. The dialysis patient is also at risk for the development of endocarditis because of AV fistula/grafts, peritoneal catheters, and use of central venous dialysis catheters. Uremic toxins may cause myocarditis or cardiomyopathy. The dysrhythmias are frequent because of the effects of hyperkalemia on the cardiac myocardium, as well as the electrolyte, osmolar, and volume shifts that occur with the dialysis procedure. Pericardial effusion with tamponade, and pericarditis may contribute to the enhanced cardiovascular morbidity and mortality rates in the renal failure patient, as well.

Hypertension is the most important risk factor for the development of cardiovascular disease in the chronic renal failure patient. Hypertension occurs in 60-70% of patients requiring dialysis. It is due primarily to chronic volume expansion. In most patients, hypertension is successfully treated with antihypertensive medications. However, the quality of treatment may be inconsistent, leading to development of left ventricular hypertrophy and coronary artery disease. Not infrequently, dialysis patients are undergoing coronary artery bypass. This therapy is leading to increased survival and should not preclude a patient from being considered for renal transplantation.

Significant abnormal pulmonary function is unusual in patients with end-stage renal disease. However, alteration in pulmonary capillary permeability resulting in pulmonary edema at atrial pressures lower than in healthy persons has been described in uremic patients. Pulmonary edema and pleural effusions are more frequent also because of increased total body fluid. These and the additive problems caused by cigarette smoking may, in some cases, be significant. In general, dialysis patients have few symptoms related to the pulmonary system with the exception of occasional pulmonary edema.

There is a high frequency of bone and joint disease in dialysis patients. Hemodialysis patients generally have low calcium levels, high phosphorus concentrations, and elevated serum PTH levels. The degree of bone disease depends on the duration of renal failure and the diligence in which the bone disease is addressed.

B. Pretransplant Evaluation

The pertinent components of a complete pretransplant recipient medical evaluation are outlined in Table 6.3. The emphasis of the evaluation should be to identify and treat all coexisting medical problems that may increase the morbidity and mortality rates of the surgical procedure and adversely impact on the posttransplant course. In addition to a thorough medical evaluation, the social issues of the patient should be evaluated to determine conditions that may jeopardize the outcome of transplantation such as financial and travel restraints and a pattern of noncompliance.

1. History of Renal Disease

There is a diverse array of diseases that destroy renal function afflicting the newborn to the aged. The overall rate of end-stage renal disease is approximately 750 persons/million population. Table 6.4 depicts the most common etiologies of renal disease leading to kidney transplantation.[5] It is important to understand the etiology of renal disease, since the primary renal pathology may influence outcome based on the propensity for recurrence of disease and the association of comorbidities. For example, patients with autosomal dominant polycystic kidney disease may have other medical problems such as intracerebral aneurysm and diverticulosis. Patients with SLE nephritis may have a lupus anticoagulant. Other valuable pieces of information derived from the history would include the clinical course on dialysis with respect to the need for blood transfusions, the occurrence of thrombotic problems with AV grafts/fistulas and control venous dialysis catheters, and the infection rate from peritoneal dialysis catheters (peritonitis) and central venous catheters. Noting the amount of urine production daily is helpful in assessing the early function of kidney allograft. A history of previous kidney transplantation is important for obtaining insight into risk of rejection, infection,

Table 6.3. Pretransplant recipient medical evaluation

1. History
   A. Etiology of renal disease
   B. Dialysis status
   C. Urine production
   D. Urologic problems
   E. Previous transplant, including complications after transplant
   F. Blood transfusions
   G. Allergies
2. Review of Systems
3. Past Medical/Surgical History
4. Physical Examination
   A. Vital signs, height, weight
   B. Abdominal pain, previous abdominal surgery, heme positive stool
   C. Vascular: carotid bruit, peripheral pulses
   D. Infection
5. Gynecologic evaluation (Pap smear)
6. Mammography (family history of, > 40years of age)
7. Dental Evaluation
8. Laboratory Studies
   A. Complete blood count, blood chemistries (including calcium, phosphorous, magnesium) liver function tests, coagulation profile, PTH level
   B. Infectious profile: CMV serologies (IgM/IgG) Epstein-Barr virus serologies (IgM/IgG), varicella-zoster serologies, Hepatitis B and C serologies, HIV, RPR (syphilis), PPD (Tuberculosis skin test with anergy panel when indicated)
   C. Urinalysis, urine culture, and cytospin
   D. Immunological profile, blood type (ABO), panel reactive antibody (PRA), HLA typing
9. Chest x-ray (PA and lateral)
10. EKG (12 lead)
11. Special procedures in selected patients
    A. Upper GI endoscopy
    B. Colonoscopy
    C. Gallbladder ultrasound
    D. Ultrasound of native kidneys
    E. Peripheral arterial Doppler studies
    F. Pulmonary function tests
    G. Abdominal x-ray
    H. Carotid Duplex study
12. Complete cardiac work-up
    A. Electrocardiogram
    B. Exercise/dipyridamole thallium scintigraphy
    C. 2D-echocardiography with Doppler (+/- dobutamine)
    D. Coronary arteriogram (if indicated)
13. Urologic assessment (in select patients)
    A. Voiding cystourethrogram
    B. Urodynamic pressure-flow studies
    C. Cystoscopy
14. Psychosocial evaluation

Table 6.4.   *Most common causes of renal failure in year 2000*

Diagnosis	Number of Cases	Percent of Cases
Hypertensive Nephrosclerosis	1,997	14.2%
Diabetes, Type II	1,836	13.1%
Retransplant/Graft Failure	1,696	12.0%
Polycystic Kidneys	1,207	8.6%
Diabetes, Type I	903	6.4%
Chronic Glomerulonephritis	662	4.7%
Malignant Hypertension	578	4.1%
IgA Nephropathy	566	4.0%
Focal Glomerularsclerosis	502	3.9%
Systemic Lupus Erythematosus	363	2.6%
Membranous Glomerulonephritis	271	2.0%
Focal Segmental Glomerulosclerosis	269	1.9%
Chronic Pyelonephritis/Reflux	248	1.8%
Other	1,028	7.3%

6

and compliance. Identification of drug allergies has obvious implications, especially with respect to perioperative and chronic postoperative antibiotic use. A review of medications should discern those that may interfere with the action of the immunosuppressive agents.

2. Review of Systems

The review of systems should focus on the medical issues pertinent to determining the candidacy or the need for additional selected interventional studies to complete the medical work-up. Gastrointestinal and hepatico-pancreatico-biliary diseases including gastroparesis, gastritis, peptic ulcer disease, cholecystitis, hepatitis, pancreatitis, diverticulitis, may lead to the need for upper and/or lower GI endoscopy, gallbladder ultrasound, liver biopsy, and other diagnostic studies. Cardiopulmonary issues including a history of angina, myocardial infarction, pericarditis, valvular disease, congestive heart failure, COPD will determine the necessary degree of imaging and interventional studies such as echocardiography, thallium scintigraphy, coronary angiography, and pulmonary function testing. Urologic issues related to recurrent urinary tract infection and other symptoms of bladder dysfunction may lead to studies such as a voiding cystourethrogram. Previously treated cancer has implications for the duration of time that should pass following curative therapy before transplantation occurs. In general, a two-year disease-free interval is necessary. This also has relevance to the screening process to detect occult recurrent malignancy. An infectious disease profile including exposure to tuberculosis and the possible need for immunizations for hepatitis B, pneumococcus, and influenza guide the work-up. Neurological problems such as stoke or transient ischemic attacks may lead to carotid duplex studies, head CT or MRI, and cerebral angiography. Patients with polycystic kidney disease may require cerebral imaging studies to rule out intracranial aneurysm. A review of medications should focus on those that interfere with the cytochrome P450IIIA system that accelerates or reduces the metabolism of the calcineurin inhibitors. Significant psychiatric illness, questionable compliance with medications, drug, alcohol, and smoking history may alter the work-up or candidacy of the patient.

3. Physical Examination

The physical examination always starts with assessment of vital signs including blood pressure. It is typical for renal failure patients to be hypertensive, however, this should be adequately controlled with medications. Poor control may lead to work-up that could result in the need for pretransplant native nephrectomy. The height and weight of the recipient may be a factor in determining the acceptability of certain cadaveric kidneys to optimize the "nephron dose". Obesity may alter the surgical approach to the kidney transplant. For example, with anticipated difficult surgical exposure, placement of the kidney allograft in the retroperitoneal right iliac fossa utilizing the proximal common iliac artery and common iliac vein may simplify the transplant procedure. Detection of carotid bruits or weak femoral pulses are indications for duplex ultrasonography, and possibly cerebral arteriography. An abdominal examination that elicits right upper quadrant or epigastric pain, or the detection of heme positive stool, would result in the need for gallbladder ultrasound and upper and lower GI endoscopy. A previous kidney transplant, abdominal surgery resulting in a right upper quadrant incision, an appendectomy incision, or the presence of a PD catheter, may influence the choice of iliac fossa to place the kidney transplant. The physical exam may also detect areas of infection particularly in patients with diabetes that may have foot ulcers or possibly an unappreciated osteomyelitis. Female transplant candidates should have a formal gynecologic examination and Pap smear, and if over 40 years, a mammogram. All patients should have a dental examination with treatment of dental caries and tooth extraction, if necessary, completed prior to transplantation.

4. Interventional Studies and Pretransplant Surgery

Laboratory studies serve to screen for abnormalities not suspected by the history and physical examination. Not all of the procedures are required as the individual examinations will differ and require a differential degree of depth of evaluation. A complete cardiac work-up including angiography is unnecessary in all patients. However, individuals with a significant history, positive review of systems, type 1 diabetes, or hypertensive renal disease, should undergo a very complete evaluation to rule out significant coronary artery disease. Coronary artery disease (CAD) is a major cause of mortality, especially in patients with type 1 diabetes. It is frequent to diagnose significant CAD in diabetic patients that present with no symptoms or history of ischemic heart disease. Some have taken the position that all patients with type 1 diabetes should undergo coronary arteriography. The rationale is that the likelihood of having significant silent disease is high. Since noninvasive tests, such as scintigraphy, have a low sensitivity the complication rate of arteriography is less than the morbidity associated with preceding with transplantation in a patient with a false negative noninvasive test for ischemic heart disease. The use of low dose, minimally toxic contrast agents, and biplanar imaging make coronary arteriography safe even in patients with pre-uremic chronic renal failure. The diagnosis of significant coronary disease logically leads to consideration for revascularization by coronary angioplasty with stent, or coronary artery bypass grafting. Successful coronary artery interventional procedures do not rule out patients for transplantation. However, it may be prudent to perform post-revascularization cardiac stress testing to confirm that ischemic myocardium does not exist, prior to proceeding with transplant surgery.

The medical work-up may reveal circumstances requiring surgical intervention to prepare the patient for kidney transplantation. The common surgical procedures and indications are outlined in Table 6.5. Pretransplant native kidney nephrectomy/nephroureterectomy is no longer a routine pretransplant procedure. The native kidneys are left in place because they may still produce significant volumes of urine and secrete erythropoietin. Nephrectomy/ nephroureterectomy is reserved for specific indications. Ultrasound evidence of gallstones would be an indication for pretransplant cholecystectomy. The mortality and morbidity of acute cholecystitis is significant in the immunosuppressed transplant recipient. Individuals with significant coronary artery disease should be seriously considered for revascularization and then reconsidered for transplantation. Splenectomy is no longer a requisite pretransplant surgical procedure.

Multiple random blood transfusions were at one time associated with improved cadaveric graft survival in the pre-cyclosporine era. Currently there is no clinical benefit to transfusion and the risk of sensitization is significant. Donor-specific transfusion therapy in the setting of living kidney transplantation has also been almost completely eliminated.

When the evaluation is completed, it is presented at the Evaluation Committee meeting. The Committee may decide that further evaluation is necessary before a final decision is made to accept or not accept the patient for kidney transplantation. When the patient is active on the waiting list, arrangements are made with the patient's dialysis unit, or if preuremic, the patient himself will submit periodic serum samples to the tissue-typing laboratory for determination of a current panel reactive antibody level that is also available to be used for crossmatching. It is usually at this point that insurance coverage is secured or confirmed.

5. The Immunological Evaluation

A recipient of a kidney transplant will undergo an extensive immunological evaluation. The primary purpose is to evaluate those variables that are associated

---

Table 6.5.   Indications for pretransplant native nephrectomy/nephroureterectomy

Large polycystic kidney disease
Chronic renal parenchymal infection
Chronic infected reflux disease
Heavy proteinuria
Intractable hypertension
Infectious nephrolithiasis

with a risk of antibody-mediated hyperacute rejection. There are four components of the immunologic evaluation: ABO blood group antigen determination, HLA typing, serum screening for humoral reactivity to HLA phenotypes, and donor/recipient crossmatching.

a. ABO blood group determination. The ABO blood group antigens are carbohydrate moieties that are expressed on endothelial cells. They are a potential target of recipient circulating preformed cytotoxic anti-ABO antibody. Transplantation across incompatible blood groups may result in humoral mediated hyperacute rejection. Therefore, transplantation is avoided in this circumstance. Several systems of blood group verification are in place at the transplant center and the organ procurement organization to ensure proper ABO matching.

An interesting exceptional circumstance is the possibility of transplanting kidneys from the blood group subtype A2 into type O, B, or AB recipients. It is confirmed that the recipient circulating preformed anti A2 antibody titer is absent. The A2 antigen is weakly immunologic and does not result in induction of antibody. There have also been experimental procedures for ABO incompatible kidney transplants.

b. HLA typing. All transplant recipients are tissue typed to determine the HLA Class I and Class II loci. Multiple alleles exist at each loci, and they are co-dominantly expressed by each chromosome. Six HLA antigens are determined. The kidney donors are also HLA typed and the degree of incompatibility between donor and recipient is defined by the number of antigens that are mismatched at each HLA locus. The implications of these results are to determine zero HLA mismatched donor/recipient pairs that allows cadaveric kidneys to be preferentially allocated to those recipients. The degree of mismatching also has implications for the number of points assigned to each transplant recipient on the waiting list for allocation. Finally, HLA matching does influence the outcome of the kidney transplant. For cadaveric kidney transplants the best outcomes are observed in recipients of a zero antigen mismatched kidney. For living related kidney transplantation, the best results are observed in recipients of an HLA- identical kidney allograft. The degree in which HLA mismatching influences outcome varies considerably from center to center, and has changed with the application of the newer immunosuppressive agents.

It has been determined that groups of HLA antigens share characteristics in their molecular composition that are related to its antigenicity. These broad specificities have been categorized into cross-reacting groups (CREG). Matching for the CREG groups, rather than the individual HLA antigens, is an alternative method to define the degree of donor/recipient matching. Recipients of CREG matched donor kidneys seldom become sensitized for those specific HLA antigens. There are also advantages to CREG matching in making cadaveric kidney allocation more equitable between races. However, any system of organ allocation using HLA matching as a criterion imposes some degree of racial bias that affects access.

c. Serum screening for antibody to HLA phenotypes. Sensitization to histocompatibility antigens is of great concern in certain populations of transplant candidates. This occurs when a recipient is sensitized as a consequence of receiving multiple blood transfusions, receiving a previous kidney transplant, or from pregnancy. Transplantation of a kidney into a recipient that is sensitized against donor Class I HLA antigens is at high risk to develop hyperacute antibody-mediated rejection. All transplant candidates are screened to determine the degree of humoral sensitization to HLA antigens. This is accomplished by testing the patient's serum against a reference panel of lymphocytes that express the spectrum of HLA phenotypes. This methodology is referred to as panel reactivity and is quantified as the percentage of the panel to which the patient has developed antibody. This measurement is expressed as percent panel reactivity (PRA). For each patient this value varies between 0-100% and may change over time. Patients on the cadaveric kidney waiting list have PRA determinations on a regular basis. Information is kept that defines the peak percentage PRA and current percent PRA. Patients highly sensitized will have a very high PRA level that can remain elevated for years. The implications are that the waiting time will be very long before a patient receives a kidney to which he/she is nonreactive. Organ allocation point systems take into consideration the PRA level for maximizing organs to those individuals with high PRA that are found to have a negative crossmatch with a particular donor.

d. Crossmatching. Crossmatching is an in vitro assay method that determines whether a potential transplant recipient has preformed anti-HLA Class I antibody against those of the kidney donor. This immunologic test is conducted prior to transplantation. A negative crossmatch must be obtained prior to considering accepting a kidney for transplantation. The patient's stored or fresh serum is reacted against donor T-cells from cadaveric lymph nodes, or cadaver/living donor peripheral blood. The relevant antibodies are cytotoxic IgG anti-HLA Class I. Occasionally, a recipient will have IgM activity to HLA antigen. This can be determined by a crossmatch assay in which the recipient's serum is treated with dithiothreitol (DTT) to denature the IgM and eliminate the IgM response. Platelet absorption and determination of autoantibody are other useful techniques to carefully characterize and interpret the results of a positive crossmatch. Anti-HLA Class II antibody is less important and is detected utilizing donor B cells.

The technique of crossmatching is referred to as a microlymphocytotoxicity test. It was developed in the late 60's by Terasaki and has been one of the most important advances in kidney transplantation to prevent hyperacute rejection.[6] There are several methodologies. The standard test is referred to as the NIH method. Purified donor lymphocytes are incubated in recipient serum in the presence of complement. Complement reacts with bound anti-HLA antibody to kill the cell. Viability staining is performed to determine if a reaction has occurred. To increase the sensitivity of the standard crossmatch, anti-IgG globulin may be added to enhance binding of complement. Flow cytometry crossmatching is even more sensitive and has gained in popularity making it the standard procedure for most transplant programs. Binding of recipient antibody to donor T-cells is determined by detection of fluorescein-labeled monoclonal mouse anti-human antibody reactive to the anti-HLA antibody. The degree of reactivity is quantified by channel shifts. A positive flow crossmatch has a channel shift above a defined threshold. By using donor T and B cells and an array of fluorescein-labeled anti-human antibody the specific characteristics of the crossmatch reaction can be defined. It is possible to determine the specificity of the reaction to the precise HLA antigen.

CADAVER AND LIVING KIDNEY DONATION

The annual number of patients that receive a kidney transplant are determined by the number of cadaveric kidneys available and the number of living kidney donors. Cadaver kidney transplants make up the majority, numbering 8,493 in 2002. The bulk of the increase in kidney transplants over the past several years is due to greater numbers of living donors. There were 6,235 living donor kidney transplants in 2002.

Numerous strides have been made to increase the total number of cadaver kidneys available by public education programs encouraging organ donation. There is also a new classification of expanded criteria organ donors, and more liberal consideration of controlled and uncontrolled non-heart-beating donors. In addition, many programs are expanding their living kidney donor experience by including distantly related donors such as spouses, cousins, aunts, uncles, close friends, and even emotionally unrelated donors. This has resulted in an increase in the proportion of all kidney transplants performed in the U.S. by living donors to almost 45%. In many transplant programs, living kidney donation accounts for nearly 75% of all the kidney transplants. There are few medical ethical issues related to accepting kidneys from living donors. The donor mortality risk is <0.01%. Life expectancy is unaffected. There is no long-term morbidity related to development of hypertension or impaired renal function (7). In fact, many patients have been benefited by the thorough medical examination during work-up by revealing unexpected medical issues.

A. EVALUATION OF THE LIVING DONOR

The pertinent aspects of the medical evaluation of the potential live kidney donor are outlined in Table 6.6. The psychosocial evaluation is necessary to confirm that the motive to donate the kidney is altruistic. Not all individuals willing to donate can be accepted as a donor because of ABO blood type incompatibilities with the recipient. This uncertainty requires the evaluation to proceed in stages so that expensive imaging studies are not performed in ABO incompatible donors. The work-up is performed in phases, beginning with determination of blood type, blood chemistry profile, complete blood count, coagulation studies, and urinalysis with culture. If the donor is one of multiple siblings willing to donate, then HLA typing is conducted. This is done to determine if an HLA identical or one-haplotype match can be found.

When the single best donor is identified with a negative crossmatch the work-up proceeds with a 24-hour collection of urine for detection of protein and creatinine clearance. Next, viral serologies are obtained, chest x-ray, EKG, and if indicated, a two-dimensional cardiac echocardiogram. Finally, the special imaging studies to evaluate the renal vasculature and collecting system are obtained. It is imperative that there is confirmation that the potential live donor has two kid-

---

Table 6.6. Medical evaluation of the potential live kidney donor

---

1. Identification of interested family or nonfamily members and orientation to live kidney donation
2. Complete history and physical examination
3. Immunological studies
   A. Blood type
   B. HLA determination
   C. Cross-matching with recipient
4. Social and psychological evaluation
5. Laboratory studies
   A. Complete blood count, serum chemistry profile including liver function tests, coagulation profile
   B. Infectious survey: hepatitis A, B, and C serologies, CMV serologies (IgG/IgM), Epstein-Barr virus serologies (IgM/IgG), HIV, RPR, urine culture and cytospin
   C. Prostate Specific Antigen (PSA) males > 50 years
   D. Pregnancy test, pap smear, mammogram (females >40 years or family history)
   E. Urinalysis and 24 hour urine for protein and creatinine clearance
   F. Chest x-ray (PA and lateral)
   G. EKG (12 lead)
   H. Special tests to evaluate renal vasculature and collecting system
      a. Renal CT with 3-D angiography (spiral CT)
      b. Renal ultrasound to determine presence of 2 kidneys and to rule out PCKD in relative of recipient with PCKD
      c. MRI/MRA (experimental)

6

neys. An excellent single imaging technique, which is capable of visualizing both kidneys and the renal arteries, veins, and collecting systems, is the renal CT with three-dimensional angiography (spiral CT). Spiral CT is becoming more popular than renal arteriography, the caveat being that mild fibromuscular hyperplasia may be overlooked with the former. Renal ultrasounds are generally not routinely utilized. However, in relatives of a recipient with autosomal dominant polycystic kidney disease, a renal ultrasound will rule out polycystic kidney disease in the donor without employing the more expensive CT examination. Once the imaging studies are performed, if there is favorable anatomy, a final crossmatch will be completed. Next, the surgery date is scheduled.

B. CADAVER DONOR EVALUATION

Potential cadaveric organ donors are referred to the organ procurement organization (OPO) by their member hospitals. The OPO screens the referrals to determine initial medical suitability and confirm that adequate documentation of brain death is present. An OPO representative should be called upon to personally discuss the situation of organ donation with the family. Permission for organ donation is obtained.

Medical management of the cadaver organ donor is typically taken over by personnel at the OPO, or a managing transplant medical or surgical specialist. The pertinent aspects of the medical evaluation to optimize care and provide useful information for the transplant centers are outlined in Table 6.7. The etiology and duration of brain death is determined, the amount of cardiac arrest time, if any,

Table 6.7. Medical evaluation of the potential cadaveric organ donor

1. Confirm diagnosis of brain death
2. Confirm consent for organ donation from family
3. History
   A. Etiology and duration of brain death
   B. High risk behaviors (e.g. drug/alcohol abuse)
   C. Pre-existing diseases (e.g. renal dysfunction, diabetes)
4. Physical examination
   A. Sites of physical trauma, intra-abdominal surgery, infection
   B. Hemodynamic stability and pressors
   C. Urine output
5. Blood Work
   A. Blood type
   B. HLA typing
   C. Infectious survey:
      i) HIV
      ii) RPR
      iii) viral serologies: CMV, EBV, hepatitis B and C
   D. Complete blood count, complete serum chemistry profile, coagulation profile
   E. Blood, urine, sputum cultures
6. Urinalysis
7. Kidney biopsy (if indicated)

noted, as well as the need for inotropic support. Significant pre-existing diseases such as renal dysfunction, diabetes, and cancer are obtained. High-risk behavior for acquired infectious diseases is noted.

Physical examination begins with description of hemodynamic stability as well as height, weight, urinary output, and presence of hematuria. Next, examination determining potential physical trauma to the intraabdominal organs, as well as sites of infection, are obtained. An extensive amount of blood work is required beginning with determination of blood type. An infectious survey with multiple viral serologies is obtained. A complete serum chemistry profile helps determine the degree of renal function, electrolyte abnormalities, and liver parenchymal dysfunction. A complete blood count and coagulation profile are obtained, the latter useful to rule out disseminated intravascular coagulation. The peripheral blood or lymph nodes are obtained and forwarded to the OPO laboratory for HLA typing and crossmatching. If the cadaveric organ donor is considered marginal there may be certain criteria that warrant a renal biopsy after procurement. The OPO is then responsible for contacting the various transplant programs that have patients on the waiting list for kidney transplantation. Each transplant program has its own criteria by which cadaveric kidneys are accepted. These decisions are made differently by different programs, each with the goal of selecting suitable kidneys for successful and safe kidney transplantation.

A new system has been instituted by UNOS to improve placement and reduce the discard rate of cadaveric kidneys from expanded criteria organ donors.

C. Organ Procurement and Preservation

The donor organs are procured by a specialist transplant surgery procurement team. In this era of multiple organ procurement, often the kidneys will be procured by surgical teams obtaining the liver or pancreas for transplantation. The donor cadaveric kidneys are removed after cold preservation solution is infused intra-arterially via the aorta followed by surface cooling with ice slush. The organs are usually obtained through an en-bloc nephroureterectomy, including the aorta and inferior vena cava. The en-bloc kidneys are separated at the backbench. Details of the anatomy including the size of the kidneys, length of ureter, numbers of arteries, veins, and ureters are recorded. Identification of large cysts or a tumor is the goal of the external examination. A renal biopsy may be taken contingent on the donor age and medical considerations.

The kidneys are then each placed in preservation solution, packed on ice, labeled as left or right, and transported to the OPO or transplant center. Some programs elect to use a machine perfusion system believing that it may decrease the incidence of delayed graft function. The preservation solution most frequently used is the University of Wisconsin preservation solution. Although this has had greatest benefit by extending preservation times for liver and pancreas allografts, it is beneficial for kidney allografts as well. Some programs are also using the HTK preservation solution. Kidney allografts will function after cold preservation times as long as 72 hours. However, the incidence of delayed graft function increases significantly after preservation times >24 hours, dependent upon on the health and age of the cadaver organ donor. In general, most cadaveric kidneys are transplanted within 36 hours. This time interval allows adequate opportunity for safe transportation of zero antigen-mismatched kidneys to the selected recipients anywhere in the country.

TRANSPLANTATION SURGERY AND POST-SURGICAL CONSIDERATIONS

A. Preoperative Transplant Care

Selection of the cadaveric kidney transplant recipient usually occurs shortly after procurement of the kidneys. The recipient is admitted to the hospital, re-evaluated, and a final decision made whether or not to proceed with surgery. The re-evaluation emphasizes work-up for infectious disease, or other medical issues that would contraindicate surgery. It is necessary to determine if dialysis is required prior to transplantation. If the patient is on peritoneal dialysis, occult peritonitis should be quickly ruled out by gram stain while the culture results of the peritoneal fluid are pending. Because patients may be on the waiting list for years, significant progression of previously insignificant medical problems may have occurred. Suspicion of cardiac disease is obtained through history and physical exam and EKG. Sometimes it is necessary to proceed with invasive tests to conclusively rule out significant coronary artery disease. It may be most prudent to proceed directly to coronary arteriography to do so. This would then require post-procedure dialysis to eliminate the contrast material. The re-evaluation admission also affords time to review the sequence of transplant events with the patient and family members. It is also during this time that informed consent may be obtained if the patient is to be included in any study protocols.

B. Special Surgical Considerations During Organ Procurement

The kidney transplant procedure begins with the organ procurement process. It is essential that the organ procurement team exhibit knowledge of the important anatomical variations of the renal vasculature and collection system. Inadvertent transection of renal vasculature or the ureter can significantly compromise the success of the transplant. Communication between the procurement team and the implantation team is valuable.

C. Kidney Transplant Surgery

Kidney transplantation is not a technically demanding procedure but it is unforgiving of even minor technical misadventures. The surgical procedure is uniform, but no two kidney transplants are exactly alike. A typical uncomplicated kidney transplant can be performed in 3 ($\pm$ 0.5) hours. Technical complications resulting in graft loss are very uncommon.

Several procedures are carried out prior to the skin incision. Patients are administered perioperative antibiotics. Intra-operative immunosuppressive induction agents may be given, including the corticosteroids and anti-lymphocyte antibody induction agents. After induction of general anesthesia, a central venous catheter may be placed. A large Foley catheter is placed in the bladder and the bladder infused with about 200cc of antibiotic fluid by gravity. Another approach is to utilize a Foley extension that allows cysto-tubing to be connected for infusion of fluid into the bladder during the case after vascular reconstruction. The Foley catheter should be very securely taped to a shaved spot on the thigh with use of benzoin and 2 inch-cloth tape. Ted hose and pneumoboots are often applied to minimize the chance for development of deep venous thrombosis. Naso/orogastric tubes are generally not used. The abdomen is then prepped and draped in a sterile manner.

Figure 6.3 illustrates the anatomic position of the heterotopically placed kidney transplant. The transplant site is the iliac fossa. Generally, the right iliac fossa is favored because the vessels are more superficial. Also, on the right, the proximal common iliac vein lays lateral to the artery and is easily accessible in obese and deep patients compared to the external iliac vein. This may be particularly important if living donor kidneys are used with short renal vessels. The exception are patients with Type 1 diabetes that may be candidates for a subsequent pancreas transplant. In this situation the left iliac fossa is used for the kidney transplant.

The skin incision is made either as a curvilinear hockey-stick type incision relatively medial compared to the alternative straighter and more diagonal and lateral incision. The incision is carried down through the external oblique aponeurosis through the oblique musculature to the peritoneum. The inferior epigastric vessels are suture ligated and divided. In females, the round ligament is divided. In males, the spermatic cord structures are preserved and mobilized medially. The

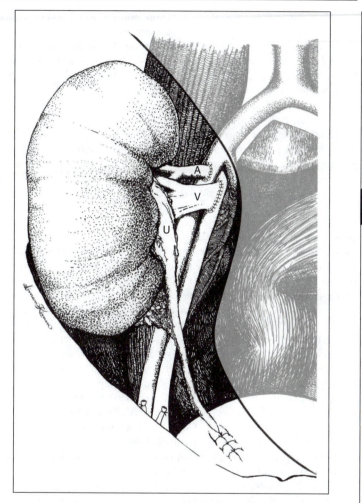

Fig. 6.3. Kidney transplant in right iliac fossa with anterior ureteroneocystostomy.

peritoneum is mobilized medial and cephalad to expose the underlying retroperitoneal iliac vessels. Often at this point, a mechanical retractor is utilized to improve exposure. The iliac artery and vein are dissected free from the surrounding soft tissues with suture ligation and division of the overlying lymphatics. This is important to minimize occurrence of a posttransplant lymphocele.

There are several alternatives for the vascularization of the renal allograft. Patients that are uremic typically do not require systemic heparinization for the vascular anastomosis. However, preuremic patients should be heparinized. Commonly, the external iliac artery and vein are utilized. There are situations that may require suture ligation and division of the hypogastric veins to mobilize the external iliac vein laterally to improve its position and to optimize the alignment of the kidney allograft. The common iliac artery may be important to use if there is atherosclerotic disease or concern about the perfusion pressure in the extra iliac artery. The internal iliac artery can be used for an end-to-end anastomosis in living donor kidneys or in kidneys that have multiple arteries. Another consideration that is used to determine the location of the arterial anastomosis is how the lie of the kidney in the iliac fossa will be affected by the relationship of its lower pole to the anteriorly rising psoas muscle.

The sutures used for anastomosis in adult kidney transplants are typically 5-0 monofilament for the vein, and 6-0 monofilament for the artery. If there is a very difficult arterial anastomosis because of intimal abnormalities, interrupted stitches are useful. If hypogastric veins need to be ligated they should be stick tied. The length of the renal artery and vein should be examined and the vein trimmed to an appropriate length relative to the artery, leaving it slightly longer when the iliac vein is medial. On right cadaveric kidneys it is very useful to utilize the inferior vena cava as an extension. There is no demonstrable disadvantage in putting a left kidney on the left side or the right kidney on right side. Placing the contralateral kidney in the iliac fossa does make for a more natural vascular alignment when the vein is mobilized lateral relative to the artery. Also, the ureteral collecting system is relatively anterior to the vessels in the hilum.

After completion of the vascular anastomoses the ureterocystostomy is performed. It is important in males that the ureter be slipped under the cord structures. The ureteral artery needs to be securely ligated. The ureter is then cut and spatulated. There are 3 common methods of ureteral anastomoses. The Ledbetter-Politano procedure requires an open cystotomy and the ureter is tunneled posteriorly near the trigone. The most common approach is the anterior ureteroneocystostomy in which the spatulated ureter is directly sutured to the bladder mucosa, followed by approximation of the muscularis to create a tunnel over the distal 2 cm of the ureter (Fig. 6.4). This approach has been modified as a single stitch procedure, whereby the ureter is invaginated in the bladder with a single stitch, followed by the approximation of the muscular layer to create a tunnel to prevent reflux (Fig. 6.5). In the unusual situation involving a duplicated collecting system, separate ureterocystotomies are performed for each ureter. Alternatively, the tips of the ureters may be fish-mouthed and sewn together creating a single ureteral orifice for anastomosis to the bladder mucosa. In very unusual cases where the bladder can not be used, urinary drainage using an ileal loop is successful. In some centers, ureteral stents are often routinely employed. This may minimize the occurrence or early urine leaks or ureteral stenosis. The ureteral stents are removed approximately 6 weeks posttransplant via flexible cystoscopy in the outpatient setting.

It is usually not necessary to place a retroperitoneal drain. However, if needed, it is perfectly reasonable to place close suction drainage which is required only for about 24-48 hours. Wound complications can be associated with significant morbidity. Careful closure of the incision that incorporates all layers of the muscle and fascia is important to prevent hernia. Keeping the wound edges moist with

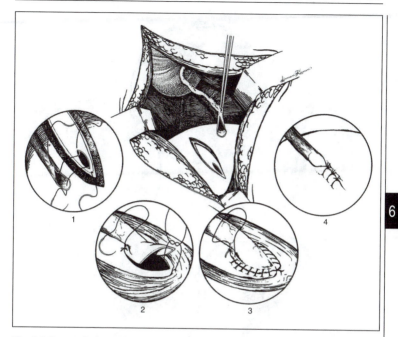

Fig. 6.4. Extravesical anterior ureteroneocystostomy demonstrating mucosa-to-mucosa anastomosis of the ureter to the bladder mucosa and approximation of the detrusor muscle to create the anti-reflux tunnel.

antibiotic solution during the course of the procedure will help minimize wound infection. The patient can usually be extubed in the OR at the completion of the case.

D. VERY EARLY POSTOPERATIVE MANAGEMENT

Immediate postoperative care begins in the post-anesthesia recovery room with airway management as highest priority, ensuring successful extubation and airway protection. Pain control is administered. Vital signs are monitored frequently. A complete chemistry profile, complete blood count, coagulation survey, chest x-ray, and EKG are typically obtained. Observation and documentation of hourly urine output is critical to determine the early degree of initial function of the kidney transplant as well as anticipating the intravenous fluid replacement necessary. Urinary output can be as low as drops or greater than 1 liter per hour. Postoperative fluid replacement must be thoughtfully approached. Assessment of volume status is important to avoid volume overload or depletion. Central venous pressure monitoring is a useful guide to intravascular volume status. If a brisk diuresis is occurring, it is not uncommon for electrolyte abnormalities to develop including hypocalcemia and hypomagnesemia. Determination of serum potassium levels is very important. When urine output is very low, hyperkalemia should be anticipated. A brisk urine output may be associated with either hyper- or

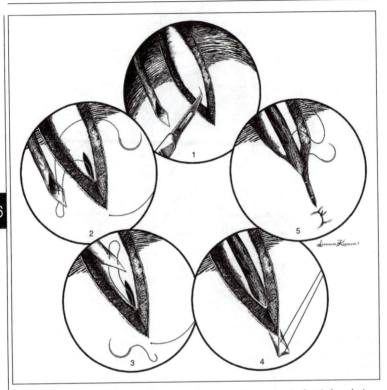

Fig. 6.5. Extravesical anterior ureteroneocystostomy utilizing the single stitch technique and anti-reflux tunnel.

hypokalemia. In the latter case, replacement of potassium in intravenous fluid must be approached with caution. In patients with voluminous urine output, the urinary concentration of potassium may be unexpectedly low. It is prudent to measure urinary potassium concentration prior to considering adding potassium to the intravenous fluids. Often replacement fluids are administered according to the rate of urine output. In that case, the potassium concentration in the intravenous replacement fluids should not exceed that in the urine.

An abrupt cessation of brisk urinary output must be quickly assessed. Suspicion that the Foley is occluded by a blood clot should prompt immediate irrigation. Importantly, an acute renal arterial thrombus will manifest as abrupt cessation of urine output. Very early vascular problems may be reversed and the kidney salvaged if acute renal artery thrombosis is suspected (usually in the recovery room) and the patient is immediately surgically re-explored.

Early significant postoperative bleeding would manifest as hypotension, tachycardia, decreased urine output, and lower than expected hemoglobin level. When the patient is stable and the early postoperative laboratory evaluation complete, the patient is typically transferred to the transplant general care unit. Routine intensive care unit observation is usually not required, however, there are individual cases in which it is sometimes required.

E. Surgical Complications

1. Wound Complications

Risk factors for, and the morbidity of, wound complications in the transplant patient are significant. Avoiding a wound complication begins with the forethought of correct surgical technique in opening the wound, dissection in correct tissue planes, gentle handling of tissue, meticulous attention to hemostasis, keeping the wound edges moist with antibiotic solution during the case; and ends with secure approximation of the deep fascia and careful approximation of the skin edges during closure. Wound complications may be a source of significant morbidity especially if ignored until deep and extensive facial necrosis and abscess develops. Wound complications initially appear as superficial drainage. It is important to differentiate between superficial and deep wound problems. Superficial wound infections must be opened and a sample of fluid submitted to microbiology for identification of the infectious organism and its sensitivity to antibiotics. Superficial infections can be treated successfully with local wound care. The fully granulated wound may then be allowed to heal by secondary intention or the patient brought back to the operating room for closure. Wound dehiscence requires urgent surgical repair.

2. Bleeding

Kidney transplantation is a vascular surgery procedure, however, it is not an operation associated with much blood loss. It is uncommon for intra-operative blood transfusions to be given. Postoperatively, a life threatening bleeding complication is very rare but could result from rupture of the arterial anastomosis from a mycotic aneurysm. Significant bleeding could also occur as a result of the loosening of a suture ligature on the inferior epigastric vessels or a branch of the renal vein. Rupture of the kidney from an aggressive, early rejection reaction, usually in a highly sensitized patient, will cause significant bleeding.

Bleeding in the retroperitoneal space is usually tamponaded by the peritoneum and the abdominal wall. However, extensive dissection of blood can occur cephalad in the retroperitoneal space accommodating several units of blood. If the kidney is placed intra-abdominally a greater amount of bleeding will occur prior to tamponade. The presence of a large organized hematoma is at risk for secondary infection and subsequent wound breakdown. Diagnosis of bleeding is manifested by hemodynamic instability, physical discomfort, reduced urine output, instability of hemoglobin levels, requirement for repeated blood transfusion, and usually detection of coagulopathy. The coagulopathy may include a prolonged protime and bleeding time. The uremic platelet dysfunction may be improved by administration of DDAVP, cryoprecipitate, or even a platelet transfusion. All antiplatelet and anticoagulation agents should be promptly stopped. Some coagulation abnormalities may be improved by administration of FFP and/or vitamin K.

3. Acute Vascular Thrombosis

Acute (<24 hours posttransplant) arterial thrombosis occurs in <1% of all kidney transplants. Salvage of the renal allograft is possible if immediately diagnosed. The typical scenario occurs in the post-anesthesia recovery room. The diagnosis is suspected during the immediate post-operative evaluation when it is noticed that a previously brisk initial diuresis suddenly stops. Acute arterial thrombosis is typically due to a technical problem or a small embolus. The technical issues may relate to the degree of pre-existing atherosclerotic vascular disease. Another consideration includes pressure on the kidney from the anterior abdominal wall in recipients that are very thin with a narrow pelvis. This occurs when the kidney is "laced in" too tightly from being positioned relatively distal on the external iliac artery with compression or tension on the underlying renal vessels resulting in vascular thrombosis.

Venous thrombosis is also a rare occurrence. If it occurs the kidney is usually unsalvageable. The cause often never satisfactorily identified. Renal vein thrombosis can extend to the external and common iliac veins proper and result in deep venous thrombosis and pulmonary embolism. Venous thrombosis manifests differently from arterial thrombsis. In the former case, sudden onset of bloody urine output often associated with unilateral swelling of the ipsilateral lower extremity, occurs. If suspected during the post-anesthesia recovery period the patient should be brought back to the operating room promptly and the transplant re-explored. There is not the luxury of time to obtain noninvasive duplex vascular studies. If venous thrombosis occurs, often too much time lapses before the problem is identified for any realistic hope of renal allograft survival, however, anecdotal reports have been described.

4. Urine Leak

Urine leaks occur at the ureterovesical junction or through a ruptured calyx, secondary to acute ureteral obstruction. It may occur days or weeks posttransplant. Often the etiology of early urine leak is due to necrosis of the tip of the ureter. Urine leaks manifest as diminished urine output, hypercreatinemia, and often lower abdominal or suprapubic discomfort. This constellation of signs and symptoms can be confused with a rejection episode. Anti-rejection therapy should never be instituted without a renal allograft biopsy, especially if a urine leak is possible.

The diagnosis of urine leak begins with a high index of clinical suspicion backed up by imaging studies. These may include an ultrasound or a CT scan showing a fluid collection. The fluid is then percutaneously accessed and the BUN and creatinine concentrations compared to that in the serum. To localize the leak, a nuclear medicine study may be useful, or a retrograde cystogram. A percutaneous nephrogram is definitive and may be useful treatment. Initial treatment of a suspected urine leak would include placement of a foley catheter. Urine leaks may be attempted to be repaired with minimal intervention with a percutaneous nephrostomy and drainage with internal stenting, or through a cystoscopic retrograde approach. More aggressive treatment of the urine leak would involve operative intervention with reimplantation of the ureter or a uretero-ureterostomy utilizing the ipsilateral native ureter.

5. Ureteral Stenosis/Obstruction

This is a relatively late complication occurring months or years posttransplant, which could result from ischemia of the ureter or a tight ureteroneocystostomy. Ureteral stenosis is manifested by elevated creatinine and hydronephrosis. Sometimes infectious pyelonephritis occurs. Diagnosis is made by several complimentary evaluations beginning with the observation of elevated creatinine followed by renal ultrasound showing moderate or severe hydronephrosis. Mild hydronephrosis is a benign common finding. To demonstrate that hydronephrosis is functional, a foley catheter should be placed at the time of ultrasound. Also, it is useful to obtain a diuresis renogram. If a functional stenosis or obstruction is present, there will be delayed excretion of nuclear material in the bladder. This result logically leads to percutaneous nephrostomy to confirm the diagnosis with contrast imaging. The percutaneous nephrostomy access is used for treatment by placement of an internal ureteral stent into the bladder, and an external drain of the renal pelvis. Those procedures solve the problem in the majority of cases; however, there are times when a surgical approach is necessary.

If surgery is required, one must be prepared to re-implant the transplant ureter, perform a ureteroureterostomy to the native ureter, or perform a ureteropyelostomy. Pre-operative retrograde stenting of the native ureter is often helpful in dissecting the native ureter in the reoperative field. Also, stenting of the transplant ureter is helpful as well. It is prudent to consult the initial operative report to understand the surgical techniques used during the initial kidney transplant procedure. It is also very useful to know if the allograft is a left or right kidney, since this has implications for understanding the anatomical relationship between the pelvic collecting system and the renal vasculature. Kidneys placed in the contralateral iliac fossa will have the collecting system relatively anterior to the renal vessels and therefore a less treacherous dissection can be carried out, especially if a ureteropyelostomy is necessary. The surgical results are very satisfactory with immediate and long-term sustainable renal allograft function.

6. Lymphocele

A lymphocele is a circumscribed collection of retroperitoneal lymph that originates from lymphatic vessels about the iliac vasculature and the hilum of the kidney. The incidence of lymphocele is greatly reduced by careful suture ligation of lymph vessels overlying the iliac vessels. The true incidence of lymphoceles is unknown because not all patients are evaluated for its presence (in otherwords, the denominator is unkown). However, significant secondary problems may arise by external compression of the iliac vein causing leg swelling and discomfort, or compression of the transplant ureter causing hydronephrosis and renal dysfunction. Significant collections are usually diagnosed during the work-up for hypercreatinemia by ultrasonography. A perinephric fluid collection that is lymph is confirmed by percutaneous access and analysis of the fluid for white blood count,

differential, BUN, and creatinine. The fluid collection is differentiated from a uri-
noma or a serum collection. If hydronephrosis is diagnosed by an imaging study,
it is critical that investigation for perinephric fluid collection is undertaken prior
to consideration for percutaneous nephrostomy or other invasive procedures.

There are multiple treatment options for a lymphocele. The standard principle
is that intraperitoneal drainage of the lymphocele should be established. This can
be accomplished surgically with either a laparoscopic approach or an open surgical approach with marsupialization of the edges of the lymphocele. Caution must
be undertaken during surgery to avoid injury to the pelvic collecting system and
the ureter of the transplant kidney. Intraoperative ultrasound may be a useful
complementary procedure if the lymph collection can not be definitively differentiated from a dilated renal pelvis. In some instances, percutaneous drainage is
undertaken. However, this has a higher risk of infection and the disadvantage of
requiring an external drain in place for extended periods of time. Patients recover
from the laparoscopic and open surgical procedures in less than 2 days. One should
expect immediate resolution of hydronephrosis, improved diuresis, and correction of hypercreatinemia. Interestingly, some have hypothesized that lymph collection may be hastened by an ongoing acute rejection episode. Therefore, if prompt
correction of hypercreatinemia does not occur, acute rejection should be suspected
and diagnosed by renal biopsy.

RENAL ALLOGRAFT PARENCHYMAL DYSFUNCTION

Renal allograft parenchymal dysfunction is an important cause of graft loss.
The clinical manifestation is uniformly that of hypercreatinemia, yet the causes
are numerous and a differential diagnosis must be approached, taking into consideration the risk of rejection, the blood concentrations of the calcineurin inhibitors, etiology of native kidney failure, and the time period following
transplantation.

A. KIDNEY TRANPLANT BIOPSY

The use of percutaneous biopsy of kidney transplant is invaluable in the prompt
accurate diagnosis of parenchymal dysfunction. Often the procedure is preceded
by an ultrasound to rule-out other nonparenchymal diseases as a cause of
hypercreatinemia. The technique is well established and is safe. The kidney transplant biopsy can be done in the outpatient office setting. Often it is done with
ultrasound guidance. Kidneys placed in the retroperitoneal iliac fossa are much
more accessible to safe biopsy than those implanted in the intra-abdominal position. The site of biopsy should be the upper pole to minimize injury to the lower
collecting system. Patients with prolonged bleeding time or coagulopathy should
not be biopsied until these abnormalities are corrected. A 16-18 gauge biopsy needle
is typically used. In most transplant programs, the renal pathologist is available to
read the results of the biopsy specimen within hours of the procedure. This facilitates rapid diagnosis and institution of appropriate therapy. Most often, the biopsy is performed to confirm the clinical suspicion of acute or chronic rejection.
However, occasionally, unexpected diagnoses are made such as hemolytic uremic

syndrome, recurrent disease, or pyelonephritis. If changes consistent with acute nephrotoxicity of calcineurin inhibitors are observed then appropriate dosing of the immunosuppression is prescribed.

B. REJECTION

1. Hyperacute Rejection

Hyperacute rejection of the renal allograft occurs when circulating preformed cytotoxic anti-donor antibodies directed to ABO blood group antigens or to donor HLA class I antigens are present. The mechanism of allograft destruction is well characterized. Antibodies bind to antigen expressed on the donor endothelium resulting in activation of the complement system, platelet aggregation, and microvascular obstruction. The frequency of hyperacute rejection is extremely low, being prevented by ruling-out transplant recipients with a positive pretransplant crossmatch. Hyperacute rejection may occur within minutes of revascularization of the allograft and observed intra-operatively, or it may occur hours later. Patients at risk for hyperacute rejection are those with a high past or current PRA level. There is no ability to salvage the renal allograft. Pathological examination will reveal significant interstitial hemorrhage, infiltration of neutrophils, and deposition of antibody on endothelium. A relatively new immunohistological technique to help determine if hyperacute rejection has occurred includes staining for complement deposition on the endothelium using an anti-c4d antibody. The ability to diagnosis hyperacute rejection histologically is often obscured because of the severe degree of kidney destruction. Other considerations would include arterial or venous thrombosis.

2. Accelerated Acute Rejection

This is a very early, rapidly progressive, and aggressive rejection reaction. It can occur within the first week of transplantation. The pathologic characteristics are massive infiltration of lymphocytes, macrophages, and plasma cells. There is injury to the renal tubules, damage of interstitial capillaries, and vascular injury of larger vessels marked by endothelial swelling. The very aggressive and rapid nature of this rejection reaction makes it difficult to reverse. Immediate therapy with anti-T-cell antibodies, in addition to pulse corticosteroids, may reverse the process. Approximately 50% of the grafts can be salvaged. It would be expected that long-term function would be compromised.

3. Acute Tubular Interstitial Cellular Rejection

This is the most common type of rejection reaction with an incidence at one year posttransplant of approximately 10 ($\pm$5)%. Typically, it occurs between 1-3 months posttransplant. It is T cell mediated and injury is directed to the renal tubules. Histopathologic examination reveals T-cell infiltration around the tubules and infiltration within the tubules, producing "tubulitis" (Fig. 6.6). The severity of rejection is defined on a continuum from mild to severe, which correlates its duration of activity.[8] The gold standard for diagnosis is renal allograft biopsy. Treatment is guided by the severity of histopathological changes. Mild rejections

Fig. 6.6. Acute renal allograft rejection with intraepithelial lymphocytes penetrating the tubular basement membrane producing tubulitis. A. Mild (PAS, x300.) B. Severe (PAS, x600). (Reprinted with permission from: Solid Organ Transplant Rejection, Editor Solez, Publisher Marcel Dekker, Inc., 1991).

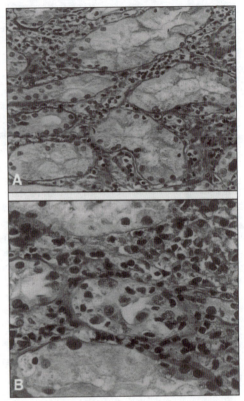

may be successfully reversed with corticosteroids alone, whereas moderate or severe rejections may require the use of anti-T-cell antibody. Acute tubular-interstitial (T-I) rejection may occur repeatedly or relatively late (1+ years posttransplant). These latter two circumstances place the patient at high risk for development of chronic allograft rejection. Acute T-I rejection is reversible in ≥95% of cases.

4. Chronic Rejection

Chronic rejection is a slow and progressive deterioration in renal function, characterized by histologic changes involving the renal tubules, capillaries, and interstitium. It is often associated with individuals with recurrent rejection or a late acute rejection episode. The dysfunction is also believed to be complicated by the nephrotoxic effects of the calcineurin inhibitors. The precise mechanisms of this disease are poorly defined and is an area of intense study. Application of conventional antirejection agents, such as corticosteroids or anti-T-cell antibodies do not appear to alter the progressive course. Unfortunately, this is a major cause of kidney allograft loss occurring >2 years posttransplant.

C. Delayed Graft Function

Delayed graft function immediately posttransplantation is usually due to acute tubular necrosis (ATN). It commonly occurs in cadaver allografts and rarely in kidneys from living donors. Delayed graft function in a living donor transplant should provoke work-up for causes other than ATN. The clinical significance of delayed graft function on allograft functional survival is significant. The frequency of delayed graft function is variable among the different transplant centers and approximates roughly 20-40% of cadaver transplants. Risk of delayed graft function in cadaveric kidneys includes prolonged cold ischemia time, the age and medical condition of the cadaveric organ donor, and excessive early use of calcineurin inhibitors. ATN is usually limited to approximately 2-6 days. It can be prolonged for up to several weeks. If persistent, a simultaneous acute rejection episode may occur, as well as acute nephrotoxicity of the calcineurin inhibitors. Successful care of the patient with delayed graft function requires good judgment on the timing of renal allograft biopsy.

D. Nephrotoxicity of Calcineurin Inhibitors

The calcineurin inhibitors, cyclosporine and tacrolimus, may induce an acute nephrotoxic effect. The mechanisms are incompletely defined but include preglomerular arteriolar vessel constriction causing reduction in renal blood flow and decreased glomerular filtration. The nephrotoxic effect is related to circulating blood levels, which appear to correlate with peak concentrations occurring 2-4 hours after dosing. The nephrotoxic effect may be severe and prolonged, mimicking an aggressive acute rejection reaction. The parenchymal dysfunction is reversible if calcineurin blood concentrations are promptly reduced. The occurrence of simultaneous nephrotoxicity and rejection are not mutually exclusive. Occasionally it is necessary to perform renal allograft biopsy if the expected correction of hypercreatinemia does not occur subsequent to reduction in calcineurin inhibitor dosing. The pathologic characteristics of cyclosporine or tacrolimus nephrotoxicity are distinguishable from acute tubulointerstitial rejection. The former demonstrates renal tubule vascularization and the latter a lymphocytic infiltrate of the renal tubules.

E. Hemolytic Uremic Syndrome

This is a glomerular injury that results in abrupt hypercreatinemia and diminished urine output. De novo HUS is different than recurrent HUS. The etiology of de novo HUS is unknown but seems to be associated with endothelial injury associated with the calcineurin inhibitors and also the occurrence of cytomegalovirus. A high index of suspicion is necessary to make a prompt diagnosis. Laboratory evaluation showing diminished platelet count, anemia, reduced haptoglobin levels, rising LDH levels, and a peripheral blood smear with schistocytes is consistent with the diagnosis. The definitive diagnosis is by renal allograft biopsy showing glomerular microthrombi. Treatment starts by discontinuing the calcineurin inhibitor, administration of gamma globulin, and possibly the application of plasmapheresis. Conversion to an alternative calcineurin inhibitor does not usually cause recurrence. If recurrent de novo HUS occurs it may be necessary to maintain immunosuppression with just an antimetabolite and corticosteroid. HUS has a high rate of renal allograft loss.

F. BK POLYOMAVIRUS ASSOCIATED NEPHROPATHY[9]

The BK virus is a polyomavirus. It is an acquired childhood pulmonary infection that transports to the kidney and uroepithelium where it remains latent. Mechanisms resulting in reactivation have not been defined, but the immunosuppressed state is a critical factor. BK nephropathy is identified as the cause of renal allograft dysfunction in 1-10% of cases. BK nephropathy is usually noted beyond 6 months post transplant. Since the late 1990s there has been an increase in the recognition of BK nephropathy. This increase in occurrence coincided with the use of the more potent induction and maintenance immunosuppressive agents. The diagnosis of BK nephropathy is suspected when the serum creatinine increases. The diagnosis of BK nephropathy is definitively made using renal histology by the identification of characteristic light microscopic or EM changes. Light microscopy of a renal biopsy specimen will reveal an interstitial infiltrate and tubulitis similar to rejection. There may also be subtle changes consistent with a virally-induced cytopathic effect of the renal tubules. Electron microscopy will confirm the presence of viral particles. Plasma and urine PCR testing is performed to detect BK polyoma genome. Treatment has generally included reduced immunosuppression however, histologic rejection may exist concurrently with BK nephropathy complicating treatment decisions. Direct antiviral therapy may be considered if there is no response to reduced immunosuppression, but there is no standard anti-viral agent being used and results are disappointing.

G. RECURRENT RENAL DISEASE

Recurrent disease in the kidney transplant accounts for <2% of all graft losses. Table 6.8 outlines the renal diseases often associated with risk of recurrent disease. The incidence of recurrent disease and the likelihood of graft loss are estimates. A few diseases have a high risk of renal allograft loss, such as focal segmental glomerulosclerosis, hemolytic uremic syndrome, oxalosis, and membranoproliferative glomerulonephritis.

TRANSPLANT NEPHRECTOMY

Transplant nephrectomy is the surgical removal of a kidney transplant. The indications for transplant nephrectomy include irreversible technical complications that result in acute failure of the transplant, hyperacute rejection, and chronic loss of renal allograft function associated with local or systemic signs of symptoms. Transplant nephrectomy may be required within days of the transplant or even years after a transplant has failed.

If transplant nephrectomy is performed within a couple months of the initial transplant procedure the kidney can be removed by taking down the easily identifiable vascular structures and ureter. For kidney allografts that have been in the retroperitoneal space for longer, the plane of dissection between the peritoneum and the renal capsule cannot be developed. The subcapsular nephrectomy tech-

Table 6.8. Recurrent diseases of the kidney transplant

A. Primary Renal Disease	Rate Loss if Present	Likelihood of Graft
Focal and segmental glomerulosclerosis (FSGS)	20-100%	30-50%
Membranous proliferative glomerulonephritis type I and type II	I: 20-70% 20-50%	II: 50-100% 10-40%
Membranous glomerulonephritis	3-10%	<2%
IgA nephropathy	50-80%	<10%
Anti-GBM disease	25-50%	<2%
B. Systemic Diseases		
Oxalosis (Type I)	80-100%	50%
Systemic lupus erythematosus	5-20%	5-30%
Hemolytic uremic syndrome	20-50%	40-50%
Diabetes mellitus	100%	<2%
Cystinosis	100%	<2%
Schonlein-Henoch purpura	75%	<1%
Amyloidosis	20%	5-20%
Mixed cryoglobulinemia	50%	50%
Alports	Common	Rare
Sickle Cell	Rare	Common
Fabry Disease	Rare	Common

6

nique is then utilized (Fig. 6.7). The transplant incision is reopened. Dissection is conducted through the fascia to the kidney capsule, avoiding entering the peritoneum. The capsule is opened and the renal parenchyma dissected from it circumferentially. The kidney is shelled out of the capsule. No attempt is made to identify the individual artery, vein, or ureter. Dissection is carried out to the hilum. A large vascular clamp is then placed upon the hilum with extreme caution to avoid occluding the iliac vessels. Confirmation that the iliac artery is opened is made by palpation of the femoral pulse. With a large vascular clamp on the hilum the broad pedicle is sharply divided and oversewn. The small amount of remaining foreign tissue left behind is not problematic. The space occupied by the kidney is obliterated by pressure from the intraperitoneal organs. Occasionally closed suction drainage is placed in the wound. Transplant nephrectomy is a shorter procedure than a kidney transplant, and the convalescence is typically a two day inpatient stay.

IMMUNOSUPPRESSION FOR KIDNEY TRANSPLANTATION

All kidney transplant recipients require life-long immunosuppression to prevent a T-cell alloimmune rejection response. Many immunosuppressive agents have been approved by the Federal Drug Administration (FDA), and several more are in phase 3 clinical trials. There are two broad classifications of immunosuppressive agents: intravenous induction/anti-rejection agents, and maintenance immunotherapy agents. There is no consensus as to the single best immunosuppressive protocol and each transplant program utilizes the various combinations of agents slightly differently. The goals of each of the programs are similar: to prevent acute and chronic rejection, to minimize the toxicities of the agents, to minimize the rates of infection and malignancy, and to achieve the highest possible rates of patient and graft survival.

A. THERAPEUTIC USE OF IMMUNOSUPPRESSION IN KIDNEY TRANSPLANTATION

1. Induction/Anti-Rejection Immunosuppressive Agents

The term "induction therapy" is generally used to describe antilymphocyte antibody pharmacologics that are parenterally administered for a short course im-

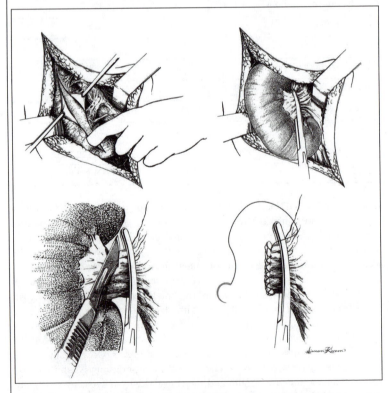

Fig. 6.7. Transplant nephrectomy utilizing the intracapsular surgical technique.

mediately posttransplant. The rationale for using induction agents pertains to its potent anti-T-cell immunosuppressive properties. In this context, induction therapy is used in conjunction with maintenance agents for the purpose of minimizing the risks of early rejection episodes, often with aims to accelerate renal allograft function, and perhaps, even inducing a tolerogenic effect to donor alloantigen.

This strategy of using induction therapy is most often applied to recipients at relatively "high risk" for rejection. Some recipient characteristics associated with a high risk of rejection include: HLA Class I antigen sensitization, elevated PRA status, re-transplantation, and African American race. In those settings, induction therapy is used to prevent or delay early acute rejection. Induction therapy is applied in other situations, as well. For example, for recipients receiving a kidney with delayed graft function, induction therapy is often used to "cover" the recipient with effective anti-T-cell therapy (using blocking or deleting agents), to delay the application of calcineurin inhibitors. This strategy serves to avoid the renal vasoconstrictive and injurious tubular effects that can occur with calcineurin inhibitors until adequate renal function occurs.

Another reason to use induction therapy is that it will provide a short course of potent immunosuppression that permits immediate and permanent elimination of one (or more) of the maintenance agents required posttransplant. This has been successfully used to immediately eliminate corticosteroid usage in kidney transplant recipients. The fourth context in which induction therapy is discussed relates to its theoretical "conditioning" effects to induce host immunological hyporesponsiveness (or tolerance) to alloantigen.

With respect to the use of induction therapy, it is estimated is that is is routinely used in approximately 50% of transplant centers for cadaveric kidney transplantation. There has not been any documented substantial effect of induction therapy on patient and graft survival rates. However, there is a significant difference in the incidence of rejection within the first six months posttransplant in recipients receiving anti-IL-2R induction agents with double or triple maintenance immunotherapy with or without azathioprine. Of note is that the incidence of rejection the first six months has also been reduced with the application of the new agents, microemulsion cyclosporine or tacrolimus in combination with MMF or sirolimus, versus the old combinations involving standard cyclosporine and azathioprine.

Trends in induction therapy. Since 1995 the proportion of kidney transplant recipients receiving induction therapy has doubled from under 30% to approximately 60%. Figure 6.8 shows the number of recipients receiving the various types of induction agents by year of transplantation. OKT3 and equine antithymocyte globulin were the predominant induction agents used through 1997. After that there has been a marked shift to the use of the anti-interleukin 2 receptor antibodies (daclizumab and basiliximab) in 1998. In 2001, 26% of the 13,109 transplants for which information is available used basiliximab and 15% used daclizumab. Rabbit antithymocyte globulin began being incorporated in 1999. The use of this agent has grown rapidly to 18% of kidney transplants in 2001. OKT3 use has dropped to <1% of transplants, and equine antithymocyte globulin, which peaked in 1997, to 2%.

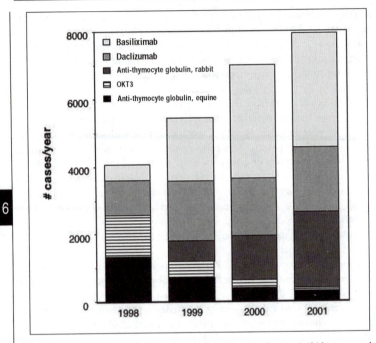

Fig. 6.8. Use of immunosuppressive agents for induction therapy in kidney transplant recipients.

2. Maintenance Immunosuppressive Agents

There are several immunosuppressive agents currently in use for maintenance immunotherapy in kidney transplant recipients. They include corticosteroids (prednisone), cyclosporine, tacrolimus azathioprine, mycophenolate mofetil (MMF), and sirolimus. The optimal maintenance immunosuppressive protocol has not been developed. The maintenance immunosuppressive agents are required life-long.

The contemporary use of transplant immunosuppression involves multi-modality therapy. Typically 2-3 agents are used in combination that together maximize efficacy and minimize toxicity rather than if used singularly as immunotherapy. The use of the various maintenance immunosuppression agents has changed significantly in the past few years. The majority of new kidney transplant recipients receive tacrolimus. The preferred antimetabolite is MMF. The most notable trends, as reported to the UNOS Scientific Renal Transplant Registry, are that microemulsion cyclosporine has replaced the use of standard cyclosporine. Tacrolimus and MMF are being used with greater frequency.

Trends in maintenance immunosuppression. For long-term immunosuppression, the majority of kidney recipients are prescribed a combination of corticosteroids a calcineurin inhibitor, and an antimetabolite. Figure 6.9 shows the trend in

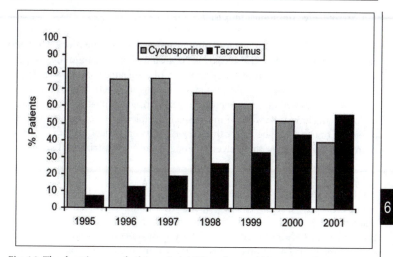

Fig. 6.9. The changing use of calcineurin inhibitors for new kidney transplants.

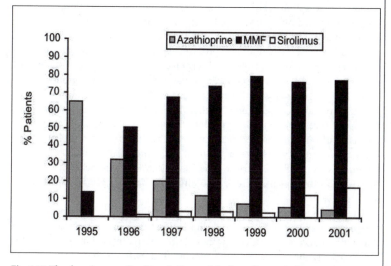

Fig. 6.10. The changing use of antimetabolites and sirolimus for new kidney transplants.

calcineurin inhibitor use in new kidney transplant recipients from 1995-2001. During this period, cyclosporine use halved from about 80% to 40%. There has been a corresponding rise in tacrolimus use from 10% to 60%. During this same period (1995-2001) Figure 6.10 shows that azathioprine use declined from 65% to 5%, while MMF use increased from 13% to 80%. In 2001, 16% of new kidney transplant recipients received sirolimus as part of their immunosuppressive medicine.

3. Steroid-Free Immunosuppression

In the past few years there has been a reassessment of the utility of including corticosteroids in the maintenance immunosuppression combinations. The beneficial contribution of corticosteroids in controlling host allo-immunity to prevent or reverse rejection is offset by a myriad of associated morbid side effects. Because the immunological risks of solid organ allograft rejection have been minimized with the newer immunosuppressive agents, greater emphasis has been placed on improving long-term wellness in transplant recipients. Major impediments to achieving optimal well-being are the erosive effects on patient health of the immunosuppressive agents themselves; and one of the most detrimental agents has been the corticosteroids.[10,11]

Previous efforts to minimize chronic corticosteroid exposure through protocols of slow weaning in kidney transplant recipients have been limited because of late rejection and graft losses. The Canadian Multicentre Transplant Study Group studied the effect of steroid withdrawal in recipients receiving dual immunosuppression with CsA and prednisone.[12] Divergent outcomes occurred at about posttransplant day 400 in recipients withdrawn from corticosteroids versus the controls maintained on CsA and prednisone. The incidence of acute rejection in both groups was similar, however, graft loss (etiology not reported) was higher in recipients weaned off corticosteroids. Schulak et al,[13] reported that renal transplant recipients receiving induction therapy, CsA and azathioprine that were randomized into a steroid withdrawal protocol experienced a higher rate and more severe rejection versus controls remaining on corticosteroids. In that study renal allograft functional survival rates were similar between treatment groups. Ahsan et al,[14] reported on a prospective, randomized study of prednisone withdrawal in kidney transplant recipients on CsA and MMF. Patient enrollment was stopped because of excess rejection in the prednisone withdrawal group. The cumulative incidence of rejection or treatment failure within 1 year posttransplant in the maintenance group was 9.8% versus 30.8% in the prednisone withdrawal group. Of note, risk was higher in African Americans (39.6%) versus the others (16.0%). At 1 year posttransplant, there was no difference between groups in patient or graft survival.

Recent trials in renal transplantation using aggressive protocols of steroid avoidance or very rapid (< 5 days) steroid elimination have been more successful than the slow withdrawal studies. Using a complete or near-complete avoidance protocol completely abrogates steroid-related side effects without affecting patient and graft survival rates or the risk of acute rejection. The Northwestern group previously reported that a very short course of corticosteroid therapy (3 days) combined with induction therapy (IL-2R anatagonist) and tacrolimus and MMF resulted in a 15% incidence of rejection and a 1-year kidney graft survival rate that exceeded 95%.[15] Interestingly, approximately 90% of rejection episodes occurred within 2 weeks of transplant. Thereafter, for up to a year, rejection was rare (3-5%). More recently, Cole et al,[16] and Vincenti et al,[17] reported short-term results of rapid steroid elimination protocols using a similar immunosuppression protocol consisting of an IL-2R antagonist, a calcineurin inhibitor with MMF. In both studies acute rejection occurred relatively earlier in recipients rapidly withdrawn from steroids, but was eventually equivalent at 1-year posttransplant to either historical controls or a control cohort. Matas et al,[18] reported short-term results of a renal transplant pilot study in which 51 recipients were given antithymocyte globulin induction, with CsA and MMF. Steroids were rapidly eliminated. Rejection rates were similar to controls with respect to frequency, timing and severity. Patient and graft survival rates were similar to historical controls. Birkeland has reported long-term results on 100 consecutive renal transplant recipients that received anti-thymocyte globulin induction, CsA and MMF with follow-up of 797-1052 days posttransplant.[19] Four-year graft survival is 82% and a cumulative rate of rejection is only 13%. Thus, there appears to be an important distinction between the earlier protocols involving slow steroid withdrawal (weaning) and rapid steroid elimination (avoidance).

6

A recent prospective, randomized study compared outcomes in recipients treated with tacrolimus (no induction) in which steroids were either rapidly eliminated (1 week) or slowly withdrawn over months.[20] The incidences of acute rejection in the first 6 months posttransplant in the rapid versus slow elimination treatment arms were 29% and 33%, respectively. Rejection occurring after stopping steroids occurred in 2 recipients in the rapid elimination group and 1 recipient in the slow taper group. Patient and graft survival rates and quality of renal allograft function were the same among treatment arms.

There is speculation why completely avoiding or rapidly eliminating steroids posttransplant does not pose a risk for rejection, and may actually be associated with a decreased rate long-term.[21-23] Although glucocorticoids decrease cytokine production, the effect may be offset by upregulation of proinflammatory cytokine receptor expression on T-cells. Hypothetically, disruption of the cytokine/cytokine receptor milieu may tip the balance toward T-cell activation and explain the enhanced susceptibility towards late rejection and graft loss in recipients in whom steroids are weaned off as opposed to the situation in which corticosteroids are essentially avoided. Therefore, as corticosteroids are slowly tapered the upregulation of cytokine receptor expression and proliferative capacity may create an immunological potential for enhanced T-cell effector function that could ensue once corticosteroid exposure falls below an individual's immunosuppression "threshold."

The alloimmune risk of graft loss in solid organ transplantation has been largely solved by application of the modern maintenance immunosuppressive agents, yet transplant recipient wellness is plagued by steroid-associated side-effects such as bone disease, cosmetic disfiguration, cataracts, gastric ulcers, increased cardiac risks, etc. The recent studies have demonstrated the feasibility of applying a prospective and flexible approach to rapid corticosteroid elimination in transplantation that gives all recipients an opportunity to have steroids permanently excluded from the maintenance immunosuppression regimen. This approach does not appear to be associated with an increased risk of rejection or graft loss in the short- or long-term.

B. Complications of Chronic Systemic Immunosuppression

Chronic systemic immunosuppression is a double-edge sword. The same immunosuppressive effects that prevent rejection of the allograft pose a risk for development of malignancy and infectious diseases. Furthermore, there are numerous possible untoward side-effects of each immunosuppressive agent differentially exhibited in each transplant recipient. For each transplant patient the proper immunosuppressive dosing and identification of the most tolerable agents and its doses are determined over the first several months after transplant. Complications result from either under immunosuppression, over immunosuppression, or the peculiar susceptibility to the side-effects of the drugs themselves. Approximately three-quarters of re-admissions posttransplant can be attributed to either infection, rejection, and/or drug toxicity.

1. Maligancy

Chronic systemic immunosuppression increases the risk of cancer after transplantation. The increased risk occurs from suppression of immune surveillance, increased incidence of viral infection leading to oncogenic stimulation, and the direct action of the immunosuppressive agents themselves. Development of significant neoplasms in the transplant population is approximately 100 times higher than that seen in the general population. Interestingly, the common malignancies seen in the general population, including lung, breast, prostate, and colon cancers, are not increased in the transplant population. The common malignancies in transplant recipients include cancers of the skin and lips, malignant lymphomas, Kaposi sarcoma, gynecologic cancer, and genital urinary cancer.

The oncogenic potential of Ebstein-Barr virus (EBV) and human herpes virus 8 (HHV-8) are important variables in the development of posttransplant lymphoproliferative disorders (non-Hodgkin lymphoma) and Kaposi sarcomas, respectively. The degree of overall immunosuppression, especially intensive immunotherapy for treatment of rejection with anti-T-cell antibodies, is a major factor in determining development of these tumors. Also the serologic profile of the organ donor and transplant recipient are related to the risk of development of these malignancies. Transplantation of organs from seropositive donors to recipients seronegative for EBV, and possibly HHV-8, have important implications for surveillance of these tumors in the posttransplant course.

2. Infectious Disease

Infectious diseases are frequent complications of systemic immunosuppression. The likelihood of infection is related to the intensity of immunosuppression and the exposure to infectious organisms. The mortality of serious infection has declined recently and is related to a reduction in the incidence of acute rejection and the consequent need for intense anti-T-cell antibody therapy. The basic principles of infectious disease management in the general population apply to the transplant population, as well. The goal is identification of the infectious disease organisms, their sensitivities to antimicrobials, and the localization of the site(s) of infection.

There are significant differences in the types of infectious organisms and the intensity of infection that the transplant recipient encounters versus the general population. There are a number of opportunistic infectious organisms that need to be considered in the transplant recipient. Also, the severity of infection is often underestimated because of the relatively benign clinical presentation. The most common clinical presentation of infection is fever. There is a relatively uniform methodology in the diagnostic approach to unexplained fever. Clinical symptoms related to specific organ systems often lead to imaging studies or interventional procedures. The work-up needs to be thorough and swift. Often broad spectrum antibacterials, antifungals, and antiviral agents need to be employed empirically while identification of the organism is being determined in the microbiology laboratory. Infectious diseases can occur in any compartment or body cavity. The time posttransplant and the particular symptoms and signs, such as the white blood count, can play important roles in anticipating the likely infectious agent. A list of the common opportunistic infections germane to the transplant recipient versus the general population is shown in Table 6.9.

Urinary tract infections are the most common bacterial infection in the kidney transplant recipient. Urinary tract infections can be relatively benign, presenting as cystitis, or rapidly progressive and potentially life-threatening if pyelonephritis with bacteremia develops. In many programs the kidney transplant recipients are prescribed prophylactic antibiotics for the first year posttransplant. The antibiotic of choice is trimethoprim-sulfamethoxazole. This has been shown to decrease the frequency of urinary tract infections and is also effective prophylaxis for reducing the opportunistic infections of Pneumocystis and Nocardia.

Table 6.9. Opportunistic infectious organisms in transplant recipients

A. Bacterial
   1. Legionella pneumophilia and micdadei
   2. Nocardia asteroides
   3. Listeria monocytogenes
   4. Mycobacteria
B. Fungal
   1. Candida albicans, tropicalis, parapsilosis
   2. Aspergillus niger, fumigatus, flavus
   3. Cryptococcus neoformans
   4. Torulopsis glabrata
   5. Mucor
   6. Rhizopus
C. Viral
   1. Cytomegalovirus
   2. Herpes simplex virus
   3. Varicella
   4. Ebstein-Barr virus
D. Protozoan and parasitic
   1. Pnemocystis carinii
   2. Toxoplasma gondii

OUTCOMES OF KIDNEY TRANSPLANTATION

The results of kidney transplantation are defined according to the specific endpoint studied. The broad endpoints include patient survival and renal allograft survival. More specific endpoints have been examined such as the incidence and severity of rejection episodes, quality of renal allograft function, hospitalizations, and even economic data. Outcomes may be reported from various sources including national databases or in multi-center trials, and single-center experiences. Over the past 15 years the results of kidney transplantation have steadily improved with appreciation of the medical nuances of each case, and the development of new immunosuppressive and antimicrobial agents.[24] The outcome of kidney transplantation is influenced by many variables (Table 6.10). Some of these that will be discussed include: donor source, degree of HLA mismatch, PRA level, race, etiology of renal disease, duration of pre-transplant dialysis therapy, delayed graft function, and the transplant center effect.

Two of the most useful national databases on kidney transplantation are the Scientific Registry of Transplant Recipients (SRTR) of the United Network for Organ Sharing, where data on all kidney transplants in the U.S. have been collected since 1987, and the United States Renal Data System (USRDS). The SRTR supports the ongoing evaluation of the scientific and clinical status of solid organ transplantation including kidney transplants. The SRTR contains information on over 200,000 transplant recipients. Funding comes from the Health Resources and Services Administration (HRSA), a division of the U.S. Department of Health and Human Services (DHHS). The SRTR is administered by University Renal Research and Education Association (URREA), a not for profit health research organization, in collaboration with the University of Michigan. The United States Renal Data System is a national data system that collects, analyzes, and distributes information about end-stage renal disease (ESRD) including renal transplantation. The USRDS is funded directly by the National Institute of Diabetes and Digestive and Kidney Diseases (NIDDK) in conjunction with the Centers for Medicare & Medicaid Service (CMS). USRDS staff collaborates with members of CMS, the

---

Table 6.10. Variables influencing outcome of kidney transplantation

Donor source
HLA match
PRA level
Race
Etiology of renal disease
Recipient age
Recipient medical status and body mass index
Expanded criteria cadveric organ donor
Delayed kidney graft function
Prior kidney transplant
Duration of dialysis prior to transplantation
CMV donor/recipient status
Clinical acute rejection
Transplant center effect

United Network for Organ Sharing, and the ESRD networks, sharing datasets and actively working to improve the accuracy of ESRD patient information.

Effect of donor source. The results are defined as patient and graft survival according to the length of time posttransplant. Definition of patient survival is obvious, loss of functional graft survival is defined for patients that have died or have returned to dialysis. Table 6.11 shows the 1-year graft survival rates of cases collected by the SRTR for the year 2000. The results are stratified according to donor source. The outcome of kidney transplantation is superior in recipients receiving a kidney from a living donor. Within this category recipients of sibling HLA-identical grafts do best. Interestingly, there is very little difference among graft survival rates in other living donor categories including 1-haplotype matches, 0-haplotype matches, and living unrelated donors.

6

Table 6.11. Kidney allograft survival according to donor source

Donor	1 Year N	%	3 Years N	%	5 Years N	%
Cadaver	15,850	88.4%	15,510	78.5%	15,186	63.3%
Living	9,862	94.4%	8,265	88.3%	7,007	76.5%

Source: OPTN/SRTR Data as of August 1, 2002. Cohorts are transplants performed during 1999-2000 for 3 month and 1 year; 1997-1998 for 3 year; and 1995-1996 for 5 year survival.

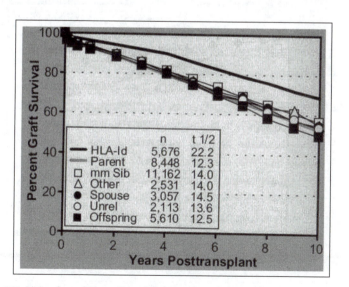

Fig. 6.11. Living donor kidney allograft functional survival according to donor relationship. (Reprinted with permission from: Clinical Transplants 2001, editor Cecka and Terasaki, Publisher UCLA Tissue Typing Laboratory, 2002).

Figure 6.11 shows the results of living donor graft survival rates by donor relationship in a cohort of transplant recipients from 1988-2000. Results are expressed in terms of half-life survival. In fact, for kidney transplants reported to the UNOS Scientific Registry during a more recent era (1996-2000), the HLA mismatched living donor grafts had superior outcomes versus the 0-antigen mismatched cadaveric grafts (Fig. 6.12). This indicates that the health of kidney donor and recipient, the elective timing of the transplant, the short cold ischemia time, the immediate function of the graft, and other factors are very important determinants of outcome over and above HLA matching by itself. Within the cadaveric transplant group, collections of large numbers of cases show subtle differences in graft survival rates according to the degree of mismatching. Only kidneys from 0 HLA antigen mismatched cadaveric donors seem to confer a survival advantage over the longer-term (Table 6.12).

Effect of PRA level. It has been recognized that highly sensitized patients have relatively poorer outcome because of greater likelihood of graft loss from immunological causes. Table 6.13 shows outcomes in recipients stratified according to pre-transplant PRA level. The higher the PRA level the worse the outcome.

Effect of race. Recipient race has an impact on renal transplant outcome in both cadaver and living donor transplants. In general, the rate of graft loss in African American recipients, especially after the first year posttransplant, is nearly double compared to Caucasians and other ethnic groups (Fig. 6.13).

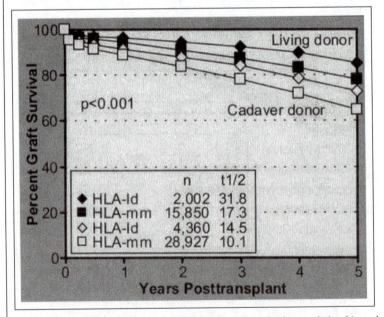

Fig. 6.12. Kidney allograft functional survival according to the donor relationship and HLA compatibility. (Reprinted with permission from: Clinical Transplants 2001, editor Cecka and Terasaki, Publisher UCLA Tissue Typing Laboratory, 2002).

Table 6.12. Cadaveric kidney allograft survival according to HLA antigenic mismatch

Mismatch	1 Year N	%	3 Years N	%	5 Years N	%
0	2,667	90.4%	2,410	83.2%	2,322	70.2%
1	481	91.4%	539	82.3%	560	63.6%
2	1,649	90.4%	1,797	81.1%	1,944	63.9%
3	3,285	89.2%	3,482	79.3%	3,484	63.2%
4	3,506	87.5%	3,549	76.6%	3,624	61.9%
5	2,860	86.4%	2,510	75.3%	2,232	60.3%
6	1,376	85.6%	1,193	73.5%	1,006	58.4%

Source: OPTN/SRTR Data as of August 1, 2002. Cohorts are transplants performed during 1999-2000 for 1 year; 1997-1998 for 3 year; and 1995-1996 for 5 year survival.

Table 6.13. Cadaveric kidney allograft survival according to PRA level

PRA at Transplant	1 Year N	%	3 Years N	%	5 Years N	%
0-19%	12,670	88.8%	12,753	79.1%	12,656	64.2%
20-79%	1,263	84.3%	1,170	73.9%	1,107	59.1%
80%+	601	84.4%	515	73.4%	533	53.1%

Source: OPTN/SRTR Data as of August 1, 2002. Cohorts are transplants performed during 1999-2000 for 3 month and 1 year; 1997-1998 for 3 year; and 1995-1996 for 5 year survival.

Effect of etiology of renal disease. The etiology of renal disease has less of an effect on outcome over the short term than in years past. Patients with diabetes used to have significantly worse outcomes a decade ago. The greatly improved results in these recipients has been primarily due to understanding the concurrent medical problems associated with diabetes, particularly cardiovascular disease. However, in recipients with diabetes, the duration of time on dialysis therapy prior to transplantation has recently been appreciated to strongly influence outcome. In general, transplant waiting time on dialysis is one of the strongest independent and modifiable risk factors for renal transplant outcomes. Much of the advantage of living versus cadaveric transplantation may relate to this phenomenon. This is because of the elective nature of living door transplantation that can be completed within months of work-up. The effect of transplant waiting time is so strong that graft survival for cadaveric renal transplant recipients with a history of renal failure of less than 6 months is equivalent to living-donor transplant recipients who wait on dialysis for more than 2 years.[25]

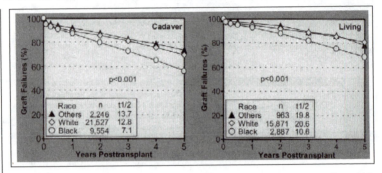

Fig. 6.13. The influence of recipient race on outcomes of cadaveric and living donor kidney transplants. (Reprinted with permission from: Clinical Transplants 2001, editor Cecka and Terasaki, Publisher UCLA Tissue Typing Laboratory, 2002).

Effect of delayed graft function. A relatively recent finding that has great impact on the outcome of transplantation is the quality of initial function of the kidney transplant.[26] If delayed graft function is significant such that dialysis is required within the first week of transplantation, outcomes at 1-year and greater are significantly inferior to immediately functioning grafts. This is greatly amplified in recipients of living kidney donors where the expected rate of delayed graft function is near zero. If delayed graft function ensues in this group, technical problems or unusual immunological problems have occurred. In the cadaveric kidney group, delayed graft function, usually because of ATN, is not unusual. However, prolonged delayed graft function results in higher rates of rejection, an event that can be obscured in a poorly functioning kidney, that is associated with a higher incidence of immuologic graft loss.

The transplant center effect. One of the strongest determinants of outcome is the transplant center effect. There are approximately 250 UNOS approved kidney transplant programs in the U.S.. Each program is required to submit outcomes data to the SRTR database. The SRTR publishes the outcomes data in various media,[27] including the internet.[28] Analyses are conducted according to national and single-center patient populations. For kidney transplantation, single-center analyses of outcomes include patient and graft survival rates for cadaver and living donor transplants over the very short- (3 months), short- (1 year), and intermediate-term (3 years). Results are expressed in actual terms as well as expected outcomes based on the patient mix of the individual institution. There are statistical analyses that determine whether the actual results are higher, the same, or lower than expected. Examination of single-center outcomes sheds light on the transplant center effect. From this it is possible to gain some insight into the extreme variability of outcomes between transplant centers.

Of the approximately 250 kidney transplant centers, 13 demonstrated 1-year graft survival rates statistically significantly higher than expected, and 12 demonstrated outcomes statistically significantly lower than expected. Cadaveric kidney transplant outcome had the strongest effect on whether overall statistical significance was reached. The group with "superior" outcomes for cadaveric transplantation included 13 centers, of which 2 also had "superior" outcomes for living donor transplantation. Table 6.14 shows the 1-year cadaveric kidney graft survival results for the 13 transplant programs that had higher than expected results. Large and relatively small transplant programs in large and small cities are represented.

The spread of actual outcomes for these centers, 97-95% 1-year graft survival, was higher than the spread of expected outcomes, 91-88%. There were 10 centers with actual 1-year cadaveric kidney transplant graft survival rates statistically significantly less than expected results. The actual graft survival spread was 59-80%, compared to the expected spread of 85-91%. Importantly, the lower graft survival rate in 3 centers in the later group was influenced by a greater rate of graft loss accounted for by death-with-function rather than immunologic or other causes of graft loss. This was reflected by the lower 1-year patient survival rates in this group (actual 80-88% versus expected 93-95%).

6

Table 6.14. One-year cadaveric kidney transplant graft survival rates in transplant centers with results statistically significantly higher than expected results.[1]

Transplant Center (City)	N	Actual Graft Survival	Expected Graft Survival[2]	P-Value
*Emory University Hosp. (Atlanta)	241	94.26	87.85	<0.01
*Tampa General Hosp. (Tampa)	332	94.02	89.08	<0.01
Univ. of Wisconsin. Hosp. (Madison)	351	92.78	87.43	<0.01
Univ. of Alabama Hosp. (Birmingham)	385	94.29	89.62	<0.01
*Univ. of California Med. Center (San Francisco)	345	93.02	89.19	0.013
*Hosp. of the Univ. of Pennsylvania (Philadelphia)	209	94.4	88.08	0.015
Northwestern Memorial Hosp. (Chicago)	117	96.15	89.95	0.020
Albany Medical Center Hosp. (Albany)	124	96.77	90.25	0.022
Washington Hosp. Center (Washington DC)	74	94.54	84.92	0.023
*Jackson Memorial Hosp. of Univ. Miami (Miami)	242	93.45	89.32	0.028
Baylor Univ. Medical Center (Dallas)	168	95.66	90.46	0.030
*Oregon Health Sciences Univ. Hosp. (Portland)	213	92.28	87.58	0.038

[1] Adult (age >18 years) recipients transplanted between 7/1/99 and 12/31/01.
[2] Based on SRTR data on U.S. graft failure rates adjusted for donor and recipient characteristics (see http//www.ustransplant.org).
* Transplant centers with results statistically significantly higher than expected results for cadaver kidney transplants at 3 years posttransplant

SUMMARY

Advances in kidney transplantation will continue to be made for many years as more specific and less toxic immunosuppressive agents, and tolerance induction protocols are developed. Other advances in improving participation in organ donation consent rates, greater use of expanded criteria cadaveric organ donors (including non-heart-beating donors), emphasis on the use of living donors, and new pre-conditioning protocols that allow transplants for sensitized recipients will help solve the problem of limited availability of organs for transplantation. Solutions to these issues will permit a greater number of ESRD patients to receive a kidney transplant, allowing them to enjoy the highest quality and duration of life that is possible.

REFERENCES

1. Wolfe RA, Ashby VB, Milford ELet al. Comparison of mortality in all patients on dialysis, patients on dialysis awaiting transplantation, and recipients of a first cadaveric transplant. NEJM. 1999; 341:1725.
2. Suthanthiran M, Strom TB. Renal transplantation. NEJM 1994; 331:365.
3. URREA; UNOS. 2002 Annual Report of the U.S. Organ Procurement and Transplantation Network and the Scientific Registry of Transplant Recipients: Transplant Data 1992-2001 [Internet]. Rockville (MD): HHS/HRSA/OSP/DOT; 2003 [modified 2003 Feb 18; cited 2003, April 3]. Available from: http://www.optn.org/data/annualReport.asp.
4. Kasiske BL, Ramos EL, Gaston RS et al. The evaluation of renal transplant candidates: clinical practice guidelines. Journal of the American Society of Nephrology 1995; 6:1.
5. U.S. Renal Data System, USRDS 2002 Annual Data Report: Atlas of End-Stage Renal Disease in the United States, National Institutes of Health, National Institute of Diabetes and Digestive and Kidney Diseases, Bethesda, MD, 2002.
6. Takemoto SK, Terasaki PI. Evaluation of the transplant recipient and donor: molecular approach to tissue typing, flow cytometry and alternative approaches to distributing organs. Current Opinion in Nephrology and Hypertension 1997; 6:299.
7. Najarian JS, Chavers BM, McHugh LE, Matas AJ. 20 years or more of follow-up of living kidney donors. Lancet. 1992; 340:807.
8. Solez K. Renal allograft histopathology. Nephrology, Dialysis, Transplantation; 10 Suppl 1995; 1:44.
9. Hirsch HH. Polyomavirus BK nephropathy: A (re-) emerging complication in renal transplantation. Am J Transplant 2002: 2; 25-30.
10. Fryer JP, Granger DK, Leventhal JR et al. Steroid related complications in the CsA era. Clin Transplant 1994; 8 :224.
11. Bialas MC, Routledge PA. Adverse effects of corticosteroids. Adverse Drug React Toxicol Rev 1998; 17:227.
12. Sinclair NR. Low-dose steroid therapy in CsA-treated renal transplant recipients with well-functioning grafts. The Canadian Multicentre Transplant Study Group. Can Med Assoc J. 1992;147:645.
13. Schulak JA, Mayes JT, Moritz CE, Hricik DE. A prospective randomized trial of prednisone versus no prednisone maintenance therapy in CsA-treated and azathioprine-treated renal transplant patients. Transplantation. 1990; 49:327.
14. Ahsan N, Hricik D, Matas A et al. Prednisone withdrawal in kidney transplant recipients on CsA and MMF—a prospective randomized study. Steroid Withdrawal Study Group. Transplantation. 1999; 68:1865.

15. Kaufman DB, Leventhal JR, Fryer JP, Abecassis MI, Stuart FP. 2000. Kidney transplantation without prednisone. Transplantation 69:S133

16. Cole E, Landsberg D, Russell D et al. A pilot study of steroid-free immunosuppression in the prevention of acute rejection in renal allograft recipients. Transplantation 2001; 72: 845.

17. Vincenti F, Monaco A, Grinyo J et al. A multicenter randomized prospective trial of steroid withdrawal in renal transplant recipients receiving basiliximab, cyclosporine microemulsion and mycophenolate mofetil. Am J Transplant 2003; 3:306.

18. Matas AJ, Rancharan T, Paraskevas S et al. Rapid discontinuation of steroids in living donor kidney transplantation: A pilot study. Am J Transplant 2001; 1:278.

19. Birkeland SA. Steroid-free immunosuppression in renal transplantation: a long-term follow-up of 100 consecutive patients. Transplantation. 2001; 71:1089.

20. Boots, JMM. Christiaans, MHL. van Duijnhoven, EM et al., Early steroid withdrawal in renal transplantation with tacrolimus dual therapy: a pilot study. Transplantation 2002; 74:1703.

21. Almawi WY, Lipman ML, Stevens AC et al. Abrogation of glucocorticoid-mediated inhibition of T cell proliferation by the synergistic action of IL-1, IL-6, and IFN-gamma. J Immunol 1991; 146:3523.

22. Almawi WY, Beyhum HN, Rahme AA, Rieder MJ Regulation of cytokine and cytokine receptor expression by glucocorticoids. J Leukoc Biol 1996; 60:563.

23. Almawi WY, Hess DA, Assi JW, Chudzik DM, Rieder MJ. Pretreatment with glucocorticoids enhances T-cell effector function: possible implication for immune rebound accompanying glucocorticoid withdrawal. Cell Transplant 1999; 8:637.

24. Hariharan S, Johnson CP, Bresnahan BA et al. Improved graft survival after renal transplantation in the United States, 1988 to 1996. NEJM 2000; 342: 605.

25. Meier-Kriesche HU, Kaplan B. Waiting time on dialysis as the strongest modifiable risk factor for renal transplant outcomes: a paired donor kidney analysis. Transplantation. 2002; 74: 1377.

26. Cecka JM, Shoskes DA, Gjertson DW. Clinical impact of delayed graft function for kidney transplantation. Transplant Rev 2001; 15: 57.

27. 2002 Scientific Registry of Transplant Recipients Report on the State of Transplantation. Am J Transplant 3(Suppl 4), 2003

28. http://ustransplant.org/center-adv.html

6

Pancreas Transplantation

Dixon B. Kaufman
Illustrations by Simon Kimm

RATIONALE OF PANCREAS TRANSPLANTATION FOR PATIENTS WITH TYPE 1 DIABETES MELLITUS

At the turn of the century a patient diagnosed with Type 1 diabetes mellitus had average life expectancy of only two years. The development of insulin as a therapeutic agent revolutionized the treatment of diabetes mellitus by changing it from a rapidly fatal disease into a chronic illness. Unfortunately this increased longevity brought to the fore serious secondary complications including nephropathy, neuropathy, retinopathy and macro- and microvascular complications in survivors 10 to 20 years after disease onset. The annual national direct and indirect costs of Type 1 and 2 diabetes in 2002 - including hospital and physician care, laboratory tests, pharmaceutical products, and patient workdays lost because of disability and premature death - exceeded $130 billion.[1]

Currently, the prevalence of Type 1 diabetes in the United States is estimated to be 1,000,000 individuals, and 30,000 new cases are diagnosed each year. Presently there is no practical mechanical insulin-delivery method coupled with an effective glucose-sensory device that could replace the function of the impaired cells to administer insulin with a degree of control to produce a near constant euglycemic state without risk of hypoglycemia. Therefore, persons with Type 1 diabetes are resigned to manually regulate blood glucose levels by subcutaneous insulin injection, and as a consequence, typically exhibit wide deviations of plasma glucose levels from hour to hour and from day to day. Since hypoglycemia is intolerable, glucose control must error on the high side and patients live with relative chronic hyperglycemia as evidenced by elevated HgbA1c levels. Hyperglycemia is the most important factor in the development and progression of the secondary complications of diabetes. These observations and the known fact that conventional exogenous insulin therapy cannot prevent the development of the secondary complications of Type 1 diabetes, has lead to a search for alternative methods of treatment designed to achieve better glycemic control to the extent that the progression of long-term complications can be altered.

The only treatments that have been demonstrated to influence the progression of secondary complications normalize or near normalize glycosylated hemoglobin levels – beta cell replacement therapy with pancreas or islet transplantation and intensive insulin therapy. Islet transplantation is discussed in Chapter 8. Pancreas transplantation is superior to that and intensive insulin therapy with regard to the efficacy of achieving glycemic control, and its beneficial effects on diabetic

Organ Transplantation, 2nd edition, edited by Frank P. Stuart, Michael M. Abecassis and Dixon B. Kaufman. ©2003 Landes Bioscience.

secondary complications. Only pancreas transplantation consistently normalizes glycosylated hemoglobin levels, and compared to intensive insulin therapy, has the added physiological properties of pro-insulin and C-peptide release. A successful pancreas transplant produces a normoglycemic and insulin-independent state. It will reverse the diabetic changes in the native kidneys of patients with very early diabetic nephropathy; prevent recurrent diabetic nephropathy in patients undergoing a simultaneous pancreas-kidney transplant; reverse peripheral sensory neuropathy; stabilize advanced diabetic retinopathy; and significantly improve the quality of life.

However, there are important considerations of pancreas transplantation that currently preclude it as therapy for all patients with type 1 diabetes. First, it is unrealistic that all patients with diabetes could receive a pancreas transplant. There are too many patients with type 1 diabetes and too few organs. Second, pancreas transplantation involves significant surgery. Third, lifelong immunosuppression is required to prevent graft rejection. Therefore, the indications for pancreas transplantation are very specific and narrow. There are three circumstances where consideration for pancreas transplantation is reasonable: i) for select medically suitable patients with type 1 diabetes that are also excellent candidates for kidney transplantation; ii) for patients with type 1 diabetes that enjoy good function of a kidney transplant and are receiving immunosuppression; and iii) for select patients with type 1 diabetes that are extremely brittle or associated with significant frequency and severity of hypoglycemic unawareness such that the risks of surgery and immunosuppression are less morbid than the current state of ill health.

INDICATIONS AND CONTRAINDICATIONS TO PANCREAS TRANSPLANTATION

Approximately 1,300 pancreas transplants are performed annually in the U.S. (Fig. 7.1). Eighty-five-90% involve a simultaneous pancreas and kidney transplant (SPK) for patients with type 1 diabetes and chronic or end-stage renal failure. These persons are excellent candidates for a simultaneous pancreas and kidney transplant from the same donor because the immunosuppressive medications needed are similar to those for a kidney transplant alone, and the surgical risk of adding the pancreas is low. The benefits of adding a pancreas transplant to ameliorate diabetes are profound – it saves lives.[2] Unfortunately, access to SPK transplantation is primarily obstructed by the donor organ shortage. Only a small proportion of patients that could benefit from an SPK transplant ever receive one. Figure 7.2 illustrates, by year, the growing size of the waiting list, the number of SPK transplants performed, and the relatively stagnant number of cadaver organ donors per year.[3]

The second category for pancreas transplantation is patients with Type 1 diabetes who have received a previous kidney transplant from either a living or cadaveric donor. This group accounts for approximately 10% of patients receiving pancreas transplants. It is the fastest growing of the three groups. The important consideration is that of surgical risk, since the risk of immunosuppression has already been assumed.

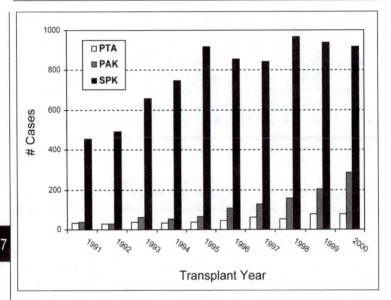

Fig. 7.1. Annual number of pancreas transplants performed in the U.S. according to category (from the International Pancreas Transplant Registry).

The third category for pancreas transplantation is in non-uremic, non-kidney transplant patients with type 1 diabetes. In this situation one assesses the risk of immunosuppression to be less than the current clinical condition with conventional exogenous insulin administration. These patients with diabetes have extremely labile disease, such that there is difficulty with day-to-day living, associated with frequent emergency room visits and inpatient hospitalizations for hypoglycemia or diabetic ketoacidosis. Other patients may have significant difficulty with hypoglycemic unawareness that results in unconsciousness without the warning. This can be a devastating condition for these select patients that affects their employment, their ability to have a license to drive, and concern about suffering lethal hypoglycemia while asleep. The indications for a pancreas transplant alone are essentially identical to those patients being considered for an islet transplant. However, in the former situation, there are fewer contraindications with respect to body mass index and insulin requirements.

An extremely interesting patient population for which the benefits of pancreas transplantation are being more thoroughly explored are those with early diabetic nephropathy. These patients show the presence of micro-albuminuria indicating the renal diseases at a stage where progression is inevitable without amelioration of the diabetic state. It is clear that either dialysis or kidney transplantation will ultimately be required. It has been established that pancreas transplant performed at this early stage of diabetic nephropathy is capable of halting and reversing the diabetic process affecting the native kidneys. It is possible that added beneficial

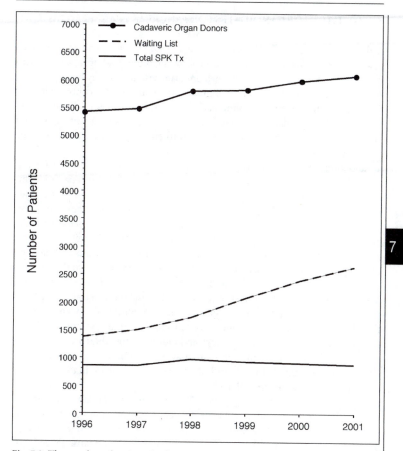

Fig. 7.2. The number of patients in the U.S. waiting for an SPK transplant, receiving an SPK transplant and the number of cadaver organ donors per year.

effects on associated secondary complications may also be achieved, but conclusive studies have not been performed.

The contraindications to pancreas transplantation include the majority of patients with type 1 diabetes that have normal renal function and do not exhibit a brittle course, or hypoglycemic unawareness, or any evidence of nephropathy. For patients that do have an indication for pancreas transplantation, it is important to rule out significant medical contraindications in a similar manner as applies to other areas of transplantation. These issues include: recent malignancy, active or chronic untreated infection, advanced forms of major extrarenal complications (i.e. coronary artery disease), life expectancy of less than 1 year, sensitization to donor tissue, noncompliance, active substance abuse, and uncontrolled psychiatric disorder.

EVALUATION OF CANDIDATES FOR PANCREAS TRANSPLANTATION

There are significant pre-existing morbidities of pancreas transplant candidates with advanced renal disease. It should be assumed that coincident extrarenal disease is present. Diabetic retinopathy is a nearly ubiquitous finding in patients with diabetes and end-stage renal disease. Significant vision loss may have occurred. Also patients may be overtly blind. Blindness is not an absolute contraindication to transplantation since many blind patients lead very independent life styles. Although rarely a problem, it should be confirmed that a patient with significant vision loss has an adequate support system to ensure help with travel and the immunosuppressive medications.

Autonomic neuropathy is prevalent and may manifest as gastropathy, cystopathy, and orthostatic hypotension. The extent of diabetic autonomic neuropathy is commonly underestimated. Neurogenic bladder dysfunction is an important consideration in patients receiving a bladder-drained pancreas-alone transplant or an SPK transplant. Inability to sense bladder fullness and empty the bladder predisposes to urine reflux and high post void residuals. This may adversely affect renal allograft function, increase the incidence of bladder infections and pyelonephritis, and predispose to graft pancreatitis. The combination of orthostatic hypotension and recumbent hypertension results from dysregulation of vascular tone. This has implications for blood pressure control posttransplant, especially in patients with bladder drained pancreas transplants that are predisposed to volume depletion. Therefore, careful re-assessment of posttransplant antihypertensive medication requirement is important. Sensory and motor neuropathies are common in patients with longstanding diabetes. This may have implications for the rehabilitation posttransplant. It also is an indicator for potential risk for injury to the feet and subsequent diabetic foot ulcers.

Impaired gastric emptying (gastroparesis) is an important consideration because of its significant implications in the posttransplant course. Patients with severe gastroparesis may have difficulty tolerating the oral immunosuppressive medications that are essential to prevent rejection of the transplants. Episodes of volume depletion with associated hypercreatinemia in patients with SPK transplants frequently occur. Patients typically require careful treatment modalities that include motility agents such as metoclopramide or erythromycin.

Advanced coronary artery disease is the most important comorbidity to consider in patients with type 1 diabetes with diabetic nephropathy. It has been estimated that uremic, diabetic patients carry a near 50 fold greater risk of cardiovascular events then the general population. The diabetic, uremic patient has several risk factors in addition to diabetes for development of coronary artery disease including, hypertension, hyperlipidemia and smoking. Because of the neuropathy associated with diabetes, patients are often asymptomatic because ischemia-induced angina is not perceived. The prevalence of significant (>50%) coronary artery stenosis in patients with diabetes starting treatment for end-stage renal disease is estimated to be 45-55%.

Uremic, diabetic patients also experience an increased rate of cerebral vascular accidents (strokes) and transient ischemic attacks. Deaths related to cerebral vascular disease are approximately twice as common in patients with diabetes versus no diabetes once end-stage renal disease has occurred. Patients with diabetes suffer strokes more frequently and at a younger age then do age and gender match non-diabetic stroke patients. Hypertension is the major risk factor for stroke followed by diabetes, heart disease and smoking.

Lower extremity peripheral vascular disease is significant in patients with diabetes. Uremic diabetic patients are at risk for amputation of a lower extremity. These problems typically begin with a foot ulcer associated with advanced somatosensory neuropathy.

Mental or emotional illnesses including neuroses and depression are common. Diagnosis and appropriate treatment of these illnesses is an important pretransplant consideration with important implications for ensuring a high degree of medical compliance.

The components of the pretransplant evaluation are very similar to that carried out in kidney-alone transplant patients with special attention to the above medical issues. The history of disease, review of systems, and physical examination are conducted in a similar focused manner. The interventional studies with respect to the workup of cardiovascular disease does require a uniform screening method because of the high prevalence of severe and often silent cardiovascular disease in the diabetic patient. Figure 7.3 illustrates an example of an algorithm for screening transplant candidates with diabetes for coronary artery disease (CAD).

The basic goal of screening is to detect significant, treatable CAD in patients not suspected to have coronary lesions. Noninvasive screening that has high sensitivity and specificity for significant coronary artery disease can be used on low risk patients. Patients considered to be at moderate or high risk for significant CAD should undergo coronary arteriography to determine the severity and location of the lesions. Patients with coronary lesions amenable to angioplasty with stenting or bypass grafting should be treated and re-evaluated and then reconsidered for transplantation. The goal of revascularization is to diminish the perioperative risk of the transplant procedure and to prolong the duration of life posttransplant. Patients that have experienced long waiting periods prior to pancreas transplantation should have their cardiac status assessed at regular intervals.

A liberal policy that virtually all diabetic, uremic patients should undergo coronary angiography is not unreasonable because the current noninvasive tests are relatively insensitive. Also, the techniques of coronary angiography have changed in the last few years, allowing for selected arteriography with very low dose, less toxic contrast agents using biplanar imaging techniques. The nephrotoxic risk of the angiography has been reduced considerably (if a left ventriculogram is omitted) in a preuremic patient with creatinine clearance >20 ml/min.

TRANSPLANT SURGERY AND SURGICAL COMPLICATIONS

The timing of allocation of the pancreas to a specific patient relative to the procurement of the organ has important implications. Determining donor HLA typing, viral serologies, and crossmatch results with patients on the pancreas transplant waiting list will permit the ideal situation of allocating the cadaveric pan-

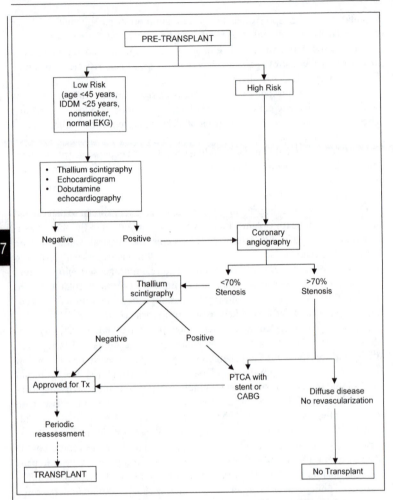

Fig. 7.3. Algorithm for screening transplant candidates with Type 1 diabetes for coronary artery disease.

creas (plus kidney with SPK transplant) prior to procurement of the organs. This sequence of events has several advantages: i) it will allow the transplant center performing the pancreas transplant the choice to also procure the pancreas; ii) it will allow patients to be admitted to the hospital and the re-evaluation process to begin simultaneously, rather than sequential, to the procurement of the organs; iii) it will minimize the cold ischemia time of the pancreas prior to implantation. Pancreas allografts do not tolerate cold ischemia as well as kidney allografts. It is ideal to revascularize the pancreas within 24 hours from the time of crossclamping at procurement. Finally, it will also allow identification of 0-antigen mismatched donor-recipient pairs to be identified prior to procurement that will minimize cold ischemia time if the organs need to be transported across country.

A. Preoperative Transplant Care

The pancreas transplant recipient is admitted to the hospital, re-evaluated and a final decision made whether or not to proceed with surgery. The re-evaluation process is similar to that for kidney-alone transplant recipients, emphasizing workup for infectious disease or other acute medical issues that would contraindicate surgery. There are several special considerations for the diabetic patient. Careful management of diabetes pretransplant is important for patients not allowed eat or drink prior to surgery. A bowel preparation is performed for patients that will undergo enteric drainage of the pancreas transplant

B. Cadaveric Pancreas Donor Selection

Identification of suitable cadaveric organ donors for pancreas transplantation is one of the most important determinants of outcome. The contraindications of cadaveric pancreas procurement for transplantation are outlined in Table 7.1.

Several anatomic and physiologic factors have been identified that affect the results of pancreas transplantation. In general, the criteria that determine an appropriate donor for pancreas transplantation are more stringent than for kidney or liver donors. Cadaveric pancreas organ donors are typically between the ages of 10 and 55 years. The lower age limits typically reflect the anticipated small size of the splenic artery that may preclude successful construction of the arterial Y-graft needed for pancreas allograft revascularization. The use of older donors has been associated with increased technical failure due to pancreas graft thrombosis, a higher incidence of posttransplant pancreatitis, and decreased pancreas graft survival rates. This may be consequent to reduced tolerability of cold ischemia time, but this has not been rigorously studied. The weight of the cadaveric organ donor is an important consideration. Obese donors over 100 kg are frequently not found to be suitable pancreas donors. Obese patients may have a history of type 2 diabetes, or the pancreas may be found to be unsuitable for transplantation because of a high degree of adipose infiltration of the pancreas. Obviously, weight alone does not exclude a donor, it is evaluated in conjunction with the height. There is also a lower weight limit that guides the decision for pancreas recovery. Recipients less than 30 kg must be carefully considered since this is also a reflection of potential small size of the splenic artery.

Importantly, pancreata from relatively older donors (age 55-65) and obese organ donors are associated with very successful islet isolation recovery required for an islet transplant. Therefore, application of beta cell replacement therapy, in general, and islet transplantation, in particular, should be considered for nearly all cadaveric organ donors.

Hyperglycemia and hyperamylasemia are very frequently observed in cadaveric organ donors. Hyperglycemia is not a contraindication to pancreas procurement for patients who are known not to have type 1 or type 2 diabetes. Hyperglycemia is generally benign and caused by a combination of factors including administra-

Table 7.1. Contraindications of pancreas procurement for transplantation

1. History of type 1 diabetes mellitus
2. History of type II diabetes mellitus
3. History of previous pancreatic surgery
4. Intra-abdominal trauma to the pancreas
5. Donor age <10 years and >55 years
6. Donor weight <30 kg and >100 kg (taken in consideration with height)
7. Intraoperative assessment
    A. Vascular supply
    B. Severe edema, significant adipose infiltration, significant fibrosis or mass
    C. Pancreatic hemorrhage or trauma

tion of pharmacologic doses of steroids to reduce brain swelling, high rate infusions of glucose-containing solutions (especially in patients with diabetes insipidus), and increased sympathetic activity associated with brain injury. Hyperamylasemia is concerning but reports have indicated that it has no meaningful influence on pancreas graft function posttransplant. The cause of hyperamylasemia due to pancreatitis, or due to pancreatic injury in the case of a donor with trauma, will be ruled out at the time of procurement.

The hemodynamic stability and need for inotropic support is an important consideration. This has more influence on the anticipated function of the kidney allograft than it does on initial endocrine function of the pancreas allograft in the case of an SPK transplant.

Perhaps the most important determinant of the suitability of the pancreas for transplantation is by direct examination of the organ during the surgical procurement. The experience of the procurement team is important for correct assessment of the suitability of the pancreas graft for transplantation. It is during procurement that judgment regarding the degree of fibrosis, adipose tissue, and specific vascular anomalies can be accurately assessed. Pancreata with heavy infiltration of adipose tissue are believed to be relatively intolerant of cold preservation, and carry with it the potential of a high degree of saponification due to reperfusion pancreatitis that follows revascularization. These organs may be more suitable for islet isolation.

The important vascular anomaly that must be evaluated during procurement is the occurrence of a replaced or accessory right hepatic artery originating from the superior mesenteric artery (SMA). The presence of a replaced right hepatic artery is no longer an absolute contraindication for the use of the pancreas for transplantation. Experienced procurement teams will be able to successfully separate the liver and the pancreas either in-situ, or on the backbench, without sacrificing quality of either organ for transplantation.

However, there are a few important caveats that determine if this is possible. First, it is important to emphasize that the pancreas is not a life-saving organ. Therefore, the highest priority must be to ensure an acceptable vascular supply to the liver allograft. The replaced right hepatic artery needs to be dissected down to the SMA. If the replaced right hepatic artery traverses deep into the parenchyma of the head of the pancreas requiring extensive dissection, this may preclude the pancreas for transplantation. The SMA is divided distal to the take-off of the replaced right hepatic artery preserving it intact on a short length of SMA with a carrel patch for the liver graft. Occasionally, there is a large inferior pancreaticoduodenal arterial branch vascularizing the head of the pancreas that originates proximal to the take-off of the replaced right hepatic artery. The inferior pancreaticoduodenal vessels are critical to vascularization of the head of the pancreas because the gastroduodenal artery is routinely ligated during the process of hepatic artery immobilization for the liver transplant. Therefore, in the case of a very proximal take-off of the inferior pancreaticoduodenal artery dividing the SMA at the appropriate location for proper liver procurement would significantly impair vascularization of the head of the pancreas and preclude its use for transplantation. Evaluation of the arterial vascularity of the pancreaticoduodenal allograft can be tested on the backbench by several methods: i) injection of Renografin® into the superior mesenteric artery or Y-graft and obtaining an x-ray; ii) intraarterial injection of fluorescein visualization with a Wood's lamp; and iii) performing a methylene blue angiogram.

The use of marginal and non-heart-beating donors for pancreas transplantation has been reported. There is a higher rate of ruling out the pancreas for transplantation at the time of procurement than in stable conventional organ donors. If the pancreas is deemed suitable, there is the added consideration of the effect of delayed kidney graft function in a uremic SPK candidate. The use of marginal and non-heartbeating donors for pancreas alone transplantation is selective, made on a case-by-case basis.

The use of living related and unrelated pancreas donors has also been described. A distal pancreatectomy is performed for a segmental pancreas transplant. Anecdotal cases of combined live donor partial pancreatectomy and nephrectomy have also been reported. These procedures are not widely performed and are confined to one or two pancreas transplant programs.

C. Procurement of the Pancreaticoduodenal Graft

There are several standard surgical methods for procurement of the pancreas for transplantation. The general principles are similar irrespective of the specific techniques utilized. The pancreas must be procured with an intact vascular supply that does not compromise the vascularity of the liver. The pancreas is procured with the spleen and duodenum intact. The organ is perfused with preservation solution and cold-stored. The donor iliac vessels and sometimes the portal vein are obtained for revascularization of the arterial supply.

There are two general methods of organ procurement. Many programs prefer to perform an en-bloc removal of the liver and pancreas together and separate the two organs at the backbench. Other programs prefer to perform a more deliberate dissection of the pancreas and liver by mobilizing the relevant vasculature prior to preservation. The liver and pancreas are separated in-situ. The relevant components of the in-situ procurement process are briefly described.

1. Long midline incision (+/- cruciate incision);
2. Mobilization of ascending colon, control of infrarenal aorta, and identification of superior mesentaric artery;
3. Control of supraceliac aorta;
4. Identification of hepatic artery (ligation of gastroduodenal artery), splenic artery, portal vein, and division of common bile duct;
5. Identification of replaced and/or accessory left and right hepatic arteries;
6. Exposure of the anterior aspect of the pancreas for visual and manual inspection;
7. Mobilization of the spleen by division of short gastric vessels and dissection of its ligamentous attachments;
8. Mobilization of head, tail, and body of the pancreas;
9. NG tube positioning into the proximal duodenum and irrigation of antibiotic solution;
10. Removal of NG tube and division of the proximal duodenum just distal to the pylorus;
11. Heparinization of the donor and infusion of intra-aortic (± intraportal) preservation solution;
12. Division of proximal jejunum, middle colic vessels, and superior mesenteric vessels distal to the pancreatic uncinate process;
13. Division of celiac, SMA, splenic arteries; and portal vein;
14. Procurement of liver, pancreaticoduodenosplenic allograft, and kidneys;
15. Procurement of donor iliac vessels;
16. Closure of incision.

D. BACKBENCH PREPARATION OF THE PANCREAS ALLOGRAFT

The backbench preparation of the pancreas allograft for transplantation requires careful and meticulous surgical technique to ensure a properly revascularized pancreas with adequate duodenum and minimal extraneous fibrotic or adipose tissue.[4] The pancreaticoduodenosplenic allograft is placed in a basin with chilled UW preservation solution. The duodenum should be opened, drained, and irrigated into a separate container. Some programs routinely culture the fluid and a small piece of duodenal tissue.

The main principles in allograft preparation are as follows (Fig. 7.4): to separate the spleen from the pancreas tail with secure ligatures on the large splenic vessels. Next it may be useful to cannulate the common bile duct with a 5F feeding tube to identify the location of the ampulla and ensure its center position as the proximal and distal duodenum are shortened to an appropriate length (the tube is removed). The staple line on the root of the mesentery is oversewn for reinforcement. The middle colic vessels are secured. A Y-graft is constructed utilizing the donor iliac artery bifurcation graft as end-to-end anastomoses on the splenic artery and superior mesenteric artery of the pancreas allograft. If a sufficient length of splenic artery can be mobilized, it is possible to perform a direct end-to-side anastomosis to the superior mesenteric artery. The portal vein is carefully mobilized to allow for appropriate length and determination if a short portal venous extension graft utilizing donor external iliac vein would be useful.

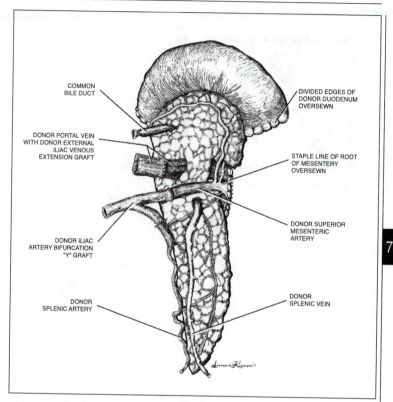

COMMON
BILE DUCT

DONOR PORTAL VEIN
WITH DONOR EXTERNAL
ILIAC VENOUS
EXTENSION GRAFT

DONOR ILIAC
ARTERY BIFURCATION
"Y" GRAFT

DONOR
SPLENIC ARTERY

DIVIDED EDGES OF
DONOR DUODENUM
OVERSEWN

STAPLE LINE OF ROOT
OF MESENTERY
OVERSEWN

DONOR SUPERIOR
MESENTERIC
ARTERY

DONOR
SPLENIC VEIN

7

Fig. 7.4. Backbench preparation of the pancreaticoduodenal allograft.

E. Pancreas Transplant Surgery

The surgical techniques for pancreas transplantation are diverse and there is no standard methodology used by all programs (Figs. 7.5-7.7).

The principles are consistent, however, and include providing adequate arterial blood flow to the pancreas and duodenal segment; adequate venous outflow of the pancreas via the portal vein; and management of the pancreatic exocrine secretions. The native pancreas is not removed. Pancreas graft arterial revascularization is typically accomplished utilizing the recipient right common or external iliac artery. The Y-graft of the pancreas is anastomosed end-to-side. Positioning of the head of the pancreas graft cephalad or caudad is not relevant with respect to successful arterial revascularization. There are two choices for venous revascularization, systemic and portal. Systemic venous revascularization commonly involves the right common iliac vein, or right external iliac vein following suture-ligation and division of the hypogastric veins. If portal venous drainage is utilized, it is necessary to dissect out the superior mesenteric vein (SMV) at the root of the mesentery. The pancreas portal vein is anastomosed end-to-side to a branch of the SMV. This may influence the methodology of arterial

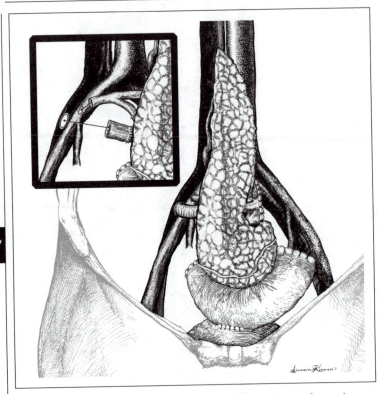

Fig. 7.5. Pancreaticoduodenal allograft with exocrine bladder-drainage and systemic venous drainage.

revascularization using a long Y-graft placed through a window in the mesentery to reach the right common iliac artery. Portal venous drainage of the pancreas is more physiologic with respect to immediate delivery of insulin to the recipient liver. This results in diminished circulating insulin levels relative to that in systemic venous-drained pancreas grafts. There has not been documented any clinically relevant difference in glycemic control.

Handling the exocrine drainage of the pancreas is the most challenging aspect of the transplant procedure. There are several methods. Pancreatic exocrine drainage is handled via anastomosis of the duodenal segment to the bladder or anastomosis to the small intestine. The bladder-drained pancreas transplant was a very important modification introduced about 1985. This technique significantly improved the safety of the procedure by minimizing the occurrence of intra-abdominal abscess from leakage of enteric-drained pancreas grafts. With the successful application of the new immunosuppressant agents, and the reduction of the incidences of rejection, enteric drainage of the pancreas transplants has enjoyed a successful rebirth.

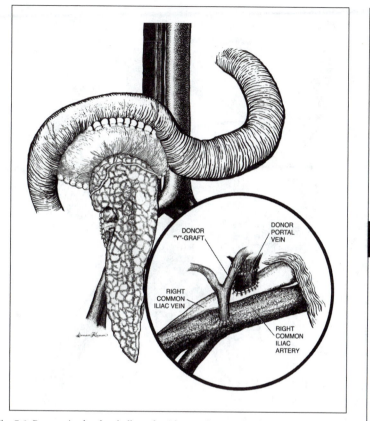

Fig. 7.6. Pancreaticoduodenal allograft with exocrine enteric-drainage and venous systemic drainage.

Enteric drainage of the pancreas allograft is physiologic with respect to the delivery of pancreatic enzymes and bicarbonate into the intestines for reabsorption. Enterically drained pancreases can be constructed with or without a Roux-en-Y. The enteric anastomosis can be made side-to-side or end-to-side with the duodenal segment of the pancreas. The risk of intra-abdominal abscesses is extremely low and the avoidance of the bladder-drained pancreas has significant implications with respect to the potential complications that include: bladder infection, cystitis, urethritis, urethral injury, balanitis, hematuria, metabolic acidosis, and the frequent requirement for enteric conversion. Currently, approximately 75% of pancreas transplants are performed with enteric drainage and the remainder with bladder drainage. Figures 7.8 and 7.9 show the annual number and relative proportions of recipients with enteric and bladder drainage of the exocrine pancreas allograft according to year and transplant category.

7

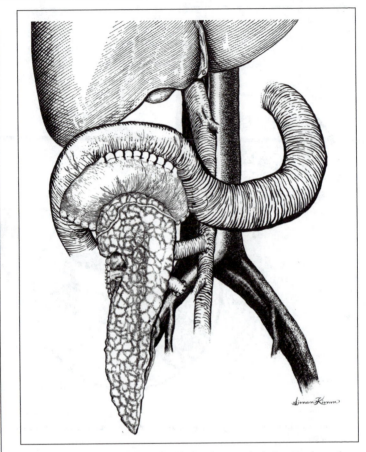

Fig. 7.7. Pancreaticoduodenal allograft with exocrine enteric-drainage and portal venous drainage.

The options of enteric versus bladder drainage depend on the choice of venous drainage and the clinical scenario of the pancreas transplant. For portally drained pancreas transplants, bladder drainage is not an option. For recipients of an SPK transplant, enteric drainage is the technique of choice because there is no urinary monitoring benefit and the morbidities as described above are significant. In the cases of PAK and PTA, bladder drainage has two important advantages: i.) urinary monitoring for rejection; and ii.) placement of the graft allowing access for percutanious biopsy for diagnosis of rejection. In the latter situations, the advantages of monitoring outweigh the morbidities associated with bladder drainage, at least in the short-term when the risk of immunologic graft loss is significant.

When the pancreas transplant is performed simultaneously with a kidney transplant, it is not uncommon for the kidney transplant to be implanted first. The

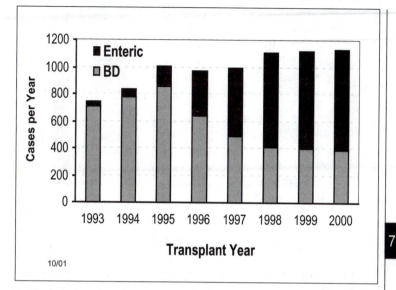

Fig. 7.8. Annual number of pancreas transplants according to exocrine drainage method.

kidney is based on the recipient left iliac vessels. Both organs may be transplanted through a midline incision and placed intraperitoneally.

Occasionally, it is necessary to consider placement of pancreas transplant based on the left iliac vessels because of a previously placed kidney transplant on the right side. In this sequential pancreas-after-kidney transplant procedure, the intra-abdominal approach is used. The pancreas is typically drained into the bladder if a pancreas transplant alone or pancreas-after-kidney transplant is performed in order to utilize measurement of urinary amylase as a method of detecting rejection. However, some programs have had good experience with enteric drainage of the pancreas transplant alone utilizing other markers for rejection, such as clinical signs and symptoms of pancreas graft pancreatitis and serum amylase or lipase levels coupled with biopsy.

F. Complications of Pancreas Transplantation

Surgical complications are more common after pancreas transplantation compared to kidney transplantation. Non-immunological complications of pancreas transplantation account for graft losses in 5-10% of cases. These occur commonly within 6 months of transplant and are as an important etiology of pancreas graft loss in SPK transplantation as acute rejection.

1. Thrombosis

Vascular thrombosis is a very early complication typically occurring within 48 hours, and usually within 24 hours of the transplant. This is generally due to venous thrombosis of the pancreas portal vein. The etiology is not entirely defined but is believed to be associated with reperfusion pancreatitis and the relatively low-flow

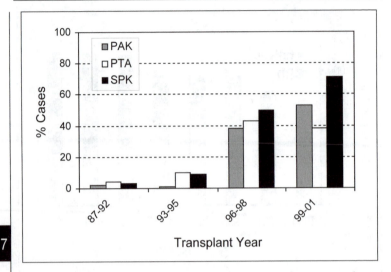

Fig. 7.9. Proportion of exocrine drained pancreas allografts according to year and transplant category.

state of the pancreas graft. To minimize graft thrombosis, prudent selection of donor pancreas grafts, short cold ischemia times, and meticulous surgical technique are necessary. Regarding the latter, it may be helpful to utilize the distal cava/proximal common iliac vein or the common iliac vein after ligation and division of the hypogastrics. Patients are often given anti-platelet agents and/or heparin during the perioperative period to minimize the occurrence of vascular thrombosis. The quality of the pancreas graft, the age of the donor, and the cold ischemia time also influence graft thrombosis rates. Approximately 3-5% of pancreas grafts will need to be removed because of portal venous thrombosis. Arterial thrombosis is less common and is usually associated with anastomosis to atherosclerotic vessels.

2. Transplant Pancreatitis

Pancreatitis of the allograft occurs to some degree in all patients postoperatively. It is common to see a temporary elevation in serum amylase levels for 48-96 hours posttransplant. These episodes are transient and mild without significant clinical consequence. Interestingly, it is common for patients receiving a simultaneous kidney-pancreas transplant to have a greater degree of fluid retention for several days posttransplant, compared to a kidney transplant alone recipient. Though not proven, this may be related to the graft pancreatitis that ensues during the perioperative period. The retained fluid is mobilized early post-operatively. It is important to minimize the risk of delayed kidney graft function by shortening cold ischemia time such that the retained third-spaced fluid may be rapidly eliminated to avoid an episode of heart failure or pulmonary edema.

3. Complications of the Bladder-Drained Pancreas Transplant

Bladder-drained pancreas transplantation is a safer procedure than enteric-drained pancreas transplantation with respect to avoiding the possibility of an intra-abdominal abscess. However, it is hampered by numerous less morbid complications. The pancreas transplant will eliminate approximately 500 cc of richly bicarbonate fluid with pancreatic enzymes into the bladder each day. Change in pH of the bladder accounts, in part, for a greater increase in urinary tract infections. In some cases, a foreign body such as an exposed suture from the duodenocystotomy acts as a nidus for urinary tract infections or stone formation.

Acute postoperative hematuria of the bladder-drained pancreas is usually due to ischemia/reperfusion injury to the duodenal mucosa or to a bleeding vessel on the suture line that is aggravated by the anti-platelet or anticoagulation protocols to minimize vascular thrombosis. These cases are self-limited but may require change in bladder irrigations and, if severe, cystoscopy to evacuate the clots. Occasionally it is necessary to perform a formal open cystotomy and suture ligation of the bleeding vessel intraoperatively. If relatively late chronic hematuria occurs, transcystoscopic or formal operative techniques may be necessary treatment.

Sterile cystitis, urethritis and balanitis may occur after bladder-drained pancreas transplantation. This is due to the effect of the pancreatic enzymes on the urinary tract mucosa. This is more commonly experienced in male recipients. Urethritis can progress to urethral perforation and perineal pain. Conservative treatment with Foley catheterization or operative enteric conversion are the extremes of the continuum of treatment. Figure 7.10 illustrates the surgical procedure of enteric conversion.

Metabolic acidosis routinely develops as a consequence of bladder excretion of large quantities of alkaline pancreatic secretions. It is necessary that patients receive oral bicarbonate supplements to minimize the degree of acidosis. Because of the relatively large volume losses, patients are also at risk of episodes of dehydration exacerbated by significant orthostatic hypotension.

Reflux pancreatitis can result in acute inflammation of the pancreas graft, mimicking acute rejection. It is associated with pain and hyperamylasemia. It is believed to be secondary to reflux of urine through the ampulla and into the pancreatic ducts. Often, the urine is found to be infected with bacteria. This frequently occurs in a patient with neurogenic bladder dysfunction. This complication is managed by Foley catheterization. Reflux pancreatitis will quickly resolve. The patient may require a complete workup of the cause of bladder dysfunction including a pressure flow study and voiding cystourethrogram. Interestingly, in older male patients, even mild hypertrophy of the prostate has been described as a cause of reflux pancreatitis. If recurrent graft pancreatitis occurs, enteric conversion may be indicated.

Urine leak from breakdown of the duodenal segment can occur and is usually encountered within the first 2-3 months posttransplant but can occur years posttransplant. This is the most serious postoperative complication of the bladder-drained pancreas. The onset of abdominal pain with elevated serum amylase, which can mimic reflux pancreatitis or acute rejection, is a typical presentation. A

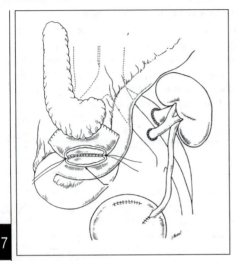

Fig. 7.10. Surgical procedure of enteric conversion. (Reprinted with permission from: Surgery, Vol 112, 1992).

7

high index of suspicion for urinary leak is necessary to accurately and swiftly make the diagnosis. Supporting imaging studies utilizing a cystogram or CT scanning is necessary to confirm the diagnosis. Operative repair is usually required with exploration. The degree of leakage can be best determined intraoperatively and proper judgment made whether direct repair is possible or more aggressive surgery involving enteric diversion or even graft pancreatectomy is indicated.

4. Complications of the Enteric-Drained Pancreas Transplant

The most serious complication of the enteric-drained pancreas transplant is that of a leak and intra-abdominal abscess. This serious problem usually occurs 1-6 months posttransplant. Patients present with fever, abdominal discomfort, and leukocytosis. A high index of suspicion is required to make a swift and accurate diagnosis. Imaging studies involving CT scan are very helpful. Percutaneous access of intra-abdominal fluid collection for gram stain and culture is essential. The flora is typically mixed with bacteria and often times fungus, particularly Candida. Broad-spectrum antibiosis is essential. Surgical exploration and repair of the enteric leak is necessary. A decision must be made whether the infection can be eradicated without removing the pancreas allograft. Incomplete eradication of the infection will result in progression to sepsis and multiple organ system failure. Peripancreatic infections can result in development of a mycotic aneurysm at the arterial anastomosis that could cause arterial rupture. Transplant pancreatectomy is indicated if mycotic aneurysm is diagnosed.

The occurrence of intra-abdominal abscess has been greatly reduced with greater recognition of the criteria for suitable cadaveric pancreas grafts for transplantation. Improved perioperative antibiosis, including anti-fungal agents, has contributed to the decreased incidence of intra-abdominal infection, as well. There is no convincing evidence that a Roux-en-Y intestinal reconstruction decreases its incidence. Perhaps the most significant contribution to reducing the incidence of intra-abdominal abscess is the efficacy of the immunosuppressive agents in reducing the incidence of acute rejection and thereby minimizing the need for intensive anti-rejection immunotherapy.

GI bleeding occurs after the enteric-drained pancreas from a combination of perioperative anticoagulation and bleeding from the suture line of the duodenoenteric anastomosis. This is self-limited and will manifest as diminished hemoglobin level associated with heme-positive or melanotic stool. Conservative management is appropriate, it is extremely unusual for reoperative exploration.

IMMUNOLOGICAL ASPECTS OF PANCREAS TRANSPLANTATION

A. IMMUNOSUPPRESSION FOR PANCREAS TRANSPLANTATION

The outcome of pancreas transplantation with respect to graft survival rates and rejection rates is most dependent upon the choice of immunotherapeutic agents employed. There is consensus that the risk of pancreas allograft rejection is much greater than that observed with kidney transplantation. The precise reasons are not well defined but likely involve greater immunogenicity of the pancreaticoduodenal graft. Recurrent autoimmune reactions are extremely rare. The majority of pancreas transplant programs are using induction therapy combined with microemulsion cyclosporine or tacrolimus, plus mycophenolate mofetil or sirolimus, and prednisone. This combination has significantly improved graft survival rates. The incidence of acute rejection has been reduced by more than half. The avoidance of induction therapy with this maintenance immunosuppression protocol is also associated with excellent patient graft survival rates but with a higher rate of acute rejection. There are steroid avoidance protocols described for pancreas transplantation, and reports of successful steroid withdrawal.

1. Trends in Induction Therapy in Pancreas Transplantation

Induction therapy is usually included in immunosuppressive protocols for recipients of whole-pancreas transplants. In fact, induction therapy is used with greater frequency in pancreas transplant recipients than for any other solid-organ recipients. One reason is the relatively higher risk of rejection observed for simultaneous pancreas-kidney (SPK), pancreas after kidney (PAK) and pancreas transplant alone (PTA) recipients, as compared with other solid organ transplants. The use of induction therapy in pancreas transplantation has been generally guided by practical experience, rather than by the results of formal randomized, prospective, multi-center trials. No FDA-approved immunosuppressive agents are on the market with a labeled indication to reduce rejection rates specifically in pancreas transplant recipients. Nonetheless, in 2001, ~ 81% of solitary pancreas (PAK and PTA) transplant recipients and over 75% of recipients of SPK transplants received induction therapy.[i]

---

[i] For comparison, the proportion of recipients of other solid organ transplants receiving induction therapy in 2001 is as follows: kidney ~ 60%; liver ~ 15%; intestine ~ 50%; heart ~ 45%; lung ~ 40%; and heart-lung ~ 75% [from SRTR (X)].

Over the past six years some interesting trends have been observed in the frequency and type of the induction therapy agent used in solitary pancreas (PAK and PTA) and SPK transplant recipients (Fig. 7.11).[5] For recipients of SPK transplants in 1996 and 1997, virtually all cases of induction therapy involved the use of either OKT3® or ALG. Beginning in 1998 the use of basiliximab increased from 6.6% to the current rate of 31%,, daclizumab from 13.3% to the current rate of 20.1%, and equine ATG from 0.1% to the current rate of 27.7%. The same trends were observed for recipients of a solitary pancreas transplant from 1996 through 1997 virtually 100% of the cases of induction therapy utilized either OKT3® or ALG (~ 50% of each). Since 1998-9, with the introduction of daclizumab, basiliximab, and equine ATG, the use of these three agents has supplanted those previous two. Among the three new agents, the proportion of solitary pancreas transplant recipients that received equine ATG has increased from 0.4% in 1998 to 55% in 2001.

2. Trends in Maintenance Therapy in Pancreas Transplantation

Maintenance immunosuppressive agents used for pancreas transplantation fall into the following categories: a) corticosteroids, b) calcineurin inhibitors (cyclosporine and tacrolimus), c) antimetabolites (azathioprine and mycophenolate mofetil), and d) other (rapamycin and Cytoxan). In 2001, solitary pancreas recipients received corticosteroids in 93% of cases, tacrolimus in 91% (cyclosporine 8%), mycophenolate mofetil in 74% (azathioprine 1%) and

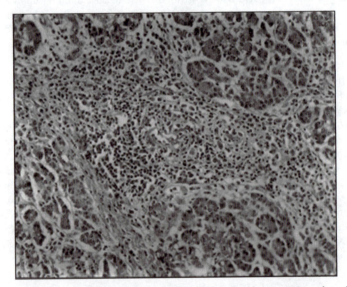

Fig. 7.11. Histopathology of acute pancreas allograft rejection demonstrating a dense inflammatory infiltrate involving septa and extending in to acinar tissue. (Reprinted with permission from: Solid Organ Transplant Rejection, Editor Solez, Publisher Marcel Dekker, Inc., 1996).

rapamycin in 19%. Therefore, in 2001, the most frequently used combination of maintenance therapy at discharge was tacrolimus, mycophenolate mofetil, and corticosteroids.

The dominant use of tacrolimus today represents a marked shift from earlier eras. In 1992-93 cyclosporine accounted for virtually 100% of the calcineurin inhibitor use in pancreas transplantation. In 2001, 92% of SPK transplant recipients received corticosteroids, 86% tacrolimus (14% cyclosporine), 82% mycophenolate mofetil, and 19% rapamycin. Based on these data, one can extrapolate that the most common maintenance immunosuppressive regimen used in SPK transplant recipients included corticosteroids, tacrolimus, and mycophenolate mofetil.

Trends in the uses of maintenance therapies over the past 7 years for SPK transplant recipients are depicted in Figures 7.12 and 7.13.[5] The use of tacrolimus rose to 86% in 2001. Because tacrolimus is used as a replacement for cyclosporine, cyclosporine usage has dropped from nearly 100% of cases in 1992 to only 14% of cases in 2001. Similar trends in the use of antimetabolites are seen with respect to azathioprine and mycophenolate mofetil. In the early 1990s azathioprine was used in nearly 100% of cases, dropping to 1% in 2001; mycophenolate mofetil usage grew from 25% in 1995 to 82% in 2001. From 2000 to 2001, sirolimus usage rose from 13% to 19% of cases.

Similar trends were observed for recipients of a solitary pancreas allogaft. The use of tacrolimus has increased yearly aince 1995 and reached 91% in 2001. The FDA approved mycophenolate mofetil for marketing for kidney transplantation in 1995, and it was used in only 14% of solitary pancreas transplant cases that year (azathioprine was used in 72% of cases). However, within one year nearly 80% of solitary pancreas transplant recipients received mycophenolate mofetil, with only 12% receiving azathioprine. The use of azathioprine has diminished yearly and dropped to 1% usage in 2001. In 1999, the FDA approved the use of sirolimus for marketing for kidney transplantation. For pancreas transplantation, this agent is usually used in combination with a calcineurin inhibitor, and as a substitute for an antimetabolite. The use of sirolimus has been relatively slow to penetrate the market, compared to the rapid spread of tacrolimus and mycophenolate mofetil usage. In 2000 and 2001, sirolimus was used for 10% and 19% of solitary pancreas cases, respectively.

B. PANCREAS ALLOGRAFT REJECTION

The early clinical presentation of pancreas allograft rejection is much different than that of kidney rejection. An understanding of the kinetics of the tissue injury during acute rejection of the pancreas allograft is essential to making a timely diagnosis. Destruction of the (beta cells occur relatively late following initial injury of the acinar tissue. Therefore, the diagnosis of pancreas graft rejection by hyperglycemia is a late and often irreversible situation. Detection of changes in acinar cell function is the basis for early suspicion of pancreas graft rejection. The graft is usually inflamed and patients experience pain and discomfort around the graft due to peritoneal irritation. This, coupled with elevation in the serum amylase or lipase, and if bladder-drained, reduction in urinary amylase, may be the

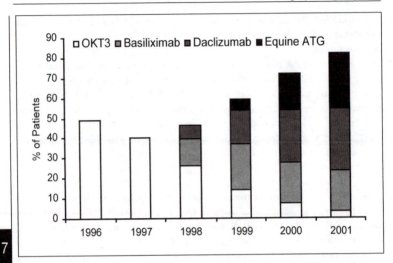

Fig. 7.12. Use of immunosuppressive agents for induction therapy in SPK transplant recipients (Source, 2002 OPTN/SRTR Annual Report).

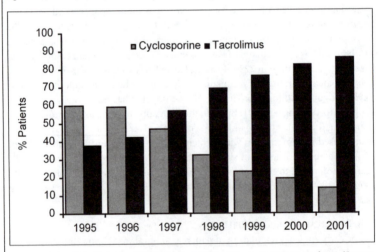

Fig. 7.13. The changing use of calcineurin inhibitors for new SPK transplants (Source, 2002 OPTN/SRTR Annual Report).

initial presentation of acute rejection. Graft pancreatitis and urinary leak of a bladder-drained pancreas can present in a similar manner. Often, patients will require bladder catheterization to differentiate graft rejection from reflux pancreatitis.

The gold standard for confirming the diagnosis of pancreas graft rejection is pancreas graft biopsy. The biopsy may be performed by several methods including the percutaneous approach, transcystoscopic biopsy in a bladder-drained pancreas, or open surgical biopsy. The usefulness of pancreas graft biopsy to confirm the clinical suspicion of rejection is so important that the surgical procedure of pancreas transplantation should include consideration of the intra-abdominal location of the pancreas to make it accessible for percutaneous biopsy. This is especially important in pancreas transplant alone and pancreas-after-kidney transplant procedures. Figure 7.14 shows the histologic features of acute pancreas graft rejection.

In the situation of a simultaneous pancreas-kidney transplant, it is the kidney allograft that is the best indicator of a rejection reaction. Rejection of the kidney allograft will manifest as a rise in serum creatinine. This will prompt ultrasound and biopsy of the kidney allograft. If rejection is present, anti-rejection therapy is instituted. If there is a concurrent pancreas graft rejection process, the anti-rejection therapy will reverse the process in both organs. It is extremely uncommon for isolated pancreas allograft rejection to occur in a setting of a simultaneous kidney-pancreas transplant. However, this may occur in 1-2% of cases and the diagnosis is made by kidney and pancreas transplant biopsies. Treatment of the pancreas alone rejection is guided by its severity and requires pulse steroids or anti-lymphocyte immunotherapy. The success rates for reversing pancreas allograft rejection are very high, in excess of 90%, if diagnosed promptly. There was a time when incidence of pancreas transplant rejection was greater than in kidney transplant-alone recipients. With the application of new immunosuppressive agents, however, the incidence of pancreas rejection has been reduced from approximately 80% to less than 30%.

RESULTS OF PANCREAS TRANSPLANTATION
The results of pancreas transplantation are typically described in terms of patient survival and pancreas graft survival. The definition of patient survival is ob-

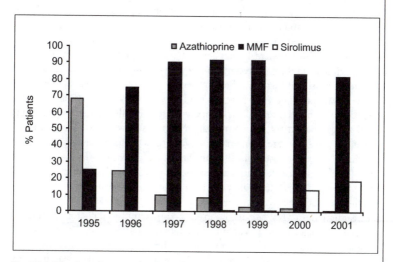

Fig. 7.14. The changing use of antimetabolites and sirolimus for new SPK transplants (Source, 2002 OPTN/SRTR Annual Report).

vious, pancreas graft losses are defined as: i) patient death with a functioning graft; or ii) loss of insulin independence irrespective of whether the pancreas allograft is in place or removed. The most valuable and complete information on the results of pancreas transplantation comes from the Scientific Registry of UNOS and the International Pancreas Transplant Registry. All cases of pancreas transplantation in the U.S. have been collected since October 1997. Single center experiences with pancreas transplantation have been valuable in reporting results of specific technical and immunosuppressive protocols. Very importantly, the effect of pancreas transplantation on secondary complications of diabetes, such as diabetic nephropathy, retinopathy, neuropathy, and quality of life have been fascinating.

A. OUTCOMES OF PATIENT AND GRAFT SURVIVAL

The results of SPK transplantation in terms of patient and graft survival have shown steady improvement over time. According to results from the SRTR and the International Pancreas Transplant Registry patient survival in SPK recipients has increased from 90% to 95% over the past 10 years. Pancreas graft functional survival has also improved over this 10-year interval from 74% to 84% at 1 year. There is no clinically significant difference in pancreas graft outcome in bladder-drained pancreases versus the enteric-drained pancreas. There is also no clinically significant difference in outcome in systemic venous drainage versus portal drainage. The immunologic risk for graft loss for technically successful cases of SPK transplantation has decreased over time. The current rate of immunologic loss is only 2% at one year. Relative risk factors for pancreas graft loss in SPK recipients have been determined and include increasing recipient age, (> 45 years), prolonged preservation time (> 24 hours), and positive effects were shown for the use of mycophenolate mofetil. The relative risk factors for pancreas graft loss in solitary pancreas recipients have been determined and include increasing recipient age, (<45 years), and prolonged preservation time (> 24 hours), and positive effects were shown for the use of mycophenolate mofetil and tacrolimus.

Figure 7.15 shows patient, kidney and pancreas graft survival rates in SPK transplant recipients in the most recent era analyzed (1998-2002) by the International Pancreas Transplant Registry. These are the best outcomes reported to date with one year patient, kidney, and pancreas graft survival rates of 95%, 92%, and 84%, respectively. Single center reports from the most active SPK transplant programs show wide variability of kidney and pancreas graft survival rates (Table 7.2).[6]

Figure 7.16 shows the comparative survival rates of the pancreas graft among the three transplant groups for the current era analyzed. The survival rates have been the highest recorded with some single center reports describing even better results. For pancreas after kidney (PAK) transplantation, pancreas graft survival has shown steady improvement over the 10-year interval 1993 through 2002 from a 1-year patient survival rate of 65% to 82%. The technique of bladder drainage is associated with a current 1-year graft survival rate of 85% versus 75% for enteric drainage. The immunologic risk for graft loss for the technically successful cases has been reduced to only 3-5% at 1 year. The relative risks for pancreas graft loss from technical failures (7% in first year posttransplant) include enteric exocrine

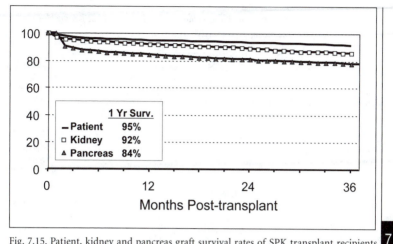

Fig. 7.15. Patient, kidney and pancreas graft survival rates of SPK transplant recipients (n=3885, 1/1/1998-7/1/2002). (Source, International Pancreas Transplant Registry, 1998)

drainage, increasing preservation time (> 24 hours), and increasing BMI (body mass index, >25 kg/m2).

For patients receiving a pancreas transplant alone (PTA) the pancreas graft functional survival rates over the past 10 years has shown significant improvement from 55% to 80%. The technique of bladder drainage is associated with a current 1-year graft survival rate of 81% versus 74% for enteric drainage. The immunologic risk for graft loss for the technically successful cases has been reduced to only 5-7% at 1 year. The relative risks for pancreas graft loss from technical failures (9% in first year posttransplant) include increasing age (> 45 years) and increasing PRA rate (>20%).

B. Effect of Pancreas Transplantation on Secondary Complications of Diabetes

Recipients of a successful pancreas transplant maintain normal plasma glucose levels without the need of exogenous insulin therapy. This results in normalization of glycosylated hemoglobin levels and a beneficial effect on many secondary complications of diabetes. The durability of the transplanted endocrine pancreas has been established with the demonstration that normalization of glycosylated hemoglobin is maintained for as long as the allograft functions. The potential lifespan of the transplanted pancreas is not precisely known since survivors with functioning pancreas transplants are now greater than 20 years posttransplant, and still going. The implications of prolonged normalization of glycemia and glycosylated hemoglobin levels are significant with respect to patients' quality of life, kidney structure, and motor and sensory and nerve function.

The quality of life of pancreas transplant recipients have been well-studied. Patients with a functioning pancreas graft describe their quality of life and rate their health significantly more favorably than those with nonfunctioning pan-

Table 7.2. One-year pancreas and kidney allograft survival rates in SPK transplant recipients at the top 5 most active centers in the U.S.[1]

PANCREAS TRANSPLANT				
Center	N	Actual Graft Survival	Expected Graft Survival[2]	P-Value
U.S.	2,244	84.96%		
Univ. of Wisconsin Hosp. (Madison)	130	88.94%	83.77%	0.115
Fairview Univ. Med. Center (Minneapolis)	95	73.08%	78.52%	0.255
Northwestern University (Chicago)	82	95.59%	83.51%	0.010
Ohio State Univ. Hosp. (Columbus)	77	87.70%	85.88%	0.716
Jackson Memorial Hosp. (Miami)	68	94.12%	86.08%	0.072

KIDNEY TRANSPLANT				
Center	N	Actual Graft Survival	Expected Graft Survival[2]	P-Value
U.S.	2,244	91.90%		
Univ. of Wisconsin Hosp (Madison)	130	95.11%	91.54%	0.197
Fairview Univ. Med. Center (Minneapolis)	95	77.95%	89.17%	0.010
Northwestern University (Chicago)	82	97.28%	91.95%	0.079
Ohio State Univ. Hosp. (Columbus)	77	93.04%	91.90%	0.900
Jackson Memorial Hosp. (Miami)	68	97.06%	92.25%	0.228

[1] Adult (age >18 years) recipients transplanted between 7/1/99 - 12/31/01.
[2] Based on SRTR data on U.S. graft failure rates adjusted for donor and recipient characteristics (see http//www.ustransplant.org).

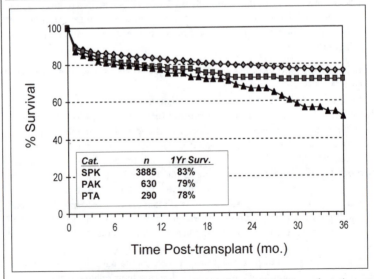

Cat.	n	1Yr Surv.
SPK	3885	83%
PAK	630	79%
PTA	290	78%

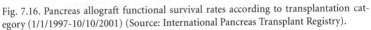

Fig. 7.16. Pancreas allograft functional survival rates according to transplantation category (1/1/1997-10/10/2001) (Source: International Pancreas Transplant Registry).

creas grafts. Satisfaction encompasses not only the physical capacities, but also relate to psychosocial and vocational aspects, as well. The functioning pancreas graft leads to even better quality of life when compared to recipients of a kidney transplant alone. Virtually all patients of a successful pancreas transplant report that managing their life, including immunosuppression, is much easier with the transplant, than prior to transplantation. Successful pancreas transplantation will not elevate all patients with diabetes to the level of health and function of the general population, but the transplant recipients consistently report a significantly better quality of life than do patients who remain diabetic.

The development of diabetic nephropathy in transplanted kidneys residing in patients with type 1 diabetes has been well-established. There is marked variability in the rate of renal pathology, including mesangial expansion and a widening of the glomerular basement membrane, in patients with Type 1 diabetes and a kidney transplant alone. The onset of pathological lesions can be detected within a few years of kidney transplantation. Clinical deterioration of renal allograft function can lead to loss 10-15 years posttransplant. A successful pancreas transplant prevents glomerular structure changes of kidney allografts in patients with type 1 diabetes. This has been observed in transplanted kidneys of patients undergoing SPK transplantation, as well as in kidneys of recipients undergoing PAK transplantation. These studies provide evidence of the efficacy of normalizing blood glucose and glycosylated hemoglobin levels to prevent the progression of diabetic glomerulopathy in renal allografts.

Furthermore, it has been established that a successful pancreas transplant will halt or reverse the pathology in the native kidneys of patients with Type 1 diabetes and very early proteinuria.[7] The pancreas transplant recipients all had persistently normal glycosylated hemoglobin values after transplantation for 5-10 years. The thickness of the glomerular and tubular basement membranes and mesangial volume steadily decrease over a 10 year interval. These early studies have important implications for the role of pancreas transplant alone in patients with type 1 diabetes and very early changes in native renal function.

Successful pancreas transplantation has been shown to halt, and in many cases, reverse motor-sensory and autonomic neuropathy 12-24 months posttransplant. This has been studied most extensively in recipients of SPK transplants. This raises the possibility that improvement of diabetic neuropathy occurs, in part, to improvement of uremic neuropathy. However, pancreas transplantation alone in preuremic patients has also been shown to result in improvement in diabetic neuropathy. Many patients express subjective improvements of peripheral sensation 6-12 months post-pancreas transplantation. Very interestingly, the effect of reversal of autonomic neuropathy in patients with type 1 diabetes with pancreas transplantation has been associated with better patient survival rates than patients with failed or no transplantation.

Pancreas transplantation does not have an immediate dramatic beneficial effect on pre-established diabetic retinopathy. Retinopathy appears to progress for at least 2 years following transplantation of the pancreas, but begins to stabilize in

3 to 4 years compared to diabetic recipients of a kidney transplant only. Longer term studies of 5-10 years, similar to that described above, have not been reported.

Other beneficial effects of the secondary complications of diabetes have been less rigorously studied, but improvement in the microcirculation and blood pressure control have been described. The effect of amelioration of diabetes by successful pancreas transplantation on the incidence, or progression of coronary artery disease is currently preliminary and inconclusive. However, it is believed that pancreas transplantation removes an important risk factor to the development of coronary artery disease, and in conjunction with control of blood pressure, hyperlipidemia, and avoidance of smoking, a beneficial impact is predicted

The overall purpose of pancreas transplantation is to improve the quality of life of patients with type 1 diabetes and end-stage renal failure above that which can be accomplished by kidney transplantation alone. The amelioration of diabetes, and with it the absence of the need for frequent blood glucose monitoring, exogenous insulin therapy, diet and exercise control, unequivocally resolves the primary complication of the disease. Many of the secondary complications are halted, and in many cases, reversed. However, it is very dependent upon the severity of the disease prior to transplantation, and the relatively subjective assessment of success. The combination of improved patient and graft survival rate that has been demonstrated in the short-term with the application of the new immunosuppression agents, will translate into greater survival of pancreas graft function long-term. This will contribute greatly to quality of life issues and facilitate ongoing research that has associated control of glycemia and normalized glycosylated hemoglobin levels with improvements in numerous secondary complications of diabetes.

REFERENCES

1. American Diabetes Association, Economic Costs of Diabetes in the U.S. in 2002. Diabetes Care 2003; 26:917.
2. Ojo AO, Meier-Kriesche HU, Hanson JA, Leichtmen A , Magee JC, Cibrik D, Wolfe RA, Port FK, Agodoa LY, Kaufman DB, and Kaplan B. Impact of simultaneous pancreas-kidney transplantation on long-term patient survival. Transplantation 2001; 71:82.
3. URREA; UNOS. 2002 Annual Report of the U.S. Organ Procurement and Transplantation Network and the Scientific Registry of Transplant Recipients: Transplant Data 1992-2001 [Internet]. Rockville (MD): HHS/HRSA/OSP/DOT; 2003 [modified 2003 Feb 18; cited 2003, April 3]. Available from: http://www.optn.org/data/annualReport.asp.
4. Mizrah S, Jones J, Bentley FR. Preparing for pancreas transplantation; donor selection, retrival technique, preservation, and and back-table preparation. Transplantation reviews. 1996; 10:1.
5. 2002 Scientific Registry of Transplant Recipients Report on the State of Transplantation. Am J Transplant 3(Suppl 4), 2003.
6. http://ustransplant.org/center-adv.html
7. Fioretto P, Steffes MW, Sutherland DER, Goetz F, Mauer MM. Reversal of lesions of diabetic nephropathy after pancreas transplantation. NEJM 1998; 339:69.

Islet Transplantation

Dixon B. Kaufman, Bernhard J. Hering
Illustrations by Simon Kimm

INTRODUCTION

The incidence of diabetes in the U.S. is increasing with at least 1 million persons affected with type 1 disease.[1] The life threatening complications[2] are preventable by maintaining euglycemia according to the Diabetes Control and Complication Trial and its follow-up studies.[3,4] Despite the development of improved means of glucose monitoring and insulin delivery, maintaining even near-normal glucose levels in most patients with diabetes is difficult and complicated by occasional severe hypoglycemic episodes. Developing a means of achieving insulin independence and good glycemic control early in the disease prior to the onset of complications would represent a major therapeutic advance in the treatment of this debilitating disease.

Pancreas transplantation is able to correct the metabolic abnormalities of diabetes. There is now great enthusiasm for developing islet transplantation as a less morbid and potentially more broadly applicable therapy. Small pilot studies have recently demonstrated the feasibility of islet transplantation to ameliorate type 1 diabetes in select patients. Future emphasis on large-scale application must confront the profound challenges of expanding the sources of insulin producing tissue and diminishing the need for chronic systemic immunosuppression. Current breakthroughs have emerged from persistent experimentation over the past 30 years to provide a solid foundation and clear vision on which the successes of tomorrow will be built.

8

RATIONALE OF ISLET TRANSPLANTATION FOR PATIENTS WITH TYPE 1 DIABETES

Pancreas transplantation is a near perfect means of normalizing glycosylated hemoglobin levels — the most important determinant of stabilizing or reversing microvascular complications.[5] Successful pancreas transplantation will result in a durable normoglycemic and insulin-independent state[6] that can reverse the diabetic changes in the native kidneys of patients with early diabetic nephropathy;[7] reverse peripheral sensory neuropathy;[8] stabilize advanced diabetic retinopathy;[9] and significantly improve the quality and duration of life.[10,11]

However, there are important considerations with pancreas transplantation that currently preclude it as therapy for all patients with type 1 diabetes. It is unrealistic that all patients with diabetes could be transplanted to allow for early intervention in the disease process. There are too few organs (only about 1400 cases are performed annually in the U.S.). Pancreas transplantation involves significant surgery that

Organ Transplantation, 2nd edition, edited by Frank P. Stuart, Michael M. Abecassis and Dixon B. Kaufman. ©2003 Landes Bioscience.

precludes consideration in patients with significant co-morbid medical conditions (i.e., cardio- and peripheral vascular disease). Lifelong immunosuppression is required to prevent graft rejection. Transplanting the whole pancreas is not necessary to achieve an insulin independent state since it is only the islets that are required for glucose homeostasis. In fact, excluding the exocrine pancreas may avoid some of the complications of pancreas transplantation.

In the early 1970s transplantation of isolated pancreatic islets for treatment of type 1 diabetes in humans was seriously considered when the technique was proven feasible in small animal models.[12-15] The demonstration that islet transplantation ameliorates the basic metabolic defects of the hyperglycemic state and stabilizes or reverses early secondary lesions provides a strong impetus and rationale for pursuit of such an approach in humans. Importantly, many of the factors that limit application of whole pancreas transplantation for treating diabetes are more likely to be overcome if isolated islets are used as the source of endocrine tissue. Therefore, islet transplantation is a treatment option that has generated great enthusiasm and is being developed for a potentially broader scope of patients.

HUMAN CLINICAL TRIALS OF ISLET TRANSPLANTATION

The pre-modern era (1990-2000) of islet transplantation might be considered to have started when the technical methods of isolating and purifying human islets on a relatively large-scale were described.[16] Using these methods, Scharp et al reported temporary insulin independence and near normal blood glucose levels after transplantation of approximately 800,000 isolated, purified islets in a patient with type 1 diabetes.[17] Shortly thereafter, the Edmonton group reported long-term (>1 year) insulin independence in 2 islet transplant recipients receiving 243,000 and 368, 000 islets, respectively.[18] These reports were among the first to demonstrate the proof-of-principle that beta cell replacement therapy by means of islet transplant could eliminate the need for exogenous insulin therapy in select patients with diabetes. During this time numerous successes in islet autotransplantation for patients undergoing total pancreatectomy for benign disease were also reported.[19] However, despite overcoming many of the technical challenges of isolating and purifying islets from the human pancreas,[20] allogeneic islet transplants for treatment of type 1 diabetes was seldom effective.

According to the International Islet Transplant Registry, of the 355 adult islet allograft transplants performed in type 1 diabetic recipients from 1990 through 1999, only 19% of recipients achieved insulin independence for >7 days, and only 11% maintained it at 1-year follow-up.[21] The failure to achieve and maintain insulin independence on a more consistent basis was ascribed to 3 obstacles: 1) the transplantation and engraftment of an inadequate mass of viable islets, 2) the use of diabetogenic immunosuppressive regimens, and 3) early and late graft loss due to nonspecific and allospecific immune mechanisms of injury, respectively, as well as autoimmune recurrence.[22]

Careful analysis of the few successes showed that islet allotransplantation could succeed if: i) the transplanted mass of islets exceeded >8000 islets/kg of body weight (usually achieved by pooling islets from several cadaver donor pancreases); ii) implantation was intraportal; iii) purity was >50%; and iv) anti-T cell induction therapy was used. The percent of patients that became insulin independent (>1 week) after receiving islet transplants fulfilling these "state of the art" criteria was closer to 35%.

Although the majority of patients with type 1 diabetes still failed to achieve insulin independence following "state of the art" islet allotransplantation procedures, in 80% of those cases C-peptide levels were above 1 ng/ml for over 1 month. Therefore, insulin dependence in the face of C-peptide secretion indicated that variables other than the mass of transplanted islets were critical factors affecting outcome. It became realized that an important variable that was different between the auto- and allotransplant settings, besides underlying autoimmunity, that accounted for the inferior results was the differential requirement for systemic immunosuppression. Some of these agents, corticosteroids in particular, impart a deleterious effect on the functional efficiency of islet grafts.[23] Application of a new combination of maintenance agents without corticosteroids involving tacrolimus and sirolimus[24,25] combined with an IL-2 receptor antagonist induction agent, and repeated infusions of purified islets, resulted in a dramatic breakthrough.

The clinical trial results reported by the University of Alberta group in Edmonton in July 2000 marked a turning point in the history of islet transplantation.[26] Shapiro et al. transplanted an adequate mass of islets by performing sequential transplants of islets from 2 to 4 donors, reduced the metabolic demand placed on transplanted islets by avoiding glucocorticoids and high-dose calcineurin inhibitors, and prevented immunologic graft loss by administering the synergistic immunosuppressants sirolimus and tacrolimus. The Edmonton protocol included all of the "state-of-the-art" techniques associated with success, while also addressing many of the previously mentioned obstacles. The outcome resulted in restored normoglycemia and insulin independence in 7 of 7 type 1 diabetic patients who previously had labile diabetes and hypoglycemia unawareness.[26] Recent publications of the Edmonton experience include 24 patients. The insulin independence rates are 87.5% at 1 year and 70% at 2 years.[27] More detailed information on clinical outcomes and metabolic test results with the Edmonton protocol was published in 2 follow-up reports.[28,29] The total experience through March 2003 exceeds 50 recipients.

Importantly, the validity of the Edmonton results has been strengthened by confirmatory findings reported by additional institutions, including diabetes reversal after islet transplants in patients with established kidney grafts[30] and after transplants of islets prepared from a non-heart-beating donor.[31] A consortium of islet transplant centers from North America and Europe, supported by the Immune Tolerance Network of the NIH, are also showing the reproducibility of the Edmonton results.

INDICATIONS AND CONTRAINDICATIONS TO ISLET TRANSPLANTATION

The indications for islet transplantation are listed in Table 8.1. Very few islet transplants are performed annually in the U.S. Those that are performed are done under rigorous study protocols at academic institutions heavily supported by clinical research resources. In the pre-Edmonton era, virtually all islet allotransplants were performed in association with kidney transplants. In the mid-1990s, solitary islet transplants began to be evaluated in patients whose type 1 diabetes was complicated by hypoglycemia unawareness.[32] Since publication of the Edmonton results in 2000,[26] the vast majority of islet allotransplants have been performed in patients with hypoglycemia, metabolic lability, or progressive microvascular diabetes complications. Today, the majority of islet transplants are performed in recipients without renal dysfunction. This is referred to as an islet-transplant-alone (ITA) procedure. The emphasis on the ITA approach for treating diabetes is in contrast to whole pancreas transplantation in which 85-90% involve a simultaneous kidney transplant.

Islet Transplantation Alone (ITA). The primary indication for islet transplantation is in patients with type 1 diabetes with extremely labile disease, such that routine daily activities are interrupted by episodes of extremely high blood glucose levels resulting in frequent emergency room and inpatient hospitalizations for diabetic ketoacidosis. They have significant difficulty avoiding and sensing extremely low blood sugar levels, referred to as hypoglycemic unawareness, that result in unconsciousness without the warning. This is a devastating condition that affects employment, the ability to have a license to drive, and concern about suffering lethal hypoglycemia while asleep. In this situation the risk of chronic immunosuppression is judged to be less than the current diabetic condition ineffectively managed by exogenous insulin therapy.

Current recipient inclusion criteria for islet transplant alone procedures include patients with type 1 diabetes mellitus for ≥5 years experiencing at least 1 of the following problems despite conscientious insulin management efforts in close cooperation with an endocrinologist: 1) Reduced awareness of hypoglycemia, or the clinical manifestation of hypoglycemia-associated autonomic failure;[33,34] 2) Metabolic lability or instability characterized by 2 or more episodes of severe hypoglycemia OR 2 or more hospital admissions for diabetic ketoacidosis during the previous 12 months; 3) Progressive secondary complications:
- progressive nephropathy, defined by a confirmed rise of microalbuminuria over at least 3 months, beginning anytime within the past 2 years, despite the use of an ACE inhibitor;
- autonomic neuropathy with symptoms consistent with gastroparesis, postural hypotension, neuropathic bowel or bladder, or persistent or progressive severe, peripheral, painful neuropathy not responding to usual management;
- a minimum of a 3-step progression using the Early Treatment Diabetic Retinopathy Study (ETDRS) grading system, or an equivalent progression as documented by an ophthalmologist familiar with diabetic retinopathy.

Thus, currently applied inclusion criteria include a small subgroup of patients with type 1 diabetes. For the subgroup of patients unable to continue intensive insulin therapy because of recurrent severe hypoglycemia, beta cell replacement therapy via islet (or pancreas) transplantation may be the only approach to achieving euglycemia.

Table 8.1. Indications for islet transplantation

Type 1 diabetes ≥ 5 years associated with at least one of the following complications:
1. Reduced awareness of hypoglycemia;
2. Metabolic lability or instability characterized by 2 or more episodes of severe hypoglycemia OR 2 or more hospital admissions for diabetic ketoacidosis during the previous 12 months;
3. Progressive secondary complications:
   i.   progressive nephropathy despite the use of an ACE inhibitor;
   ii. autonomic neuropathy with symptoms consistent with gastroparesis, postural hypotension, neuropathic bowel or bladder, or persistent or progressive severe, peripheral, painful neuropathy not responding to usual management;
   iii. progression of diabetic retinopathy
4. Kidney transplant recipient with adequate renal function on immunosuppression

---

The contraindications and relative contraindications for islet transplantation are listed in Tables 8.2 and 8.3, respectively. The contraindications to islet transplantation include the majority of patients with type 1 diabetes that have normal renal function and do not exhibit a brittle course, or hypoglycemic unawareness. For patients that do have an indication for islet transplantation, it is important to rule out significant medical contraindications in a similar manner as applies to other areas of transplantation. These issues include: recent malignancy, active or chronic untreated infection, advanced forms of major extrarenal complications (i.e., coronary artery disease), life expectancy of less than 1 year, sensitization to donor tissue, noncompliance, active substance abuse, uncontrolled psychiatric disorder.

The criteria regarding body weight, body mass index, and pre-transplant insulin are important considerations. The critical number of human allogeneic islets required to achieve insulin independence is generally perceived to be 8,000 islet equivalents per kilogram recipient body weight. The number of islet equivalents, using currently available techniques, that can be prepared from a high-quality human cadaver donor pancreas deemed suitable for single-donor islet allotransplants, rarely exceeds 500,000. The number of islet equivalents retrievable from a less-than-optimal cadaver pancreas accepted for a 2-donor islet transplant protocol is typically 300,000. For these reasons, islet transplants are restricted to recipients with body weights less than 75 kg (expected to require a more than 600,000 islet equivalents for diabetes reversal).

Increasing evidence suggests that insulin resistance is more likely in type 1 diabetic individuals with a high body mass index and high insulin requirements. Insulin resistance increases the metabolic demand placed on transplanted islets. That fact may not matter in the setting of a vascularized whole organ pancreas transplant, but insulin resistance can spoil the ability of a marginal mass of transplanted and engrafted islets to reverse type 1 diabetes. Therefore, islet transplants are currently restricted to a subgroup of type 1 diabetic individuals with a higher probability of insulin independence.

Table 8.2. Contraindications for islet transplantation age less than 18 years

Body weight >75 kg

Body mass index >26 (female), >27 (male) kg/m^2

Insulin requirement of >0.7 IU/kg/day or >50 IU per day (whichever is less)

Positive C-peptide response ($\geq$0.2 ng/mL) to oral or intravenous glucose tolerance testing

Creatinine clearance <80 ml/min/1.73 m^2

Positive pregnancy test or failure to follow effective contraceptive measures

Active infection including hepatitis C, hepatitis B, HIV, or tuberculosis

Invasive aspergillus infection during the previous 12 months

History of malignancy (except for treated squamous or basal cell carcinoma of the skin)

Active alcohol or substance abuse

History of nonadherence to prescribed medical regimens

Psychiatric disorder that is unstable or uncontrolled on current medication

Inability to provide informed consent

Severe coexisting cardiac disease, characterized by any 1 of these conditions: recent myocardial infarction (within past 6 months); angiographic evidence of noncorrectable coronary artery disease; evidence of ischemia on functional cardiac exam; left ventricular ejection fraction <30%

Baseline liver function tests outside of normal range or history of significant liver disease

Gallstones or hemangioma in liver, on baseline ultrasound examination

History of coagulopathy or medical condition requiring long-term anticoagulant therapy

Active peptic ulcer disease

Severe gastrointestinal disorders potentially interfering with the ability to absorb oral medications

Islet After Kidney (IAK) Transplantation. The recent success of islet transplantation-alone for non-uremic patients has demonstrated the feasibility and practicality of the procedure using a new corticosteroid-free immunosuppressive protocol referred to as the "Edmonton Protocol".[26] Individuals with type I diabetes and a successful kidney allograft are especially appropriate as candidates for a sequential islet transplant procedure because they already receive chronic immunosuppression. The added risk of a subsequent islet transplant is minimal compared to non-uremic patients with diabetes receiving new immunosuppression in addition to the transplant procedure.

Candidates for IAK transplants are distinguished from non-uremic individuals with type I diabetes in several respects. First, IAK transplant recipients may have a greater degree of diabetes-related co-morbidities such as cardio-, cerebro- and peripheral vascular diseases that preclude them from consideration of operative whole pancreas after kidney (PAK) transplantation because of excessive operative

Table 8.3. Relative contraindications for islet transplantation

Serum creatinine >1.3 mg/dL (female), >1.5 mg/dL (male)

History of panel-reactive anti-HLA antibodies >20%

Negative screen for Epstein-Barr virus (EBV) by an EBV nuclear antigen (EBNA) method

Active cigarette smoking (must be abstinent for 6 months)

Baseline hemoglobin <11.7 g/dL (female), <13 g/dL (male); lymphopenia (<1,000/mL), leukopenia (<3,000 total leukocytes/mL), or platelets <150,000/mL

Severe allergy requiring acute (within 4 weeks of baseline) or chronic treatment, or hypersensitivity to protocol regulated treatment products

Hyperlipidemia (fasting LDL cholesterol >130 mg/dL; and/or fasting triglycerides >200 mg/dL)

Addison's disease

Current treatment for a medical condition requiring chronic use of systemic steroids

risk. That is not a trivial consideration for islet transplantation since those same co-morbidities may make the "minimally" invasive radiological procedure of portal venous access and islet infusion considerably more risky in the IAK cohort than in the "healthy" ITA recipient. There simply has not been a sufficient experience with the interventional radiological experience of islet transplantation to fully understand its risk profile in patients with significant vascular diseases. Therefore, patients at prohibitive risk for surgical pancreas transplantation cannot be automatically considered for islet transplantation at this time. Consideration for islet transplantation must be made on a case-by-case basis with full informed consent of the patient of the possible risks of even the less invasive radiological procedure.

Second, recipients that have a successful kidney transplant may be prescribed an immunosuppressive regimen that differs from the corticosteroid-free protocol described by the Edmonton group. Therefore, immunosuppression conversion to the Edmonton protocol prior to islet transplantation may be required. This will be discussed in greater detail in the section covering immunosuppression.

Third, maintenance of optimal kidney transplant function becomes paramount. This difference requires consideration of treatment protocols for islet transplantation to take a backseat to those needed to maintain optimal kidney transplant function. Conflicts involving approaches to immunosuppression could appear when application of tacrolimus and sirolimus immunosuppression suited for islet transplantation unexpectedly compromises renal allograft function due to nephrotoxicity. For example, an individual who has enjoyed good and stable kidney transplant function for years while receiving cyclosporine, azathioprine and prednisone for immunosuppression is converted to tacrolimus and sirolimus at levels appropriate for the subsequent islet transplantation begins to demonstrate deteriorating renal function. In that case, it may not be possible to achieve the target levels of tacrolimus and sirolimus that have been established in the Edmonton protocol. Moving ahead with the subsequent islet transplantation in the face of "subtherapeutic" immunosuppression may result in inferior outcomes and at the risk of worsening renal function.

Fourth, IAK transplant recipients would be pre-immunosuppressed prior to islet transplantation. This may result in more efficient islet engraftment such that insulin independence could be achieved with a mass of islets significantly less than the threshold of approximately 8-10,000 islet equivalents/kilogram body weight that usually requires two or more cadaveric donors. The Edmonton group and others have observed that insulin requirements could be virtually immediately stopped following the second islet transplant. There are at least two issues to consider regarding the circumstances of the second transplant: the greater mass of islets and the pre-immunosuppressed state. Since both may contribute to the success it raises the theoretical possibility that the pre-immunosuppressed state could result in more efficient early islet survival such that insulin independence may be achieved with a mass of islets significantly less than the threshold of approximately 8-10,000 islet equivalents/kilogram body weight that usually requires two or more cadaveric donors. The caveat is that the patient selection criteria for IAK transplants are similar to that as applied in ITA candidate selection. Consideration of body mass index and insulin requirements are objective measures that can be standardized. The biggest difference may be in insulin sensitivity where less objective criteria are obtained and the best indicator may be the frequency and degree of hypoglycemic episodes. Because IAK transplant candidates have already accepted the risk of immunosuppression, looser criteria for proceeding with IAK with respect to hypoglycemia may take place. That is not necessarily an unreasonable approach, but how that difference affects outcome will be an important consideration.

HUMAN ISLET PROCESSING, PRODUCTION TESTING AND TRANSPLANTATION

Regulatory Aspects. Treatment of type 1 diabetes by transplanting human allogeneic islets is an investigational procedure. It is ensconced by layers of quality and regulatory oversight to safeguard public health and monitor the development of the new procedure. In the United States, transplantable allogeneic pancreatic islets meet the definition of a "drug" in the Federal Food, Drug, and Cosmetic Act (FD&C Act), 21 USC 321(g), and are subject to certain requirements of the FD&C Act. Therefore, before the initiation of any studies in humans of allogeneic islet transplantation, an investigational new drug (IND) application must be approved by the FDA (Regulations for Biological Products Title 21, Code of Federal Regulations Part 312—Investigational New Drugs). Appendix 1 lists the pertinent documents published by the FDA concerning the regulation of cellular and tissue-based products intended for transplantation, including allogeneic islets. The basis for this regulatory order is to ensure that a safe, quality product is used for transplantation.

For biological products, safety is ensured by control of the "manufacturing" process. This requires methodology to adequately characterize and demonstrate that the final therapeutic product can be "manufactured" consistently. To successfully obtain islet preparations that are safe to implant and of high quality, each stage of the production process is defined by standard operating procedures and quality control checks implemented during the course of the isolation and purification procedure. The "manufacturing" process of islets begins at the time of organ procurement.

Pancreas Donor Criteria, Procurement and Preservation. An ideal donor will have a favorable medical, sexual, and social history; pass the physical examination requirements; and clear all standard laboratory tests used in multiorgan donor workups to show low risk of disease transmission. The impact of pancreas donor criteria on the results of islet isolation and purification has been studied retrospectively by a number of groups.[35-38] Older donor age, a local procurement team, and high body mass index are positively correlated with successful islet isolations. Hyperglycemia, increased duration of cardiac arrest, and increased duration of cold storage are negatively correlated.

The surgical technique for pancreas recovery for islet transplantation follows the principles established for immediately vascularized, whole-organ pancreas transplantation.[39,40] The competence and commitment of the surgical team procuring a pancreas for islet transplantation is just as crucial for success of the islet transplant as it is in ensuring successful whole pancreas transplantation. One particularly important consideration of pancreas procurement for islet transplantation concerns the care in maintaining the integrity of the pancreas capsule. Since distention of the pancreas by intraductal injection of collagenase represents a crucial step in subsequent islet isolation, any breach of the pancreatic capsule will compromise this aspect of the isolation procedure. A moderately firm, hard, edematous, or fatty pancreas is not a contraindication to procurement for islet transplantation. In fact, the visual criteria to rule-in a pancreas for islet transplantation are more liberal than for whole pancreas transplantation.

Adequate perfusion of the pancreas with cold preservation solution is accomplished by aortic cannulation and flush through the splenic and superior mesenteric arteries. Venous hypertension in the pancreas should be avoided during in situ flush. If the cannula for in situ portal perfusion is located and secured in the portal vein, the pancreatic portion of the portal vein must be transected to allow continuous drainage of pancreatic fluid outflow. Thus, excessive perfusion pressure or restriction of venous outflow of the pancreas is avoided.

Continuous and effective surface cooling of the pancreas is of paramount importance: warm ischemia is detrimental to subsequent islet isolation.[41] Topical cooling is accomplished by widely opening the lesser sac after dividing the gastrocolic omentum and placing ice slush on the anterior aspect of the pancreas immediately after aortic crossclamping and vascular flush. The pancreas is kept cool while liver procurement occurs. Next the pancreas is procured and stored in cold preservation solution. Finally the kidneys are procured.

Pancreata to be processed for islet isolation are less tolerant of cold ischemia than those used in whole pancreas transplantation. Cold preservation time should be kept less than 9 hours to improve islet isolation yield and function. Simple cold storage in University of Wisconsin (UW) solution has been the standard preservation method in clinical islet transplantation.[42] The two-layer (perfluorochemical and UW solution) pancreas preservation method holds great promise for islet transplants.[43] The perfluorochemical is water insoluble. Its high density ensures that the pancreas floats at the interface of the perfluorochemical and preservation solution. The perfluorochemical, after being saturated with oxygen, provides ample oxygen to the pancreas during preservation, thereby allowing the oxygenated pancreas to produce adenosine triphosphate necessary for maintaining tissue integrity. It has very recently been applied to preservation of human pancreata before pancreas and islet transplantation.[44-46]

Preparation of Islets. On arrival at the islet isolation facility, the transport container is opened and inspected for package integrity. The pancreas is then removed and briefly exposed to an antibiotic and antifungal solution. The pancreas is placed in a cooling pan and the extraneous fat and nonpancreatic tissue carefully dissected and discarded. In preparation for the distension (enzyme loading) procedure, the pancreas is divided at the neck. Two cannulas are inserted into the main pancreatic ducts at the divided surface, one directed to the head and the other to the body and tail, and secured in place (Fig. 8.1). The pancreas is weighed and then perfused under controlled conditions using the perfusion protocol developed by Lakey et al.[47] The cooled perfusion solution consists of a purified enzyme blend containing collagenase[48] and serine-protease inhibitor.[49] The collagenase solution is loaded retrograde into the ducts to distend the pancreas under control of a roller pump with pressure monitoring.[47] The distended panceas may be cut into a few pieces and then placed into the Ricordi digestion chamber. Pancreatic dissociation is accomplished when the collagenase solution is circulated through the Ricordi chamber at a temperature of 37° to 38° C (Fig. 8.2). The chamber is agitated manually or using an automated system.[16] Samples are taken from the circuit at regular intervals to monitor the breakdown of the pancreas by visual inspection of tissue via the inverted microscope. When the amount of tissue liberated from the chamber increases and intact islets are observed, and it is determined that most or all of the islets are free of the surrounding acinar tissue, the recirculation reservoir and the heating circuit are bypassed. The islet isolation continues with the temperature progressively decreased to 15° to 20° C and the collagenase diluted with tissue culture medium. The digest containing the free islets is collected in containers pre-filled with tissue culture medium supplemented with 10% human serum albumin. The pancreatic digest is washed and accumulated for the purification step.

The pancreatic digest containing endocrine and exocrine tissue is purified by placing it on a continuous gradient of sodium diatrizoate ficoll[50] or with iodixanol[45] using a Cobe 2991 cell separator[51] usually under cooling conditions (5-10° C). Fractions with adequate islet purity are combined for immediate transplantation or for pre-transplant tissue culture.[52]

If islet culture is performed pre-transplant, the purified islets are placed in tissue culture flasks containing tissue culture medium. The flasks are placed in an incubator in an atmosphere of 95% air and 5% CO_2. The islets are cultured overnight at 37° C and for an additional 24 to 48 hours at 22° C.

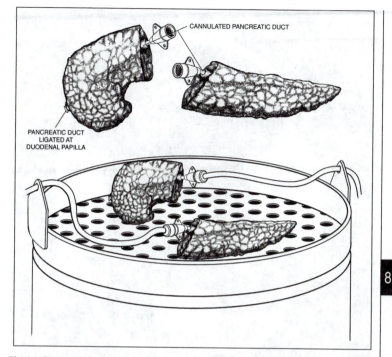

Fig. 8.1. Pancreatic intraductal cannulation and collagenase perfusion.

When it is time to perform the transplant, the islets are collected from the tissue culture flasks and washed. The final islet product is suspended in transplant medium at an approximate concentration of 25-50 ml of medium/ml of tissue.

Islet Product Testing. Islet product testing in the setting of clinical transplantation should follow the regulatory framework of the U.S. Food and Drug Administration (FDA) Center for Biologics Evaluation and Research for cellular and tissue-based products.

Four regulatory requirements have been established for the manufacture of cellular and tissue-based products: 1) product safety, 2) product characterization, 3) control of the manufacturing process, and 4) reproducibility and consistency of product lots. To ensure product safety, specific tests must be established to determine sterility (aerobic, anaerobic, and fungal cultures), pyrogenicity and endotoxin content, and absence of mycoplasma or adventitious agents. Product characterization requires the design and implementation of batch production records and standard operating procedures to test cell and tissue identity, purity, potency, stability, viability, and cell number or amount of tissue. Table 8.4 summarizes current assays for islet product safety and characterization. Table 8.5 lists islet product release and post-release criteria.

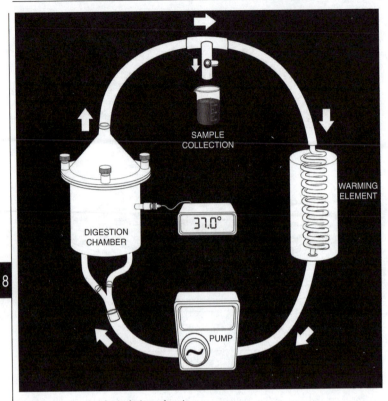

Fig. 8.2. Pancreatic islet isolation schemia.

Islet Transplantation. Intraportal infusion of islets has emerged as the most common technique. Any history of abdominal surgery or liver abnormalities (hemangioma) is considered when deciding how to best access the portal vein. Percutaneous transhepatic catheterization is the most common access route in clinical islet transplantation.[53,54] Alternative approaches to access the portal vein include mini-laparotomy and cannulation of an omental or mesenteric vein, and transjugular intrahepatic portosystemic access.

Access to the portal vein via transhepatic catheterization is provided by the interventional radiologists. Either the left or right intrahepatic portal system is accessed percutaneously. The position of the tip of the infusion catheter is guided to the main portal vein. Position is confirmed with contrast dye, the opening portal pressure is obtained, and a formal portogram is performed. If the portal pressure is <20 mm Hg, and no abnormalities are noted, the islet infusion bag is connected with the portal vein catheter using a standard intravenous infusion set. The islet are infused intraportally, over a period of 15 to 60 minutes, using gravity. Portal vein pressure is recorded halfway through the infusion process, and more often if deemed appropriate. Islet infusion is halted if the portal pressure exceeds

22 mm Hg; it is restarted only if the pressure drops below 18 mm Hg. If the pressure remains elevated, infusion is stopped. After completion of the islet infusion, and after a final rinse, the closing portal pressure is obtained and recorded. No additional intraportal contrast agents are given after islet infusion. The sheath is withdrawn and coils or gelatin-sponge (Gelfoam) pledgets are deployed in the puncture tract to prevent bleeding and augment hemostasis.

POSTTRANSPLANT MANAGEMENT

Insulin and Glycemic Control Immediately Post-transplant. Before the injection of islet cells, both an IV insulin infusion and an IV glucose (5%) infusion are started. Plasma glucose levels are targeted to 80 to 120 mg/dL. Insulin is administered, as needed, to achieve and maintain the plasma glucose levels in the target range. Thereafter, plasma glucose levels are measured every 1 to 2 hours until the recipient is discharged on day 1 or 2 posttransplant. The recipient is asked to test blood glucose several times per day (fasting, before lunch, 2 hours after lunch, before supper, and at bedtime). Exogenous insulin is withdrawn or adjusted, as needed. Recipients able to maintain fasting blood glucose levels below 126 mg/dL, and 2-hour postprandial levels below 180 mg/dL, after insulin discontinuation has been attempted are considered insulin-independent. Insulin-independent recipients are advised to continue to measure and record fasting blood glucose levels daily and postprandial blood glucose levels intermittently. In the event insulin independence is not achieved within the first 6 weeks post-transplant, an additional transplant may be considered.

8

Table 8.4. Current assays for islet product safety and characterization

Category	Assay
Safety	Sterility (aerobic, anaerobic, and fungal cultures)
	Mycoplasma
	Pyrogenicity and endotoxin
Identity	Diphenylthiocarbazone (DTZ)-positive cells
	Insulin content
Cell number	Enumeration of islets and islet equivalents (IE) in DTZ-stained aliquots
	Insulin content
	Volume of tissue pellet
Purity	Percent of DTZ-positive cells Immunoreactive insulin/DNA ratio
	Percent of beta, alpha, delta, ductal, acinar, and other cells, per analysis of cellular composition using immunohistochemistry
Viability	Percent of IE recovery after 48-hour culture
	Microfluorometric membrane integrity test using fluorescent dyes
Potency	Insulin secretory response to glucose challenge in vitro
	Diabetic immunodeficient mouse bioassay
Stability	Studies repeated before and after islet storage in culture and shipment

*IE = 1 IE is equal to 1 150-µm islet.

Table 8.5. Current human islet product release criteria

Sample	Product Test	Specification
Islets	Glucose-stimulated insulin release	Stimulation index >1
	Viability	Must be ≥70%
	Islet enumeration	>4,000 IE/kg recipient body weight
	Purity	Volume of islet prep ≤10 g of tissue
Islets in medium	Mycoplasma	Negative
Islet supernatant	Gram stain	Negative
	Endotoxin	≤5 EU/kg
	Aerobic culture	Negative
	Anaerobic culture	Negative
	Fungal culture	Negative

*IE = 1 IE is equal to 1 150-μm islet.

Baseline efficacy assessment includes the determination of insulin requirements, HbA1c levels, and the number of episodes of severe hypoglycemia, both before and at intervals after the first and final islet transplant. The basic metabolic testing profile includes periodic mixed meal testing (Ensure High Protein, formerly Sustacal, Abbot, Abbot Park, IL) to assess glucose and the C-peptide responses. The intravenous glucose tolerance test (IVGTT) allows calculation of the acute insulin response to insulin, glucose disposal (Kg), and areas under the curve for insulin and C-peptide;[55,56] it has remained the mainstay assessment of islet graft function.[29] Glycemic control in islet recipients has recently been evaluated using a subcutaneous continuous glucose testing system (CGMS, Minimed, Sylmar, CA), which measures the glucose every 5 minutes.[30,57] Other metabolic tests include: the homeostasis model assessment (HOMA) to assess insulin resistance;[58,59] the Sustacal stimulation index for evaluating posttransplant islet function.[17,60,61] Acute C-peptide and insulin responses to intravenous arginine are frequently measured per the protocol of Teuscher et al.[62]

Ryan et al, recently compared metabolic tests in a large series of patients successfully transplanted under the Edmonton protocol.[30] His data indicates that the acute insulin response to arginine provides the best measure of islet mass posttransplant, and that insulin response to glucose stimulation correlates better with the measures of glycemia. An important finding was that the area under the curve for insulin derived from the IVGTT was lower in islet recipients who eventually became C-peptide deficient. Further studies are clearly needed to identify, standardize, and validate measures of islet mass and function in islet recipients and to assess metabolic markers predictive of nonimmunological islet graft failure.

Advanced metabolic tests performed in islet recipients include glucose potentiation of arginine-induced insulin secretion as a measure of insulin secretory reserve and islet mass;[62] euglycemic hyperinsulinemic clamps, with infusion of labeled glucose, to quantify peripheral insulin sensitivity and hepatic glucose production;[63] the frequently sampled intravenous glucose tolerance test (FSIVGTT, minimal model), as modified by Finegood et al[64] to quantify glucose tolerance, insulin response, insulin sensitivity, and glucose effectiveness; and stepped hypoglycemic clamp tests, to measure hormonal glucose counterregulation, autonomic, and neuroglycopenic symptoms in response to hypoglycemia.[32]

Key to any analysis of the benefits of islet transplants are the assessments of the development, progression, and reversal of microvascular and macrovascular complications; health-related quality of life; cost-utility; and life expectancy. The first pilot study to address health-related quality of life in islet recipients used telephone interviews; the survey instruments were the Health Utilities Index Mark 3, SF-32 version 2, Hypoglycemia Fear Survey, and Audit of Diabetes Dependent Quality of Life Survey[65] and showed a marked improvement in studied parameters.

Immunosuppression. Shapiro et al developed a glucocorticoid-free immunosuppressive protocol, which markedly improved islet transplant outcomes. Their protocol consisted of the IL-2 receptor blocker daclizumab for induction, then sirolimus and low-dose tacrolimus for maintenance immunosuppression.[26] Daclizumab induction therapy was given intravenously at a dose of 1 mg/kg every 14 days for a total of 5 doses over a 10-week period, thus allowing an extended period for a supplemental islet transplant procedure. If the second islet transplant procedure occurred more than 10 weeks after the first, the course of daclizumab was repeated. No glucocorticoids were given at any time. Sirolimus was dosed to achieve and maintain trough levels of 12 to 15 ng/mL for the first 3 months and of 7 to 10 ng/mL thereafter. Tacrolimus was administered at an initial dose of 1 mg twice daily, then adjusted to maintain a trough concentration at 12 hours of 3 to 6 ng/mL. Type 1 diabetic islet allograft recipients reliably achieved and maintained freedom from the need of exogenous insulin after transplantation of an adequate mass of islets prepared from 2 to 4 donor organs, suggesting that the protocol by Shapiro et al protected against alloimmune and autoimmune reactivity.[26,66]

The success reported by the Edmonton group using glucocorticoid-free immunosuppression involving sirolimus has been confirmed by other institutions.[45,67-69] Immunosuppression with daclizumab, sirolimus, and reduced-dose tacrolimus has evolved as the gold standard for type 1 diabetic islet transplant recipients. A multicenter trial (with 9 participating islet transplant centers in North America and Europe) is currently underway to evaluate, in more detail, the safety and efficacy of the Edmonton immunosuppressive protocol.

The acute, and in particular, long-term risks associated with novel immunosuppressive regimens are unknown. The preliminary results on about 300 islet recipients since 1995 provide only incomplete information as to the risks of immunosuppression. Since publication of the Edmonton trial, most islet transplants are performed as solitary islet transplants in nonuremic recipients whose diabetes is complicated by hypoglycemia unawareness. Invasive CMV disease, opportunistic infections, post-transplant lymphoproliferative disorders, and other malignancies have not been reported in this recipient category. These encouraging results are presumably related to the small CMV viral load transferred with islet products,[70] the exclusion at most institutions of EBV-negative patients from participation in islet transplant trials, and the very low incidence of treated rejection episodes in islet recipients immunosuppressed with sirolimus and reduced-dose tacrolimus.

For recipients of an islet-after-kidney transplant procedure, there are special considerations for immunosuppression. Recipients that have a successful kidney transplant may be prescribed an immunosuppressive regimen that differs from the corticosteroid-free protocol described by the Edmonton group. Conversion of immunosuppression to the Edmonton protocol prior to islet transplantation may be necessary to improve results over those previously reported.[60,71,72] In those three series, IAK recipients were continued on cyclosporine-based immunosuppression with an antimetabolite adjuvant. At the time of IAK, prednisone boluses were administered—typically 500-1000 mg of prednisolone, with or without an antibody induction agent. The detrimental effects of steroids, and the beneficial effects of induction therapy have been well documented in preclinical large animal models of islet transplantation.[23] Therefore, modifications and consistent application of the newer immunosuppression approaches to islet transplantation may be needed to achieve insulin independence following IAK.

There are at least two situations that describe approaches to kidney transplant immunosuppression that have implications with respect to preparation for a subsequent IAK transplant. The first scenario includes renal transplant recipients with Type I diabetes that were initially prescribed a "conventional" immuno-suppression regimen including corticosteroids at the time of kidney transplantation prior to any consideration of an IAK procedure. A typical immunosuppression protocol might entail cyclosporine, azathioprine (or mycophenolate mofetil) and prednisone. This situation in which immunosuppression was initially prescribed in consideration of the kidney transplant only is referred to as a "casual approach." In this circumstance the patient would undergo immunotherapy conversion to that resembling the Edmonton protocol of tacrolimus and sirolimus without corticosteroids prior to the IAK transplant.

The second scenario includes renal transplant recipients with type I diabetes that are prescribed an immunosuppression regimen in which the corticosteroids are avoided or immediately rapidly eliminated following the kidney transplantation in anticipation of the subsequent islet transplant. This approach is referred to as "expectant immunosuppression." The combined use of tacrolimus and MMF with an IL-2 receptor antagonist allows corticosteroids to be withdrawn within 3 days of renal transplantation.[73] The risk of a renal allograft rejection episode is approximately 13%, (85-90% of occurring within 3 weeks of transplantation), and 100% were reversible with appropriate anti-rejection therapy. One of the criti-cisms of steroid avoidance protocols is that the long-term results are not known. The long-term outcome of steroid avoidance was addressed by Birkeland.[74] A 5-year follow-up of 100 kidney transplant recipients indicated that renal allograft sur-vival was not compromised by omitting chronic steroid exposure.

The ability to coordinate the approach of renal transplant immunotherapy with subsequent medical conversion including possible corticosteroid withdrawal in preparation of the islet transplant requires that there is integration of the kidney and islet transplant programs. A functionally integrated program also assumes that decisions regarding the application of whole pancreas transplantation would have been considered for some recipients of a functioning kidney allograft.

Host Autoimmune Responses to Transplanted Islets. The immune response to transplanted islets and autoantigen has been measured in recipients with functioning and failing islet allografts. One-year islet allograft survival in recipients with positive autoantibodies (GAD65 or ICA) before or after their transplant compared unfavorably with survival in recipients who remained autoantibody-negative.[75] The clinical relevance of autoimmunity after islet transplants was confirmed by another study: insulin independence was achieved in >60% of recipients without, but not in any recipients with, autoantibody elevations.[76] The study by Bosi et al[76] also identified glutamic acid decarboxylase (GAD) as the key autoantigen in the reexposure of patients with autoimmune diabetes to islet beta cells. Recurrent autoimmunity was independent of donor-recipient HLA matching and autoantibody titer at the time of the transplant; autoimmunity also recurred in the absence of alloimmunity.

FUTURE DIRECTIONS

After almost 3 decades of persistent trials, islet transplants are only beginning to contribute in a clinically significant fashion to the treatment of a select group of people with type 1 diabetes. While the success of the Edmonton trial has been considered a turning point, challenges remain. With currently available technology, only half of all pancreata processed for islets now meet release criteria for clinical transplants. Islets from 2 donor pancreata are still required to reliably reverse diabetes. It is reasonable to assume that 4 pancreata are actually required to reverse diabetes, making islet transplants much less effective than vascularized whole-organ transplants, at least for now. The next critical step is to enhanced the viable islet mass retrievable from 1 donor pancreas by optimizing the entire islet isolation process until consistent diabetes reversal after transplants of islets prepared from a single donor pancreas is achieved. This achievement would have a marked impact on pancreas allocation for islet transplants. It would represent a significant boost to the field. Achieving this outcome will also require application of innovative therapies that enhance islet engraftment and early function in combination with new approaches of immunosuppression that minimze the metabolic demand on transplanted islets.

With the development of new immuosuppressive preconditioning and maintenance agents, the immunosuppressive risks now associated with islet transplants will continue to diminish. This will permit a greater number of individuals to be transplanted earlier in the course of their disease without unacceptable immunosuppressive risk. Tissue availability will be the limiting factor in determining the magnitude of the impact of islet transplants on treating diabetes. Until the day that islet preparations can be taken down off the shelf for transplantation, only a small proportion of the millions of persons with diabetes can be treated by cellular replacement. Intense research in developmental and stem cell biology and xenotransplantation aims to achieve those goals.

Appendix 8.1.

Center for Biologics Evaluation and Research (CBER)
www.fda.gov/cber
1-800-835-4709

A Proposed Approach to the Regulation of Cellular and Tissue-based Products, (62 FR 9721) – available at http://www.fda.gov/cber/gdlns/CELLTISSUE.pdf

Guidance for Industry: Guidance for Human Somatic Cell Therapy and Gene Therapy (63 FR 36413) – available at http://www.fda.gov/cber/gdlns/somgene.pdf

Transcript of discussion of allogeneic pancreatic islets by FDA Biologic Response Modifier Advisory Committee– available at http://www.fda.gov/ohrms/dockets/ac/cber00.htm

Federal Register. Establishment Registration and Listing for Manufacturers of Human Cellular and Tissue-Based Products; Final Rule (1/2001) - available at http://fda.gov/cber/rules.htm

Federal Register. Current Good tissue Practice for Manufacturers of Human Cellular and Tissue-Based Products; Inspection and Enforcement; Proposed Rule (1/2001) - available at http://fda.gov/cber/rules.htm-gtp

8

It is generally anticipated that pancreas transplants will someday be largely replaced by islet-cell transplants. Just as pancreas transplants set the stage for islet transplants, the real value of islet transplants will be to create and build momentum for the development of surrogate beta cells that will then make cell replacement therapy routine and commonplace. Then beta-cell replacement will become the premier treatment option for people with type 1 diabetes. The current breakthroughs of today are providing an exciting and solid foundation on which tomorrow's success will be built.

REFERENCES
1.   Diabetes in America (2nd edition), NIH Publication No. 95-1468, 1995.
2.   Portuese E, Orchard T 1995 Mortality in Insulin-Dependent Diabetes. In: Harris MI, Cowie CC, Stern MP, Boyko EJ, Reiber GE, Bennett PH (eds) Diabetes in America. US Govt Printing Office, Washington DC, pp 221-232.
3.   Epidemiology of Diabetes Interventions and Complications (EDIC) Research Group. Effect of intensive diabetes treatment on carotid artery wall thickness in the epidemiology of diabetes interventions and complications. Epidemiology of Diabetes Interventions and Complications (EDIC) Research Group. Diabetes 1999; 48: 383-90.
4.   DCCT/EDIC Research Group. Effect of intensive therapy on the microvascular complications of type 1 diabetes mellitus. JAMA 2002; 287:2563-9.
5.   The Diabetes Control and Complications Trial Research Group. The effect of intensive treatment of diabetes on the development and progression of long-term complications in insulin-dependent diabetes mellitus. NEJM 1993; 329:977-86.
6.   Morel P, Goetz FC, Moudry-Munns K, et al. Long-term glucose control in patients with pancreatic transplants. Annals of Internal Medicine 1991;115:694-699.
7.   Fioretto p, Steffes MW, Sutheralnd DER, et al. Reversal of lesions of diabetic nephropathy after pancreas transplantation. NEJM 1998; 339:69-75.

8.   Kennedy WR, Navarro X, Goetz FC, et al. Effects of pancreatic transplantation on diabetic neuropathy. NEJM 1990; 322:1031-7.

9.   Ramsay RC, Goetz FC, Sutherland DER, et al. Progression of diabetic retinopathy after pancreas transplantation for insulin-dependent diabetes mellitus. NEJM 1988; 318:208-14.

10. Zehrer CL, Gross CR. Quality of life pancreas transplant recipients. Diabetologia 1991;34(Suppl 1):S145-S149.

11. Ojo AO, Meier-Kriesche HU, Hanson JA, et al. Impact of simultaneous pancreas-kidney transplantation on long-term patient survival. Transplantation 2001; 71; 82-90.

12. Younoszai R, Sorenson RL and Lindall AW. Homotransplantation of isolated pancreatic islets (abstract). Diabetes 1970; 19 (Suppl):406.

13. Ballinger WF and Lacy PE. Transplantation of intact pancreatic islets in rats. Surgery 1972; 72:175-82.

14. Reckard CR, Zeigler MM, Barker CF. Physiological and immunological consequences of transplanting isolated pancreatic islets. Surgery 1973; 74:91-95.

15. Panijayanond P, Soroff HS and Monaco AP. Pancreatic islet isografts in mice. Surg Forum 1973; 24:329-30.

16. Ricordi C, Lacy PE, Finke EH, Olack BJ, Scharp DW. Automated method for isolation of human pancreatic islets. Diabetes 1988; 37:413-420.

17. Scharp DW, Lacy PE, Santiago JV, et al. Insulin independence after islet transplantation into type 1 diabetic patient. Diabetes 1990; 39:515-8.

18. Warnock GL, Kneteman NM, Ryan EA, et al., Long-term follow-up after transplantation of insulin-producing pancreatic islets into patients with type 1 (insulin dependent) diabetes mellitus. Diabetologia 1992; 35:89-95.

19. White SA, Robertson GS, London NJ, et al. Human islet autotransplantation to prevent diabetes after pancreas resection. Dig Surg 2000;17:439-50.

20. London NJ, Robertson GS, Chadwick DR, et al. Human pancreatic islet isolation and transplantation. Clin Transplant. 1994; 8:421-59.

21. Brendel MD, Hering BJ, Schultz AO, Bretzel RG. International Islet Transplant Registry Newsletter #9. Department of Medicine, Justus-Liebig-University of Giessen. 2001; Volume 8 (Newsletter No. 9):1-20.

22. Hering BJ, Ricordi C. Islet transplantation for patients with type 1 diabetes: Results, research priorities and reasons for optimism. Graft 1999; 2:12-27.

23. Kaufman DB, Morel P, Condie R, Field MJ, Roney M, Tzardis P, Stock P Sutherland DER: Beneficial and detrimental effects of RBC-adsorbed antilymphocyte globulin and prednisone on purified canine islet autograft and allograft function. Transplantation 1991; 51:37-42.

24. Kneteman NM, Lake JR, Wagner T, et al. The metabolic impact of rapamycin (sirolimus) in chronic canine islet graft recipients. Transplantation 1996; 61:1206-1210.

25. McAlister VC, Gao Z, Peltekian K, Dominguas J, Mahalati K, MacDonald AS. Sirolimus-tacrolimus combination immunosuppression. Lancet 2000; 355:376-377.

26. Shapiro AMJ, Lakey JR, Ryan EA, et al. Islet transplantation in seven patients with type 1 diabetes mellitus using a glucocorticoid-free immunosuppressive regimen. NEJM 2000; 343:230-238.

27. Shapiro AMJ, Ryan EA, Paty B, Korbutt GS, O'Kelley K, Kneteman NM et al. Human islet transplantation can correct diabetes. Transplantation 2002; 74[4 (Suppl.)]:119.

28. Ryan EA, Lakey JR, Rajotte RV, Korbutt GS, Kin T, Imes S et al. Clinical outcomes and insulin secretion after islet transplantation with the Edmonton protocol. Diabetes 2001; 50:710-719.

29. Ryan EA, Lakey JR, Paty BW, Imes S, Korbutt GS, Kneteman NM et al. Successful islet transplantation: continued insulin reserve provides long-term glycemic control. Diabetes 2002; 51:2148-2157.

30. Kaufman DB, Baker MS, Chen X, Leventhal JR, Stuart FP. Sequential kidney/islet transplantation using prednisone-free immunosuppression. Am J Transplant 2002; 2:674-677.

31. Markmann JF, Deng S, Huang X, Desai NM, Velidedeoglu E, Liu C et al. Evaluation of islets from non-heart beating donors for human transplantation. Transplantation 2002; 74[4(Suppl.)]:142.

32. Meyer C, Hering BJ, Grossmann R, Brandhorst H, Brandhorst D, Gerich J et al. Improved glucose counterregulation and autonomic symptoms after intraportal islet transplants alone in patients with long-standing type 1 diabetes mellitus. Transplantation 1998; 66:233-240.

33. Cryer PE. Hypoglycemia: the limiting factor in the management of IDDM. Diabetes 1994; 43:1378-1389.

34. Cryer PE. Iatrogenic Hypoglycemia as a Cause of Hypoglycemia-Associated Autonomic Failure in IDDM: A Vicious Cycle. Diabetes 1992; 41:255-260.

35. Zeng Y, Torre MA, Karrison T, Thistlethwaite JR. The correlation between donor characteristics and the success of human islet isolation. Transplantation 1994; 57:954-958.

36. Benhamou PY, Watt PC, Mullen Y, et al. Human islet isolation in 104 consecutive cases. Transplantation 1994; 57:1804-1808.

37. Brandhorst H, Brandhorst D, Hering BJ, Federlin K, Bretzel RG. Body mass index of pancreatic donors: a decisive factor for human islet isolation. Experimental & Clinical Endocrinology & Diabetes 1995; 103 Suppl 2:23-26.

38. Lakey JR, Warnock GL, Rajotte RV, Suarez-Alamazor ME, Ao Z, Shapiro et al. Variables in organ donors that affect the recovery of human islets of Langerhans. Transplantation 1996; 61:1047-1053.

39. Marsh CL, Perkins JD, Sutherland DER, Corry RJ, Sterioff S. Combined hepatic and pancreaticoduodenal procurement for transplantation. Surgerey, Gynecology & Obstetrics 1989; 168:254-258.

40. Sollinger HW, Vernon WB, D'Alessandro AM, Kalayoglu M, Stratta RJ, Belzer FO. Combined liver and pancreas procurement with Belzer-UW solution. Surgery 1989; 106:685-691.

41. Lakey JR, Kneteman NM, Rajotte RV, Wu DC, Bigam D, Shapiro AM. Effect of core pancreas temperature during cadaveric procurement on human islet isolation and functional viability. Transplantation 2002; 73:1106-1110.

42. Belzer FO, Ploeg RJ, Knechtle SJ, D'Alessandro AM, Pirsch JD, Kalayoglu M et al. Clinical pancreas preservation and transplantation. Transplant Proc 1994; 26:550-551.

43. Kuroda Y, Kawamura T, Suzuki Y, Fujiwara H, Yamamoto K, Saitoh Y. A new, simple method for cold storage of the pancreas using perfluorochemical. Transplantation 1988; 46:457-460.

44. Matsumoto S, Kandaswamy R, Sutherland DE, Hassoun A, Hiraoka K, Sageshima J et al. Clinical application of the two-layer [University of Wisconsin Solution (UW)/ Perfluorochemical (PFC) plus O_2] method of pancreas preservation before transplantation. Transplantation 2000; 70:771-774.

45. Hering BJ, Kandaswamy R, Harmon JV, Ansite JD, Clemmings S, Sakai T et al. Insulin independence after single-donor islet transplantation in type 1 diabetes with hOKT3gamma1 (Ala-Ala), sirolimus, and tacrolimus therapy. Amer J Transplantation 2001; 1(Suppl. 1):180.

46. Ricordi C, Fraker.C., Szust J, al-Abdullah I, Poggioli R, Kirelew T et al. Towards making every pancreas count: significant improvement in human islet isolation from marginal (older) donors following addition of oxygenated perflourocarbon to the cold storage solution. Amer J Transplantation 2002; 2(Suppl. 3):229.

47. Lakey JRT, Warnock GL, Shapiro AM, Korbutt GS, Ao Z, Kneteman NM et al. Intraductal collagenase delivery into the human pancreas using syringe loading or controlled perfusion. Cell Transplant 1999; 8:285-292.

48. Linetsky E, Bottino R, Lehmann R, Alejandro R, Inverardi L, Ricordi. Improved human islet isolation using a new enzyme blend, Liberase. Diabetes 1997; 46(7):1120-1123.

49. Lakey JR, Helms LM, Kin T, Korbutt GS, Rajotte RV, Shapiro AM et al. Serine-protease inhibition during islet isolation increases islet yield from human pancreases with prolonged ischemia. Transplantation 2001; 72:565-570.

50. Brandhorst H, Brandhorst D, Brendel M, Hering BJ, Bretzel RG. Assessment of intracellular insulin contents during all steps of human islet isolation procedure. Cell Transplant 1998; 7:489-495.

51. Lake SP, Bassett PD, Larkins A, Revell J, Walczak K, Chamberlain J et al. Large-scale purification of human islets utilizing discontinuous albumin gradient on IBM 2991 cell separator. Diabetes 1989; 38 Suppl 1:143-145.

52. Hering BJ, Bretzel RG, Hopt UT, Brandhorst H, Brandhorst D, Bollen et al. New protocol toward prevention of early human islet allograft failure. Transplant Proc 1994; 26:570-571.

53. Alejandro R, Mintz DH, Noel J, Latif Z, Koh N, Russell E et al. Islet cell transplantation in type I diabetes mellitus. Transplant Proc 1987; 19(1 Pt 3):2359-2361.

54. Weimar B, Rauber K, Brendel MD, Bretzel RG, Rau WS. Percutaneous transhepatic catheterization of the portal vein: A combined CT- and fluoroscopy-guided technique. Cardiovasc Intervent Radiol 1999; 22:342-344.

55. Bingley PJ, Colman P, Eisenbarth GS, Jackson R, McCulloch DK, Riley WJ et al. Standardization of IVGTT to predict IDDM. Diab Care 1992; 15:1313-1316.

56. Chase HP, Cuthbertson DD, Dolan LM, Kaufman F, Krischer JP, Schatz DA et al. First-phase insulin release during the intravenous glucose tolerance test as a risk factor for type 1 diabetes. J Pediatr 2001; 138:244-249.

57. Geiger MC, Ferreira JV, Froud T, Caulfield A, Baidal D, Rothenberg L et al. Evaluation of metabolic control by continuous subcutaneous glucose monitoring system in patients with type 1 diabetes mellitus after islet cell transplantation. Amer J Transplantation 2002; 74[4(Suppl.)]:110.

58. Haffner SM, Gonzalez C, Miettinen H, Kennedy E, Stern MP. A prospective analysis of the HOMA model. Diab Care 1996; 19:1138-1141.

59. Levy JC, Matthews DR, Hermans MP. Correct homeostasis model assessment (HOMA) evaluation uses the computer program. Diab Care 1998; 21:2191-2192.

60. Scharp DW, Lacy PE, Weide LG, Marchetti P, McCullough CS, Flavin K et al. Intraportal islet allografts: the use of a stimulation index to represent functional results [published erratum appears in Transplant Proc 1991;23:2702]. Transplant Proc 1991; 23:796-798.

61. Alejandro R, Lehmann R, Ricordi C, Kenyon NS, Angelico MC, Burke G et al. Long-term function (6 years) of islet allografts in type 1 diabetes. Diabetes 1997; 46:1983-1989.

62. Teuscher AU, Kendall DM, Smets YF, Leone JP, Sutherland DE, Robertson et al. Successful islet autotransplantation in humans: functional insulin secretory reserve as an estimate of surviving islet cell mass. Diabetes 1998; 47:324-330.

8

67. Keymeulen B, Ling Z, Gorus FK, Delvaux G, Bouwens L, Grupping A, Hendrieckx C, Pipeleers-Marichal M, Van Schravendijk C, Salmela K, Pipeleers DG. Implantation of standardized beta-cell grafts in a liver segment of IDDM patients: graft and recipient characteristics in two cases of insulin-independence under maintenance immunosuppression for prior kidney graft. .Diabetologia1998; 41:452-59.

63. Luzi L, Hering BJ, Socci C, Raptis G, Battezzati A, Terruzzi I et al. Metabolic effects of successful intraportal islet transplantation in insulin-dependent diabetes mellitus. J Clin Invest 1996; 97:2611-2618.

64. Finegood DT, Warnock GL, Kneteman NM, Rajotte RV. Insulin sensitivity and glucose effectiveness in long-term islet-autotransplanted dogs. Diabetes 1989; 38 Suppl 1:189-191.

65. Johnson JA, Nanji SA, Supina A, Ryan EA, Shapiro AMJ. Islet transplantation and health related quality of life in type 1 diabetes: results of an initial pilot study. American Journal of Transplantation 2002; 2[Suppl. 3]:229.

66. Ryan EA, Lakey JR, Shapiro AM. Clinical results after islet transplantation. J Investig Med 2001; 49:559-562.

67. Alejandro R, Ferreira JV, Caulfield A, Froud T, Baidal D, Geiger M et al. Insulin independence in 7 patients following transplantation of cultured human islets. Transplantation 2002; 2[Suppl. 3]:227.

68. Rother KI, Hirschberg B, Gaglia JL, Chang R, Wood B, Kirk AD et al. Islet transplantation in patients with type 1 diabetes. NIH experience in 6 patients. Transplantation 2002; 2[Suppl. 3]:228.

69. Markmann JF, Deng S, Huang X, Liu C, Velidedeoglu E, Desai NM et al. Reversal of type 1 diabetes by transplanting isolated pancreatic islets from single donors. Transplantation 2002; 2[Suppl. 3]:228.

70. Preiksaitis JK, Lakey JRT, LeBlanc BA, Fenton JM, Ryan EA, Bigam DL et al. Cytomegalovirus (CMV) is not transmitted by pancreatic islet transplantation. Amer J Transplantation 2002; 2[Suppl. 3]:308.

71. Secchi A, Socci C, Maffi P, Taglietti MV, Falqui L, Bertuzzi F, DeNittis P, Piemonti L, Scopsi L, Di Carlo V, Pozza G: Islet transplantation in IDDM patients. Diabetologia 1997; 40:225-231.

72. Keymeulen B, Ling Z, Gorus FK, Delvaux G, Bouwens L, Grupping A, Hendrieckx C, Pipeleers-Marichal M, Van Schravendijk C, Salmela K, Pipeleers DG. Implantation of standardized beta-cell grafts in a liver segment of IDDM patients: graft and recipient characteristics in two cases of insulin-independence under maintenance immunosuppression for prior kidney graft. .Diabetologia 1998; 41:452-59.

73. Kaufman DB, Leventhal JR, Fryer JP, Abecassis MI, Stuart FP. Kidney transplantation without prednisone. Transplantation 2000; 69: S133.

74. Birkeland SA. Steroid-free immunosuppression in renal transplantation: a long-term follow-up of 100 consecutive patients. Transplantation 2001; 71:1089-91.

75. Jaeger C, Brendel M, Hering BJ, Eckhard M, Bretzel RG. Progressive islet graft failure occurs significantly earlier in autoantibody-positive than in autoantibody-negative IDDM recipients of intrahepatic islet allografts. Diabetes 1997; 46:1907-1910.

76. Bosi E, Braghi S, Maffi P, Scirpoli M, Bertuzzi F, Pozza G et al. Autoantibody response to islet transplantation in type 1 diabetes. Diabetes 2001; 50:2464-2471.

8

Liver Transplantation

*Michael Abecassis, Andres Blei, Alan Koffron, Steven Flamm,
and Jonathan Fryer*

INTRODUCTION

Orthotopic liver transplantation is the accepted therapeutic option of choice for acute and chronic end-stage liver disease. The indications and contraindications to liver transplantation have become standardized, as has the operative and post-operative management. This chapter will address the evaluation and management of patients with acute and chronic liver failure with particular emphasis on recipient selection, operative and postoperative management, and will consist of a practical approach to patients undergoing liver transplantation. Our goal is to provide helpful guidelines to caregivers involved in the care of these complex patients.

Liver failure can present as either acute (fulminant and subfulminant failure) or chronic (advanced cirrhosis). The term decompensated cirrhosis reflects the presence of one or more complications. Each disease etiology presents unique features and it is therefore important to recognize these distinctions. In the pre-transplantation era, liver failure was associated with an almost universal fatal outcome, with a spontaneous survival in fulminant hepatic failure of 10-20% and a 1 year mortality in decompensated cirrhosis of >50%. In contrast, liver transplantation patient survival outcomes are presently >85% at one year and >70% at five years, underlining the application of liver transplantation as the standard of care in patients with both acute and chronic liver failure. In addition, the advent of both split liver transplant and live-donor liver transplantation offers additional hope to patients with liver failure in the presence of an ever-growing cadaveric organ shortage.

A. LIVER TRANSPLANTATION FOR PATIENTS WITH ACUTE LIVER FAILURE

Acute liver failure (ALF) is often used synonymously with fulminant liver failure. ALF is defined as an acute hepatic deterioration not preceded by evidence of chronic liver disease, which has progressed from the onset of jaundice to the development of hepatic encephalopathy in less than 8 weeks.[1]

Subsequent refinements include a division between fulminant (<2 weeks) and subfulminant hepatic failure (>2weeks), a difference that reflects the greater predominance of brain edema and intracranial hypertension in patients with a shorter interval between the onset of jaundice and the development of encephalopathy. More recently, a differentiation between hyperacute (< 1week), acute (1-4 wks) and subacute failure (>4 wks) has been suggested. Both drug-induced hepatic failure and an indeterminate etiology appear to be more commonly associated with a longer interval. There are also geographic differences in the etiology of

Organ Transplantation, 2nd edition, edited by Frank P. Stuart, Michael M. Abecassis
and Dixon B. Kaufman. ©2003 Landes Bioscience.

fulminant hepatic failure. Hepatitis E is a common cause of ALF during pregnancy in India but is not seen in the United States. A recent survey of 295 cases in the U.S. (Table 9.1) showed acetaminophen intoxication as a leading cause, followed by nonA-nonE (also termed cryptogenic) and drug-induced failure.[2] Acetaminophen toxicity was associated with the best spontaneous survival (60%), and it is important to recognize its etiologic role in patients with either underlying alcohol consumption or with poor food intake, in whom lower daily doses (4 grams rather than 10-12 grams) may induce severe liver injury. The cause of nonA-nonE fulminant hepatitis remains elusive. Although a transmissible agent has been implicated, hepatitis C, hepatitis G or TTV (transfusion-transmitted virus) have been shown not to be the culprits. Drug-induced hepatic failure has a particularly poor prognosis and spontaneous survival is rare once encephalopathy develops.

Clinical evidence of intracranial hypertension include, hyperventilation, opisthotonus, hyperpronation-adduction of the arms, cardiac arrhythmia, myoclonus, seizures, poorly reactive pupils.

Patients with ALF present initially with vague symptoms, such as anorexia and malaise. Attention by patients and their caregivers may not focus on the diagnosis of liver failure until jaundice is evident. Patients often describe a syndrome suggestive of and consistent with a viral illness. When jaundice is identified, liver function tests typically reveal massive elevations in AST and ALT, elevated bilirubin, significant elevation in the PT, and, in some patients, metabolic acidosis. If Tylenol overdose is suspected, acetaminophen levels should be obtained and the patient should be started on IV acetyl cysteine (Mucomyst). A delay in diagnosis may lead to referral of a patient with ALF late in the clinical course, resulting in advanced cerebral edema.

9

Table 9.1. Etiology of fulminant hepatic failure in the United States[2]

		Spontaneous		
		N	%	Survival (excludes death or transplantation)
Acetaminophen	60	20%		60%
Hepatitis nonA-E	44	15%		10%
Drug-induced	33	12%		10%
Hepatitis B	30	10%		15%
Hepatitis A	21	7%		35%
Miscellaneous	122	Wilson's disease, acute fatty liver of pregnancy, Budd-Chiari Syndrome mushroom intoxication, ischemic injury, tumor infiltration, autoimmune hepatitis, rare viruses (herpes, adenovirus).		

Total of series   295

Fulminant hepatic failure typically affects young individuals who had previously been in a perfect state of health and, prior to the availability of liver transplantation, was associated with 80 to 90 percent mortality, especially in patients who progressed to grade 3 or 4 hepatic encephalopathy. With successful transplantation, >90% of patients survive.[3] Although many factors contribute to the deterioration and death of these patients, the terminal event is typically brainstem herniation as a result of progressive brain swelling. Hepatic encephalopathy is typically divided into 4 stages. Furthermore, coma in stages 3 and 4 is subdivided into 4 grades.(Table 9.2).

Table 9.2a. Hepatic encephalopathy

Stage 1	Slowing of consciousness
Stage 2	Drowsiness
Stage 3	Confusion, reactive only to vocal stimuli
Stage 4	Presence of deep coma with absence of reaction to vocal stimuli

Table 9.2b. Grading of coma in stages 3 and 4

Grade 1	Reactivity to vocal stimuli
Grade 2	Absence of reactivity to vocal stimuli, but with a coordinated response to painful stimuli
Grade 3	Absence of reactivity to vocal stimuli with a incoordinated response to painful stimuli
Grade 4	Brain death

It is essential that these patients be admitted and monitored closely in a specialized liver unit where frequent surveillance of their LFTs, PT, CBC, blood gases, blood sugars, electrolytes, and neurological status is performed. With Tylenol overdose, liver transplantation can be prevented if therapy is initiated early. With progression of encephalopathy to stage 3 or 4, the patient should be intubated for airway protection, as these patients have a very high incidence of aspiration as they deteriorate neurologically. An NG tube should be placed at this time and lactulose initiated. The patient should be started on an H2-blocker to prevent ulceration. A Foley catheter should be placed, as well as an arterial line. Central venous monitoring should be entertained if there is a deterioration in renal function or hemodynamic instability. An intracranial pressure monitor should be placed if the patient's neurologic status cannot be followed clinically, in order to accurately assess progressive brain swelling.[4] Cerebral perfusion pressures determined by subtracting the intracranial pressure from the mean arterial pressure provides a marker for cerebral perfusion. In the case of sustained untreatable cerebral hypoperfusion, the patient may no longer be considered a transplant candidate since irreversible brain injury may occur. If there is evidence of ongoing brain swelling, hyperventilation and/or mannitol may help temporarily.

Prior to the availability of liver transplantation, many non-surgical approaches were attempted in patients with acute liver failure including exchange transfusions, steroids, hemodialysis, and charcoal hemoperfusion. Unfortunately, none of these approaches have been particularly successful. There is new evidence that hypothermia may help to delay brain swelling which is often the terminal complication, but further assessment of this approach is needed. Presently, liver transplantation is considered the best therapeutic option for acute liver failure not thought to be reversible. The criteria for determining whether a patient will need liver transplant or not include factor V level less than 30%, pH less than 7.3, INR >6.5, stage 3 or 4 encephalopathy, and lack of response to medical therapy within 20 to 48 hours. (Table 9.3)

Early referral to a liver transplantation center is essential since: a) it is difficult to predict which patients will recover spontaneously; b) deterioration can occur very suddenly; c) there is a shortage of donor organs and the chance of receiving a transplant is greater with early placement on the waiting list; and d) once brainstem herniation has occurred, patients are not salvageable by liver transplant or by any other means.

It is important to recognize etiologies of fulminant hepatic failure in which transplantation is contraindicated. These include diffuse infiltration of the liver by lymphoma or extensive liver metastases as an initial manifestation of malignancy. Hepatic ischemia can be a manifestation of left-sided ventricular failure without signs of congestive heart failure. Acute hepatic vein thrombosis, with fulminant failure as a result of venous outflow block, is best treated with a decompression procedure (side-to-side portacaval shunt) rather than organ replacement.

OPTIONS FOR HEPATIC SUPPORT

Due to the severe shortage of human donors, many patients with acute liver failure die waiting for a suitable organ. For this reason, these patients should be referred to centers which are not only capable of liver transplantation, but which are also capable of supporting such patients until an organ becomes available. In addition to standard medical supportive measures, several strategies are being developed to provide temporary hepatic support.(Table 9.4) These options are discussed in the ensuing section.

Charcoal hemoperfusion systems have been evaluated as artificial liver support devices. Although some studies have suggested a survival advantage with fulminant hepatic failure of certain etiologies[7], most patients do not appear to benefit. Other forms of artificial liver support have included dialysis-like systems coupled with absorbant technology. One system in this category, which is currently under-

---

Table 9.3.   Criteria for transplantation of acute liver failure

---

Kings College Criteria[5]
- Acetaminophen toxicity
  - ph < 7.30 (after hydration and regardless of degree of encephalopathy)
                              or
    INR >6.5
    creatinine >3mg/dl
    Encephalopathy III-IV
- Non-acetaminophen etiology
  - INR >6.5 irrespective of degree of encephalopathy
                  or 3 of the following five criteria
    Age<10, >40
    Etiology: nonA-E hepatitis, drugs
    Duration of jaundice before encephalopathy >7 days
    INR >3.5
    Serum bilirubin >17.5 mg%.

Clichy Criteria[6]
- Factor V <20% (age <30 years) or 30% (age >30 years)
- Confusion and/or coma

Table 9.4. Options for hepatic support

Artificial liver support devices
Bioartificial livers
Hepatocyte transplantation
Extracorporeal liver perfusion
Artificial liver support systems

going clinical trials, utilizes dialysis fluid containing charcoal and a cation exchange resin to bind toxic substances in the blood. A pilot study performed in acute liver failure patients showed that this system was well tolerated and could produce biochemical improvements, although its ability to reverse the progression to terminal brain swelling has not be demonstrated. A second support device, the molecular absorbents recirculating system (MARS), consists of a dialysis system where the polysulphone membrane is impregnated with albumin and the dialysate enriched with albumin to facilitate the removal of toxic metabolites.[8] A third artificial liver support system, the microsphere based detoxification system (MBS), involves plasma recirculation at very high flow rates with all flow being exposed to particle size absorbents, which provide a large surface area for absorption.[9]

Bioartificial Liver Support Systems

Another approach consists of a Bioartificial Liver (BAL). In this system, plasma obtained with a centrifugal plasma separator is subsequently perfused through microcarrier bound porcine hepatocytes.[10] This device has been studied clinically with some promising early results. It is difficult to determine what role the hepatocytes played in these instances since a charcoal column is also included in the circuit. An alternative approach, the Extracorporeal Liver Assist Device (ELAD), utilizes blood perfusion through hollow fiber membranes surrounded by cells of a human tumor cell line (C3A).[11]

Hepatocyte Transplantation

More recently, hepatocyte transplantation has been used successfully to treat certain metabolic disorders[12] and preliminary data indicate that it may also be effective in acute liver failure. However, the number of cryopreserved hepatocytes required to achieve success may limit the utility of this approach.

Extracorporeal Liver Perfusion

This approach overcomes many of the problems associated with the previous approaches including: a) the inability to support all the functions provided by the liver and b) inability to provide enough hepatic support to overcome the derangement associated with fulminant hepatic failure. Both human and porcine livers have been used successfully with this approach. Since the shortage of human livers remains the essential problem in patients with fulminant hepatic failure, the only human livers which will be available for this technique will be those of poor quality and that are not usable for transplantation.

With this approach, an extracorporeal circuit perfuses blood from the femoral vein, incorporates a centrifugal pump and a tissue oxygenator which lead to the porcine liver (which is kept in a sterile temperature controlled environment at the bedside), and then returns the blood to the patient through the jugular or axillary vein.(Fig. 9.1) This approach has been successful in the past in providing both biochemical and neurological improvement in patients. More recently, successful 'bridging' to successful liver transplantation has been achieved.[13] The limiting factor with porcine livers has been a vascular rejection that occurs within 2 to 4 hours of perfusion due to preformed human antibodies to porcine endothelium.

Because of severe organ shortages, recent interest in xenotransplantation has led to strategies which have overcome the early rejection associated with pig to primate transplantation.[14] The most exciting of these approaches has been the development of pigs which are transgenic for human complement regulatory proteins (CD55 and CD59). In this setting, complement activation does not occur in pig endothelium and early rejection can be potentially avoided. Transplantation of organs from transgenic pigs to non-human primates extends kidney graft survivals from hours to weeks when compared to organs from non-transgenic pigs. Therefore, it is anticipated that prolongation of survival will provide a period of hepatic support, which will clearly exceed that experienced with non-transgenic pig livers.

B. LIVER TRANSPLANTATION FOR PATIENTS WITH CHRONIC LIVER DISEASE

COMPLICATIONS OF CIRRHOSIS

Cirrhosis can arise from two major categories of disease: hepatocellular and cholestatic. Within both groups, further subclassifications can be delineated.(Table

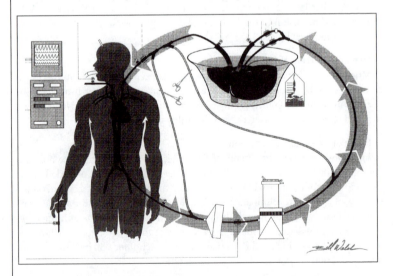

Fig. 9.1. Extracorporeal liver perfusion circuit (see text).

9.5) While all etiologies share common features of liver failure once an advanced stage is reached, unique aspects of each etiology influence management during and following transplantation.

Liver transplantation is also indicated for patients with certain metabolic diseases that can present with liver failure in the absence of cirrhosis.(Table 9.6) This is more common in the pediatric population, but can occasionally extend into young adulthood. Other congenital abnormalities (urea cycle enzyme deficiencies, familial hypercholesterolemia, familial amyloidosis) can present with extra-hepatic manifestations that are so severe that liver transplantation is recommended in the absence of hepatic disease. Finally, a miscellaneous group of chronic disorders may require transplantation in the absence of both cirrhosis and hepatic failure.

PATHOPHYSIOLOGY OF CHRONIC LIVER DISEASE

The pathophysiology of advanced liver disease results in two cardinal pathophysiological abnormalities: hepatocellular failure and portal hypertension. In acute liver failure, portal hypertension is seldom a clinical problem while in cirrhosis, an increased portal pressure may give rise to complications while hepatocellular function is preserved. The importance of these two factors is recognized in the Child-Turcotte-Pugh classification, a prognostic tool in patients with cirrhosis.(Table 9.7)

PORTAL HYPERTENSION

Portal pressure rises as a result of both a high hepatic vascular resistance and an increased portal venous inflow. The anatomical site of the increased vascular resistance in the liver will vary with different etiologies of cirrhosis[15], the hepatic sinusoids being the critical site for alcoholic cirrhosis. A functional component to this resistance may also be present, as transformed stellate cells in the sinusoids may respond to vasoconstrictive stimuli, such as endothelin. Once a critical level of portal hypertension is reached (hepatic venous pressure gradient of 10-12 mmHg, defined by the pressure gradient between the portal vein and the hepatic vein), portal-systemic collaterals form in an attempt to decompress the portal system. Portal hypertension is sustained by the development of increased portal venous inflow.

This increase in portal flow is part of a generalized hemodynamic abnormality of both acute and chronic liver failure consisting of a hyperdynamic circulation. The mechanisms which contribute to the arteriolar vasodilatation are under investigation, but an increased production of nitric oxide in the vascular endothelium and hence low systemic vascular resistance may explain the levels of circulating cytokines (such as TNFa) that are present in patients with both acute and chronic liver disease. The hyperdynamic state has repercussions on other organs, such as lung and kidneys, which pose specific problems in the management of the patient before, during and after liver transplantation.(Fig. 9.2)

Hepatocellular Failure

The "intact hepatocyte" theory of hepatocellular failure postulates that a critical number of viable hepatocytes is needed to maintain liver function. The "sick hepatocyte" theory suggests a generalized malfunction of individual cells. There

Table 9.5. Cirrhosis and liver transplantation

	Special Considerations for Liver Transplantation (OLT)
Hepatocellular Diseases	
Chronic hepatitis	
Hepatitis B	Virus should be non-replicating (HBV-DNA negative)
Hepatitis D	Co- or superinfects Hepatitis B. Rare in the US
Hepatitis C	Important to exclude alcohol as comorbid factor
Autoimmune	Pre-OLT medication may affect post-OLT bone disease
Drug-induced	Examples: nitrofurantoin, alphamethyldopa
Steatohepatitis	
Alcohol	Abstinence and social support critical for OLT.
Obesity	Increasing prevalence of cirrhosis. Rate of recurrence.
Drug-induced	Example: Amiodarone.
Vascular disease	
Chronic Budd-Chiari syndrome	Acute occlusion is amenable to decompressive surgery. R/O myeloproliferative syndrome, thrombotic tendency.
Inborn errors of metabolism	
Hemochromatosis	Cardiac involvement results in increased OLT morbidity.
Alpha-1-antitrypsin deficiency	Lung disease is rare in the presence of liver cirrhosis
Wilson's disease	OLT for acute disease not amenable to medical therapy
Glycogen storage disease type I/III	Can present in early adulthood.
Cholestatic Diseases	
Disease of intrahepatic bile ducts	
Biliary atresia	Kasai procedure may offer relief for a few years before OLT.
Primary biliary cirrhosis	Bone disease can be especially problematic post-OLT.
Drug-induced disease	Examples: Chlorpromazine, tolbutamide.
Familial cholestasis	Byler's syndrome, arteriohepatic dysplasia.
Cystic fibrosis	Insipissated bile syndrome leading to cirrhosis.
Disease of extrahepatic bile ducts	
Primary sclerosing cholangitis	Secondary cholangiocarcinoma may contraindicate OLT.
Secondary biliary cirrhosis	Requires Roux-en-Y anastomosis at OLT.

9

may be elements of both theories in advanced of cirrhosis. On a practical level, the 3 biochemical tests used in the Child-Turcotte-Pugh classification have not been superseded by more sophisticated tests, such as those that arise from tests of drug metabolism (e.g., lidocaine, caffeine).

Recipient Evaluation

A thorough evaluation of the subject's candidacy for liver transplantation must include an assessment of the need, urgency and technical feasibility of OLT. The acuity and extent of the investigation is frequently determined by the severity of liver disease. In patients with fulminant hepatic failure, in whom therapeutic decisions need to be made over a short interval, the evaluation phase may need to be

Table 9.6. *Liver abnormalities without cirrhosis*

Congenital abnormalities
  Urea cycle enzyme   Severe hyperammonemia may cause
    deficiency   neurological deficits.
  Homozygous   Important to assess status of coronary
    hypercholesterolemia   arteries pre-OLT.
  Primary hyperoxaluria   May also require renal transplantation.
    type I
  Familial amyloidotic   Need to assess cardiac status. Disease may be
    polyneuropathy   too advanced.
Developmental abnormalities
  Polycystic liver disease   OLT indicated for symptoms from massive
                             hepatomegaly
  Caroli's disease   Chronic biliary sepsis can be an indication
                             for OLT.

Table 9.7. *Prognosis in cirrhosis. Child-Turcotte-Pugh classification*

	Points	1	2	3
Reflecting Portal Hypertension				
Ascites		None	Controlled with meds	Not controlled
Hepatic encephalopathy		None	Controlled with meds	Not controlled
Reflecting Hepatocellular Failure				
Bilirubin (mg%)		0-2	2-3	>3
Prothrombin time (secs prolonged)		0-3	3-6	>6
Albumin (g%)		>3.5	2.8-3.5	<2.8

Minimum score: 5. Maximal score: 15
CTP Class: A: 5-6
                     B: 7-9
                     C: ≥ 10

streamlined and accelerated. The recipient evaluation includes the investigation of four major areas:

i) Assessment of Etiology of Liver Disease

This aspect requires an adequate history, physical examination and laboratory testing.(Table 9.8) Radiologic imaging of the liver and endoscopic evaluation of the GI tract are also needed. A liver biopsy, obtained either percutaneously or via the transjugular route in patients with ascites and severe coagulopathy, can provide a definitive diagnosis and may be critical in selected patients with acute liver failure and for others in whom alcoholic hepatitis is suspected.

ii) Assessment of the Complications of Cirrhosis

Several complications of cirrhosis signal the need to proceed with liver transplantation and require selective diagnostic tests. The tools to complete such workup are delineated in Table 9.9.

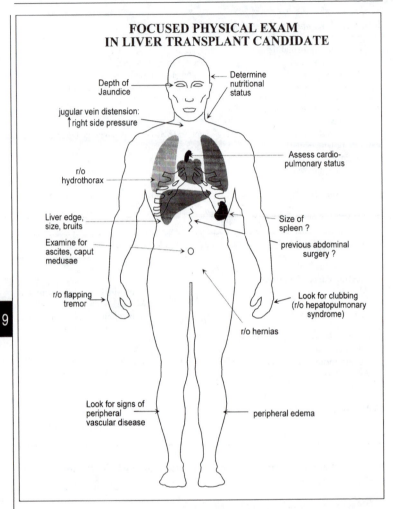

**FOCUSED PHYSICAL EXAM
IN LIVER TRANSPLANT CANDIDATE**

Depth of Jaundice

Determine nutritional status

jugular vein distension: ↑ right side pressure

Assess cardio-pulmonary status

r/o hydrothorax

Liver edge, size, bruits

Size of spleen ?

Examine for ascites, caput medusae

previous abdominal surgery ?

r/o flapping tremor

Look for clubbing (r/o hepatopulmonary syndrome)

r/o hernias

Look for signs of peripheral vascular disease

peripheral edema

Fig. 9.2.

iii) Assessment of Exclusion Criteria (Contraindications)

Older recipients are increasingly referred for evaluation. Although there is no absolute chronological limit for age above which transplantation is contraindicated, evaluation of physiological age requires a thorough clinical assessment. Adequate evaluation of cardiac function is critical. Obese and diabetic individuals are also at risk of atherosclerotic vascular disease and require full cardiovascular evaluation. While non-invasive cardiac testing may be adequate in the younger, otherwise healthy candidate, this will be insufficient for patients at risk (Table

---

Table 9.8. Testing to assess etiology of liver disease (blood tests)

Hepatitis B, HBV-DNA, HBeAg, anti-HBe, and anti-Delta Abs.
Hepatitis C, HCV-RNA, HCV genotype
Autoimmune: Anti-smooth muscle Ab (ASMA), Antinuclear Ab (ANA),
Antimitochondrial Ab (AMA).
Alpha-1-antitrypsin level/phenotype.
Wilson: Ceruloplasmin, 24 hr urine copper, liver copper.
Hemochromatosis: Iron saturation, ferritin, HFE gene test.
Blood group (for listing purposes)

Table 9.9. Testing to assess the complications of liver disease

Arterial blood gases: r/o hypoxemia/hepatopulmonary syndrome
Liver imaging: r/o hepatocellular carcinoma (HCC)
Serum alpha-fetoprotein, Ca19-9: r/o HCC, cholangiocarcinoma
Doppler ultrasound: r/o portal vein thrombosis (PVT)
Upper gastrointestinal endoscopy: Assess portal hypertension
Bone densitometry: Selected patients
Neuropsychological testing: Selected patients

Table 9.10. Testing to exclude contraindications

Infectious disorders:	HIV, syphilis, CMV, EBV, toxoplasmosis
Malignancy:	Colonoscopy in primary sclerosing cholangitis (ulcerative colitis)
	ERCP in primary sclerosing cholangitis (cholangiocarcinoma)
	In HCC: bone scan, lung CT (metastatic workup)
	Screening (colon, breast, cervical, prostate cancer)
Cardiopulmonary status:	CXR, EKG, 2D-Echo (routine)
	Thallium stress test, coronary angiography (patients at risk)

9.10). Other organs also require attention. Bone disease post-transplantation is affected by the pre-transplantation bone status (especially in older patients, those receiving corticosteroids pre-transplantation and those with cholestatic liver disease) and post-transplant medications. Bone densitometry is required in such individuals for adequate evaluation and follow-up.

Pulmonary Function Tests

Co-existing medical conditions need to be ascertained. Uncontrolled infection outside the biliary tree is an absolute contraindication to transplantation. In the case of malignancy, metastatic hepatobiliary and extrahepatic malignancy are also absolute contraindications. For other neoplasias, a waiting period of 5 years after treatment of a solid organ tumor and 2 years for a hematological disorder is recommended. The presence of AIDS is a contraindication to transplantation, as post-transplant immunosuppression accelerates the course of the disease. Irreversible brain damage and multiorgan failure also preclude the liver transplant procedure.

iv) Psychosocial Assessment

It is important to predict the ability of the candidate to exhibit discipline and responsibility during his post-transplant care. Assessment of the patient's life style, psychological stability (including his/her perception of disability) and extent of family support require interaction with Psychiatry/Social Work support services. This evaluation is critical for patients with alcoholic liver disease, in whom the ability to abstain from alcohol post-transplant can be assessed by the ability to abstain before transplantation (at least 6 months), employment history and a support structure (family, friends). Patterns of drug abuse need to be explicitly discussed. Emergency psychiatric assessment is needed for acute hepatic failure from ingestion of acetaminophen with suicidal intent, as an interview should occur before the patient develops an altered mental state. If the latter is present, the team needs to rely on the individual history (e.g., previous suicidal attempts) and a family interview to reach a decision.

SELECTION CRITERIA AND LISTING PROCESS

The decision to proceed with transplantation requires a careful assessment of the etiology and staging of liver disease, the complications of cirrhosis, potential contraindications, and a comprehensive psychosocial evaluation. The results of the work-up may not be clear-cut and a determination to exclude a candidate can be difficult, especially when the alternative outcome to the patient is certain death. A Multidisciplinary Transplant Review Board, composed of all individuals involved in the different aspects of care of the transplant recipient, needs to weigh dispassionately the pros and cons of each candidate in order to reach a rational decision. Input from consulting physicians, psychiatry, ethicists and social workers is critical to resolve specific situations. Each candidate must have an advocate who presents his/her case to the selection committee and the vote to proceed must be unanimous.

The patient needs to meet minimal listing criteria before placed in the waiting list (Child-Turcotte-Pugh score of at least 7 for most causes of cirrhosis). Once listing is approved, the patient is awarded a priority based on the current UNOS organ allocation scheme, the Model for End Stage Liver Disease (MELD). This scheme, based on predicted three-month mortality of patients awaiting liver transplant, uses laboratory values to generate a score which determines priority. The MELD equation incorporates serum creatinine, serum bilirubin, and international normalized ration (INR) as illustrated in Table 9.11.

Table 9.11. Model for end stage liver disease (MELD)

MELD Score	$= 0.957 \times \text{Log}_e(\text{creatinine mg/dl})$ $+ 0.378 \times \text{Log}_e(\text{bilirubin mg/dl})$ $+ 1.120 \times \text{Log}_e(\text{INR})$ $+ 0.643$

CLINICAL MANAGEMENT WHILE AWAITING LIVER TRANSPLANTATION

With the increasing waiting times, maintaining the patient in an acceptable medical condition in order to undergo a successful liver transplant is a challenge for the managing team. Both prophylactic measures and therapeutic interventions are needed to deal with the numerous complications that can arise.

TIMING OF LIVER TRANSPLANTATION

A sound knowledge of the natural history of disease is essential in the decision making process vis-à-vis the timing of transplantation. The development of complications typically results in an upgrade of the priority status for transplantation in exchange for a higher surgical mortality and a large increase in cost. This apparent paradox cannot be resolved given the current organ shortage. In acute liver failure, prognostic criteria have been developed to assess the necessity for urgent liver transplantation. In patients with chronic liver disease, the prevention and management of potential complications requires an inordinate amount of attention and comprehensive care on the part of the clinician.

PROPHYLAXIS OF COMPLICATIONS

Patients in the waiting list are at risk for developing HCC. Screening with ultrasound and alpha-fetoprotein level determination every 6 months is performed by most transplant centers. Screening upper endoscopy to rule out the presence of medium/large varices with red wheals is also recommended, as these patients may benefit from prophylaxis of variceal hemorrhage with beta-blockers. Hepatitis B vaccination is seldom useful in advanced stages of liver disease, but is recommended by some centers. Hepatitis A vaccination has become recently available and its utility in patients with liver disease is currently being evaluated.

THERAPY OF COMPLICATIONS

The rationale for each therapy is beyond the scope of this handbook and the reader is referred to standard references.[16,17] Each of the four major complications has a management protocol.(Table 9.12) However, the development of one complication can trigger additional problems. GI hemorrhage and infection have the potential of aggravating liver and renal function, while intractable ascites impairs respiratory function and aggravates malnutrition. Overt hepatic encephalopathy can result in aspiration pneumonia and may require prophylactic tracheal intubation. Fluid overload in the setting of renal failure and severe hypoalbuminemia requires extracorporeal measures for correction, such as CVVH (continuous venovenous hemofiltration). These patients require extensive and intensive support to overcome these problems.

RECIPIENT OPERATION

When a suitable donor is identified for a recipient, a rapid evaluation of the recipient is done so that any potential contraindications that may have arisen during the waiting period are noted and appropriately investigated. Please refer to the appropriate chapters for donor and anesthetic issues. The ensuing section will address the technical aspects of liver transplantation.

Table 9.12. Treatment of complications of cirrhosis

A. Variceal Hemorrhage

Initial hemostasis
- Pharmacological therapy
  Vasopressin (0.1-0.4 U/min) and nitroglycerin (start with 1 mg/kg/min iv).
  Octreotide [100 ucg bolus, 50 ucg/hr infusion (still unproven when given alone)]
- Endoscopic therapy
  Variceal band ligation preferred over endoscopic sclerotherapy.
  Fundic varices not amenable to endoscopic therapy in the US.
- Mechanical tamponade
  Sengstaken-Blakemore tube requires knowledge of potential complications.

Prevention of early rebleeding
- Octreotide infusion for 5 days
- Treatment of bacterial translocation: Norfloxacin 400 mg/day.

Maintenance therapy
- Pharmacological therapy
  Propranolol, to reduce portal pressure by 20%, start with 20 mg bid (requires hepatic vein catheterization) or
  Maximal dosage that reduces heart rate to 25% of baseline or not<55 beats/minute.
  If portal pressure reduction not attained, add isosorbide mononitrate 5 mg bid.
- Endoscopic therapy
  Continue variceal band ligation until erradication of varices (achieved with 4-5 sessions in 40-50% of patients).

Failure of therapy
- Shunt surgery, especially distal splenorenal shunt
  For patients with good liver function (Child 5-7 and no ascites).
- Transjugular intrahepatic portal-systemic shunt (TIPS)
  Rescue therapy, for patients with poor liver function

B. Hepatic Encephalopathy

1. *Correct precipitating event*
   Cleansing enemas for GI bleeding
   Volume expansion/electrolyte correction
   Treatment of infection, (without aminogylcosides !)
   Antagonism of sedatives (flumazenil, Narcan)
2. *Diet*
   Protein intake should be at least 0.75-1 g/kg (counteract catabolic state).
3. *Non-absorbable disaccharides*
   Lactulose po 20-30 cc q 8-12 hours (via NG in ICU)
4. Zinc sulfate, 300 mg q 12 hours (to increase urea synthesis in liver)
5. *Antibiotics on intestinal flora*
   Neomycin (3-6 g/day) for short periods (to avoid toxicity)
   Metronidazole, start at 250 mg bid.
6. In stage III-IV encephalopathy, Endotracheal intubation to prevent aspiration

C. Ascites

1. *Diet and fluid balance*
   Bed rest and low sodium diet (2-4 g/d)
   Fluid restriction (1L/day) for serum sodium <130 mEq/l
   Daily weight, urinary output and fluid balance
2. *Diuretics*
   With no response to a low sodium diet and a low U_{Na} (r/o dietary non-compliance)→
   Spironolactone (100-400 mg/d) alone or with furosemide (20-160 mg/day)
   Restrict weight loss to not > 1kg/d when no peripheral edema
   Careful with diuretic complications
     Renal impairment
     Hepatic encephalopathy
     Hyperkalemia with renal failure (Spironolactone)

cont'd on next page

Table 9.12, cont'd.

3. *Large-volume paracentesis*
   Indicated for tense ascites that impairs respiration, for refractory ascites.
   Albumin administered after paracentesis (6g/L removed) to avoid post-paracentesis circulatory dysfucntion.
   Diuretics continued after procedure if possible.
4. *TIPS*
   Poor outcome (worsening liver failure) in patients with Child class C cirrhosis.
5. *Hepatorenal syndrome*
   Assure volume expansion with central pressure monitoring
   Experimental therapy: Vasoconstrictors, TIPS

D. Spontaneous bacterial peritonitis
1. *Choice of antibiotics*
   Initial therapy with cefotaxime 3-6 g/d (or equivalent) until culture results.
   Repeat paracentesis after 48 hours to assure response (ØPMN of 50%).
2. *Culture-negative neutrophylic ascites*
   Repeat paracentesis critical
3. *Prevention of renal failure*
   Discontinue diuretics until satisfactory microbiological response
   Experimental therapy: iv albumin.
4. *Prophylaxis. Several regimens proposed*
   Norfloxacin 400 mg/d, Bactrim 5 days/week, Ciprofloxacin 1/week.

STANDARD SURGICAL TECHNIQUE

The recipient operation consists of hepatectomy of the native liver followed by implantation of the donor liver. The native hepatectomy can be difficult, especially in patients with previous upper abdominal operations and severe portal hypertension. The ligamentous attachments of the liver are systematically taken down followed by skeletonization of the hilar structures, namely the bile duct, hepatic artery, and portal vein, in preparation for implantation of the new liver. The retroperitoneal (bare) area is taken down last since most of the blood loss can result from this dissection. Finally, the inferior vena cava (IVC) is encircled below the liver having divided the adrenal vein, and above the liver allowing enough room between the diaphragm and the origin of the hepatic veins for a vascular clamp to be comfortably placed. At this 'point of no return', the bile duct is ligated and divided, as is the hepatic artery. Vascular clamps are then placed on the portal vein and the IVC below and above the liver and the liver is removed by transecting the portal vein and the IVC and removing the retrohepatic IVC with the liver.

At this point, hemostasis is achieved as well as possible. Occasionally, the bare area may require coagulation with the argon beam coagulator and a few hemostatic sutures. Depending on the degree of coagulopathy the new liver may need to be implanted while there is ongoing bleeding from the bare area. The donor liver is prepared for implantation on the back table by removing its diaphragmatic attachments including ligation of phrenic veins, removing the adrenal gland and ligation of the adrenal vein, and preparing the arterial and portal venous structures. The donor liver is then brought onto the operative field and end-to-end anastomoses are constructed using running non-absorbable monofilament suture between donor and recipient suprahepatic IVC first, then the infrahepatic IVC. Prior to completion of the infrahepatic IVC anastomosis, the liver is flushed with 500 cc of cold Ringer's lactate solution until the effluent from the infrahepatic IVC is clear and, at this point, the IVC anastomosis is completed. Next the portal vein anastomosis is performed end-to-end with running non-absorbable monofilament suture leaving a 'growth factor' in order to prevent a narrowing of the anastomosis. In the case of a thrombosed or inadequate portal vein, a donor iliac vein conduit is anastomosed preferably to the confluence of the splenic and superior mesenteric veins (SMV) or alternatively to any patent branch of the portal venous system including the SMV. SMV-to-portal vein grafts are tunneled through the transverse mesocolon. Once the portal vein anastomosis is completed, the clamps are removed in sequence and the liver is thus perfused with portal venous inflow.

Venous-venous bypass (VVP) is occasionally used, prior to completion of the hepatectomy, in order to decompress the splanchnic venous system as well as venous return from the lower extremities. Some centers use VVP routinely, whereas in other centers it is not used at all.[18] Most centers use VVP in selected patients, especially when the hepatectomy has been difficult and bloody, or when significant portal hypertensive bleeding is evident especially from the bare area. VVP requires cannulation of both a lower extremity vein, typically the saphenofemoral vein, and an upper extremity or neck vein. This can be achieved either via cutdown or by a percutaneous approach. Partial VVP can also be used, consisting of lower extremity to upper extremity bypass alone, as compared to full VVP which includes a portal venous line in order to decompress the portal vein. The decision to use or not to use VVP may depend on the hemodynamic stability of the recipient upon clamping, especially of the IVC. Rapid infusion can be used to offset some degree of hemodynamic instability during the clamping phase, but if the patient does not tolerate clamping without significant hemodynamic instability, then VVP should be considered.(Table 9.13)

Upon reperfusion of the liver with portal venous inflow, patients can develop a "reperfusion syndrome" consisting of right-sided ventricular failure associated with high filling pressures and systemic hypotension, significant arrhythmias can also occur. This syndrome is usually transient in nature and thought to be secondary to infusion of potassium or acid load from the preserved liver, and from splanchnic and lower extremity venous congestion. Expert anesthetic management and correction of electrolyte abnormalities are needed during this transient period.

Table 9.13. Potential indications for venous-venous bypass

1. Severe retroperitoneal collateralization
2. Poor preoperative renal function
3. Hypotension following test clamping of the vena cava despite adequate volume loading
4. Intestinal or mesenteric edema
5. Fulminant hepatic failure
6. Inexperience with the procedure

(For more details on anesthetic considerations of liver transplantation including monitoring of coagulation please refer to Chapter 13.)

The hepatic artery anastomosis is typically performed between the recipient hepatic artery, at the junction of the gastroduodenal artery, and the donor celiac axis using a Carrel patch. Approximately 15 to 20 percent of the time, abnormal arterial anatomy is identified in the donor liver consisting of either an aberrant left hepatic artery emanating from the left gastric artery of the donor, which does not require any particular reconstruction, or an aberrant right hepatic artery originating from the superior mesenteric artery. This latter type of arterial anatomy requires arterial reconstruction on the back bench which most commonly consists of implanting the origin of the aberrant vessel onto the donor splenic artery so that the celiac axis can be used as a single inflow. Occasionally, the inflow from the recipient hepatic artery is inadequate either because of inadequate flow or as a result of abnormal arterial anatomy in the recipient. Donor iliac arteries are routinely harvested as part of the donor procedure and these can be used to construct a conduit between the recipient infrarenal aorta and the donor hepatic artery or celiac axis. This conduit can also be made to originate from the supraceliac aorta, although infrarenal reconstruction is more commonly used. The conduit can be brought to the hilum by creating a tunnel behind the pancreas, but can also be placed anteriorly through the transverse mesocolon.

Once the liver is arterialized and the hepatic artery demonstrates satisfactory flow, hemostasis is achieved, and the bile duct reconstruction is performed using end-to-end choledochocholedochostomy over a T-tube stent. Several variations of this anastomosis have been used. Recently the necessity for a T-tube has been

9

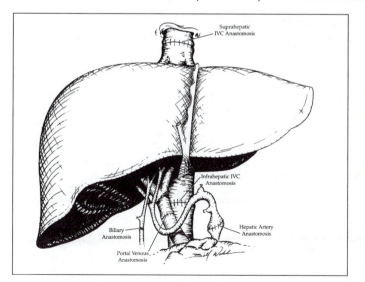

Fig. 9.3. Standard technique. This figure illustrates a completed liver transplantation with vascular and biliary anastomoses.

questioned[19] and some centers have elected not to use T-tubes primarily because of an unavoidable rate of biliary leaks following removal of the T-tube, as well as other technical problems associated with the T-tubes. Therefore, an end-to-end choledochocholedochostomy is performed using absorbable interrupted monofilament suture without stenting.(Fig. 9.3) If the recipient bile duct is not appropriate for end-to-end reconstruction, a Roux-en-Y choledochojejunostomy is performed in standard fashion with or without internal stenting.(Table 9.14)

ALTERNATIVE TECHNIQUES

"PIGGYBACK PROCEDURE"

The recipient hepatectomy can be altered to leave the recipient retrohepatic IVC in situ. Thus, during the hepatectomy, the caudate venous branches are ligated and divided individually as the IVC is separated from the liver. Occasionally accessory hepatic veins are encountered particularly to the right lobe and eventually the liver remains attached to the IVC only by the hepatic veins. The hepatic veins can then be either clamped and the ostia used for the IVC anastomosis (Fig. 9.4A), or suture ligated and another site on the recipient IVC used for anastomosis. The donor IVC is then anastomosed to the recipient IVC in a piggyback fashion by performing either an end-to-side or side-to-side IVC-to-IVC anastomosis. Once the IVC anastomosis is completed, the infrahepatic IVC is used as outflow of portal venous blood (instead of cold Ringer's lactate) in an effort to wash out preservation solution from the liver and following this, the infrahepatic IVC is ligated. The remaining structures are anastomosed in standard fashion (Figure 9.4B).

"SPLIT LIVER PROCEDURE"

Recently, the use of split livers has become routine for selected donor livers for most liver recipients. The liver is typically 'split' along the falciform ligament separating the left lateral segment (Couinaud segments II and III) from the remaining liver. The main hilar vascular and biliary structures are retained with the right side of the liver. The left lateral segment is typically transplanted into a child and the remaining liver transplanted into an adult. The transplant procedure for a split liver is identical to that for a whole liver with the exception that hemostasis at

---

Table 9.14. Indications for choledochojejunostomy

1. Donor-recipient bile duct size discrepancy
2. Diseased recipient bile duct
   a) Secondary biliary cirrhosis
   b) Primary sclerosing cholangitis
   c) Choledocholithiasis
   d) Biliary atresia
3. Presence of biliary duct malignancy
4. Poor blood supply to recipient bile duct
5. Inability to pass biliary probe through ampulla

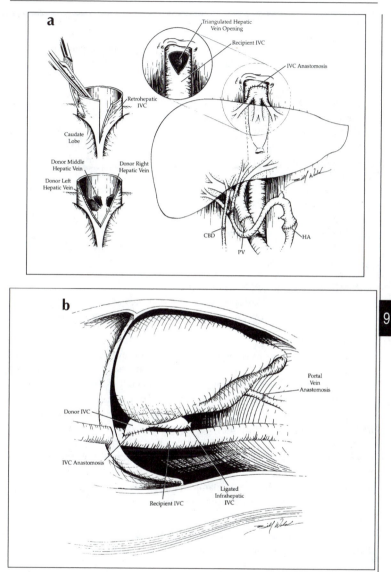

Fig. 9.4A and B. A) Piggyback technique. This figure illustrates the preparation for the piggy-back cavo-cavo plasty. First, the donor suprahepatic IVC is viewed from the back where a vertical slit is made in the middle of the back wall. This is triangulated to match the triangulated hepatic vein opening on the recipient side. Finally, the liver is viewed after all of the anastomoses have been completed showing an end on view. B) Side view of the piggyback procedure. Showing the triangulated cavo-cavo plasty of the donor suprahepatic IVC and the ligated infrahepatic IVC.

the cut surface needs to be secured and a careful check for biliary leaks in the raw surface needs to be carried out. Split liver transplant, when performed on proper recipients using suitable donor organs has survival results comparable to whole livers, but is associated with a higher rate of surgical complications.[20,21]

"AUXILIARY PROCEDURE"

In selected recipients with either metabolic diseases or acute liver failure, auxiliary transplants have been performed. A left lobe resection of the native liver is carried out and a donor left lateral segment or left lobe is transplanted orthotopically by anastomosing the donor left hepatic vein to recipient IVC end-to-side and portal vein hepatic artery and bile duct connections constructed in standard fashion. Nuclear studies are used to follow uptake/function by the donor/recipient liver, and in cases where the native liver recovers, the donor liver is either allowed to atrophy following withdrawal of immunosuppression, or is removed. Alternatively, the donor liver is treated like any other transplanted liver and ultimately becomes the predominantly functioning liver. Differential portal venous flow between recipient and donor liver segments may be responsible for preferential function and hypertrophy.

IMMEDIATE POSTOPERATIVE CARE

There are three major considerations in the immediate postoperative period: 1) liver function, 2) postoperative bleeding, and 3) general considerations.

1. LIVER FUNCTION

One of the most disastrous complications following liver transplantation is primary nonfunction (PNF). PNF needs to be differentiated from graft dysfunction which encompasses a spectrum ranging from mild graft dysfunction, manifested by elevated liver enzymes and poor early synthetic function, to severe dysfunction manifested by prolonged synthetic dysfunction, some degree of hemodynamic instability, and associated multiorgan dysfunction.[22] This end of the dysfunction spectrum along with PNF require consideration of urgent retransplantation, whereas mild to moderate dysfunction require close observation and supportive therapy. The appearance of the liver following reperfusion, the production of bile intraoperatively, and the hemodynamic status of the recipient provide intraoperative evidence of liver function (Table 9.15).

Table 9.15. Helpful signs of hepatic function in the intraoperative period

1. Restoration of hemodynamic stability
2. Good renal function evidenced by adequate urine output
3. Stabilization of acid-base status
4. Normalization of the coagulation system
5. Normalization of body temperature
6. Maintenance of proper glucose metabolism
7. Adequate bile production
8. Good texture and color of the liver

However, the first 6 to 12 hours (immediate postoperative period) provide more definitive evidence of liver function. The best indicators of early graft function include normalization of Factor V levels, prothrombin time, and transaminases. In addition, clearance of lactic acidosis, awakening from the anesthetized state, and good renal function provide further affirmation of liver function (Table 9.16).

2. POSTOPERATIVE BLEEDING

Significant coagulopathy can be present following revascularization of the liver and can be attributed to fibrinolysis, heparin-like effect, and platelet and coagulation factor deficiencies. Under normal circumstances, with a functioning graft, coagulopathy is reversed by the time of abdominal closure. However, complete hemostasis may not be fully achieved at time of closure despite the best of efforts, especially if the recipient is hypothermic and if the operation has been long, difficult, and bloody. This scenario has become uncommon, but can nevertheless occur, especially in the setting of a dysfunctional graft. Under these circumstances, it may be preferable to place appropriate drains or even packs, close the abdomen, and return the patient to the intensive care unit. Close attention to ongoing bleeding despite correction of coagulopathy is essential. This can be achieved with a combination of hemodynamic monitoring, serial hematocrit determinations, and overall condition of the patient including urine output and measuring of drainage output. It may also be helpful to perform hematocrit determinations on the drain fluid. If ongoing bleeding, despite correction of coagulopathy and rewarming of the patient, is suspected, especially if hemodynamic instability and oliguria are present, the patient should be returned to the operating room for evacuation of hematoma and identification of ongoing bleeding. At this time, generalized oozing may have improved so that specific bleeding sites can be more easily identified and oversewn, especially in the bare area. The presence of a dry operative field at the time of abdominal closure however should not be viewed as evidence that postoperative bleeding cannot occur. Postoperative bleeding should be considered highly in the differential diagnosis of hypotension and oliguria in the immediate postoperative period even in patients in whom a dry field was achieved intraoperatively.

3. GENERAL CONSIDERATIONS

Hemodynamic stabilization is guided by the usual clinical assessments of adequate organ and tissue perfusion. Of note, patients with cirrhosis typically ex-

Table 9.16. Helpful signs of hepatic function in the immediate postoperative period

1. Hemodynamic stability
2. Awakening from anesthesia
3. Clearance of lactate
4. Resolution of hypoglycemia
5. Normalization of coagulation profile
6. Resolution of elevated transaminases
7. Bile of sufficient quantity and golden brown in color

hibit hemodynamic parameters consistent with those of a septic patient including high cardiac output and low systemic vascular resistance. These hemodynamic conditions may persist for several weeks following transplantation, and may require vasoconstrictive agents for optimal management.

Pulmonary management consists of appropriate ventilatory support with manipulation of respiratory rate, tidal volume, positive end expiratory pressure, and optimal oxygenation. Serial blood gases are used to monitor progress. The patient is typically extubated as soon as he/she is awake and exhibits a good inspiratory effort with adequate vital capacity. Early extubation leads to speedier recovery. However, massive fluid shifts and preoperative generalized debilitation may delay extubation. Once the patient is extubated, careful attention to incentive spirometry and the liberal use of chest physical therapy can help prevent the development of atelectasis and pneumonia. The nature of the incision combined with the state of debilitation of the patient are likely reasons why pulmonary complications are common in the postoperative period. In addition, the common presence of a right-sided pleural effusion in these patients may further delay pulmonary recovery. The importance of pulmonary care following extubation cannot be overstated.

Laboratory testing includes careful attention to glucose levels and electrolyte status. In addition to the usual attention to sodium and potassium, magnesium levels are typically low and magnesium supplementation is required. Ionized calcium determinations should be frequent and ionized calcium should be normalized. In addition, normalization of transaminases and prothrombin time or Factor V levels should be expected in the first 24 hours. If a T-tube is used, the quality of the bile can provide a helpful hint of good liver function. Finally, a baseline doppler ultrasound to assess patency of the hepatic artery in particular should be performed within the first 24 hours of transplantation.

INVESTIGATION OF LIVER FUNCTION TEST ABNORMALITIES

Liver function test abnormalities may consist of elevations in liver transaminases suggestive of hepatocellular necrosis or alkaline phosphatase and bilirubin suggestive of cholestasis. These two patterns of liver function abnormality are not mutually exclusive and can, therefore, occur simultaneously. However, the pattern of liver function abnormality may determine the most appropriate investigation algorithm by suggesting a cause for the laboratory abnormalities. In addition, the timing of the abnormalities may render some causes more suspect than others. The differential diagnosis of abnormal liver function tests include graft dysfunction, technical complications (vascular and biliary), immunological complications (rejection), infectious complications, and finally, recurrence of native disease (Table 9.17).

Graft dysfunction encompasses a wide spectrum ranging from mild to severe dysfunction. Mild dysfunction is manifested by a significant rise in transaminases postoperatively (above 2,500 IU) as a result of preservation injury. In addition, there may be a second peak in transaminases within 24 hours which is thought to be secondary to reperfusion injury. Regardless of the peak transaminase level, it is important that the trend in transaminase levels be downward. If transaminases continue to rise beyond 12 to 24 hours following transplantation, a more complete evaluation including assessment of mental status, coagulation profile, renal function, and hemodynamic stability should be carried out (Table 9.18).

A diagnosis of severe dysfunction or primary nonfunction must be differentiated from that of technical vascular complications including hepatic artery thrombosis, portal vein thrombosis, and hepatic congestion secondary to venous outflow obstruction. Preservation injury is generally associated with improving mental status and stable or improving prothrombin time which is easily correctable. In contrast, primary nonfunction is manifested by a patient who does not awaken and has progressive deterioration of mental status, a worsening coagulation profile which is not correctable, renal dysfunction, and hemodynamic instability. The treatment of severe hepatic dysfunction is primarily supportive. Intravenous prostaglandin E_1 has been shown to be beneficial.[23] Bioartificial liver support has been also used as a "bridge" until the liver either recovers or a suitable donor liver is located for urgent retransplantation. In cases of less severe dysfunction, the transaminases normalize over time as do the coagulation parameters. These patients, however, become severely cholestatic in the recovery period, likely as a result of impaired bile transport mechanisms and liver biopsy in these patients may reveal extensive bile plugging with ballooning hepatocyte degeneration consistent with severe cholestasis.

9

Table 9.17. Causes of hepatic dysfunction

Immediate
1. Primary allograft nonfunction
2. Primary allograft dysfunction
3. Hepatic artery thrombosis
4. Portal vein thrombosis
5. Hepatic vein and caval thrombosis
6. Biliary tract obstruction/leak

Delayed
1. Rejection
2. Infection
3. Biliary tract obstruction
4. Recurrent disease
5. Graft Dysfunction

Table 9.18. Signs of primary non-function

1. Failure to regain consciousness
2. Hemodynamic instability
3. Poor quality and quantity of bile
4. Increasing prothrombin time
5. Renal dysfunction
6. Rise in transaminases and bilirubin
7. Acid-base imbalance
8. Persistent hypothermia

VASCULAR COMPLICATIONS

Hepatic artery thrombosis can present with a variety of liver test abnormalities including very subtle elevations in transaminases and, therefore, may go undiagnosed in the early period and become manifest later with biliary complications such as bile leaks, bilomas, liver abscess, and biliary strictures.(Table 9.19)

Therefore, any abnormal trend in liver function tests should be investigated immediately with ultrasound/doppler and, if the hepatic arterial signal is not clearly seen, then an angiogram should be performed. The role of lytic therapy and/or urgent reoperation for thrombectomy remains controversial. Retransplantation may be necessary especially if liver function is severely compromised in the early postoperative period. Hepatic artery thrombosis is usually related to technical complications and, therefore, a satisfactory pulse in the hepatic artery should be obtained before leaving the operating room at the time of transplantation. There is increasing data to suggest that the use of flow probes and the measurement of hepatic artery flow may predict the risk of hepatic artery thrombosis.[24]

Portal venous thrombosis is less common, but can occur in the setting of significant portal vein stenosis or previous portal vein thrombosis in the recipient, especially in the pediatric recipient. Typically, severe elevations in transaminases are observed in the early period and ascites is a manifestation in delayed portal vein thrombosis. Also, acute portal hypertension manifested by variceal bleeding should alert the surgeon to the possibility of acute portal vein thrombosis. In the acute setting, thrombectomy should be attempted in an effort to save the graft, although retransplantation may be necessary especially if the graft is compromised.

Finally, venous outflow obstruction causing a Budd-Chiari-like congestion of the liver can be seen either following standard hepatic transplantation with end-to-end SVC anastomosis, but has been more commonly described in the setting of piggyback operations. Several innovative techniques have been advocated for repair. In the early postoperative period, a significant elevation in transaminases results from the acute congestion, whereas delayed manifestations consist primarily of ascites and evidence of portal hypertension.

BILIARY TRACT COMPLICATIONS

Anastomotic biliary leaks may occur early in the postoperative period resulting in either localized or generalized peritonitis. Biliary output from the drains and elevation in serum bilirubin out of keeping with elevation in the other liver function tests should raise this diagnostic possibility. These biliary leaks can occur either as a result of technical problems or as a result of hepatic artery thrombosis

Table 9.19. Manifestations of hepatic artery thrombosis

1. Elevation of the transaminases and bilirubin
2. Fulminant hepatic failure
3. Sepsis with hepatic abscesses or gangrene of the liver
4. Biliary anastomotic disruption
5. Biliary tract strictures

with ischemic compromise of the bile duct. These early leaks are best treated by reoperation and revision to a Roux-en-Y choledochojejunostomy. Localized leaks may be treated with endoscopic retrograde cholangiography (ERCP) and sphincterotomy with stenting of the bile duct leak.

Biliary leaks from the raw surface of split livers can be treated conservatively, especially if the leak is contained and adequately drained. If the leak continues, ERCP with sphincterotomy may be necessary. Delayed complications include stenoses of the bile duct anastomosis and intrahepatic biliary strictures which may or may not be related to hepatic artery thrombosis. These are typically managed by skilled ERCP intervention with dilatation and stenting. Where these fail, biliary reconstruction with a Roux-en-Y choledochojejunostomy may be necessary. Finally, dysfunctional motility of the bile duct and of the Sphincter of Oddi may result in functional obstruction in the absence of mechanical obstruction.[25] These types of problems manifest later on in the postoperative period. Also, biliary casts and stones can form, especially in the presence of longstanding T-tubes and may result in biliary obstruction requiring ERCP intervention.

The use of either endoscopic or percutaneous (transhepatic) techniques in the management of biliary complications is dictated by the availability of skilled interventional endoscopists and radiologists at the particular institution. In our opinion, endoscopic (ERCP) intervention is preferred and percutaneous transhepatic procedures are used, when for technical reasons, endoscopic access to the involved biliary tract is not possible.

REJECTION

Rejection can occur in the first days following transplantation, especially if induction immunosuppressive therapy is not used. The pattern of liver function test abnormalities varies and can be hepatocellular or cholestatic in nature. Diagnosis is made by liver biopsy since clinical signs and symptoms of rejection are extremely variable, non-specific, and unreliable (Table 9.20).

Rejection is a common phenomenon with at least 60 percent of liver transplant recipients having at least one episode. Acute cellular rejection usually occurs between the fourth and fourteenth day posttransplant with most episodes occurring within three months of transplantation. Some patients are asymptomatic while others may experience profound symptoms due to a failing liver allograft. The diagnosis of allograft rejection is confirmed by histologic examination of a liver biopsy. Classic histologic findings of acute cellular rejection include a portal infiltrate consisting of mixed inflammatory cells, where the presence of eosinophils

---

Table 9.20. Signs and symptoms of rejection

1. Fever
2. Decreased quality and quantity of bile
3. Elevation of the bilirubin and/or transaminase levels
4. Sense of ill being
5. Increased ascites

can be diagnostic, as well as lymphocyte-mediated bile duct injury, and endothelialitis.(Fig. 9.5, Table 9.21)

INFECTION

Abnormality of liver function tests secondary to infection is most commonly secondary to viral infections which include cytomegalovirus (CMV) hepatitis, as well as recurrence of previous viral hepatitides. Recurrence of disease will be covered in the next section. CMV hepatitis is diagnosed by the presence of inclusion

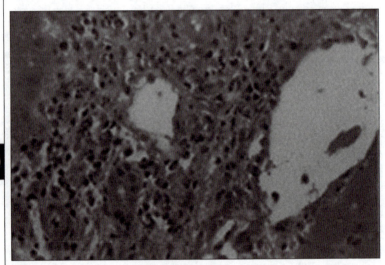

Fig. 9.5. Hematoxylin and eosin stain of acute cellular rejection demonstrating a mixed portal infiltrate with many eosinophils, endothelialitis, and evidence of cellular-mediated bile duct disruption.

Table 9.21. Histologic determinants of acute cellular rejection

1. Portal infiltrate with mixed inflammatory cells
2. Bile duct injury
3. Endothelialitis

Grading of Acute Liver Allograft Rejection – Banff Criteria[24]

Grade	Criteria
I (mild)	Cellular infiltrate in a minority (< 50%) of the triads, that is generally mild, and confined within the portal spaces.
II (moderate)	Cellular infiltrate, expanding most (> 50%) or all of the triads.
III (severe)	As above for moderate, with spillover into periportal areas and moderate to severe perivenular inflammation that extends into the hepatic parenchyma and is associated with perivenular hepatocyte necrosis.

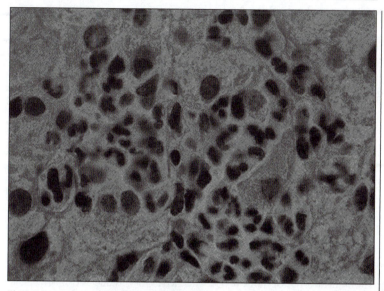

Fig. 9.6. Immunoperoxidase stain of CMV hepatitis demonstrating an inclusion body with intranuclear staining for CMV and a surrounding cluster of polymorphonuclear cells.

9

bodies with clusters of polymorphonuclear cells (Fig. 9.6). These "clusters" represent the "footprints" of CMV. This evidence of tissue invasive disease is often associated with symptoms of fever, general malaise, myalgias, and diagnosis is corroborated with shell vial culture or positive antigenemia tests. Treatment consists of reduction in immunosuppression and antiviral agents such as ganciclovir. In addition to CMV infection, other bacterial and fungal systemic infections may result in a secondary abnormality in liver function tests associated primarily with a cholestatic pattern. These elevations are difficult to sort out and may require multiple diagnostic efforts. Finally infection of the liver secondary to abscess formation may occur resulting in abnormal liver function tests typically as a result of hepatic artery thrombosis. These can be bacterial and fungal in nature and can be diagnosed first with ultrasound, then CT scan, and finally angiography. ERCP may be helpful in delineating the extent of biliary duct disruption. If severe enough, these biliary tract complications may require retransplantation of the liver.

RECURRENCE OF NATIVE DISEASE

Recurrence of native disease consists most frequently of recurrence of viral infection such as hepatitis B and hepatitis C, as well as non-A, non-B, non-C hepatitis. Recurrence of hepatitis B is easy to diagnose with either serum markers or stains for surface antigen and core antigen on the biopsy. In contrast, recurrence of hepatitis C may be more difficult to differentiate histologically from other causes of liver abnormality such as rejection. Although there is some evidence that there may be a role for immune modulators and antiviral agents such as interferon and ribavirin in the prophylaxis and treatment of recurrent hepatitis C, the data are inconclusive. Similarly, recurrence of non-A, non-B, non-C hepatitis can be extremely difficult to diagnose and these patients are often treated mistakenly for acute cellular rejection, which may initially improve liver number abnormalities, but eventually these abnormalities recur. In addition, recurrence of other diseases such as primary biliary cirrhosis, non-alcoholic steatohepatitis, and to a lesser degree, primary sclerosing cholangitis and autoimmune hepatitis have been described. These disease recurrences are typically diagnosed with a combination of liver biopsy and imaging of the biliary tree. The management of recurrence in these diseases can be difficult and for the most part, they are treated much in the same way as in the native liver.

POSTOPERATIVE CARE

The aspects and specifics of postoperative care are best delineated according to the particular postoperative period. These periods include: i) the immediate postoperative, ii) early postoperative inpatient, iii) early outpatient, and iv) long-term outpatient periods.

I) IMMEDIATE POSTOPERATIVE INPATIENT CARE

The bulk of the specifics of the immediate postoperative care are discussed above. The immediate postoperative period is defined by the postoperative intensive care unit stay. However, since immunosuppression is usually instituted in this period, a discussion of immunosuppression as it applies to liver transplantation follows.

IMMUNOSUPPRESSION

Induction therapy, traditionally in the form of anti-lymphocyte preparations (MALG, ATG, OKT3), have for the most part, not been widely used in liver transplantation. More recently, a resurgence of interest in induction therapy has resulted from the introduction of humanized IL-2 receptor antibodies (Zenapax and Simulect). The role of these and other newer induction agents in liver transplantation remain to be elucidated.

Baseline immunosuppression is instituted in the immediate postoperative period and typically consists of a calcineurin inhibitor (either Neoral (cyclosporine) or Prograf (tacrolimus)) and steroids. There are very few indications for intravenous administration of calcineurin inhibitors. Steroids are administered initially as intravenous Solu-Medrol and, once the patient is tolerating oral intake with sips of fluids, prednisone is used. Some centers advocate the use of a third agent, historically Imuran (azathioprine). Cellcept (mycophenolate mofetil) which has largely replaced Imuran in kidney and kidney/pancreas transplantation is being used increasingly either as a third agent or in an attempt to obviate the use of steroids, and in some patients the use of calcineurin inhibitors. The role of Cellcept in baseline immunosuppression for liver transplantation remains to be better defined. Rapamycin is presently being evaluated as an additional agent for baseline immunosuppression.

II) EARLY POSTOPERATIVE INPATIENT CARE

Patients are transferred out of the intensive care unit onto the transplant ward as soon as they are extubated and hemodynamically stable. This period is typically 24 to 48 hours and, upon transfer, the patients are encouraged to ambulate. Often, the patients' pretransplant debilitated state does not allow for early ambulation and these patients require special rehabilitation requiring transfer to acute rehabilitation units. However, if the patients are doing well and do not need long-term rehabilitation care, their diet is advanced as tolerated. Standard wound care is administered and the drains are removed, especially if no biliary leak is evident. Of note, the presence of large volumes of ascites in the drains should not result in delay in removing the drains.

In addition to immunosuppressive agents, prophylaxis against Pneumocystis carinii (PCP) is achieved with Bactrim. In patients with an allergy to sulfa-containing compounds, pentamidine inhalation and dapsone have been used successfully. Cytomegalovirus (CMV) prophylaxis is achieved with ganciclovir therapy. Most centers have transitioned from the use of intravenous ganciclovir preparations to the recently available oral preparations of ganciclovir. Newer preparations of oral ganciclovir appear to have better absorption and bioavailability kinetics and are likely to replace intravenous ganciclovir for prophylaxis. Of concern, increasing resistance to ganciclovir may dictate the use of anti-cytomegalovirus cocktails in the future especially for preemptive therapy rather than prophylaxis.

Standard antibacterial prophylaxis necessitates coverage of gram negative and anaerobic agents typically present in bile. Gram positive coverage appears to be less important. Finally, antifungal prophylaxis is achieved with swish and swallow of nystatin suspension or other such topical antifungal. In addition, agents such as fluconazole and itraconazole are used in the early postoperative period as prophylaxis against systemic fungal infections. Of note, these latter agents can result in dramatic increases of calcineurin inhibitor levels due to competition with cytochrome P450, and therefore, levels need to be monitored closely.

Most patients also receive peptic ulcer disease prophylaxis especially when receiving high-dose steroids in the form of either H_2 blockers or proton pump inhibitors. Magnesium supplementation is often necessary in patients who exhibit hypomagnesemia.

In addition to consideration of immunosuppression and prophylaxis, close attention to liver function tests and hematology and biochemistry laboratory values is essential in the first few days following transplantation. Typical problems of thrombocytopenia and mild renal dysfunction may require intervention such as platelet transfusion and optimization of central filling pressures, respectively. Liver function test abnormalities are investigated as outlined above for the immediate postoperative period.

In the case of inability to tolerate oral feedings, enteral feedings via nasoduodenal tube or intravenous hyperalimentation may be important. There are no convincing data to show that routine use of hyperalimentation, either intravenous or enteral, is beneficial in the majority of patients.

Common infections following liver transplantation include urinary tract, pulmonary, intra-abdominal, central venous catheter, and wound infections. Any fever or leukocytosis needs investigation for possible infection in these systems (Table 9.22).

If a T-tube is used, a T-tube cholangiogram is obtained at approximately the fifth postoperative day and in the absence of leak or obstruction, the T-tube is clamped. Clamping of the T-tube can result in a transient elevation in liver function tests. If the patient develops any abdominal pain following clamping of the T-tube, the house staff should be instructed to unclamp the T-tube and attach it to a drainage bag to gravity immediately. Following this, a repeat cholangiogram or HIDA scan should be obtained to rule out a biliary leak.

III) EARLY OUTPATIENT CARE

As soon as the patients are tolerating a diet and able to ambulate, they can be discharged to the outpatient setting and followed closely in the outpatient clinic. Typically, blood work is obtained three times weekly and the patients are seen and examined on a weekly basis. A standard protocol for the frequency of laboratory investigations and clinic visits is established (Table 9.23). Clinic visits are used to evaluate the patient and to review their medications to avoid errors.

Any elevation in liver function tests or any lab work abnormality is investigated further. Standard algorithm for elevation in liver function tests includes an ultrasound doppler examination of the liver looking for patency of the hepatic artery, portal vein, and hepatic veins. Also, the ultrasound will detect any dilatation of the biliary tree and any abnormalities within the parenchyma such as liver abscess formation. If the ultrasound is unremarkable, the next step usually consists of a percutaneous liver biopsy to rule out rejection and infection. In the early postoperative period, especially in patients undergoing transplantation for diseases other than chronic viral hepatitis, elevation in liver numbers can be treated empirically with steroid boluses without a need for biopsy. When needed, biopsies can be performed as outpatients and rejection can also be treated in the outpatient setting. In the case of steroid-resistant rejection which must be documented by a liver biopsy, treatment consists of anti-lymphocyte preparations (OKT3, ATG) typically for two weeks. OKT3 can be administered via peripheral vein, but a cytokine release syndrome may be associated with injection of OKT3 and, therefore, the first two to three doses of OKT3 need to be given in the inpatient setting

---

Table 9.22. Common causes of bacterial infection following liver transplantation

1. Line sepsis
2. Infected peritoneal fluid
3. Pneumonia
4. Intra-abdominal abscess
5. Biliary anastomotic leak
6. Cholangitis secondary to biliary tract obstruction
7. Urinary tract infection

Table 9.23. Frequency of outpatient visits and laboratory investigations

Outpatient Visits		Laboratory Investigations	
0 – 4 weeks	Twice weekly	0 – 1 months	Mon, Wed, Fri
5 – 8 weeks	Weekly	1 – 2 months	Twice weekly
9 – 12 weeks	Every other week	2 – 3 months	Weekly
3 – 6 months	Monthly	3 – 6 months	Every 2-3 weeks
6 – 9 months	Every 2 months	6 – 12 months	Monthly
9 months – 1 year	Every 3 months	12 – 18 months	Every 2 months
After 6 months, patients are returned		18 – 24 months	Every 3 months
to their referring physician.		Over 2 years	Every 6 months

and, therefore, require readmission. These reactions can be mild consisting of fever, diarrhea, and general feeling of malaise with myalgias. Alternatively, OKT3 treatment can be associated with violent reactions consisting of shaking chills, dyspnea, pulmonary edema requiring intubation especially in patients who are volume overloaded, and other manifestations of cytokine release.

ATG administration requires a central venous catheter and does not tend to be associated with overt cytokine release syndrome. When anti-lymphocyte therapy is used, ganciclovir prophylaxis intravenously administered concomitantly has been associated with an decreased incidence of CMV infection. More recently, treatment with high-dose Prograf has been used to reverse acute cellular rejection.

If the biopsy does not reveal the cause for elevation in liver function tests, visualization of the biliary tree is imperative, in order to rule out obstruction. If a T-tube was used, a T-tube cholangiogram can be obtained in order to visualize the biliary tree even in the absence of a dilated biliary duct by ultrasound evaluation. Some centers use cystic duct stents and these also can be injected with radiopaque material in an attempt to visualize the biliary tree. If a T-tube or cystic duct stent is not used, visualization of the bilary tree requires ERCP for both diagnosis and intervention if necessary.

On occasion, despite these diagnostic maneuvers, the reason for elevation of liver function tests remains elusive. In the case of recurrent hepatitis C, the biopsy can be misleadingly normal or show the occasional 'Councilman Body' or apoptosis of hepatocytes despite significant elevations in liver function tests. In these cases, conservative management and close observation will eventually reveal the cause for liver function tests abnormality. Not infrequently, repeat biopsies are required before a diagnosis can be established.

The use of routine protocol biopsies is controversial. Although some centers use protocol biopsies in every patient, the occasional findings of histologic rejection in a patient with normal liver function tests and no clinical evidence for rejection can pose a management dilemma. Consequently, most transplant centers have abandoned the use of routine protocol biopsies and rely on either laboratory or clinical abnormalities as a stimulus for liver biopsy and other investigations.

In addition to abnormalities in liver function tests, patients are encouraged to report any potential signs of infection such as fever or chills. If a patient experiences a fever, this is quickly investigated with pancultures for bacterial, fungal, and viral infections including CMV. A chest x-ray is also part of the routine fever workup and, if there is an indwelling central venous catheter, retrograde cultures are also used. Antibiotic therapy is instituted empirically in the immunosuppressed patient while awaiting the results of the cultures especially if the patient appears septic or toxic. Low-grade fevers can be investigated and managed in an outpatient setting, whereas high fevers, especially in a toxic patient, require urgent readmission to the hospital and may require more thorough investigation such as a CT scan of the abdomen to rule out intra-abdominal sepsis.

Side effects of the immunosuppressive agents need to be considered. The most common drugs which result in significant side effects are the calcineurin inhibitors. These side effects include nephrotoxicity, neurotoxicity, hyperkalemia, hypomagnesemia, hypertension, and tremor. Prograf has the additional side effect of inducing new onset diabetes and is more prone to result in GI symptoms of abdominal pain and diarrhea. Both drugs are metabolized through the cytochrome P450 system and, therefore, drugs which increase the effective level include erythromycin and antifungal agents ketoconazole, fluconazole, and itraconazole, as well as calcium channel blockers such as diltiazem, verapamil, and nicardipine. Drugs which decrease levels are primarily the anti-seizure medications in general (phenytoin, phenobarbital, and carbamazepine) and most of the anti-tuberculosis medications such as isoniazid, rifampin, and rifabutin. Twelve-hour serum trough levels are measured and monitored closely and the dosage of these agents is guided by these levels.

Imuran and Cellcept primarily cause leukopenia and, when used, these agents must be adjusted according to the white blood cell count. If the white blood cell count is below 3,000, as a rule, these agents should be held. The use of GCSF and GM-CSF have made leukopenia in these patients much easier to manage. The use of these agents has not resulted in increased rejection.

In patients undergoing transplantation for hepatitis B-related chronic liver disease, human hepatitis immunoglobulin (HBIg) preparations are administered in high doses during the perioperative period. Typically, 10,000 U are administered intravenously during the anhepatic phase and then daily for six to seven days. Titers of antibody are measured and are maintained above 300. At one week following transplantation, the HBIg are administered intravenously weekly at first and then monthly. Eventually HBIg can be administered intramuscularly at monthly intervals always maintaining titers above 300 IU (Table 9.24). In addition, antiviral agents have been used, particularly in patients with HBV DNA positivity prior to transplant. DNA positivity is considered a contraindication to transplant unless patients can be rendered DNA negative with the use of antivirals such as lamivudine. In patients who are rendered HBV DNA negative with lamivudine, over time, lamivudine-resistant mutants arise and, therefore, the combination of HBIg and lamivudine is thought to provide better recurrence prophylaxis than either agent alone. Lamivudine is continued in the posttransplant period and the optimal combination regimen for HBIg and lamivudine in the long term remains to be worked out.

Table 9.24. Liver transplant protocol for HBV DNA positive patients

---

1. Pretransplant
   HBV DNA positive patients are to be started on lamivudine 100 mg po q day (available in elixir). If unable to tolerate elixir, it is available in 150 mg tablets; may take 150 mg q day. Recheck Hepatitis B surface antigen, HBV DNA, and Hepatitis Be antigen and antibody at one month intervals for three months. Following two successive negative results, repeat every two to three months. Due to reported cases of pancreatitis using other related drugs, also check amylase periodically. Patient remains on lamivudine.

2. Intraoperative
   Stage II (Anhepatic) – 10,000 IV of HBIg given IV.

3. Postoperative – Inpatient
   A. 10,000 IU HBIg IV daily x 6 days.
   B. Blood samples at trough point (just prior to next dose). Samples to be done on days 1, 2, and 6 and more frequently if necessary based on results (desired level is > 300 IU).

4. Postoperative – Outpatient
   A. If the trough HBIg level on day six is > 300 IU, begin giving 10,000 IU IV q week x 4 weeks. At this point, if desired levels are maintained, can switch to 5cc HBIg IM q week x 4 weeks. If desired levels are still maintained, can switch to 5cc HBIg IM q month.
   B. Laboratory monitoring:

Hep B surface antigen	q month x 3 months
Hep Be antigen	then, q 3 months x 12 months
Hep Be antibody	
HBV DNA	
HBIg levels (trough)	q week x 3 months
	then, q 2weeks x 3 months
HBV DNA quantitative	q 3 months x 12 months

5. Long-Term Monitoring
   If patient remains HBV DNA negative after 6 months, obtain HBIg levels q month. Continue administering 5 cc HBIg IM q month for at least the first year. Longer term dosing and laboratory monitoring will depend on patient response and broader experience data. Patient is to remain on lamivudine for lifetime.

Patients who undergo transplantation and are found to have hepatocellular carcinoma (HCC) need close monitoring posttransplantation for recurrence. A large proportion of these patients have elevated alpha-fetoprotein (AFP) levels prior to transplantation and, in those patients, serial AFP determinations can be used to monitor recurrence. Unfortunately, recurrence of HCC is associated with poor long-term outcome. Despite the use of pretransplant adjuvant therapy in the form of either chemoembolization or local therapy, the use of adjuvant chemotherapy following transplantation is not used universally. The treatment of recurrent HCC is not very satisfactory. On occasion, local recurrence at a site of needle biopsy can be excised with no negative impact on survival. However, intra-abdominal recurrence is usually associated with poor long-term survival.

Long-Term Outpatient Care

Once the patient has overcome the problems that characterize the first three to six months after transplantation, such as acute rejection and infection, new diagnostic and therapeutic goals become important. Some are directly related to transplantation, such as the progressive reduction in the dose of immunosuppressive agents and the periodic surveillance of liver function tests to screen for late complications. Others require a focus on the potential for complications that may arise from medications including the immunosuppressive agents. This longer term follow-up requires special attention to six major areas.

i) Liver function tests: Disease recurrence is possible following transplantation for certain indications. Reinfection of the liver after transplantation is common with hepatitis C, which may progress slowly to a fibrotic/cirrhotic stage in a smaller group of patients (10%) over a 2-5 year period. Reinfection with hepatitis B can progress more rapidly to a potentially fatal course with a picture of fibrosing cholestatic hepatitis; patients require active measures to prevent reinfection (e.g., hyperimmuneglobulin) and viral replication (e.g., lamivudine). Recurrence of primary biliary cirrhosis has been documented, although histological overlap with chronic rejection may confuse interpretation.

ii) Arterial blood pressure: The administration of calcineurin inhibitors and corticosteroids may result in arterial hypertension in up to 70 % of transplant recipients. Sympathetic stimulation with attendant vasoconstriction and volume expansion are thought to be responsible. In addition, liver disease prior to transplantation, characterized by systemic vasodilatation may "protect" the hypertensive patient. The need for calcium channel blockers (with attention to possible interactions with cyclosporin/tacrolimus), selective and non-selective beta blockers and diuretics is commonly seen in patients following liver transplantation. Decreasing the doses of calcineurin inhibitors and/or steroids may help reduce the need for antihypertensive agents.

Renal function: Deterioration of renal function is common after liver transplantation, mainly as a result of the sympathetic stimulation, renal vasoconstriction and decrease in glomerular filtration rate induced by calcineurin inhibitors. A reversible reduction in creatinine clearance of approximately 50% can be seen at one year. Over a more prolonged period, permanent reductions in renal function may occur, with variable degrees of proteinuria. Histology reflects both ischemic injury to the glomerulus as well as tubular damage. The use of prostaglandins has not been shown to diminish calcineurin-induced nephrotoxicity. In addition, the use of non-steroidal antiinflammatory agents and the use of drugs which affect the metabolism of calcineurin inhibitors can both result in impaired renal function.

Metabolic issues: Chronic liver disease is an insulin-resistant state and the administration of corticosteroids after the transplantation may result in overt diabetes. In addition, tacrolimus has been demonstrated to be diabetogenic irrespective of steroids. Elevations of both serum cholesterol and triglycerides can occur with cyclosporin, which may be somewhat less pronounced with tacrolimus. Drugs which effect cholesterol production have been used effectively in patients with elevated lipids. Hyperuricemia with gouty attacks reflect an effect of cyclosporin. Weight gain after transplantation can be substantial, especially after the 2nd month. Hyperphagia, high steroid dosage and central effects of calcineurin inhibitors contribute to this effect. Excessive weight gain has adverse repercussions on the control of blood pressure and diabetes. Development of atherosclerotic vascular disease is of concern and every effort should be made to control these metabolic effects, especially hyperlipidemia and diabetes..

Bone disease: Prior to liver transplantation, patients with chronic liver disease often have underlying bone disease, especially those individuals with cholestatic liver disease and those receiving steroids for autoimmune hepatitis. Bone loss occurs primarily in the first six months after transplantation and active measures to prevent this deterioration should be instituted. Baseline bone densitometry is of assistance to guide replacement. The latter includes supplementation with calcium and vitamin D. Anti-resorptive therapy with biphosphonates should also be considered.

Screening: It is important to be proactive in the search for potential complications. This includes periodic ophthalmological exams to rule out glaucoma and cataracts. Screening for malignancy proceeds under similar protocols to the general population, including mammography, gynecological examination and colonoscopy. Some additions exist as a result of the immunosuppressive state. Careful dermatological examination is important at every visit, as there is an increased incidence of squamous cell carcinoma of the skin. Patients should avoid bright sunlight hours and use sun-protecting lotions. Patients with primary sclerosing cholangitis and ulcerative colitis may be at high risk for development of colon carcinoma. Post-transplant lymphoproliferative disorder has a wide spectrum of pathology and can respond well to a reduction of immunosuppression. Knowledge of the rates of de novo malignancies following transplantation can be a useful guide to screening in these patients.

Psychological well-being: Well adjusted patients should resume their pre-transplant activities, retain gainful employment, and maintain normal social interactions. The transplant evaluation process should detect any warning signals which suggest a lack of motivation. Depression should be diagnosed as should other important factors such as personality disorders. Issues of compliance and recidivism, especially in patients with alcoholic liver disease cannot be overemphasized. This evaluation should include nursing, social work, and psychiatric services as needed. Vocational rehabilitation may be necessary. Finally, quality of life assessments should routinely be performed by transplant centers.

LIVE-DONOR LIVER TRANSPLANTATION (LDLT)

Overview

Over the past decade, the gap between the number of adult patients in need of liver transplantation and the number of organs donated has increased greatly. This discrepancy has increased both the mean waiting time to undergo transplant and mortality from complications of end-stage cirrhosis for patients on the waiting list. Over the past several years, attempts to address the inadequate supply of organs for transplant have included the use of marginal donors (age, hemodynamics, viral infection). More recently, living donors have been used to address this need.

The concept of using a living donor developed in pediatric transplantation more than a decade ago[26], waiting list mortality declined, and the procedure was shown to have excellent recipient results and low risk for morbidity and mortality in the donor. This concept was extended to adult live-donor liver transplant (LDLT). The LDLT procedure involves transplantation of the right hepatic lobe from one adult donor to another, with the first series in the United States presented in 1998.[27]

Live-Donor Liver Transplant Recipient

LDLT is considered for those patients likely to experience mortality while awaiting a cadaveric organ donor. Table 9.26 outlines those patients who are candidates for LDLT.

Table 9.26. LDLT candidate recipients

A. Pre-MELD
   Hepatocellular carcinoma (T_4 and T_2)
   Fulminant hepatic failure
   Patients not likely to receive cadaveric organ with life expectancy less than 6 months
B. Post-MELD
   Hepatocellular carcinoma (exceeding T_2 criteria)
   Complications of cirrhosis, low MELD score
      GI bleeding
      Hepatic encephalopathy
      Intractable pruritus
      Recurrent cholangitis
      Fulminant hepatic failure

Donor Candidacy and Evaluation

Potential donors are evaluated by a donor advocate team, must be complete healthy, and have hepatic size and anatomy compatible with right lobe transplantation (Table 9.27).

Table 9.27. Right lobe donor evaluation

History and physical exam (donor advocate physician)
Psychosocial evaluation (social work, psychiatry)
Laboratory assessment
   CBC, chemistry, coagulation profile
   Thrombophilia screening, viral serologies (HIV, HBV, HCV, etc.)
   ECG, chest radiograph
   Cardiac stress testing, if indicated
   Liver imaging (MRI, MRA, MRV, MRCP, or CT scan/ERCP)
   Liver biopsy, if indicated
   Family agreement/consent, no evidence of compensation/coercion

---

Donor and Recipient Procedure

The donor procedure consists of a formal right hepatic lobectomy with extreme care to avoid injury to those structures servicing the residual liver, or left lobe. Intraoperative cholangiogram and ultrasound is often necessary in this regard. Once harvested, the lobe is flushed with preservative solution and, if necessary, vascular reconstruction is completed on the back table in preparation for implantation. The recipient operation involves an IVC-sparing hepatectomy with anastomosis of the donor right-sided structures (vascular, biliary) to the corresponding recipient structures.[28] LDLT provides an alternative which may reduce the waiting-list mortality in selected patients. Ongoing studies will determine the true risk to the donors and whether recipient outcomes are comparable to whole liver transplant.

LIVER TRANSPLANTATION – A NEW ERA

Approximately 10,000 liver transplants have been performed to date, mostly in the post-cyclosporine era. For the most part, one-year and five-year patient survival rates are 90 percent and 75 percent, respectively. Graft survival rates may be slightly lower reflecting an incidence of retransplantation. Quality of life studies have shown that most patients have an excellent quality of life following transplantation, although the long-term care of the immunosuppressed patient is an evolving field which presents many interesting challenges. Certainly, chronic side effects of immunosuppressive therapy, de novo malignancies, and recurrence of native disease continue to present significant problems. These important clinical entities form the basis for present and future research in transplantation.

There has been a dramatic shift in the paradigm of liver transplantation in the last decade. Long-term results are unequivocally excellent and there is no longer a need to convince other clinicians that liver transplantation is a worthwhile therapeutic entity. Currently, our most significant hurdle includes a prohibitive organ shortage with resulting ongoing disagreements about allocation. Although living donor transplants have become increasingly utilized in both pediatric and adult recipient, the discrepancy between the need and the supply of organs continues to widen. Until xenotransplantation becomes a clinical reality, live donors will be used increasingly. The inherent risk to the donor requires a meticulous assessment of both clinical and ethical issues. Therefore, it behooves the transplant community to monitor closely the results of adult-to-adult living donor liver transplantation, as well as donor morbidity and mortality. This effort will require funding from the Federal Government so that appropriate registries can be supported. Finally, the resulting longer waiting times will necessitate more aggressive and innovative management algorithms for the complications of cirrhosis.

REFERENCES

1. Benhamou, JP. Fulminant and sub-fulminant hepatic failure; definition and causes. In: Williams, R, Hughes RD, ed. Acute Liver Failure: Improved Understanding and Better Therapy. London: Mitre Press, 1991:6-10.
2. Schiodt FV, Atillasoy E, Shakil AO, et al. Etiology and outcome for 295 patients with acute liver failure in the United States. Liver Trans Surg 1999; 5:29-34.

3. Ascher NL, Lake JR, Emond JC, et al. Liver transplantation for fulminant hepatic failure. Arch Surg 1993; 128:677-82.

4. Blei AT, Olafsson S, Webster S, Levy R. Complications of intracranial pressure monitoring in fulminant hepatic failure. Lancet 1993; 341:157-8.

5. O'Grady JG, Alexander GJ, Hayllar KM, et al. Early indicators of prognosis in fulminant hepatic failure. Gastroenterology 1989; 97:439-45.

6. Bernuau J, Rueff B, Benhamou JP. Fulminant and subfulminant liver failure: definitions and causes. Sem in Liver Disease 1986; 6:97-106.

7. O'Grady JG, Gimson AE, O'Brien CJ, et al. Controlled trials of charcoal hemoperfusion and prognostic factors in fulminant hepatic failure. Gastroenterology 1988; 94:1186-92.

8. Stange J, Ramlow W, Mitzner S, et al. Dialysis against a recycled albumin solution enables the removal of albumin-bound toxins. Artif Organs 1993; 17:809-13.

9. Weber C, Rajnoch C, Loth F, et al. The microspheres based detoxification system (MDS). A new extracorporeal blood purification procedure based on recirculated microspherical absorbent particles. Int J Artif Organs 1994; 17:595-602.

10. Rozga J, Podesta L, LePage E, et al. A bioartificial liver to treat severe acute liver failure. Ann Surg 1994; 219:538-46.

11. Sussman NL, Gislason GT, Conlin CA, Kelly JH. The Hepatix extracorporeal liver assist device: initial clinical experience. Artif Organs 1994; 18:390-96.

12. Fox, IJ, Langnas AN, Fristoe LW, et al. Successful application of extracorporeal liver perfusion: a technology whose time has come. Am J Gastroenterol 1993; 88:1876-81.

13. Chari RS, Collins BH, Magee JC, et al. Brief report: treatment of hepatic failure with ex vivo pig-liver perfusion followed by liver transplantation. N Engl J Med 1994; 331:234-7.

14. Harland RC, Platt JL. Prospects for xenotransplantation of the liver. J Hepatol 1996; 25:248-58.

15. Coy D., Blei AT. Portal hypertension. In: Berk J, Schaffner F, eds. Bockus Textbook of Gastroenterology. Philadelphia: Saunders, 1994:1955-87.

16. Cordoba J, Blei AT. Treatment of hepatic encephalopathy. Am J Gastroenterol 1997; 92:1429-39.

17. Chari RS, Gan TJ, Robertson KM, Bass K, Camargo CA Jr, Greig PD, Clavien PA. Venovenous bypass in adult orthotopic liver transplantation: routine or selective use? J Am Coll Surg 1998; 186:683-90.

18. Ferraz-Neto BH, Mirza DF, Gunson BK, et al. Bile duct splintage in liver transplantation: is it necessary? Transplant International 1996; 9 Supp.1:S185-7.

19. Clavien PA, Harvey PR, Strasberg SM. Preservation and reperfusion injuries in liver allografts. An overview and synthesis of current studies. Transplantation 1992; 53:957-78.

20. Reyes J, Gerber D, Mazariegos GV, et al. Split-liver transplantation: a comparison of ex vivo and in situ techniques. J Ped Surg 2000; 35:283-9.

21. Ghobrial RM, Yersiz H, Farmer DG, et al. Predictors of survival after in vivo split liver transplantation: analysis of 110 consecutive patients. Ann Surg 2000; 232:312-23.

22. Greig PD, Woolf GM, Sinclair SB, et al. Treatment of primary liver graft nonfunction with prostaglandin E_1. Transplantation 1989; 48:447-53.

23. Abbasoglu O, Levy MF, Testa G, et al. Does intraoperative hepatic artery flow predict arterial complications after liver transplantation? Transplantation 1998; 66:598-601.

24. Douzdjian V, Abecassis MM, Johlin FC. Sphincter of Oddi dysfunction following liver transplantation: screening by bed-side manometry and definitive manometric evaluation. Dig Dis Sci 1994; 39:253-6.

25. Demetris AJ, Batts KP, Dhillon AP, et al. Banff schema for grading liver allograft rejection: an international consensus document. Hepatology 1997; 25:658-63.
26. Broelsch CE, Emond JC, Whitington PF, et al. Application of reduced-size liver transplants as split grafts, auxiliary orthotopic grafts, and living related segmental transplants. Ann Surg 1990; 212:368-75, discussion 375-7.
27. Wachs ME, Bak T, Karrer FM, et al. Adult living donor liver transplantation using a right hepatic lobe. Transplantation 1998; 66:1313-6.
30. Abecassis M, Koffron A, Fryer J, Superina R. Surgical techniques in living liver donors. In: Adult Living-Donor Liver Transplantation, Current Opinion in Organ Transplantation, Volume 6, Issue 4: Philadelphia, Lippincott William & Wilkins, 2001.

9

Intestinal Transplantation

Jonathan P. Fryer

HISTORY

Transplantation of vascularized organs, such as the intestine, was first conceptualized by Alexis Carrel at the turn of the century, who recognized the potential for such procedures with the establishment of a reliable method of performing vascular anastomoses.[1] However, the feasibility of intestinal transplantation was not demonstrated until 1959 when Richard Lillihei, at the University of Minnesota, reported success in a canine model.[2] This inspired the first human intestinal transplants, which were performed by Ralph Deterling in Boston in 1964 (unpublished). The first reported human intestinal transplant was performed by Lillihei in 1967, and included the entire small bowel and right colon, with the superior mesenteric vessels being anastomosed to the left common iliac vessels.[3] Unfortunately, these and other early attempts which followed were uniformly unsuccessful.[4]

When the effectiveness of cyclosporine was established in other organ transplants in the early 1980's, there was renewed interest in intestinal transplantation. Although the first intestinal transplant using cyclosporine, performed in 1985 by Zane Cohen in Toronto[5] was also unsuccessful, in 1988 Deltz in Kiel, Germany performed what is considered to be the first successful intestinal transplant.[6] The recipient of this living-related allograft remained TPN-free for 4 year before the graft was lost to chronic rejection. Soon after, other successful outcomes were reported by the groups headed by Goulet in Paris,[7] and Grant in London, Canada[8] who had established the first intestinal transplant programs. The successes of these groups inspired other institutions to establish similar programs in the early 1990's.[9] There are now over 50 centers worldwide which have performed intestinal transplants, with close to 700 transplants performed to date.[10]

INDICATIONS

The indication for intestinal transplant is intestinal failure. This is defined as an inability to maintain greater than 75% of essential nutrition through the enteric delivery of nutrients[11] and is commonly the result of previous extensive small bowel resections, although severe malabsorption or dysmotility syndromes can also produce this situation. The short bowel syndrome, which manifests in these individuals, consists of massive diarrhea or stomal output, electrolyte abnormalities, fat malabsorption, gastric hypersecretion, Vitamin B12 deficiency, hyperbilirubinemia, and hepatic steatosis.[12]

While in the past patients with intestinal failure would not survive, these patients can now be kept alive with parenteral nutrition. Over the long term parenteral nutritional support can be provided at home, and many individuals with intesti-

Organ Transplantation, 2nd edition, edited by Frank P. Stuart, Michael M. Abecassis and Dixon B. Kaufman. ©2003 Landes Bioscience.

nal failure have done very well for many years with home parenteral nutrition (HPN). However, HPN is a very expensive therapy, costing $250 to $500 US dollars a day. Furthermore, HPN can be associated with potentially life-threatening complications such as catheter-related sepsis, catheter-related thrombosis, metabolic derangements, liver dysfunction, and bone disorders. In the pediatric population and in adults with extremely short guts (i.e., < 50 cm with colon, <100 cm without colon), gross impairment in liver function is seen in up to 50% of patients. Because central venous access is required for administration of TPN, and recurrent central line placements often lead to venous stenosis or occlusion, long term HPN often results in a loss of sites for vascular access.[13]

Since in some patients HPN may only be needed temporarily, before considering intestinal transplantation attempts at establishing enteral feeding should be pursued since there can be significant adaptation in intestinal function. Adaptation of the intestine is a result of both an increased absorptive surface due to hypertrophy and an increase in the efficiency of absorption. Generally, if an individual with intestinal failure remains HPN-dependent after 1 year, intestinal transplantation should be considered. If life-threatening complications of HPN develop prior to 1 year, intestinal transplantation should be considered earlier. If during the intestinal transplant assessment evidence of irreversible liver disease [cirrhosis, fibrosis, portal hypertension] is discovered, a liver/intestine transplant should be performed. If the underlying disease process compromises the organs supplied by both the mesenteric and celiac arterial systems, or if it mandates replacement of other sections of the alimentary tract, a multivisceral transplant (i.e., stomach, duodenum, pancreas, liver, small intestine, and colon) should be considered.

Although no specific disease entity, in and of itself, is an indication for intestinal transplant, in the intestinal transplants performed to date the primary diseases which have most commonly led to consideration of an intestinal transplant are, in adults: mesenteric thrombosis, Crohn's disease, trauma, volvulus, desmoid tumor, Gardner's syndrome/familial polyposis; and in children: volvulus, gastroschisis, necrotizing enterocolitis, pseudo-obstruction, intestinal astresia, and Hirschsprung's disease.[10]

CONTRAINDICATIONS

In general, intestinal transplants should not be performed in individuals who have significant co-existent medical conditions that have no potential for improvement following transplantation, and which would negate any potential benefit provided by an intestinal transplant in terms of life expectancy or quality of life. If the patient has active infection, malignancy, or HIV, transplantation is contraindicated. If there is substantial evidence to indicate that a potential recipient or the primary care givers are not willing or able to reliably assume the responsibilities of the day-to-day management of the potential recipient following the transplant, transplantation is contraindicated.

PRETRANSPLANT RECIPIENT EVALUATION

All individuals under consideration for intestinal transplant should be seen and evaluated by a multidisciplinary intestinal failure team including transplant surgery, gastroenterology, nutritional services, psychiatry, social work, anesthesia, and financial services. Further consultation with other specialties [i.e., cardiology, hematology, chest medicine, infectious disease, chemical dependency, dentistry, etc], will be required in some cases. Baseline laboratory investigations including routine blood work, ABO blood group determination, HLA status, and panel reactive antibody status will be performed. If not done previously, the GI tract should be assessed both radiologically and endoscopically to accurately determine the length and condition of the remaining bowel. It is also important to establish which large veins are available for vascular access, as many of the patients will have limited options. Living related donor transplantation can be discussed as an option if a potential living related donor is availabl.[14]

If after these evaluations there is consensus that the patient is a good candidate for intestinal transplantation, the patient will be listed. While waiting for a donor to become available the stable patient should be reassessed every three months to determine whether there is any change is their PRA status, deterioration in liver function, or development of other medical problems. Furthermore while waiting for intestine only transplantation, the HPN administration should be monitored very closely to ensure that it does not contribute further to the development of hepatic steatosis and fibrosis since optimal balancing of carbohydrates and lipids in the HPN solutions can minimize the development of hepatic pathology. These patients will also need ongoing maintenance of their central lines to minimize line-related complications such as infections and thrombosis. Furthermore while waiting for transplantation close attention must be paid to fluid and electrolyte disturbances which are common due to the often-excessive output from the residual GI tract, particularly in individuals who continue to eat or drink. In some instances patients who have dysfunctional intestine [i.e., dysmotility or malabsorption syndromes] or a blind loop, which result in stasis of intestinal contents, will develop severe problems with bacterial overgrowth and translocation resulting in recurrent, bacteremia and life threatening sepsis. Surgical revision to eliminate blind loops including, in extreme situations, total enterectomy of dysfunctional small bowel are sometimes warranted to keep these patients alive until transplantation can be performed.

DONOR EVALUATION AND MANAGEMENT

CADAVERIC DONORS

All cadaveric donors are potential intestine donors. The cadaveric donor needs to be ABO compatible with the recipient and because of the risk of graft versus host disease ABO identical combinations should be used in most circumstances. In most cases, extensive prior bowel resection has significantly reduced the size of the recipient peritoneal cavity and therefore a donor that is 50 to 75% the size of the recipient is needed. In certain circumstances segments of the intestine from a larger donor may be considered.

Donors should have no previous history of significant intestinal pathology. As with all organs donors there should be no significant hemodynamic instability, sepsis, history of malignancy or chronic infection, severe hypoxia, severe acidosis, and they must have negative serology for HIV, hepatitis B and hepatitis C. A cross match should be performed either using a standard cytotoxicity assay or flow cytometry. In certain circumstances, if the cross match results are not available, but the patient has had no evidence of presensitization based on pre- transplant serologic surveillance, it may be reasonable to proceed without the cross match results. Because of the need to minimize the intestinal cold ischemia time (<6 hours),[15,16] it may not always be possible to obtain the cross match results in time. Although HLA matching has not been studied extensively in small bowel transplantation it is also useful to know the HLA status of both donor and recipient, particularly if the recipient in known to be sensitized to certain HLA antigens.

Two other important considerations are the CMV and EBV serologic status of the donors and recipients. Transplantation of a serologically positive donor into a serologically negative recipient for either of these viruses can have serious consequences.[17] In addition to the risk of a systemic CMV infection, a CMV enteritis can occur which can lead to graft loss. A new EBV infection combined with posttransplant immunosuppression puts the patient at high risk for developing a post transplant lymphoproliferative disorder (PTLD).[18]

If a donor is considered suitable, an NG tube should be placed and oral antibiotics administered to try and decrease bacterial counts in the donor gut. Amphotericin B, Neomycin, and Erythromycin base are typically administered immediately after the decision is made to go ahead with the procurement and then again at initiation of the multi organ procurement. A formal bowel prep should not be performed in most circumstances because, with the time constraints involved, the bowel will end up severely distended making it difficult to transplant. In the rare circumstance that there will be 12 to 24 hours between the identification of a donor and the donor procurement, a formal bowel prep may be considered. Some programs also consider administering OKT3 to the donor to decrease the numbers of lymphocytes in the allograft prior to transplantation,[19] although the merit of this has not yet been determined.

Because the optimal cold ischemia time for intestinal grafts is less than 6 hours, careful attention must be given to the timing of the donor and recipient procedures to prevent prolonged cold ischemia. Consideration should also be given to what other organs are going to be procured, as this may influence the length of the donor procedure and the approach used by the small bowel procurement team.

LIVING DONORS

If a living donor is being evaluated, it is important that the potential donor be evaluated by a multidisciplinary team that includes transplantation surgery, GI medicine, psychiatry, nutritional services, and social work. To avoid a conflict of interest, it is imperative that the physician who is in charge of working up the donor not be an active part of the transplant team. As with any living donor procedure, the potential complications should be explained in great detail to the prospective donor on multiple occasions. It should also be made quite clear to the patient that other options besides using a living donor are available. Time must also be taken to fully understand the nature of the relationship between the donor and the recipient. Living donation should not be pursued if coercion or financial incentive appear to be the primary motivation for donation.

If a number of potential living donors are available, particularly among family members, then careful consideration should be given to the best available HLA match. The donor-recipient size discrepancy must also be considered but since, in a living donor, only a segment of the intestine is transplanted, size limits are less restrictive. As with cadaver donors, the donor and recipient should be ABO identical, although in some circumstances ABO compatible combinations can be considered.

As with cadaveric donors, living donors must be free of significant pathology involving the GI tract. Any potential living donor must be in good health with no previous significant medical problems, including diabetes, malignancy, or chronic infection. There should be no history of substance abuse or other high-risk activities in the donor, and no significant psychiatric history. Serology in the living donor must also be negative for HIV, Hep C and Hep B. Obese donors should be avoided. As with cadaveric donors, the CMV and EBV status of the donor and recipient must be carefully considered and the combination of positive donors to negative recipients should be avoided. The living donor should be worked up completely including CBC, electrolytes, liver function tests, EKG, chest x-ray. The GI tract should be evaluated endoscopically and if any concerns exist, GI contrast studies should be performed. A mesenteric angiogram with selective study of the SMA and its venous phase should be performed to ensure that the terminal SMA and SMV are adequate.

One day prior to surgery the potential donor should be kept on clear fluids and administered neomycin 1 gram and erthyomycin base 1 gram PO at 1300 and 1400 and 2300 hours. The potential living donor should also undergo a formal bowel preparation using GoLYTELY (4L) the day prior to surgery.

DONOR PROCUREMENT

It is important for all procurement teams to work closely in coordinating their various roles in the procurement process. The small bowel team must work most closely with those teams that are procuring other intra-abdominal organs. Prior to initiating the procurement there must be an agreement as to where the portal vein, or superior mesenteric vein will be divided. If no pancreas is being procured, then the portal vein is usually divided at least 2 centimeters superior to the splenic vein take off. If the pancreas is going to be used then the superior mesenteric vein must be taken immediately below the uncinate process. With regards to the artery, if the pancreas is not being used then typically the entire superior mesenteric artery will be taken along with a long tube of adjoining aorta extending up into the chest, to provide additional length for the artery should it be necessary. If this is done, great care must be taken in preserving a small Carrel patch at the origin of the celiac artery for the liver procurement team, if requested. If the pancreas is going to be used then the proximal superior mesenteric artery will need to be preserved for the head of the pancreas and the artery will have to be divided immediately below the uncinate process. In all of these circumstances, extra segments of donor artery, preferably the iliac artery; and donor vein, preferably the iliac vein, should be taken in case vascular extension grafts are needed when the small bowel allograft is revascularized.

In general, after the abdomen is open, the first step in small bowel procurement is to perform a gross visual inspection of the small bowel. If all appears well, the omentum should be taken off the right side of the transverse colon to approximately the mid transverse colon. Care must be taken not to transect the transverse mesocolon. At this point, after identifying the middle colic vessels, a site immediately to the donor's left of the middle colic vessels is chosen as the distal extent of the small bowel graft. A small hole is made in the transverse mesocolon at this site in preparation for transsection of the bowel. Next after entering into the lesser sac along the greater curvature of the stomach near the pylorus, the pylorus is encircled taking care not to injure the arteries going to the liver. An NG tube is then manipulated into the duodenum where the Amphotericin/Neomycin/Erythromycin base solution is infused. Once the solution has been infused [250-500cc], and the NG tube is withdrawn into the stomach, the pylorus is divided using a GIA stapler. After a few minutes are given for the solution to pass through the small bowel and into the colon, the jejunum just distal to the ligament of Treitz is encircled and divided using a GIA stapler. Next the transverse colon should be divided at the previously selected site. If any solid stool is palpated in the right colon it should be milked distally prior to transection so that it is not included in the graft. Therefore, the intestinal segment to be removed extends from the ligament of Trietz to the mid-transverse colon. After this segment has been completely mobilized, attention is diverted to the arterial and venous supply, which are isolated as described previously.

For procurement of a liver-intestine graft, the portal vein is not divided but is procured in continuity with the liver after ligating all posterolateral branches in the head of the pancreas. The correct orientation of the portal vein should be made apparent using small clips or indelible ink to avoid twisting during implantation. The hepatic arteries are also not divided but are procured in continuity with the celiac artery, SMA and a long, adjoining segment of thoracic aorta.

For a multivisceral transplant, all organs to be transplanted are removed en bloc with their blood supply procured in continuity with the celiac artery, SMA and a long, adjoining segment of thoracic aorta.[3]

When the organs are ready for removal a cannula is placed in the distal aorta, which is flushed retrogradely with University of Wisconsin solution. Simultaneous with initiation of the flush, the supra-hepatic vena cava is partially divided in the chest cavity to facilitate extravasation. The thoracic aorta is also clamped in the chest. After the small bowel graft has been extravasated and completely flushed with cold preservation solution, it is removed and placed in sterile bags which are placed in a cooler for transport.

It is very important that the small bowel procurement is done in close coordination with the preparation of the recipient. The two procedures should be timed so that when the donor team arrives back at the recipient hospital, all is ready for the graft revascularization.

RECIPIENT PROCEDURE

When a donor is first identified the recipient must be notified immediately so that their surgery can be coordinated with the donor procedure. The waiting recipient should at all times be prepared to transport themselves to the hospital within a couple of hours of notification. Preoperative blood work and other mandatory preoperative tests should be obtained immediately upon arrival to the hospital and broad spectrum antibiotics should be administered approximately 15 minutes prior to the opening incision. Since most intestinal transplant recipients have limited vascular access, the current TPN line may be utilized. The surgical team should inform the anesthesia team of potential available sites for other I.V. access, so that futile attempts to establish IV access are avoided.

The recipient is taken to the OR at an appropriate time dictated by the amount of surgery that is anticipated to be necessary to prepare for implantation of the donor graft. In some circumstances residual segments of diseased bowel will need to be removed from the potential recipient. Furthermore, a decision will have to be made as to which vessels the donor bowel will be anastomosed to. Ideally, if they are not diseased and are of satisfactory caliber, the recipient superior mesenteric artery and vein can be used. Alternative choices would be the infrarenal aorta for arterial input and the portal vein or inferior mesenteric vein for venous drainage. If the portal venous system is not accessible or useable, the inferior vena cava can also be used. Although anastomoses between a donor portal venous branch and the recipient cava are not physiological, in the instances where they have been performed, patients have had no adverse consequences.

For a liver-intestine graft, the caval anastomoses are performed as with a liver-only transplant. The recipient portal vein, which will still be draining the residual recipient visceral organs can either be anastomosed end-to-side to the recipient cava or to the donor portal vein. The aortic segment with its celiac and SMA trunks intact is then anastomosed end to side to the infrarenal aorta.

For a multivisceral graft, if the liver is included, the caval anastomoses are performed followed by the donor aortic segment to recipient infrarenal aortic anastomosis. If the liver is not included, the donor portal vein is anastomosed to the recipient portal vein or cava.[20]

In addition to preparing sites for the vascular anastomoses, appropriate sites for the proximal and distal intestinal anastomoses should also be identified. Ideally, the proximal end of the donor intestine will be anastomosed to the most distal and accessible segment of the recipient's remaining small intestine, which typically is at or distal to the ligament of Treitz. If in the pretransplant evaluation the recipient has been shown to have severe gastric dysmotility with delayed gastric emptying, consideration of what to do with the stomach must be included in the overall surgical plan. The management of the stomach in these circumstances is somewhat controversial. The options include:

  a. Doing nothing at the time of transplant and following the patient to see if gastric emptying remains a problem post transplant.
  b. Performing a gastrojejunostomy—anastomosing proximal donor intestine to the stomach.
  c. Performing a partial gastrectomy and gastrojejunostomy.
  d. Performing a multivisceral transplant which would include stomach, duodenum, pancreas, intestine and, if necessary, liver.

Another area of controversy is whether a segment of colon should be transplanted with the small intestine or not. The primary advantage of transplanting the colon is that it helps to control the severe fluid and electrolyte imbalances which can occur posttransplant. The disadvantage is that it may predispose to a higher incidence of bacterial translocation and infectious complications.[21]

If the recipient has remaining healthy colon, its proximal end would be the ideal site for anastomosis to the distal end of the donor intestine. If the recipient has had a proctocolectomy, the distal end of the donor intestine can be brought out as an end colostomy or ileostomy. In certain circumstances it may be preferable to perform a pelvic pull-through with a colo-anal anastomosis, but this if often better left for a second operation. If an end-ileostomy is not created, a site for a loop ileostomy must be selected. An ileostomy of some form is essential to provide direct vision and direct endoscopic access to the small bowel for surveillance following the transplant. Some centers perform a Bishop-Koop type of ileostomy rather than a loop ileostomy.

Another important consideration in the recipient operation is the placement of a feeding jejunostomy tube. Because early establishment of enteral feeding is essential, and since the establishment of oral feeding is less predictable a feeding jejunostomy should be placed at the time of transplant. The safest approach is often to put a percutaneous gastrojejunal tube into the native stomach, passing it into the proximal jejunum of the intestinal allograft. This precludes any allograft-related problems compromising the integrity of the tube insertion site. In some circumstances, however, it may be preferable to place a jejunostomy tube directly into the donor jejunum.

Upon arrival of the donor team at the recipient hospital, implantation of the graft must begin as soon as possible. The patient should be fully heparinized prior to the vascular anastomosis. Overall the total cold and warm ischemia time should be kept less than 6 hours. The warm ischemia time should ideally be less than 30 minutes. After completion of the vascular anastomoses and reperfusion of the graft, if all segments are perfused well the proximal and distal intestinal anastomoses should be performed followed by the ileostomy. The patient can then be closed after the feeding jejunostomy is placed. The recipient should be left with a tube or combination of tubes that will both decompress the stomach and allow feeding in the jejunum.

10

POSTOPERATIVE MANAGEMENT

The recipient should be established on immunosuppression immediately following surgery. For the first several days posttransplant, only select medications, included Tacrolimus should be administered via the GI tract. In circumstances where Tacrolimus absorption via the GI tract has been questionable, sublingual administration can be utilized. In most circumstances Tacrolimus is the main immunosuppressive drug. However, if the patient is intolerant of Prograf, consideration can be given to other immunosuppressive regimens based on Neoral. Sirolimus, especially in combination with Tacrolimus, has improved patient and graft survival and is now being incorporated into most immunosuppressive protocols. Steroids are also included in the postoperative immunosuppressive regimen. While induction with OKT3 or ATGAM has generally been avoided because of the higher incidence of PTLD associated with intestinal transplantation,[18] some centers have been reevaluating their role. Alemtuzumab (CAMPATH-1H)‘, an anti-CD52 mAB, has also been used by some centers although its safety and efficacy in intestinal transplantation has not yet been clearly established. Monoclonal anti-IL2 receptor antibodies (Basiliximab, Daclizumab) are currently being used for most intestinal transplants, as they appear to provide benefit. While some programs have included mycophenolate mofetil,[22] others have avoided it because of its association with gastrointestinal side effects. Prostaglandin E1 is commonly administered intravenously while the patient is in the hospital, both for its ability to improve the small bowel microcirculation and its potential immunosuppressive effects. Broad-spectrum intravenous antibiotics are usually continued for at least 1 week following the transplant.

It is imperative to maintain prophylaxis for cytomegalovirus (CMV) and Epstein Barr, virus (EBV) infections post operatively particularly where the donor is positive for CMV or EBV and the recipient is negative. CMV prophylaxis is best accomplished with Gancyclovir, although CMV immune globulin (Cytogam) has also been used. Acyclovir, which is less effective than Gancyclovir for CMV, is effective prophylaxis for EBV. Intravenous immune globulin (IVIG) is also used by some centers as EBV prophylaxis.

In the immediate postoperative period it is essential to check hemoglobins regularly for evidence of bleeding. It is also important to monitor serum pH and lactate levels to detect any evidence of intestinal ischemia or injury. Prograf levels should be followed daily and doses adjusted to achieve a serum level of 20-25 ng/ml in the early posttransplant period.

Approximately 5 days post transplant, if all is stable, an upper GI contrast study should be performed to ensure that there is no leakage or other gross abnormality in the newly established gastrointestinal tract. If the upper GI contrast study reveals no contraindication, tube feed should be initiated slowly but can usually be advanced to provide full nutritional support within a couple of days. The ideal features of an enteral feeding solution to be established in a new intestinal transplant recipient are that it: (a) provides maximum calories with minimal volume without being hyperosmolar; (b) has minimal or no complex fatty acids [medium chain triglycerides are ok]; and (c) is supplemented with glutamine and/or arginine.

POSTOPERATIVE SURVEILLANCE

In the post operative period several potential complications need to be closely watched for, including the following.

REJECTION

Acute cellular allograft rejection is unlikely to occur within the first few weeks following the transplant, provided immunosuppression is adequate. Subsequently, rejection can occur at any time but is most common in the first year, particularly the first 6 months. Unfortunately, as of yet, there is no single blood test, which will detect an early rejection. Therefore, suspicion of rejection must be based on clinical evaluation. Although no single sign, or combination of clinical signs is entirely reliable, in most instances rejection is associated with fever, a significant increase in stomal output, and GI symptoms such as abdominal pain, cramping, nausea, vomiting and diarrhea.

Although most lab tests are not helpful in confirming the diagnosis, chromium EDTA,[23] or Technetium DPTA[24] isotope studies have been useful in identifying increased intestinal permeability which correlates well with, but is not specific for, rejection. If rejection is suspected, endoscopic evaluation of the intestinal graft must be performed. The endoscopic evaluation should include as much of the small bowel as possible and biopsies from numerous sites (at least 6) should be obtained, since rejection can often be segmental. The loop ileostomy greatly facilitates this type of assessment and for that reason the ileostomy is usually kept in place for 6 months to a year following the transplant. Although the endoscopic appearance of rejecting small bowel is often abnormal with evidence of inflammation and ulceration, in early rejection it can be quite normal. Zoom-endoscopy appears to provide more a valuable endoscopic identification of acute rejection in the small bowel. The gold standard for diagnosing rejection is histologic evaluation of the biopsies. Typically early rejection is associated with increased apoptotic figures [normal less than 2 to 3 per high power field]. Other histologic findings associated with rejection include: the presence of activated lymphocytes in the lamina propria; loss of goblet cells; loss of villus height, and ulceration.[25]

When a diagnosis of rejection is made, the patient should be treated with Solumedrol 500mg IV for 3 days. Prograf levels should be rechecked and doses increased accordingly. If there is persistent evidence of rejection following treatment with steroids, the patient should be treated with OKT3 or Thymoglobulin. If, despite maintaining adequate immunosuppressive levels, rejection episodes continue to occur, consideration should be given to adding additional drugs, such as Sirolimus to the immunosuppressive regimen. Because escalation of immunosuppression can be complicated by life threatening infections or malignancies, such patients should be carefully monitored.

INFECTION

Patients who undergo small bowel transplant are even more susceptible to infectious complications than other transplant recipients. There are primarily two reasons for this:

1. The intestinal allograft is transplanted with a significantly higher load of microorganism than any other organ allograft. Therefore, any process which compromises the intestinal allograft will influence the containment of these microorganisms within the graft and contribute to their spread to various areas of the body.

2. Because intestinal rejection is difficult to detect and because severe rejection can often lead to life threatening sepsis, these patients are maintained on higher degrees of baseline immunosuppression than recipients of other organ transplants.

Bacterial Infections

When bacteria translocate from the compromised intestinal allograft, there are commonly two places where they go initially. Since the lymphatics are divided in the procurement of the intestinal allograft it is common that there is leakage of intestinal lymph into the peritoneal cavity. This often contains bacteria. While typically the peritoneal cavity is capable of handling a moderate load of bacteria, in the immunocompromized state—particularly when significant ascites is present—bacterial peritonitis can occur. The second route by which bacteria can spread is by direct translocation into the portal circulation and subsequent dissemination to other sites. Particularly common infections resulting from bacterial translocation are central line infections and pneumonias. The typical organisms are consistent with those, which are found in the GI tract and include E.Coli, klebsiella, enterobacter, staphylococci enterococci, etc. Because of the degree of immunosuppression used, other typical and atypical postoperative infections are more likely to occur.[26]

Viral Infections

A primary concern with intestinal transplantation is the development of a CMV infection, which can manifest as CMV enteritis that can be severe and lead to graft loss. In general, transplantation of a graft from a CMV positive donor to a CMV negative recipient is avoided. The clinical manifestations of CMV enteritis are not unlike that of rejection with fever, increased stomal output and GI symptoms. Other important clues which may sway the clinical diagnosis more towards CMV enteritis include: the CMV status of the donor and recipient, the degree of immunosuppression at the time symptoms developed, and a positive CMV antigenemia assay. Also with CMV infections there is typically a decrease in the white blood cell count and flu-like symptoms. Endoscopy should be performed and multiple biopsies taken if there is a clinical enteritis. While the histologic picture of CMV can sometimes be similar to that of rejection, with CMV enteritis the presence of CMV inclusion bodies is diagnostic. If CMV is diagnosed, the patient should be treated with therapeutic doses of Gancyclovir. If there is evidence of Gancyclovir resistance, Foscarnet or CMV immune globulin (Cytogam) should be considered.

Furthermore, immunosuppression should be reduced until the CMV infection is controlled.[27]

Epstein Barr virus (EBV) associated infection can initiate an entire spectrum of disease. Those particularly at risk are recipients who are EBV negative and who receive an EBV positive graft. An acute EBV virus infection is typically associated with severe malaise and fever and flu-like symptoms i.e., infectious mononucleosis. Other evidence of EBV infection can include an increase of liver function tests, splenomegaly and lymphadenopathy. In certain instances an EBV infection can progress to a post transplant lymphoproliferative disorder (PTLD) which can develop into a malignant lymphoma. Surveillance for PTLD should therefore began immediately following the transplant particularly in EBV negative recipients who have received EBV positive grafts. PCR has been utilized to semiquantitatively monitor EBV replication by quantitatively determining the amount of EBV encoded RNA (EBER) in the serum as an early warning of an impending PTLD.[28] Other approaches using in situ hybridization have also been described.

While there is no standardized strategy for preventing PTLD, two basic approaches have evolved. One approach is to give long term prophylaxis with recipients maintained on ganciclovir and/or IVIG for 3 to 12 months following the transplant. The other approach is to have a shorter period of prophylaxis (2 to 6 weeks,) followed by surveillance as described above and preemptive therapy should surveillance identify increased EBV replication. Similar strategies are also used or CMV surveillance.

POSTTRANSPLANT FUNCTION

Typically the transplanted intestine will initiate peristalsis immediately after reperfusion. However, in the process of procuring the donor intestine all extrinsic innervation to the bowel is disrupted. This and other factors contribute to a less orderly peristalsis than is seen in a normal intestine. Often a more significant problem is the dysfunction of residual native intestine in a patient with a primary dysmotility syndrome. In some instances the stomach, duodenum, and colon, etc, will be left in place to best approximate re-establishment of normal gastrointestinal continuity. Sometimes these retained native segments function adequately while in other instances they do not. It remains controversial whether such patients are best served by isolated intestinal transplants, or by multivisceral transplants which would provide a new stomach, duodenum and colon if necessary.

The absorptive capacity of the transplanted intestine is typically good. While there may be some initial malabsorption of carbohydrates, for the most part carbohydrate absorption appears to normalize within the first several months as determined by d-xylose absorption.[29] Clearly, absorption of immunosuppressive drugs, particularly Prograf, is instantaneous and some transplant programs initiate oral immunosuppressive drugs immediately following surgery. While drug malabsorption has been described,[30] difficulty in obtaining levels is often associated with inability to retain ingested drugs because of nausea or vomiting, or noncompliance. Although very little has been done to measure amino acid absorption in intestinal transplantation, this also appears to be adequate quite early as determined by nonspecific markers of protein nutrition such as pre-albumin. Fat absorption on the other hand is impaired for several months following intestinal transplantation. Because the intestinal lymphatics are unavoidably disrupted in the procurement process, intestinal lymphatic drainage is not re-established for several months following the transplant. Absorption of dietary lipids, which primarily are made up of long chain triglycerides, depends on lymphatic drainage. Medium chain triglycerides (MCTs), i.e., those consisting of 8 to 12 carbon fatty acids, can be absorbed directly into the portal circulation. For these reasons it is essential to supplement enteral feeds with MCTs for several months following transplantation. Use of more complex fatty acids will lead to malabsorption of fat with increased ileostomy output and possible dehydration. To avoid an essential fatty acid deficiency, it may be necessary to intermittently supplement with intravenous fats, until the intestinal lymphatics are reestablished. Because of the obligatory fat malabsorption, there can also be malabsorption of the fat-soluble vitamins [Vitamin D, E, A, K]. Despite this, 72% of adults and 93% of children gain weight, and essentially all achieve their ideal body weight range.[31]

Because of the abnormal intestinal motility and malabsorption, associated with the early posttransplant period, the ileostomy output can be unpredictable and often excessive. Even in the best of circumstances, high ileostomy output can be anticipated early once full enteral nutrition has been established. Very close attention must be made to the overall fluid and electrolyte balance to prevent severe dehydration and/or electrolyte imbalances. It is imperative, in addition to accurate monitoring of daily in and outs, to follow daily weights and electrolytes. Once enteral nutrition is found to be providing all nutritional requirements, TPN is discontinued. If weight is maintained or weight gain occurs, and there is no significant evidence for protein malnutrition, TPN can be permanently discontinued. After a brief period of adjustment, ostomy output should become quite predictable over a given period of time. Dramatic changes in ostomy output should be investigated, as this can be an early indicator of rejection or other pathology. Overall, 70-80% of patients who undergo successful transplantation can be completely removed from TPN.[10]

PATIENT AND GRAFT SURVIVAL

The one-year graft survival for intestinal transplants performed since 1991 is approximately 60%. Despite early trends, there appears to be no difference in long-term graft survival when other organs are transplanted with the intestine. With regards to patient survival, overall 1-year survival for intestine-only transplants has been approximately 70%, while for intestine plus liver, or multi visceral transplants 1-year patient survivals have been 62 % and 52%, respectively. However, evaluation of the most recent cohort of transplants performed at the most experienced centers suggests that patient survival in two of these three groups is improving with 77%, 69% and 62% one year patient survivals in intestine-only, multivisceral, and intestine plus liver transplants respectively. While meaningful data on long term graft and patient survival is not yet available, it appears that a plateau in survival may begin to occur at approximately the two-year mark. As has been the case with all newly established organ transplants, there appears to be learning curve phenomena with improved patient survivals observed in the most experienced centers.[32]

MORBIDITY

Acute rejection has occurred in 79% of patients undergoing intestine-only transplants. Once again the liver, and perhaps other organs, may have a protective effect since the acute rejection rates for liver/intestine and multivisceral transplants have been 71% and 56% respectively. Similarly, chronic rejection, which has been demonstrated in 13% of intestine-only transplants, has been uncommon in liver/intestine (3%) and multivisceral transplants (0%). Despite the fact that most centers avoid transplanting intestinal grafts from cytomegalovirus (CMV) positive donors, CMV infections occurred in 24% of intestine-only grafts, 18% of liver/intestine grafts, and 40% of multivisceral grafts.

Post-transplant lymphoproliferative disorders (PTLDs) have been seen in 8.3% of intestine-only, 13.3% of liver/intestine, and 15.8% of multivisceral grafts.[10] PTLDs often manifest as fever and lymphadenopathy or lymphoproliferation in either donor or recipient tissue. Lymphoma can also manifest with gastrointestinal symptoms including nausea, vomiting, diarrhea, bowel obstruction, GI bleeding, or perforation.

The incidence of PTLD in intestinal transplant recipients is higher than in other organ transplant recipients. The occurrence of PTLD clearly correlates with the intensity of immunosuppression. Significant increases in the incidence of PTLD are noted in-patients who receive OKT3 or ATGAM, especially if their total antibody course exceeds 21 days. While PTLD tends to first manifest between 2 weeks and 6 months after a transplant, it can appear at any time.

The diagnosis of PTLD usually requires a biopsy. Often this is most easily obtained from an enlarged superficial lymph node or from clinically or radiologically involved tissue. If the suspected organ is the intestine graft itself, it can sometimes be difficult to differentiate PTLD from rejection, or CMV infection. When this is the case it is often useful to obtain further studies including EBER staining of suspicious tissue. It is often also useful to evaluate the serum for a typical monoclonal or polyclonal immunoglobulin bands which can sometimes be present. Gene studies are often helpful to identify abnormal karyotypes which can aid in diagnosis and prognosis (C-myc, N-ras, p 53) is polyclonal or monoclonal. It should also be determined whether the abnormal lymphocytes sites are primarily B cells or T cells. T cell lymphomas are less common than B cell lymphomas in post-transplant PTLDs.

If the diagnosis of PTLD is made, immunosuppression should be reduced to approximately half of what it had been. In approximately one third of cases, this will result in a remission of the PTLD. Anti B-cell mAB (Rituximab) therapy is initiated. If after 2 weeks there is no evidence of improvements, all immunosuppression should be discontinued and serious consideration should be given to additional therapeutic measures including chemotherapy and/or adoptive immunotherapy. If necessary, an intestine-only graft can also be removed.

10

MORTALITY

Overall, the most significant cause of morbidity and mortality has been infectious complications. Over half (51%) of the deaths in intestinal transplant patients have been clearly attributed to sepsis. Other causes of death have included rejection (10%), technical complications (7%), PTLD/Lymphoma (7%), and respiratory causes (7%).[10]

FUTURE DIRECTIONS

Intestinal transplantation provides unique and difficult challenges. Because of the delicate balance that must be maintained to provide adequate immunosuppression without over immunosuppression, it is imperative that a simple marker be developed which will alert clinicians that an early rejection is brewing. Another goal is to develop strategies, which eliminate or minimize the risk of rejection. To this end many researchers are attempting to develop strategies for inducing tolerance. Several groups have attempted to induce a state of microchimerism and tolerance by transplanting bone marrow along with the intestinal allograft.[32] To date, this approach has not been shown to be effective. Other groups have administered donor specific transfusions simultaneous with implantation of the intestinal graft.[33] While there are some preliminary animal studies suggesting that this approach might be effective, its benefit has not yet been proven in humans. Another approach, which has been effective in kidney transplantation, is HLA matching. Although due to time constraints this may not always be practical in the realm of cadaveric intestinal transplantation, it is possible with living related donors. While the experience with living related donor intestinal transplantation has been very limited to date, some of the longest surviving intestinal grafts from the pre cyclosporine era were achieved when living related donors were utilized. More recent experiences with modern immunosuppression have shown that graft survival with living donors is at least comparable to that achieved with cadaveric donors.[10,14] The potential advantages of using living donors are: (a) opportunity for better HLA matching; and (b) better control over ischemia times. The potential disadvantages are that: (a) the donor, who does not need a surgical procedure, is put at risk; (b) the allograft will consist of a shorter segment of bowel with smaller blood vessels.

SUMMARY

Intestinal transplantation is an option for individuals who are otherwise committed to a life of HPN because of intestinal failure. Intestinal transplantation is a fairly new procedure, which is still evolving, and at this time is still associated with significant risks. Rejection in intestinal transplantation is controllable with current immunosuppressive drugs, provided it is identified early. Infectious complications are the most significant cause of morbidity and mortality with intestinal transplantation. Post transplant lymphoproliferative disorders are also more common after intestinal transplantation, particularity when multivisceral transplants are performed. New strategies for detecting rejection and preventing infection are needed for intestinal transplantation to achieve the level of success that has been achieved with other solid organ transplants.

REFERENCES

1. Carrel A. The transplantation of organs. A preliminary communication. J Am Med Assoc 1905; 45:1645.
2. Lillehei R, Goott B, Miller F. The physiological response of the small bowel of the dog to ischemia including prolonged in vitro preservation of the bowel with successful replacement and survival. Ann Surg 1959; 150:543.
3. Lillehei RC, Idezuki Y, Feemster JA et al. Transplantation of stomach, intestine, and pancreas: experimental and clinical observation. Surgery 1967; 62:721.
4. Margreiter R. The history of intestinal transplantation. Transpl Rev 1997; 11(1):9.
5. Cohen Z, Silverman R, Wassef R et al. Small intestine transplantation using cyclosporine. Report of a case. Transplantation 1986; 42:613.
6. Deltz E, Schroeder P, Gebhardt H et al. Successful clinical small bowel transplantation: report of a case. Clin Transplant 1989; 3:89.
7. Goulet O, Revillon, Jan D et al. Two and one-half-year follow-up after isolated cadaveric small bowel transplantation in an infant. Transplant Proc 1992; 24:1224.
8. Grant D, Wall W, Mimeault R et al. Successful small-bowel liver transplantation. Lancet 1990; 335:181.
9. McAllister VC, Grant DR. Clinical small bowel transplantation. In: Grant DR, Woods RFM, eds. Small Bowel Transplantation. Great Britain: Edward Arnold, 1994:121.
10. Grant DR. Intestinal Transplantation: 1997 report of the international registry. Transplantation 1999; 67(7):1061.
11. Scott N, Irving M. Intestinal Failure—the clinical problem. Dig Dis 1992; 10(5):249.
12. Nightingale J. The short bowel syndrome. Eur J Gastroenterol Hepatol 1995; 7(6):514.
13. Howard L, Ament M, Fleming R et al. Current use and clinical outcomes of home parenteral and enteral nutrition therapies in the United States. Gastroenterology 1995; 109:355.
14. Gruessner RW, Sharp HL. Living-related intestinal transplantation: first report of a standardized surgical technique. Transplantation 1997; 64(11):1605.
15. Schweizer E, Gassel A, Deltz E et al. A comparison of preservation solutions for small bowel transplantation in the rat. Transplantation 1994; 57(9):1406.
16. Scholten E, Manek G, Green C. A comparison of preservation solutions for long term storage in rat small bowel transplantation. Acta Chirurgica Austriaca 1992; 24:5.
17. Kusne S, Manez R, Frye B et al. Use of DNA amplification for diagnosis of cytomegalovirus enteritis after intestinal transplantation. Gastroenterology 1997; 112(4):1121.
18. Reyes J, Green M, Bueno J et al. Epstein Barr virus associated posttransplant lymphoproliferative disease after intestinal transplantation. Transplant Proc 1996; 28(5):2768.
19. Langnas A, Shaw B, Antonson D et al. Preliminary experience with intestinal transplantation in infants and children. Pediatrics 1996; 97(4):443.
20. Todo S, Tzakis A, Abu-Elmagd et al. Abdominal Multivisceral Transplantation. Transplantation 1995; 59(2):234.
21. Spada M, Alessiani M, Fabbi M et al. Bacterial translocation is enhanced in pig intestinal transplantation when the colon is included in the graft. Transplant Proc 1996; 28(5):2658.
22. Tzakis A, Weppler D, Khan M et al. Mycophenolate mofetil as primary and rescue therapy in intestinal transplantation. Transplant Proc 1998; 30(6):2677.
23. Grant D, Lamont D, Zhong R et al. 51Cr-EDTA: a marker of early intestinal rejection in the rat. J Surg Res 1989; 46(5):507.

10

24. D'Alessandro AM, Kalayoglu M, Hammes R et al. Diagnosis of intestinal transplant rejection using technetium-99m-DTPA. Transplantation 1994; 58(1):112.

25. Lee RG, Nakamura K, Tsamandas AC et al. Pathology of human intestinal transplantation. Gastroenterology 1996; 110(6):1820.

26. Kusne S, Furukawa H, Abu-Elmagd et al. Infectious complications after small bowel transplantation in adults: and update. Transplant Proc 1996; 28(5):2761.

27. Bueno J, Green M, Kocoshis S et al. Cytomegalovirus infection after intestinal transplantation in children. Clin Infect Dis 1997; 25(5):1078.

28. Rowe DT, Qu L, Reyes J et al. Use of quantitative competitive PCR to measure Epstein-Barr virus genome load in the peripheral blood of pediatric transplant patients with lymphoproliferative disorders. J Clin Microbiol 1997; 35(6):1612.

29. Kim J, Fryer J, Craig RM. Absorptive function following intestinal transplantation. Dig Dis Sci 1998; 43(9):1925-30.

30. Fryer J, Kaplan B, Lown, K et al. Low bioavailability of cyclosporine microemulsion and tacrolimus in a small bowel transplant recipient. Transplantation 1999; 67(2):333.

31. Abu-Elmagd, Todo S, Tzakis A et al. Three years clinical experience with intestinal transplantation. J Am Coll Surg 1994; 179(4):385.

32. Todo S, Reyes J, Furukawa H et al. Outcome analysis of 71 clinical intestinal transplantations. Ann Surg 1995; 222(3):270.

33. Gruessner RW, Nakhleh RE, Harmon JV et al. Donor-specific portal blood transfusion in intestinal transplantation: a prospective, preclinical large animal study. Transplantation 1998; 66(2):164.

10

Heart Transplantation

Keith A. Horvath and David A. Fullerton

INTRODUCTION

On December 3, 1967, Mr. Louis Waskansky underwent the first successful human cardiac transplant, performed by Dr. Christian Barnard in Cape Town, South Africa. This milestone was reached after the technical aspects of orthotopic cardiac transplantation had been described in 1959 by Ross M. Brock of Guy's Hospital in London. The following year Shumway published the seminal paper on orthotopic cardiac transplantation in which the technical aspects, recipient support and donor organ preservation were integrated into a single approach. The initial experience with heart transplantation in the ensuing twelve months after Barnard's first operation was dismal: 71 of the first 100 recipients died. The introduction of cyclosporine provided the next breakthrough that allowed hospital mortality to drop below 10 percent and five-year survival rate to approach 80 percent.

There are currently 143 heart transplant centers in the United States and the number of transplantations has plateaued at an annual rate of approximately 2300 per year in the United States and 3400 per year worldwide. These plateaus are due to limitations of donor availability and as a result approximately 30 percent of patients on the waiting list will die before a suitable organ is available.

PRETRANSPLANT MANAGEMENT OF THE RECIPIENT

Most patients are evaluated for cardiac transplantation due to symptoms of heart failure. Some patients, however, are considered primarily because of low left ventricular ejection fractions with or without ventricular arrhythmias, severe angina refractory to medical therapy or end-stage coronary artery disease. Despite ejection fractions that may be considered low enough to be an indication for transplantation, many patients can be managed medically. Their arrhythmias may be treated with amiodarone, radio frequency ablation or an implantable defibrillator. Intractable angina due to severe coronary artery disease may be amenable to less conventional methods like transmyocardial laser revascularization. Unfortunately the number of patients who can be treated by these methods is small, and the majority are referred with severe symptoms of heart failure.

The initial medical therapy for these patients is exercise within their level of tolerance, restrictions on fluid (2 L/day), and sodium (2 gm/day). First line medications typically include digoxin, loop diuretics, and ACE inhibitors. When this medical regimen fails it is most commonly due to a direct failure to recognize fluid overload. This may lead to recurrent hospitalizations. During these hospitalizations optimization is usually possible with fluid balance and vasodilatation with hemodynamic monitoring. For example, intravenous nitroprusside and diuretics

Organ Transplantation, 2nd edition, edited by Frank P. Stuart, Michael M. Abecassis and Dixon B. Kaufman. ©2003 Landes Bioscience.

are used to obtain the following hemodynamic measurements as obtained by pulmonary artery catheterization. The pulmonary capillary wedge pressure should be < 15 mm Hg, the systemic vascular resistance < 1200 dynes/s/cm², the right atrial pressure < 8 mm Hg and the systolic blood pressure > 80 mm Hg. Conversion from intravenous medications to oral vasodilators and diuretics may then be feasible. Additional medications such as hydralazine, beta-blockers, amiodarone, anticoagulants and ultrafiltration may be useful as a next line of therapy or to stabilize the situation. If intravenous inotropic support is required, particularly if the patient then becomes dependent on such medications as dobutamine, or milrinone, then mechanical ventricular assist devices may be used as a next line of therapy. Mechanical assistance is particularly valuable to prevent irreversible failure in other organ systems. The simplest form of mechanical support is an intraaortic balloon pump (IABP). The intraaortic balloon pump may be very useful but should be considered as temporary support. This becomes an issue as patients supported on inotropes or with the IABP have experienced increasingly long waits on the transplant list. The indications for mechanical support are a cardiac index less than 2.0, mean arterial pressure less than 60 mm Hg and worsened hepatic and/or renal function. If a patient is unable to be weaned from an IABP in two to three weeks, then ventricular assist device (VAD) (see Table 11.1) or total artificial heart (TAH) may be indicated.

At present several total artificial heart devices are being investigated, including the Penn State heart, and the Abiomed TAH. Several axial flow pumps to be used as VAD and TAH are in development. Additional surgical approaches fall short of transplantation, such as cardiomyoplasty, which entails the wrapping of the latissimus dorsi muscle around the heart after the muscle has been preconditioned by artificial stimulation. Another procedure, based on the principles of volume reduction to improve stroke work and decrease wall tension, is the Dor procedure. This operation entails resection of the left ventricle, frequently with mitral valve replacement or repair. This is primarily reserved for patients with largely dilated left ventricles. While the early results are encouraging, the incidence of sudden death post procedure remains high.

INDICATIONS

Cardiac transplantation is an accepted therapy for many forms of end-stage heart disease. According to the Registry of the International Society for Heart and Lung Transplantation (ISHLT), the principle diagnoses among heart transplant recipients are ischemic cardiomyopathy (44%) and idiopathic cardiomyopathy (43%) (Fig. 11.1).

There are no strict criteria dictating when a patient with severe heart disease should be listed for heart transplantation. Characteristics of patients for whom heart transplantation should be considered include:

1. end-stage heart disease refractory to maximal medical management,
2. an estimated survival without transplantation of less than approximately 6-12 months,
3. not amenable to any other conventional therapy.

Table 11.1. Ventricular Assist Devices (VADs)

Type of Device	IABP	ECMO	Centrifugal Pumps	Abiomed	Thoratec	Novacor	HeartMate
Ventricular Support	Partial Left	Complete Cardio-pulmonary	Left, Right, or Both	Left, Right, or Both	Left, Right, or Both	Left only	Left only
Position	Intra-Aortic	External	External	External	Internal	Internal	Internal
Average Duration	Short	Short	Short	Intermediate	Intermediate to Long	Long	Long
Power Source	Pneumatic	Electric	Electric	Pneumatic	Pneumatic	Electric	Electric or Pneumatic
Cannulation Site	Peripheral Arterial	Peripheral Arterial and Venus	Arterial	Arterial	Arterial or Ventricular	Ventricular	Ventricular
Anti-coagulation	Not necessary	Yes	Yes	Yes	Yes	Yes	No
Patient Ambulation	No	No	No	No	Yes w/assistance	Yes	Yes
Patient Discharge	No	No	No	No	No	Yes	Yes (electric)

11

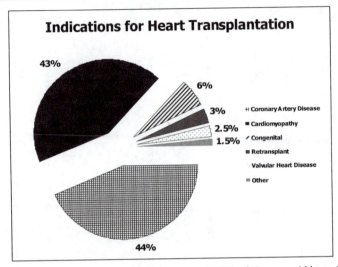

Fig. 11.1. Indications for heart transplantation. From: ISHLT website—www.ishlt.org, 2/03.

Once a patient is recognized to have those characteristics, a rigid evaluation process must be undertaken to determine if the individual patient is a suitable candidate for transplantation. The following must be considered in the evaluation process.

PATIENT AGE

Establishing an upper age limit for recipients within a given program is highly controversial. During the decade of the 1980s, most programs limited transplantation to patients under age 55. By the late 1980s and early 1990s, successful outcomes with heart transplantation, coupled with many more patients above age 60 and 65 with end-stage heart disease, prompted many programs to offer heart transplantation to older patients. It is now apparent that patient selection by strictly age limit alone is inappropriate. Selection must instead be based on the patient's physiological rather than chronological state.

Nonetheless, it must be noted that one-year survival following heart transplantation decreases with age. According to the ISHLT Registry data, patient age 50-59 years is a significant risk factor for mortality within the first year following transplant (odds ratio 1.23). Recipient age > 60 years has an odds ratio of 1.73. Age seems to have its greatest impact by creating a significantly higher operative mortality associated with transplant procedure: by three months posttransplant the survival curves for patient

< 65 years and > 65 years are parallel (Fig. 11.2). Older patients in otherwise good physiological condition may benefit greatly from heart transplantation, with improved survival and quality of life. But because the risk of heart transplantation increases with age, patients above age 60 must be very carefully evaluated to exclude any other important risk factors.

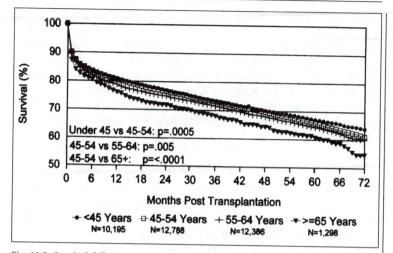

Fig. 11.2. Survival following heart transplantation as it relates to recipient age. From: Hosenpud JD et al. The registry of the International Society for Heart and Lung Transplantation: fifteenth official report. J Heart Lung Transplant 1998; 17:656-668. Reprinted by permission from Mosby Year-Book, Inc.

PATIENT SUPPORT SYSTEMS

Potential recipients must demonstrate an ability to comply with the complex medical regimens associated with posttransplant care. A support system of family and/or friends is extremely important in this regard. A thorough psychosocial evaluation is required to exclude the personality traits or interpersonal relationships, which will preclude successful care during the pre- and posttransplant time periods.

SEVERITY AND PROGRESSION OF PATIENT ILLNESS

A spectrum of illness exists among patients with end-stage heart failure. Heart failure may be acutely life threatening following an acute myocardial infarction, myocarditis, or failure to wean from cardiopulmonary bypass. Without inotropic infusions and/or support from an intraaortic balloon pump or mechanical ventricular assist device, such patients may die. Although acutely ill patients may survive to enter a more chronic condition, a decision must often be made during the acute phase whether or not the patient should be listed for transplantation.

Patients with more chronic heart failure may typically be managed as outpatients with strict attention to their medical regimen; this affords further stratification of the severity of their disease. Among patients with good control of their symptoms and satisfaction with their quality of life, continued medical therapy is indicated. Alternatively, repeated hospitalizations for heart failure indicate a poor prognosis and heart transplantation should be considered.

Peak Oxygen Consumption (VO₂)

Peak VO_2 consumption has become an important tool in the decision regarding the timing of transplantation. It helps quantify the degree of cardiac dysfunction and offers an objective prognosis of the patient. Serial determinations of peak VO_2 may be used to assess response to therapy and to track progression of disease. A peak VO_2 greater than 14 ml/kg/min is associated with one-year survival greater than 90% and suggests that transplantation may be deferred. A peak VO_2 less than 10-12 ml/kg/min is associated with a poor prognosis and patients should be considered for transplantation.

Systemic Diseases

Cardiac dysfunction may be a manifestation of some systemic diseases. Unfortunately, some systemic disease may preclude transplantation. Patients with diabetes mellitus often have renal dysfunction, peripheral vascular disease and retinal vascular disease, which may be exacerbated by the steroid therapy needed following transplantation. Such patients are typically unsuitable candidates. Systemic amyloidosis typically has multiorgan involvement and may recur in the transplanted heart. Therefore most programs recommend against heart transplantation for amyloidosis.

Pulmonary Vascular Resistance (PVR)

Determination of PVR is extremely important. Patients with left ventricular failure typically have elevated pulmonary artery pressures derived from hydrostatic pressure transmitted retrograde from elevated left-heart pressures. Once left-heart pressures are normalized following transplantation, the pulmonary artery pressures typically normalize. But long-standing left ventricular failure may also produce pulmonary vascular remodeling and a "fixed" increase in PVR; such "fixed" PVR is unresponsive to pulmonary vasodilator therapy.

Because cardiac donors almost invariably have normal pulmonary arterial pressure, the right ventricle of the donor heart is not conditioned to pump against a high resistance: the PVR of the recipient must be low enough to allow the transplanted heart to support the recipient's circulation. Heart transplantation into a recipient with increased PVR unresponsive to vasodilator therapy predictably leads to right ventricular failure of the transplanted heart, and death.

A potential heart transplant recipient must have a right-heart catheterization in order to accurately measure pulmonary arterial pressures, determine the transpulmonary gradient and calculate the PVR. If the patient is found to have increased PVR, provocative testing with vasodilator therapy is indicated in the cardiac catheterization suite. Using an infusion of sodium nitroprusside, an attempt should be made gently to vasodilate the systemic circulation without lowering systolic arterial blood pressure below 90 mm Hg. Such reduction in left ventricular afterload increases forward cardiac output. In turn, left atrial pressure (and pulmonary capillary wedge pressure) decrease, thereby lowering the hydrostatic component of the patient's pulmonary arterial pressure. During such provocative testing, the patient's hemodynamic variables typically mimic the situation

Table 11.2. Contraindications to cardiac transplantation

- Advanced age
- Irreversible hepatic, renal or pulmonary dysfunction
- Severe peripheral vascular disease
- Cerebrovascular disease
- Insulin-requiring diabetes mellitus with end-organ damage
- Active infection
- Malignancy
- Inadequate psychosocial support systems
- Limited life expectancy from systemic disease
- Pulmonary vascular resistance > 6 Wood units
- Transpulmonary gradient > 15 mm Hg

which will be created by a new heart. Specifically, pulmonary capillary wedge pressure is < 12-14 mm Hg with a systemic systolic arterial pressure > 100 mm Hg.

If under the conditions of such provocative testing the patient's systolic blood pressure falls below 90 mm Hg, and/or the PVR remains > 4 Wood units and/or transpulmonary gradient remains > 15 mm Hg, the patient's pulmonary circulation affords too much resistance to the donor heart's right ventricle. Heart transplantation should not be performed, and consideration should be given to heart-lung transplantation.

CONTRAINDICATIONS

Evaluation of potential heart transplant recipients is very difficult and must be thorough. Established criteria of eligibility within a given program may vary somewhat with the maturity and local experiences within a given transplant program. Adherence to such criteria produces predictable results following transplantation. Conditions considered by most programs to contraindicate heart transplantation are listed in Table 11.3.

LISTING STATUS CODES

Provided the patient meets the aforementioned criteria and has no contraindications they are listed for transplantation. In 1999 the urgency status codes were modified to provide better guidance with allocation. The present status codes are listed in Table 11.2.

TRANSPLANT OPERATION

DONOR SELECTION AND PROCUREMENT

In addition to the scarcity of donor organs, cardiac transplantation is further limited by a short ischemic time (4-6 hours). While this has been lengthened through numerous experimental studies clinically, extension of the preservation time has not been well tolerated. As a result the management of the donor before procurement is critical, and this is especially true as older donors are being used. With brain death there are often significant changes in the patient's hemodynamics, metabolism, and temperature control. Myocardium may be damaged from

Table 11.3. Current medical urgency status codes for heart allocation

1A   Adult—Registrant at least 18 years of age, admitted to listing hospital with at least
     one of the following: (a) mechanical circulatory support for acute hemodynamic
     decompensation with VAD 30 days or less, TAH, balloon pump, or ECMO; (b)
     mechanical circulatory support for more than 30 days with objective medical
     evidence of significant device-related complications; (c) mechanical ventilation; (d)
     continuous infusion of a single high-dose intravenous inotrope or multiple
     intravenous inotropes, in addition to continuous hemodynamic monitoring of left
     ventricular filling pressures; or (e) meets none of the criteria specified above but
     admitted to the listing hospital with a life expectancy without a heart transplant of
     less than seven days.
     Pediatric—Registrant less than 18 years of age and meets at least one of the
     following criteria: (a) requires assistance with a ventilator; (b) requires assistance
     with a mechanical assist device; (c) requires assistance with a balloon pump; (d) is
     less than 6 months old with congenital or acquired heart disease exhibiting reactive
     pulmonary hypertension at greater than 50% of systemic level; (e) requires infusion
     of high dose or multiple inotropes; or (f) meets none of the criteria specified above
     but has a life expectancy without a heart transplant of less than 14 days.
1B   Adult—A registrant who (a) has a left and/or right ventricular assist device
     implanted for more than 30 days; or (b) receives continuous infusion of intravenous
     inotropes.
     Pediatric—A registrant who (a) requires infusion of low dose single inotropes, (b) is
     less than 6 months old and does not meet the criteria for Status 1A, or (c) exhibits
     growth failure (see OPTN policies for definition).
2   A patient of any age who does not meet the criteria for Status 1A or 1B.
7   Temporarily inactive.

the changes in blood pressure. Electrolyte imbalances may lead to dysrhythmias
and myocardial edema. To stabilize the blood pressure, vasoactive drugs are often
instituted. Ideally there should be no need for significant inotropic support. Typically the donor can be managed with dopamine or dobutamine. Maintenance of a
mean arterial pressure near 80-mm Hg is necessary. Unfortunately, diabetes insipidus can also lead to instability, electrolyte imbalance and acid base abnormalities.

Ideally donors should be less than 40 years old. Those in the fourth and
fifth decades of life need careful evaluation for coronary artery disease.
Echocardiography must be performed on all potential donors. The echocardiogram
should demonstrate normal cardiac anatomy, normal valve function, and normal
ventricular function.

In patients greater than 50 years of age, cardiac catheterization should be performed to exclude coronary artery disease. If a pulmonary artery catheter has
been placed, the CVP should be in the 10-12 mm Hg range, the pulmonary capillary right pressure less than 15 mm Hg and the cardiac index greater than
2.5 l/min/m2. Allocation is based on ABO blood type and body size. Typically
an organ from a donor that is within ten percent of the recipient's weight is acceptable.

Procurement of the donor organ is performed via median sternotomy. The pericardium is opened and the heart is suspended in a pericardial cradle. The heart is

inspected and palpated. Plaque in the coronary arteries may be identified by this method and valvular insufficiency or stenosis may be noted by thrills. The pericardial reflection is dissected free of the aorta and the pulmonary artery. The reflections at the SVC and IVC are similarly dissected free to allow adequate exposure. In freeing the superior vena cava the azygos vein may be isolated, ligated and divided. This is particularly helpful if a caval-to-caval anastomosis is planned in the recipient. A cardioplegia cannula is placed high in the ascending aorta, permitting room for a cross-clamping below the innominate artery. A 14-gauge angiocath is acceptable for cardioplegic delivery. The SVC is ligated, the aorta is cross-clamped, and a liter of cold (4°C) cardioplegic solution is administered. Once this has been commenced the IVC is clamped and partially transected to allow decompression of the right heart. The heart is elevated, and the left pulmonary vein is also partially transected to decompress the left heart. Iced slush is then placed in the pericardium for topical hypothermia. Hypothermia is the most important component of cardiac preservation, since it provides a profound reduction in myocardial oxygen consumption and demand. Once cardioplegic delivery is complete the IVC is completely transected and each of the pulmonary veins is transected at the pericardial reflection. The SVC is divided and the aorta is transected at or above the level of the innominate artery. The pulmonary artery is transected at its bifurcation. If concomitant lung procurement is not being done then the pulmonary artery division should include the bifurcation and portions of the main pulmonary arteries. Once excised the heart is inspected through the great vessels looking at the aortic and pulmonary valves. Similar inspection of the tricuspid and mitral valve is achieved via the cavae and pulmonary veins. The heart is then placed in a plastic bag with cold saline. An additional bag of cold saline is then wrapped around the first, and both bags are then placed in a bucket of cold saline, which is in turn placed in a cooler filled with ice.

Technical aspects of the recipient operation in an orthotopic transplantation are largely unchanged from the original description by Lower and Shumway as reported in 1960 (Figs. 11.3A and 11.3B). A median sternotomy is performed and the patient is placed on cardiopulmonary bypass. If the heart appears to be adherent to the underside of the sternum as a result of previous operations, cardiopulmonary bypass may be instituted by cannulation of the femoral artery and vein. Otherwise standard aortic cannulation at the level of the innominate artery and bicaval venous cannulation are used. The heart is excised at the level of the atrial ventricular groove and excess donor atrium is excised as well. The donor left atrial cuff is created by making an incision that connects all of the pulmonary veins. The right atrial cuff is reestablished by making an incision extending from the IVC up towards the right atrial appendage. If caval-to-caval anastomosis is to be completed, the IVC and SVC are trimmed and beveled to prevent stenosis of the anastomoses. Using a 3-0 monofilament suture, the left atrial anastomosis is performed first. Before closure of the septum, the left atrium is filled with saline to eliminate as much air as possible. The right atrial anastomosis is then completed. The aortic anastomosis is completed next using a 4-0 monofilament suture. A

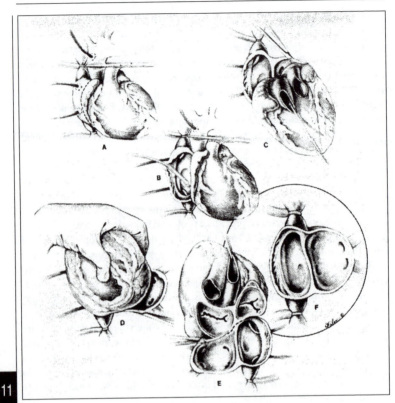

Fig. 11.3A. Explantation of the native heart. A. Initial incision is made in the right atrium close to the AV groove. B. Right atrial cuff completed down to the septum. C. Aorta and pulmonary artery transected, roof of the left atrium entered. D. With heart elevated, lateral left atrial cuff developed; pulmonary veins can be seen. E. As viewed from above, incision completed along AV groove. F. Final appearance—SVC and IVC entering RA cuff; pulmonary veins entering LA cuff.

vented cardioplegia cannula is placed proximal to the suture line to allow evacuation of air and delivery of cardioplegia if necessary. If the ischemic time has been prolonged at this point the cross-clamp can be removed and the heart reperfused. The pulmonary artery anastomosis can then be completed. If the ischemic time has been short, the pulmonary artery anastomosis can be completed before removal of the cross-clamp. The recipient, having been cooled to 30°C, is rewarmed and weaned from cardiopulmonary bypass. Temporary atrial and ventricular patient wires are placed.

Technical pitfalls include redundant atrium, which can lead to stasis and thrombus formation. Bleeding from the suture lines, particularly the atrial anastomosis, has also been reported. Because of the double atrium anastomosis, mitral and tricuspid insufficiencies can occur. As a result, separate anastomosis of the supe-

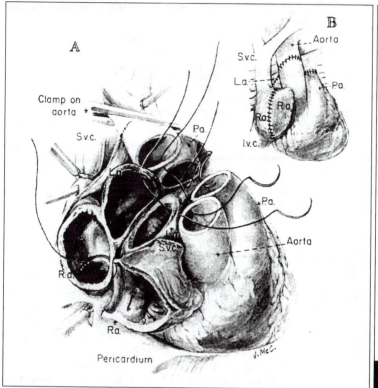

Fig. 11.3B. Orthotopic heart transplantation. A. Four end-to-end anastomoses performed: RA, LA, aorta, pulmonary artery. B. Appearance of completed transplantation (Welch KJ, Randolph JG, Ravitch MM et al, eds. Pediatric Surgery. Chicago: Mosby-Year Book, 1986, Fig. 41-4, p. 385. Reprinted by permission.)

11

rior and inferior vena cava and individual pulmonary venous implantation have been proposed.

Heterotopic heart transplantation is being considered useful as 1) a biologic-assist device; 2) in patients with elevated pulmonary vascular resistance; 3) in cases where the donor heart is too small. The technical aspects include opening of the pericardium on the right side down to the phrenic nerve to allow for the donor heart to sit in the right chest. An incision is made in the left atrium of the recipient and is then sewn to the left pulmonary veins of the donor. The donor right pulmonary veins are ligated. Caval anastomoses are completed by sewing the superior vena cava of the donor to the recipient, and the inferior vena cava is ligated. The aorta of the donor is sewn in an end-to-side fashion to the recipient, and the pulmonary artery is similarly connected in an end-to-side fashion using prosthetic graft material as needed.

Table 11.4. Cardiac drugs

Agent	Effects: Compared to normal hearts
Isoproterenol	Unchanged or increased inotropic and chronotropic effect
Dobutamine	Unchanged inotropic and chronotropic effect
Epinephrine	Unchanged inotropic and chronotropic effect
Norepinephrine	Unchanged inotropic and chronotropic effect; no reflex brachycardia with increase in blood pressure
Dopamine	Diminished inotropic response
Ephedrine	Diminished inotropic response
Neosynephrine	No reflex brachycardia
Atropine	No effect on atrial ventricular conduction
Digoxin	No effect acutely; may exert mild effect chronically
Amrinone	Unchanged inotropic effect

POSTOPERATIVE MANAGEMENT

As a result of the ischemic insult, the transplanted heart exhibits depressed systolic function and impaired contractility. To assure adequate cardiac output, preload must be maintained with right atrial pressures of 10-15 mm Hg and left atrial pressures of 15-20 mm Hg. Aside from recovering from the preservation ischemia, the transplanted heart is totally denervated. As a result, there are significantly altered responses to cardiovascular drugs (Table 11.4). As a result of the denervation, the most commonly used inotropic agent is isoproterenol in doses of 0.25 to 5.0 mcg/min. If further inotropic support is needed dobutamine or epinephrine may be used. Dopamine is primarily used to enhance renal perfusion. Inotropic support is usually required for 2-3 days postoperatively.

Denervation also alters the use of antiarrhythmic agents. Digoxin is of little use since its antiarrhythmic properties are vagally mediated. Quinidine or procainamide are typically used for supraventricular and ventricular tachyarrhythmias. Verapamil is also used to manage supraventricular tachyarrhythmias. Lidocaine is effective for ventricular arrhythmias.

IMMUNOSUPPRESSION

The mainstay of immunosuppression is triple drug therapy, including cyclosporine, azathioprine and corticosteroids (Table 11.5).

CYCLOSPORINE

Cyclosporine is administered postoperatively in a dose of 2.5-5.0 mg/kg/day and may be given intravenously or orally. This dose is then used as maintenance therapy and adjusted based on levels and interaction with other medications. The side effects of cyclosporine include nephrotoxicity, which is dose-related. Hypertension is a frequent side effect, as are tremor and paresthesias.

AZATHIOPRINE

Azathioprine (Imuran) is administered orally or intravenously in the postoperative period in a dose of 2 mg/kg/day. The dose is adjusted to keep the WBC count greater than 4000. Drug interactions are not as common, although given in

Table 11.5. Immunosuppression

Preoperative	
Azathioprine	4 mg/kg IV
Intraoperative	
Methylprednisolone	500 mg
Postoperative	
Cyclosporine	2.5-5.0 mg/kg/day
Azathioprine	2 mg/kg/day
Methylprednisolone	125 mg every 8 hours x 3-4 doses
Prednisone	1 mg/kg/day tapering over 1 week to 0.5 mg/kg/day

combination with allopurinol can cause significant bone marrow toxicity. Leukopenia and/or thrombocytopenia require dose reduction. Pancreatitis and hepatitis can occur.

CORTICOSTEROIDS

Methylprednisolone is administered intraoperatively with reperfusion of the heart in a dose of 500 mg intravenously. This is repeated postoperatively in 125 mg/8hrs until the patient is able to take oral medications. Prednisone is then administered in a dose of 0.3-0.15 mg/kg/day as an initial dose for the first 3-6 months. Corticosteroids can cause pituitary adrenal suppression, glucose intolerance, hypercholesterolemia, peptic ulcer disease, osteoporosis, hypertension and behavioral changes. Prednisone may be tapered over time as determined by endomyocardial biopsies.

Induction of immunosuppression with antilymphocyte antibodies has also been used. They include antithymocuite globulin (ATGAM), rabbit antithymocyte globulin (RATG) and monoclonal antibody OKT3. These have been shown to prevent acute rejection and have primarily been used for treatment of acute and chronic rejection. Milder or moderate rejection is typically treated with a steroid pulse of methylprednisolone in a dose of 1 gm IV x 3 days. More severe rejection, or recurrent rejection, invokes the use of ATGAM or OKT3 for 7-14 days.

New immunosuppressor drugs include tacrolimus (FK506) and mycophenolite mofetil. FK506 restricts T-cell proliferation similar to the mechanism of action of cyclosporine and initially it was hoped that FK506 could replace cyclosporine. Clinical trials have not indicated a major advantage to FK506. It has been found to be an effective rescue therapy when regimens including cyclosporine were not effective at preventing rejection.

Mycophenolite mofetil is a lymphocyte-specific inhibitor of purine synthesis with impact proliferative effects on T and B lymphocytes. It possesses many of the properties of azathioprine. Early reports indicate it may be superior to azathioprine in cardiac patients but long-term results are pending.

Other drugs such as rapamycin, deoxysperdualin and leflunomide are being evaluated. Additional treatments such as total lymphoid irradiation and photopheresis have also been tried with limited success.

11

Table 11.6. The International Society for Heart Transplantation biopsy grading system

Grade	Findings	Rejection Severity
0	No rejection	None
1	A = Focal (perivascular or interstitial) infiltrate without necrosis	Mild
	B = Diffuse but sparse infiltrate without Necrosis	
2	One focus only with aggressive infiltrate and/or myocyte damage	Focal moderate
3	Myocyte damage	Moderate damage
	A = Multifocal aggressive infiltrate and/or myocyte damage	
	B = Diffuse inflammatory process with necrosis	
4	Diffuse aggressive polymorpholeukocyte infiltrate with edema, hemorrhage, vasculitis and necrosis	

REJECTION

Clinically the most frequent symptom associated with rejection is fatigue. Examination may reveal a relative hypotension, increased jugular venous distension, and the presence of an S3. These findings should prompt an emergent biopsy. Similarly atrial ventricular arrhythmias may be the early warning signs of rejection and biopsy is warranted.

Treatment of rejection depends on the severity as assessed by histologic grading and allograft function. Mild rejection may require increased immunosuppression if accompanied by significant cardiac dysfunction. The diagnosis of rejection is made by endomyocardial biopsy. Traditionally it is performed via percutaneous internal jugular or femoral vein puncture with fluoroscopic guidance. An internationally accepted grading scale for reporting cardiac allograft rejection has been adopted (Table 11.6).

Recommended frequency for performing surveillance right ventricular biopsy varies, but it is typically performed weekly for the first month, then every other week for another month, then monthly until six months postoperatively and then every three months until the end of the first postoperative year. At that time additional biopsies as surveillance have not shown to be of clinical significance. Aside from these routine biopsies additional biopsies are performed after treatment of rejection.

REJECTION TREATMENT

There are a number of protocols for the treatment of rejection. After the initial biopsy demonstrates rejection, echocardiography is performed to evaluate function. Cyclosporine and azathioprine dosages are optimized and the patient receives a steroid pulse. Resolution may be seen by follow-up echocardiography and confirmed by rebiopsy. For more severe rejection ATGAM and/or OKT3 may be used with rebiopsy in 3-5 days and the steroid pulse may be continued. For severe rejection antilymphocyte treatment is extended and hemodynamic support, both pharmacologically and mechanically, is implemented as needed. Relisting and consideration for retransplantation are the recommended plans for progressive rejection despite escalating immunosuppression.

ANTIBIOTIC PROPHYLAXIS

As a result of immunosuppression, infection can be a major problem postoperatively. Prophylaxis against cytomegalovirus, with gancyclovir or acyclovir, should be administered if the recipient or the donor is positive. Toxoplasmosis, Pneumocystis carinii, Candida albicans and Herpes Simplex are seen as opportunistic infections.

OUTCOMES

PHYSIOLOGY OF THE TRANSPLANTED HEART

Compared to the normal heart, differences in the physiology of the transplanted heart derive largely from the fact that the transplanted heart is denervated. Data suggesting that some degree of reinnervation occurs include case reports of angina and the presence of cardiotonic reflexes among some transplant recipients. But functionally the heart remains denervated. Loss of vagal tone results in a somewhat higher resting heart rate. Carotid baroceptor reflexes are typically absent. Increases in heart rate and contractility must rely upon circulating catecholamines. Therefore heart rate increases slowly with the onset of exercise and remains elevated longer after cessation of exercise, which parallels the changes in circulating catecholamine levels. In addition to an increase in β-adrenergic receptor density, the myocardial β-andrenergic receptors of the denervated heart are more sensitive to catecholamines to provide normal cardiac output over a broad range of total body oxygen requirements (Fig. 11.4).

SURVIVAL FOLLOWING TRANSPLANTATION

The operative mortality rate associated with cardiac transplantation in most centers is typically between 3% and 5%. The principal cause of death within the first 30 days is primary allograft failure (25%). Primary allograft failure typically results from inadequate myocardial preservation of the donor heart or prolonged ischemic time. Other leading causes of early mortality include infection (15-20%), acute right ventricular failure usually resultant to increased PVR in the recipient (15%) and rejection (15%).

Risk factors for death within the first year are shown in Table 11.7. As with other surgical procedures, patient-specific risk factors reflecting greater illness of the patient or increased patient age increase the operative risk of the transplant procedure. Transplant centers performing fewer than nine transplants per year have lower survival rates. Donor-specific risk factors include female gender of the donor and increased donor age. Ischemic time of the donor heart is a well recognized risk factor and adds approximately 10% risk of death for each one hour ischemia.

Data from the ISHLT Registry indicate that the one-year survival following the cardiac transplantation is 82% worldwide. Individual programs, however, have

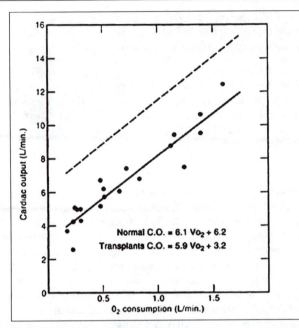

Fig. 11.4. Relationship between cardiac output and oxygen consumption in the transplanted heart. From: Perloth MG, Reitz BA. Heart and heart-lung transplantation. In: Braunwald E, ed. Heart Disease, 5th Ed.n, Philadelphia: WB Saunders Company, 1997: 527.

11

achieved one-year survival rates of 90% or higher. Beyond the first year, the annual mortality is approximately 4%. Thus, the ISHLT data indicate three-year survival is 77% and the five-year survival is about 67% (Fig. 11.5). The average survival is nine years.

QUALITY OF LIFE

The vast majority of transplant recipients achieve long-term survival and excellent functional recovery. In most transplant centers, at least 90% of survivors achieve complete rehabilitation and are classified as New York Heart Association

Class I. Nonetheless, little more than 40% return to work full-time. The reasons more recipients don't return to work full-time include insurance issues and chronic medical conditions. Those factors, more than inability to work, reflect the difficulties patients have with employment.

CARDIAC ALLOGRAFT VASCULOPATHY

Cardiac allograft vasculopathy (CAV) is the achilles heel of cardiac transplantation. In fact, the first long-term survivor of heart transplantation died one and one-half years posttransplant with severe CAV in 1969. Today CAV is the leading cause of death among patients > one year posttransplant. Unlike atherosclerotic coronary artery disease, which usually creates localized areas of eccentric stenosis in the epicardial coronary arteries, CAV is manifested by an accelerated form of

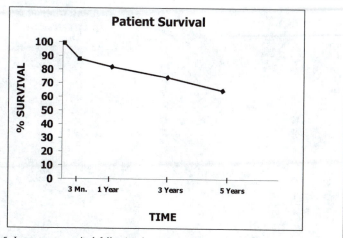

Fig. 11.5. Long-term survival following heart transplantation. From: ISHLT website— www.ishlt.org, 2/03.

Table 11.7. Risk factors for one year mortality after heart transplantation

Negative Recipient Factors

Variable	Odds Ratio
Ventilator	2.66
Retransplant	2.33
Ventricular Assist Device	1.49
Center volume < 9/yr	1.3
Female donor	1.22
Ischemic time (each time)	1.1
Recipient age > 60	1.19
Recipient age > 70	1.5
Donor age 40	1.18
Donor age 50	1.48
Donor age 60	1.99

From: Hosenpud JD, et al: The registry of the International Society for Heart and Lung Transplantation: fifteenth official report. J Heart Lung Transplant 1998;17:656-668.

11

concentric coronary artery disease that diffusely affects the larger epicardial and smaller intramyocardial coronaries. Cardiac veins are affected as well. Histologically CAV seems to begin as concentric intimal thickening and progresses with time to complex atherosclerotic plaque.

The incidence of CAV is very high. By one year posttransplant, 11% of patients are diagnosed with CAV by coronary angiography; it is found in 50% by five years. Most cardiac transplant recipients cannot experience angina because allograft reinnervation is incomplete. Thus, the clinical presentation of CAV is not angina, but is instead usually heart failure or sudden death. Occasionally CAV will present as ventricular dysrhythmias.

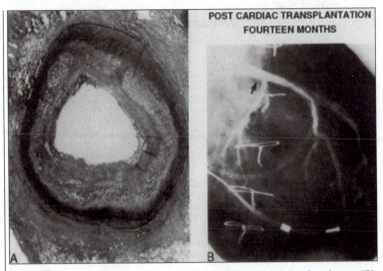

Fig. 11.6. Coronary allograft vasculopathy. Histological section (A) and angiogram (B). Despite being the best diagnostic imagining modality, angiography is relative insensitive. From: Johnson DE. Transplant coronary artery disease: histopathologic correlations with angiographic morphology. J Am Coll Cardio. 1991; 17:449. Reprinted with permission from the American College of Cardiology.

The pathogenesis of CAV remains unclear. Following coronary vascular injury associated with ischemia/reperfusion at the time of transplantation, further coronary vascular endothelial and smooth muscle cell injuries are promoted by the local inflammatory response to rejection. Even subclinical allograft rejection may contribute to the development of CAV. Common risk factors for the development of atherosclerosis such as hyperglycemia, hypertension and hyperlipidemia are commonly noted following heart transplantation and may contribute to CAV. Furthermore, some investigators suggest an infectious etiology since human CMV infection has been associated with CAV.

All heart transplant recipients must be followed indefinitely for evidence of CAV. Unfortunately, noninvasive tests for myocardial ischemia such as radionuclide scintigraphy and exercise electrocardiography are neither sensitive nor specific for detection of CAV. Therefore most programs rely upon annual coronary angiography for follow-up. Because CAV is a diffuse rather than a localized process, it may be difficult to detect with certainty using angiography. Unfortunately, coronary angiography is quite specific but a relatively insensitive test (Fig. 11.6). Intravascular coronary ultrasound (IVUS) is becoming more widespread as a screening modality for the detection of CAV. Whether it will supplant coronary angiography as the study of choice is unknown.

Because of the diffuse nature of the disease, coronary artery bypass grafting and catheter-based revascularization procedures have limited effectiveness in the treatment of CAV. Following coronary balloon angioplasty, the restenosis rate is extremely high: 55% at 8 months. Isolated case reports describing successful coronary artery bypass surgery as treatment for CAV have been published, but such successes are rare. Because of its diffuse nature, transmyocardial laser revascularization (TMR) may be an alternative to either bypass surgery or angioplasty. In a limited number of patients, the results of TMR have been encouraging.

The ultimate treatment for CAV is retransplantation. The indications for retransplantation include 1) the development of severe CAV, 2) the treatment of severe acute early injection, and 3) treatment of early acute right heart failure. The morbidity and mortality rates figures of retransplantation operations are greater than are typically seen with primary transplantation. A retransplant patient should meet the same indications as patients who undergo an initial transplant. Several reports have demonstrated that long-term survival after retransplantation is less than for primary transplantation. There is, however, a significant range in the survival results reported. Data from some institutions indicate that survival rates after retransplantation are comparable to those seen after primary transplantation. As a result, it is often difficult to determine an appropriate algorithm for retransplantation. This is a particularly controversial issue due to the scarcity of available organs. Fortunately, the results for patients undergoing retransplantation for CAV are better than those for patients retransplanted for early rejection or early right heart failure. Unfortunately, by definition the patients with CAV are older and have been chronically immunosuppressed and therefore may be less tolerant of a reoperation.

Selected Readings

1. Carrier M, Ferraro P. Immunosuppression in heart transplantation: History and current and future regimens. In: Emery RW, Miller LW, eds. Handbook of cardiac transplantation. Philadelphia: Hanley & Belfus, Inc. 2000:89-100.
2. Emery RW, Arom KV. Techniques in cardiac transplantation. In: Emery RW, Miller LW, eds. Handbook of cardiac transplantation. Philadelphia: Hanley & Belfus, Inc. 1996:61-86.
3. Hunt SA. Current status of cardiac transplantation. JAMA 1998; 280:1692-1698.
4. Perlroth MG, Reitz BA. Heart and heart-lung transplantation. In: Braunwald E, ed. Heart disease. Philadelphia: W. B. Saunders Co., 1997:520-529.
5. Stevenson LW. Medical management before cardiac transplantation. In: Emery RW, Miller LW, eds. Handbook of cardiac transplantation. Philadelphia: Hanley & Belfus, Inc., 1996:1-10.

11

Lung Transplantation

Alberto de Hoyos and Matthew Blum

INTRODUCTION

The initial attempts at human lung transplantation began forty years ago and were met with no long-term success. However, in the last twenty years lung transplantation has enjoyed increasing success and has become the mainstay of therapy for most forms of end-stage lung disease. Improved donor and recipient selection, technical advances, superior immunosuppressive strategies, and newer antibiotic regimens have improved results significantly. The operative mortality rate is now in the range of 9%. One, 2 and 5-year survival rates are 80%, 70% and 50% respectively. The Registry of the International Society for Heart and Lung Transplantation reported in 2002 that almost 15,000 lung transplants have been performed worldwide and that more than 1,500 lung transplants are performed annually.[1]

This chapter provides an overview of the pre-operative evaluation as well as the intra-and post-operative management of the lung transplant recipient.

INDICATIONS

The primary indication for lung transplantation is irreversible end-stage pulmonary disease expected to result in death within 1to 2 years and for which there are no other treatment options. Clinical conditions for which lung transplantation is indicated include the following:

1. Obstructive lung disease/Emphysema
    Chronic obstructive pulmonary disease (COPD)
    Alpha-1-antitrypsin deficiency
2. Restrictive lung disease
    Idiopathic pulmonary fibrosis (IPF)
3. Pulmonary vascular disease
    Primary pulmonary hypertension (PPH)
    Eisenmenger's syndrome (ES)
4. Septic lung disease
    Cystic fibrosis (CF)

Other "miscellaneous" causes of end-stage lung disease that have been treated with lung transplantation include cystic lung diseases such as sarcoidosis, lymphangioleiomyomatosis, and histiocytosis-X; chemotherapy or radiation therapy induced pulmonary fibrosis, idiopathic bronchiectasis, and obliterative bronchiolitis (OB) as a manifestation of chronic rejection following lung transplantation. The leading indications for lung transplantation between 1995 and 2001 were COPD (39.4%), IPF (16.9%), CF (16.1%), alpha-1-antitrypsin deficiency emphysema (9.3%) and PPH (4.6%).[1]

Organ Transplantation, 2nd edition, edited by Frank P. Stuart, Michael M. Abecassis and Dixon B. Kaufman. ©2003 Landes Bioscience.

Table 12.1. Recipient selection: general guidelines

- Clinically and physiologically severe lung disease
- Limited life expectancy (12 to 24 months)
- Other medical or surgical treatment modalities are ineffective, unavailable, inappropriate and prognosis is poor without lung transplant
- Ambulatory with rehabilitation potential
- Satisfactory nutritional status
- Appropriate mental state
  - Satisfactory psychosocial profile and good support system
  - Comprehend and accept procedure, risks, complications
- Well motivated and compliant with treatment
- Adequate financial resources for medications and follow-up
- Absence of contraindications

RECIPIENT SELECTION

A summary of the general selection criteria is shown in Table 12.1. These have been described in detail by Maurer and associates.[2] Only those patients believed to have a limited life expectancy (1 to 2 years) as a result of their underlying pulmonary disease are listed. Absolute and relative contraindications to lung transplantation are listed in Table 12.2. Once a referral is made to the lung transplant center, an initial screening evaluation of medical records, chest-x-rays, and CT scans is done. If this information is deemed satisfactory, then a formal evaluation of the patient takes place at the lung transplant center by a multidisciplinary team. The scheme for this formal on-site evaluation is outlined in Table 12.3. The data gleaned from the formal evaluation allows the lung transplant team to determine whether the patient is a suitable candidate for lung transplantation. Once patients are listed for transplantation (except those with PPH or ES), they are enrolled in a progressive, monitored cardiopulmonary rehabilitation program. Virtually all patients experience a significant improvement in strength, exercise tolerance and well being without any measurable change in pulmonary function. The average wait for a lung transplant in the United States was 19 months as of 2002.

12

SPECIFIC PRE-OPERATIVE ISSUES

Ventilator Dependency

Progression to ventilator dependency and even death during the waiting period is not uncommon because of the shortage of donor lungs. Patients who are ventilator dependent at the time of referral are not considered for transplantation. However, patients that develop ventilator dependency while on the waiting list and who remain stable are still considered candidates for transplantation.

Corticosteroid Therapy

Patients who are receiving high-dose corticosteroid therapy (prednisone >40 mg/day) are not considered eligible for transplantation due to the well-documented negative effect on bronchial healing. Low to moderate doses of corticosteroids (prednisone £ 0.2 mg/kg/day) does not contraindicate transplantation.

Table 12.2. Contraindications to lung transplantation

Absolute	Relative
Acutely ill/unstable	Systemic diseases with failure of nonpulmonary vital organs
Uncontrolled tuberculosis	Cardiac disease (coronary artery disease, ventricular dysfunction
Uncontrolled sepsis	Ongoing high-dose corticosteroids
Uncontrolled neoplasm	Age >65 years
Current smoker	Unsatisfactory nutritional status
Psychosocial problems	Osteoporosis
Inadequate resources	Medical noncompliance
HIV infection	Mechanical ventilation
Irreversible CNS injury	

Table 12.3. Scheme for evaluation of potential lung transplant recipients

- Medical history and physical examination
- Chest radiograph, electrocardiogram, and routine blood chemistries
- ABO blood typing, HLA typing and panel of reactive antibodies
- Serologic tests for hepatitis A, B, and C; HIV, cytomegalovirus
- Pulmonary studies
  - Standard pulmonary function testing, arterial blood gases
  - Quantitative ventilation/perfusion scanning
  - Cardiopulmonary exercise testing
  - CT of the chest
- Cardiovascular Studies
  - Radionuclear ventriculography
  - Doppler echocardiography (with saline contrast)
  - Right heart catheterization with angiography
  - Transesophageal echocardiography
- Rehabilitation Assessment
  - Six-minute walk test
  - Determination of supplemental oxygen requirements
- Psychosocial Evaluation
- Nutritional Assessment

12

Previous Thoracic Surgery

Previous thoracic surgery or pleurodesis is not a specific contraindication to lung transplantation. Some patients with emphysema are candidates for lung volume reduction before transplantation. Adhesions and anatomic distortion from previous thoracic surgical procedures complicate the conduct of the explantation and transplantation procedures and must be taken into consideration in planning the operative procedure.

Malignancy

History of malignant disease within the previous 5 years is usually a contraindication to transplantation. A patient judged to be cured of a more recent malignancy might be considered. An exception to this is the rare patient with bronchoalveolar carcinoma without metastatic disease. These patients have been successfully transplanted in some programs.[3]

Smoking
Patients who continue to smoke are not candidates for transplantation.

DONOR SELECTION
The principle criteria by which donor lungs are matched with recipients are ABO blood group compatibility and size matching (based upon the predicted total lung capacity and vital capacity; which in turn, are determined by the age, height, and gender). Donor and recipient histocompatibility antigen (HLA) matching is not currently performed.

Criteria to determine suitability of donor lungs for transplantation have traditionally been rigorous and at most 20% of otherwise suitable multiple organ donors have lungs that meet the traditional criteria for lung transplantation, Table 12.4. Most of the conditions that result in brain death (trauma, spontaneous intracerebral hemorrhage) are associated with significant pulmonary injury that precludes transplantation (lung contusion, aspiration, infection, and neurogenic pulmonary edema).

Satisfactory gas exchange is imperative for donor lungs. This can be confirmed by a partial oxygen pressure (PaO_2) that is greater than 300 mmHg with a ventilator delivering a fraction of inspired oxygen (FIO_2) of 1.0 and 5 cmH_2O positive end-expiratory pressure (PEEP). A PaO_2 to FIO_2 ratio of 300 or greater provides adequate evidence of satisfactory gas exchange. A donor chest x-ray must reveal clear lung fields. Bronchoscopic evidence of aspiration or frank pus in the airway is a definitive contraindication to transplantation.

A strategy that has been employed to overcome the shortage of donor lungs is the use of "marginal " lungs (defined as donor lungs that do not meet all of the traditional criteria). Relaxation of the normally strict criteria has been shown not to adversely affect outcome under carefully selected circumstances.[4-10] A minor degree of parenchymal infiltration can be accepted in a donor who is being used for a bilateral lung transplant. Judicious use of the contralateral lungs from donors with unsatisfactory gas exchange and radiographic or bronchoscopic findings confined to one lung also helps to increase the pool of donor lungs.

Other strategies to increase the donor pool include the use of living related donors for lobar transplantation and pulmonary bipartitioning.[11,12] Although innovative and exciting, these procedures are unlikely to have any meaningful impact in the overall shortage of lung allografts.

Crucial issues to successful management of the potential lung donor once brain death is declared are pulmonary toilet and fluid management. Because of its exposure to the external environment, the donor lung is at greater risk of infection than other organs. Measures such as frequent positioning change, chest physiotherapy and sterile endotracheal suctioning are recommended to protect the potential lung allografts.

Table 12.4. Traditional criteria for donor lung suitability

Preliminary Assessment
- Age <55 years
- ABO compatibility
- Chest radiograph: clear, allows estimate of size match
- History
  - Smoking <20 pack-years
  - No significant chest trauma
  - No aspiration, sepsis
  - Gram stain and culture data if prolonged intubation
  - No prior cardiac/pulmonary operation
- Oxygenation
  - PaO_2 >300 mmHg on FIO_2 1.0, PEEP of 5 cm H_2O
- Adequate size matching

Final Assessment
- Chest radiograph shows no unfavorable changes
- Oxygenation has not deteriorated
- Bronchoscopy shows no purulent secretions, aspiration
- Visual/manual assessment
  - Parenchyma is satisfactory
  - No masses

Before lung transplantation became commonplace, large volumes of fluid were used to maintain mean arterial blood pressure in multiple organ donors, to maintain urine output in kidney donors, to avoid inotropes in heart donors, and to avoid the use of vasopressin in liver donors. This volume overload can lead to pulmonary edema, thereby compromising the pulmonary allografts. General guidelines for managing the multiple organ donors are shown in Table 12.5. These guidelines should be followed in close cooperation with the other organ procurement teams.

CHOICE OF PROCEDURE

The type of lung transplant procedure to be performed depends primarily upon the clinical condition responsible for the end-stage pulmonary process. Both single lung transplantation (SLT) and BLT have been used to treat COPD.[13-15] Patients with CF or other diffuse septic pulmonary processes must undergo BLT to eliminate the risk of transmitting infection from the remaining native lung to the newly transplanted pulmonary allograft.[16] Single lung transplantation is the procedure of choice in patients with IPF since the diminished compliance and increased vascular resistance of the native lung will lead to preferential ventilation and perfusion of the transplanted lung.[17] Bilateral lung transplantation may also be offered to patients with IPF, especially in very young or large individuals. For patients with PPH, heart-lung transplantation has traditionally been the procedure of choice. However, both SLT and BLT lead to satisfactory recovery of right ventricular function in these patients. Recent evidence indicates that for patients with ES

Table 12.5. General guidelines for managing the multiple organ donor

- Blood pressure (mean) >70 mmHg
- Pulmonary capillary wedge pressure <12 mmHg
- Judicious fluid replacement
    - Replace electrolyte losses
    - Maintain urine output at 1 to 2 ml/kg/hr
    - Use DDAVP for treatment of documented diabetes insipidus
- Treat hypotension with a combination of fluid replacement and dopamine infusion up to 10 mcg/kg/min
- Maintain normothermia
- Maintain PEEP at 5 cm H_2O and PaO_2 >100 mmHg on lowest FIO_2
- Strict pulmonary toilet
- Measure arterial blood gases every 2 hours
- Elevate head of bed if possible

---

secondary to a ventricular septal defect, heart-lung transplantation should be considered the procedure of choice.[18]

According to the most recent Registry report, the majority of SLT's have been performed for either COPD (54.4%) or IPF (23.8%), and most of the BLT's have been performed for either CF (33%) or COPD (22.5%).[1]

DONOR LUNG EXTRACTION AND PRESERVATION

Currently, the parameters used to assess donor lungs are based on donor history, arterial blood gases, chest-x-ray appearance, bronchoscopic findings, and physical examination of the lung at the time of retrieval.[19] Recent evidence suggest that a bolus dose of corticosteroids (methylprednisolone 15 mg/kg) administered do organ donors after brain death declaration can improve PaO_2 and increase lung donor recovery.[10]

The following procurement procedure will allow a single donor to provide thoracic organs for up to three recipients. Both lungs are routinely extracted en block using a procedure described in detail in a previous report.[19] After the preliminary evaluation of the chest radiographs and fiberoptic bronchoscopy, the final assessment is made by gross inspection of the lungs once they are exposed by a median sternotomy in conjunction with the midline laparotomy for the extraction of the abdominal organs. The three basic components of the thoracic dissection are:

1. Venous inflow: The intra-pericardial superior and inferior vena cavae (SVC, IVC) are isolated and the SVC is encircled with heavy silk ligatures.
2. Arterial exposure: The ascending aorta and main pulmonary artery (PA) are separated from one another and encircled with umbilical tapes.
3. Airway dissection: The posterior pericardium (between the aorta and SVC) is incised, exposing the distal trachea.

On completion of the thoracic and abdominal dissection, the donor is heparinized to permit cannulation, which can be performed by all teams simultaneously, or sequentially if the donor maintains a stable condition. A cardioplegia cannula is inserted into the ascending aorta. A large bore pulmonary flush cannula is then

12

placed in the main pulmonary artery immediately proximal to its bifurcation. Occasionally the inter-atrial groove must be developed to increase the size of left atrial cuff on the right pulmonary veins. Lung procurement proceeds as follows:

1. A bolus injection of prostaglandin E_1 (500 µg) is administered directly into the main pulmonary artery alongside the cannula;
2. Double ligation of the SVC and clamping of the IVC at the diaphragm achieve inflow occlusion;
3. Cross-clamping of the ascending aorta and administration of cardioplegic solution;
4. Venting of the right heart (cardioplegic solution is vented through the IVC by transecting the IVC above the clamp; this necessitates a prior request to the abdominal procurement team to cannulate the abdominal segment of the IVC, so that their effluent flush can drain off the table);
5. Pulmonary flush is initiated: with the lungs continuously ventilated, pulmonary artery flushing is achieved with 50 ml/kg of modified Euro-Collins solution delivered at a pressure of 30 cmH$_2$0. Alternative preservation solutions include Perfadex and UW solutions.
6. Venting the left heart by amputating the tip of the left atrial appendage allows drainage of the pulmonary flush into both pleural spaces
7. Topical lung hypothermia is supplemented saline slush and ice.
8. The lungs are gently ventilated throughout to prevent atelectasis.

After completion of cardioplegia administration and lung flush, the heart is extracted first. The SVC is divided between the previously placed ligatures. The aorta is divided distal to the cardioplegia cannula. The main pulmonary artery is divided through the cannulation site, typically just proximal to the bifurcation. The heart is then elevated and retracted to the right. The left atrium is opened midway between the coronary sinus and inferior pulmonary veins. The left atrial incision is then continued toward the right. The right side of the left atrial wall is then divided, taking care to preserve a rim of atrial muscle on the pulmonary vein side. This completes the cardiac excision.

The lungs are extracted en block by:

1. Digitally encircling the trachea and dividing it between two applications of the TA-30 stapling device well above the carina keeping the lungs moderately inflated;
2. Division of the great vessels at the apex of the chest
3. Division of the esophagus superiorly and inferiorly by sequential application of the GIA stapler; and
4. Transection of the descending thoracic aorta at the level of the diaphragm.

The lung allografts are then immersed in cold crystalloid solution in the semi-inflated state. If the two lungs are to be used at separate centers, they can be divided at the donor hospital. This is done by dividing the posterior pericardium, the middle of the left atrium separating the pulmonary vein cuffs, transecting the pulmonary artery at its bifurcation, dividing the residual mediastinal tissue, and finally dividing the left main bronchus at its origin with a cutting stapling device. The double lung block or two separate lungs are then triple-bagged and transported in an ice chest (at 0-1°C).

Over the past decade, numerous studies have been performed to optimize the technique of lung preservation. Several strategies for the prevention and treatment of ischemia-reperfusion-induced lung injury have been introduced into clinical practice and have translated into a reduction in the incidence of severe ischemia-reperfusion injury from approximately 30% to 15% or less.[20-22]

TECHNIQUE OF LUNG TRANSPLANTATION

RECIPIENT ANESTHESIA

Lung transplantation requires active involvement of anesthesiologists with expertise in complex cardiothoracic anesthesia techniques, bronchoscopy and cardiopulmonary bypass. Full hemodynamic monitoring is required and includes the following: Foley catheter, central venous pressure line, pulmonary artery catheter, and radial and femoral arterial lines. Transesophageal echocardiographic monitoring is also routinely employed and is critical in patients with severe pulmonary hypertension and right ventricular dysfunction.

The airway is routinely intubated with a left-sided double-lumen endotracheal tube. This enables lung ventilation of either or both lungs. A single lumen tube with a bronchial blocker provides independent lung ventilation as required, but intraoperative maneuvering can be troublesome and lacks the reliability offered by the double-lumen tube. In patients of small stature and children, a single-lumen tube must be used. Aprotinin has been shown to decrease perioperative blood loss in patients with extensive pleural adhesions requiring cardiopulmonary bypass.[23,24]

CHOICE OF SIDE

One consideration in SLT is to try to avoid the side of a prior thoracotomy or pleurodesis if possible. Otherwise, in SLT for obstructive or restrictive pulmonary disease, the approach is to transplant the side with the least pulmonary function as demonstrated by pre-operative nuclear perfusion lung scans. In patients with Eisenmenger's syndrome, the right side is preferred to facilitate closure of the co-existing atrial or ventricular septal defects. If cardiopulmolnary bypass is anticipated, as in patients with pulmonary hypertension or severe pulmonary fibrosis with associated pulmonary hypertension, the right side is preferred. Bilateral lung replacement is accomplished by bilateral sequential single lung technique, in which the side with the least function is transplanted first.

INCISION AND APPROACH

SLT is accomplished through either a fifth intercostal space posterolateral thoracotomy or a muscle sparing incision as proposed by Pochettino.[25] Alternatively, in patients with Eisenmenger's syndrome, a median sternotomy can be utilized to accomplish simultaneous cardiac repair along with right single lung transplantation or heart-lung transplant. When the thoracotomy approach is used, the ipsilateral groin is always prepped and draped within the field, so that femoral cannulation can be performed if necessary. Bilateral lung transplantation was initially performed through a median sternotomy and later by a bilateral transverse thoracosternotomy incision ("clam shell" approach, shown in Fig. 12.1A and B)

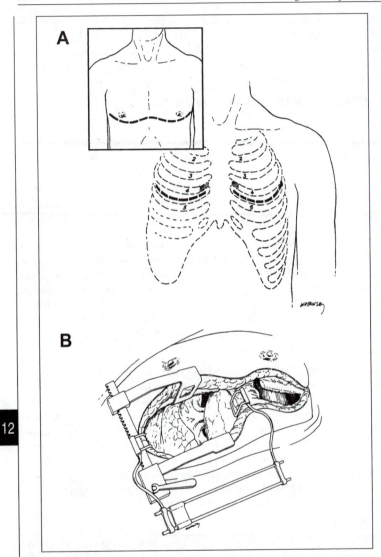

Fig. 12.1. A) The skin incision runs along the infra-mammary crease and crosses the sternum at the level of the fourth interspace. The intercostal incision in made along the upper surface of the fifth rib. The costal cartilage of the fourth rib (shaded in diagram) is resected to allow upward mobility of the fourth rib with retraction. The transverse sternotomy is omitted. B) A Finochietto chest retractor is used to spread the ribs vertically, while a Balfour retractor is placed with one jaw on the sternum and one jaw on the muscle and skin of the lateral chest. The intercostal muscle division is carried far more lateral and posterior than the skin incision to maximize rib spreading. The combined efforts of these two retractors typically result in excellent exposure without sternal division.

through the fifth intercostal space. It is now appreciated that separate bilateral anterior thoracotomies, without sternal division, provide superb exposure for BLT, and minimize the morbidity associated with transverse sternotomy.[26] With experience, intrathoracic cannulation for cardiopulmonary bypass can be accomplished avoiding a groin incision and vascular repair following decannulation.

RECIPIENT PNEUMONECTOMY AND USE OF CARDIOPULMONARY BYPASS

Except for lung transplants performed for pulmonary vascular disease, cardiopulmonary bypass (CPB) is used in a selective fashion.[27] Cardiopulmonary bypass is warranted for refractory pulmonary hypertension, poor right ventricular function as measured by transesophageal echocardiography, hemodynamic instability, hypoxemia, or worsening hypercarbia during temporary pulmonary artery occlusion. In about 10-20% of bilateral sequential lung transplants, CPB is required to facilitate implantation of the second lung, due to dysfunction of the first allograft implanted. CPB is used routinely for all patients with pulmonary vascular disease. In SLT, CPB is performed through the chest or alternatively through femoral cannulation. In BLT, standard cannulation techniques are employed utilizing the right atrial appendage and ascending aorta.

Once the pleural space has been widely exposed, the inferior pulmonary ligament is divided. The pulmonary veins and the pulmonary artery are encircled outside the pericardium. At this time the need for CPB is determined.[28] The ventilation to the contralateral lung and occlusion of the ipsilateral pulmonary artery determine whether the contralateral lung provides adequate gas exchange and hemodynamics to tolerate pneumonectomy and allograft implantation without CPB. Assessment of right ventricular function with the transesophageal echocardiography probe is especially useful at this time.

The upper lobe pulmonary artery branches are ligated and divided. This maneuver increases the length of pulmonary artery available for the subsequent anastomosis. The pulmonary artery is then stapled, a clamp is placed distally and the artery divided in between. The pulmonary veins are divided between the stapled lines or between silk ligatures placed around each branch at the hilum. This increases the size of the atrial-pulmonary vein cuff for the subsequent anastomosis.

The bronchus is identified and the bronchial arteries are secured with ligatures. The bronchus is transected just proximal to the upper lobe origin and the lung is excised. The recipient bronchus is trimmed back into the mediastinum taking care to avoid devascularization. The pericardium around the pulmonary veins is opened widely and hemostasis is achieved.

LUNG IMPLANTATION

Topical cooling of the graft during implantation is critical, and is accomplished by wrapping the allograft in a gauze sponge soaked in ice slush. The lung is kept cold with additional application of crushed ice. This provides an extended period of cold preservation and gives additional time for meticulous anastomoses.

The bronchial anastomosis is performed first. The membranous posterior wall is first closed using a continuous suture of 4-0 absorbable monofilament suture (Fig. 12.2A and B). The anterior cartilaginous airway is then closed by using an

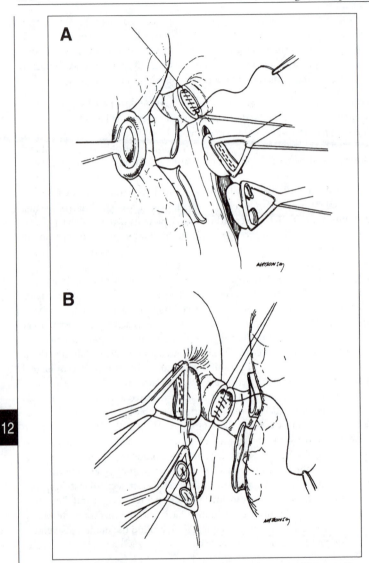

Fig. 12.2. A) The bronchial anastomosis of the right lung is depicted, showing the running closure of membranous bronchus. Self-retaining retractors, fashioned with Duval lung retractors suspended by heavy silk ties, are shown suspending the recipient pulmonary artery and pulmonary vein medially and anteriorly to expose the bronchus. B) The bronchial anastomosis of the left lung is depicted, showing the running closure of membranous bronchus. Self-retaining retractors, fashioned with Duval lung retractors suspended by heavy silk ties, are shown suspending the recipient pulmonary artery and pulmonary vein anteriorly and medially to expose the bronchus.

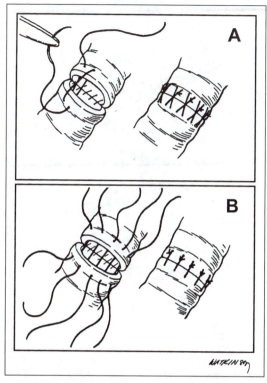

Fig. 12.3. Interrupted figure of-eight sutures. A) are used for bronchial closure in normal or larger sized airways. No attempt is made to intentionally intussuscept the smaller of the bronchial ends into the larger. In smaller airways, particularly left bronchial anastomoses in small recipients, a simple interrupted anastomosis. B) is performed to enhance accuracy and minimize stenosis.

interrupted suture technique. The donor and recipient cartilaginous arches are approximated using interrupted figure-of-eight or horizontal mattress sutures of a similar suture material with no attempt to intussuscept the smaller bronchus (Fig 12.3). The use of simple interrupted suture to perform an end-to-end anastomosis, as originally described is utilized for small caliber bronchi, since a figure-eight technique can result in bronchial narrowing. Loose peribronchial nodal tissue around the donor and recipient bronchi is used to cover the bronchial anastomosis. Rarely a pedicle flap of pericardium or thymic fat may be necessary. Bronchial omentopexy is virtually never used today.[29]

The pulmonary artery anastomosis is performed next. A vascular clamp is applied proximally on the ipsilateral main pulmonary artery. The donor and recipient arteries are trimmed to appropriate size and an end-to-end anastomosis is created using 5-0 polypropylene suture interrupted at two sites (Fig. 12.4). Care must be taken to avoid excessive lengths that may result in kinking of the pulmonary artery.

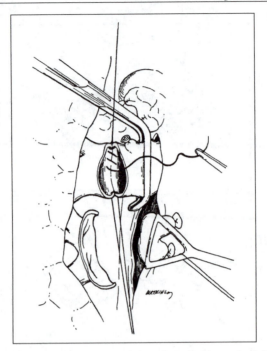

Fig. 12.4. The pulmonary artery anastomosis is performed with running 5-0 prolene suture. In the example depicted, the recipient pulmonary artery has been divided distal to the first branch to the right upper lobe. If a larger donor artery requires a larger recipient vessel, the recipient artery can be divided centrally to improve size matching.

The left atrial anastomosis is performed last. Lateral traction on the pulmonary vein stumps facilitates application of a Satinsky clamp centrally on the recipient left atrium. A large recipient atrial cuff is created by amputating the pulmonary vein stumps and excising the bridge of tissue between the two stumps (Fig. 12.5). The anastomosis is then created using a continuous 4-0 polypropylene suture, which is also interrupted, at two sites around the anastomosis (Fig. 12.5). Following completion of this anastomosis, but before tightening and tying the final stitch, the lung is gently inflated while the pulmonary artery clamp is temporarily removed, enabling the lung to be de-aired through the open left atrial anastomosis. All suture lines are then secured and the vascular clamps removed. Before restoring perfusion of the graft, 500-1000 mg of methylprednisolone is administered. Two pleural drains are inserted into the pleural space, and a standard closure is performed. Finally, the double lumen endotracheal tube is exchanged for a regular single lumen tube. Fiberoptic bronchoscopy is then performed to inspect the bronchial anastomosis, and to evacuate the airway of any blood and secretions.

Bilateral lung transplantation is performed utilizing bilateral anterior thoracotomies without sternotomy. The anastomoses for the second allograft are constructed in the same way as for the first allograft.

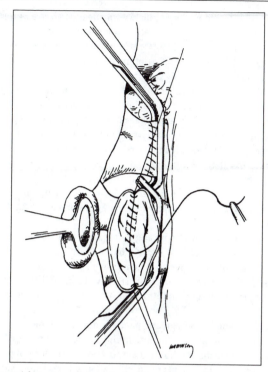

Fig. 12.5. The atrial anastomosis is performed last using running 4-0 prolene. The graft is flushed thoroughly through this partially completed suture line before the last sutures are tightened and tied.

POST-OPERATIVE MANAGEMENT

12

INTENSIVE CARE UNIT MONITORING

All lung transplant patients are admitted post-operatively to the intensive care unit, where monitoring includes the following: Electrocardiogram, oximetric evaluation of arterial and mixed venous oxygen saturation, and continuous monitoring of systemic and pulmonary arterial pressures.

VENTILATION

Most patients arrive in the intensive care unit mechanically ventilated through a single lumen endotracheal tube, which is preferable for pulmonary toilet. After BLT or SLT for IPF, standard ventilatory parameters are used. The FIO_2 is kept at a level to maintain a PaO_2 greater than 70 mmHg. A tidal volume of 8 to 12 ml/kg is usually sufficient. Extubation is performed once satisfactory gas exchange and lung mechanics are accomplished. Most of these patients are extubated between 24 and 48 hours following transplantation.

Patients who have undergone SLT for COPD or pulmonary hypertension are managed differently.[30] In patients with COPD, PEEP is avoided and tidal volumes are usually smaller to reduce hyperinflation of the excessively compliant native lung and compression of the less compliant transplanted lung.

After SLT for PPH, a PEEP of 5 to 10 cm H_2O is used for at least 36 hours, as it is believed that this minimizes the development of edema in the allograft. These patients are typically mechanically ventilated for 48-72 hours and are kept heavily sedated and often paralyzed in an effort to decrease the occurrence of pulmonary hypertensive crisis. They are also maintained with the native lung in a dependent position to promote inflation and drainage of the transplanted lung.

Prolonged periods of post-operative ventilatory requirement may occasionally be seen in lung transplant recipients. There are many potential reasons for inability to tolerate extubation but important processes to consider include: ischemia-reperfusion injury, early graft dysfunction, gas trapping, and phrenic nerve injury. Ischemia-reperfusion injury is an area of intense investigation and appears to be related to anoxic injury to the lung sustained during the period of preservation.

Gas trapping can occur in a native emphysematous lung following SLT and can be recognized radiographically as hyperinflation of the native lung and simultaneous compression of the allograft. When recognized this can be treated by positioning the patient with the allograft side up, bronchodilator therapy, and occasionally a double lumen endotracheal tube to allow to independent lung ventilation.

Clinically detectable diaphragmatic paralysis due to phrenic nerve injury is a rare complication of lung transplantation but should be considered in patients who require prolonged mechanical ventilation. Long term effects of phrenic nerve injury are uncommon. In patients requiring prolonged ventilatory support, early tracheostomy improves patient comfort, facilitates mobilization, allows oral nutrition and expedites liberation from the ventilator.

ISCHEMIA-REPERFUSION-INDUCED LUNG INJURY

Ischemia-reperfusion-induced lung injury remains a significant cause of early morbidity and mortality after lung transplantation. The syndrome typically occurs within the first 72 hours after transplantation and is characterized by nonspecific alveolar damage, lung edema, and hypoxemia. Neutrophil-mediated oxidant injury to the lung sustained at reperfusion appears to play a role. The clinical spectrum can range from mild hypoxemia associated with few infiltrates on chest x-ray to a picture similar to severe acute respiratory distress syndrome requiring mechanical ventilation, moderate levels of PEEP and FIO_2, pharmacologic therapy and occasionally nitric oxide or extracorporeal membrane oxygenation.[31,32] In addition, this syndrome can also be associated with an increased risk of acute rejection that may lead to graft dysfunction in the long term.[21]

HEMODYNAMIC AND FLUID MANAGEMENT

The lung allograft is extremely sensitive to volume. Therefore, minimizing fluid administration and careful use of diuretics are crucial in these patients. Most patients are maintained on low dose dopamine infusion (1-3 μg/kg/min) for the first 24 to 48 hours. Fluid management is facilitated by determination of pulmonary diastolic pressure or capillary wedge pressure and daily weight. Overly aggressive diuresis resulting in hypotension and prerenal azotemia should be avoided.

Pulmonary hypertension and elevated pulmonary vascular resistance can be controlled with prostaglandin E_1 infusion 10-100 ng/kg/min. Nitric oxide (at 20-60 parts per million) has also been found to be useful in decreasing pulmonary artery pressures and in improving oxygenation.[31]

PAIN CONTROL

Pain relief is effectively achieved with the use of an epidural catheter. This is placed pre-operatively in virtually all patients, except those who are anticoagulated, or in patients with pulmonary vascular disease, in whom systemic heparinization and cardiopulmonary bypass is needed. In these cases, when the epidural catheter is not placed pre-operatively, it is placed as soon after the transplant as possible. After several days the epidural catheter is removed, and patient controlled analgesia (PCA) is initiated as soon as the patient can cooperate.

POSTURAL DRAINAGE AND PHYSIOTHERAPY

Optimization of pulmonary toilet expedites weaning and extubation. SLT recipients are maintained in a lateral position with the allograft side "up" for the first 24 hours to enhance ventilation and adequate drainage of the allograft. BLT recipients are maintained supine as much as possible for the first 12 hours, then rotated from side to side as tolerated. Physiotherapy consists of vigorous chest percussion and postural drainage, and includes early mobilization of the patient after extubation. These aggressive maneuvers are withheld for the first 36 hours in SLT recipients for PPH to avoid the occurrence of pulmonary hypertensive crisis.

BRONCHOSCOPY

Bronchoscopy is useful for clearance of airway secretions, inspection of the integrity of the anastomosis, and obtaining washings to guide antimicrobial therapy. Bronchoscopy is performed in the operating room at the end of the transplantation, then on the first post-operative day and again immediately prior to extubation, and also whenever indicated by the clinical situation.

12

PLEURAL DRAINAGE

Two thoracostomy tubes are inserted into the pleural space at the time of the transplantation. They are removed as soon as there are no air leaks and drainage is minimal (<200 ml per 24 hours). Pleural space complications after lung transplantation have been reviewed in a previous report.[33]

NUTRITION

Intravenous alimentation is started within 24 hours of transplantation. In most patients, an oral diet is started within 3-7 days of the procedure, but if prolonged ventilatory support is required, a feeding tube is placed to provide enteral nutrition. Optimization of the nutritional state in these frequently malnourished patients is encouraged.[34]

Sepsis Prophylaxis

Bacterial

Prophylaxis for bacterial infection is guided by donor specimens (washings obtained during the donor bronchoscopy, as well as swabs from the donor bronchus at the time of implantation) and swabs from the recipient bronchus at the time of recipient pneumonectomy. If no organisms are identified in these specimens, then a cephalosporin is utilized for 3 or 4 days postoperatively.

If gram positive organisms are identified, vancomycin 1 g q12hrs is the regimen of choice. If gram negative organisms are seen, ceftazidime 1 to 2 g intravenously q8hrs can be used. However, if there is a progression of pulmonary infiltrates despite use of intravenous ceftazidime, then conversion to imipenem 500 mg intravenously q6hrs should be considered.

In patients with chronic septic lung disease (for example cystic fibrosis), initial antibiotics are directed at the recipient's organisms based on pre-operative culture and sensitivity data.[35] Aerosolized colistin or tobramycin can also be utilized.

Viral

Routine use of acyclovir has eliminated herpes infection as a postoperative complication. However, CMV infection remains a significant problem in lung transplant recipients. The reported incidence of CMV infection and disease following lung transplantation in the post-ganciclovir era ranges from 35% to 86% with an associated mortality rate of 2% to 12%.[36] The highest risk occurs in the case of the CMV mismatch. The seronegative recipient of a CMV-positive organ is at the highest risk of developing severe, sometimes fatal disease. It is critical to define CMV infection and disease.[37] Cytomegalovirus infection is defined by isolation of the virus or by demonstrating its presence by immunologic or molecular techniques or by seroconversion. Currently available techniques include the shell vial assay, pp65 antigenemia, polymerase chain reaction or hybrid capture assays for CMV DNAemia. Cytomegalovirus disease is diagnosed by histologic evidence of tissue invasion or a characteristic syndrome after exclusion of other etiologies in the presence of CMV infection.

Three potential mechanisms of CMV infection have been recognized: transmission by the donor organ; transmission by blood products; or reactivation of latent virus in the recipient. In the absence of endogenous antibody protection, primary CMV infections, particularly CMV pneumonitis or gastrointestinal disease, may be quite severe with mortality rates of 2% to 20%.

Based on the potential mechanisms on infection, four strategies to prevent CMV infection have been utilized:

    a. matching the donor-recipient pair by CMV serologic status

    b. use of CMV-negative blood products

    c. use of antiviral agents to suppress viral replication

    d. use of immunoglobulin (Ig) preparations to provide passive immunization.

However, to date, primarily due to the lack of randomized, controlled trials comparing regimens, the "ideal" strategy for the prevention, monitoring and treatment of CMV remains controversial. Several protocols have been utilized for prophylaxis and include monotherapy and combination regimens.[38,39] A recent recommendation is as follows:

a. seropositive recipients receive:
   CMV-Ig 150 mg/kg on postoperative day (POD) 1 and 100 mg/kg on PODs 15 and 30
   Ganciclovir IV for 30 days
   Acyclovir 3.2 g/day for 5 months

b. CMV donor positive/recipient negative receive:
   CMV-Ig for seven doses
   Ganciclovir for 100 days
   Acyclovir 3.2 g/day for 3 months

Utilizing this regimen, no evidence of CMV infection was seen in 69% (96 of 140) of patients by the end of the first year post-transplant. The incidence of symptomatic CMV infection was only 20%. The overall mortality rate was 1.4%.

It may be difficult to distinguish CMV pneumonitis from acute rejection in patients who develop cough, fever, and radiographic infiltrates. Bronchoscopy with transbronchial biopsies may be useful in differentiating acute rejection from CMV pneumonitis by providing additional specimens for culture, immunostaining for CMV, and histologic identification of acute rejection or cytomegalic inclusions. Open lung biopsy may be necessary to make the diagnosis in rare cases after failure of empiric therapy.

Standard treatment of CMV disease consists of 2 to 3 weeks of IV ganciclovir at a dose of 5 mg/kg twice daily adjusted for renal function. Although no controlled trials are available for the use of CMV-Ig for therapy, many centers add it to their ganciclovir for the treatment of tissue-invasive CMV pneumonia or colitis.

Resistance of CMV to antiviral agents has been well recognized in the laboratory and clinical setting. The incidence of ganciclovir resistance ranges from 2% to 10%, depending on the intensity of the immunosuppressive regimen and the specific organ transplanted.[40] Risk factors for the development of resistance include the intensity of the immunosuppression, particularly the utilization of anti-thymocyte globulin or OKT3 for induction therapy or treatment of rejection episodes, or prolonged exposure to either IV or oral ganciclovir. Furthermore, patients with resistant isolates have decreased overall survival and an earlier onset of BOS. Foscarnet is typically effective for the treatment of ganciclovir-resistant CMV isolates but cross-resistant strains are emerging that may require the use of Cidofovir or a combination of agents.

12

Fungal

Fungal prophylaxis is not routinely employed. Fungal organisms such as *Candida albicans* and *Aspergillus* species may colonize the respiratory tract in lung transplant patients but localized or disseminated disease is uncommon. The diagnosis of infection with these organisms is based on respiratory symptoms, isolation of the organism, and radiographic changes that cannot be explained otherwise. Invasive or disseminated disease should be treated with amphotericin B. To prevent oropharyngeal candidiasis, nystatin 500,000 units in the form of a mouthwash four times a day can be used.

PNEUMOCYSTIS CARINII

Pneumocystis carinii pneumonia can cause significant morbidity and mortality in transplant recipients. Effective prophylaxis against this organism is achieved with trimethoprim-sulfamethoxazole-DS, one tablet on Monday, Wednesday, and Friday once oral intake starts or nebulized pentamidine (in sulfa allergic patients). Active disease rarely occurs in patients receiving prophylaxis. The diagnosis of Pneumocystis carinii infection is established by bronchoscopic washings and is treated with parenteral trimethoprim-sulfamethoxazole, trimethoprim-dapsone, or pentamidine in sulfa allergic patients.

IMMUNOSUPPRESSION

Most programs rely on a "triple-drug" protocol that combines cyclosporine, azathioprine and corticosteroids.[41] Recent additions to the armamentarium include tacrolimus, sirolimus and mycophenolate mofetil. A typical strategy consists of:
   a. Pre-transplant:
      azathioprine 1.5-2 mg/kg intravenously just prior to transplantation
   b. Post-transplant
      • Cyclosporine 3-5 mg/hr intravenously, later converted to an oral dose (twice daily to maintain blood levels in the range of 250-300 ng/ml. Some centers prefer to avoid early intravenous cyclosporine due to its potential for renal toxicity.
      • Azathioprine 2 mg/kg intravenously daily (initially), later converted to 2 mg/kg orally daily, adjusted to maintain a white blood cell count greater than 3500/dl.
      • Corticosteroids:
         - Methylprednisolone 10-15 mg/kg intravenously before graft reperfusion.
         - Then 0.5 mg/kg intravenously daily,
         - Then, convert to Prednisone, (0.5 mg/kg orally daily, tapered to 15 mg p.o. daily at 1 year.
      • Anti-thymocyte globulin (ATGAM) 15 mg/kg intravenously over 8-24 hrs for 1 week (usually from the 1st-8th post-operative days).
   Primary immunosuppression trials comparing cyclosporine with tacrolimus in lung transplant recipients have shown fewer acute rejection episodes in patients taking tacrolimus.[42,43]

BRONCHIAL ANASTOMOTIC COMPLICATIONS

Airway complications were formerly a major cause of morbidity and mortality.[44] Failure of the bronchial anastomosis is generally an ischemic problem. Using standard implantation techniques, the donor bronchus is rendered ischemic and relies on collateral pulmonary flow during the first few days following transplantation. Keeping the length of donor bronchus short—two rings proximal to the upper lobe takeoff—reduces the length of donor bronchus at risk for ischemia. Due to better graft preservation (thereby maintaining superior microcirculation in the graft and better preservation of collateral flow to the bronchial arteries),

12

better techniques of airway anastomosis and routine use of corticosteroids peri-operatively, the incidence of bronchial anastomotic complications has been reduced drastically.[45] When a bronchial anastomotic complication is suspected, the patient should have a chest x-ray, fiberoptic bronchoscopy, and a chest CT scan to document the presence of any mediastinal pathology such as collections of fluid or air.

Ischemia of the airway can be seen on bronchoscopy as patchy zones of necrosis of the donor airway mucosa. This is a common finding and if limited to the superficial layers, has no clinical significance and ultimately heals.[15] However, ischemic necrosis seen at the suture line is usually associated with some degree of dehiscence. Membranous wall defects generally heal without any airway compromise, whereas cartilaginous defects usually result in some degree of late stricture. Significant dehiscence involving greater than 50% of the circumference of the airway may result in compromise of the integrity of the airway. Early post-transplantation, this form may manifest with air leak or the features of sepsis due to mediastinal infection. This problem should be managed expectantly by gentle mechanical debridement to maintain a satisfactory airway patency.[15] If extensive dehiscence occurs, a massive air leak may result, and rarely, fistulization can occur between the pulmonary artery and

bronchial anastomosis. Massive dehiscence with an uncontrolled leak or mediastinal contamination has been treated with successful retransplantation in a small number of cases. The first priority of treatment during the acute phase of an anastomotic dehiscence is to provide adequate drainage, either by chest tube, percutaneous drainage under CT guidance, or even mediastinoscopy with transcervical drainage. In the intermediate phase, periodic rigid brochoscopy and dilatation of the airway may be necessary.

Late sequelae involve the development of stricture or the development of a malacic segment of bronchus in which symptoms may include dyspnea, stridor or wheezing. Chronic airway stenoses can present significant management problems.[15,30] A right main bronchial anastomosis is usually easily managed by repeated dilatation and ultimate placement of an endobronchial stent to maintain patency.[15,45] On the left side, strictures are somewhat more difficult to manage. Dilatation is technically more difficult to accomplish because of its angulation. Furthermore, the lobar bifurcation just distal to the stricture site does not provide an adequate length of bronchus for the placement of large-caliber dilating bronchoscopes. A stent across a stricture in this location may result in occlusion of one of the lobar orifices.

Silastic endobronchial stents are preferred over other types. These stents can be inserted easily over a rigid bronchoscope and are tolerated exceptionally well. However, patients require daily inhalation of N-acetylcystine to keep the stents patent. Stents have resulted in dramatic improvement in pulmonary function.[46] Fortunately, most of these stents have been required only temporarily. After several months, most patients are able to maintain satisfactory airway patency without the stent.

12

Wire-mesh stents can only be used for malacic strictures that are completely lined by epithelium. If a wire stent is placed in a granulating stricture, the wire becomes embedded by granulation, and a stricture develops within the stent. Furthermore, removal of the wire stent may prove unsuccessful.

Distal bronchial strictures that are unmanageable by dilatation or stent insertion may require retransplantation.[47]

FOLLOW-UP STRATEGIES AND PROCEDURES REQUIRED IN LUNG TRANSPLANT RECIPIENTS

SURVEILLANCE

Lung transplant recipients have an ongoing risk of acute and chronic rejection as well as septic complications. Therefore, an approach to routine surveillance (that is, monitoring of the clinically and physiologically stable patient) seems appropriate. Different lung transplant centers vary in their approach to surveillance, but the fundamental components include:

Clinical Follow-up

Patients are "released" from the local geographic region of the lung transplant center to return "home" at around three months after transplantation, as long as their post-operative course has been satisfactory. Thereafter, they return to the care of their referring pulmonologist, ensuring that close contact and dialogue must be maintained with the lung transplant center.

PULMONARY FUNCTION TESTS

The International Society for Heart and Lung Transplantation[7] has recommended that spirometry in lung transplant recipients should be performed with equipment conforming to the American Thoracic Society Standards, without bronchodilator therapy, and ideally monthly in the first year post-transplantation.[48] This would allow a baseline value of the FEV_1 (the average of the two previous highest consecutive measurements taken 3 to 6 weeks apart) to be established. This baseline value may of course continue to increase over time. Significant allograft dysfunction would then be based on a fractional decline in the FEV_1 relative to the baseline value. At the Washington University Lung Transplant Program, spirometry is performed weekly in the first three months, then monthly between 3 and 12 months post-transplantation, and then every 2 to 3 months beyond 1 year.[49]

CHEST RADIOGRAPH

Chest radiographs are obtained on a schedule similar to that of pulmonary function testing and, of course whenever clinically indicated. Chest CT scanning is not routinely done for surveillance, but can be obtained if clinical, radiographic, or bronchoscopic findings suggest the occurrence of an intra-thoracic complication following the transplantation, such as bronchial dehiscence, pleural effusion or empyema.

FIBEROPTIC BONCHOSCOPY WITH TRANSBRONCHIAL LUNG BIOPSY (TBLB)

Fiberoptic bronchoscopy (FOB) is performed liberally in the early post-operative period, as stated previously. Subsequently, surveillance bronchoscopies are performed at 3 to 4 weeks, at around 3 months, at around 6 months, then at one year, and then annually thereafter.[50,51] FOB is also performed for clinical indications including symptoms (dyspnea, cough), signs (fever, adventitious breath sounds), the presence of radiographic infiltrates, and declining spirometry and/or oxygenation. TBLB is obtained using fluoroscopic guidance with a 2 mm fenestrated biopsy forceps. Taking 10 to 12 TBB specimens ensures a high diagnostic yield and rarely fails to provide adequate diagnosis.[52] If there is a discrete infiltrate, the majority of the biopsies are taken there, with a few being taken from uninvolved areas. If there is a diffuse infiltrate or normal chest radiograph then, specimens are taken from all several bronchopulmonary segments. These samples should then be sent in formaldehyde for routine hematoxylin and eosin (H and E) staining. Gomori methenamine silver stains are obtained to detect the presence of fungi or Pneumocystis carinii, and acid fast staining to detect the presence of mycobacteria. In addition, immunoperoxidase staining (for detection of CMV infection), and connective tissue stains (for detection of OB) are obtained if clinically indicated.

Open Lung Biopsy

When FOB with TBLB is inconclusive in the face of continuing clinical and physiological deterioration despite empiric therapy, open lung biopsy may be necessary to determine the underlying pathology and guide specific therapy.[44]

REJECTION IN THE LUNG TRANSPLANT RECIPIENT

The lung allograft is more susceptible to rejection than other solid organ allografts. Potential reasons for this include: the lung is one of the largest transplanted organs; its extensive vasculature is exposed to the entire cardiac output; unlike other allografts, it is routinely exposed to antigens from external environment via the airways; and the lung contains large populations of immunologically active cells (such as macrophages, dendritic cells, and lymphocytes). Furthermore, immunologic matching is crude (ABO blood type only) and immunosuppressive strategies remain imperfect.

ACUTE REJECTION

Virtually all lung transplant recipients sustain an episode of acute rejection within the first month post-transplantation. Patients are susceptible to acute rejection anytime beginning 3 to 5 days post-transplantation to several years later, but the risk appears to decrease with time. Fortunately, most episodes of acute rejection rarely represent a serious problem. The clinical manifestations of acute rejection are nonspecific. In addition, the manifestations early after lung transplantation (within the first month) can differ substantially from those in the late phase (that is several months or even years after the transplant).[50] In the early post-operative period, patients may experience a vague sensation of feeling unwell; they may also experience fever (more than 0.5°C above previous stable baseline), a drop in oxygenation (pO_2 dropping by more than 10 mm Hg below

Table 12.6. Working formulation for classification and grading of pulmonary rejection

A. Acute rejection
   0. Grade 0-no significant abnormality
   1. Grade 1-minimal acute rejection
   2. Grade 2-mild acute rejection
   3. Grade 3-moderate acute rejection
   4. Severe acute rejection
B. Airway damage without scarring
   1. Lymphocytic bronchitis
   2. Lymphocytic bronchiolitis

C. Chronic airway rejection
   1. Bronchiolitis obliterans, subtotal
     a. active
     b. inactive
   2. Bronchiolitis obliterans
     a. active
     b. inactive
D. Chronic vascular rejection
E. Vasculitis

previously stable baseline), and pulmonary infiltrates may develop on chest roentgenograms (in particular a peri-hilar or basal haziness). The main differential diagnosis in the early post-operative period includes acute rejection, bacterial sepsis, and pulmonary edema (resulting either from reperfusion injury or fluid overload). Later in the post-transplant period, a similar constellation of clinical features may be noted, but the chest x-ray is often not abnormal.

The main concern during these later episodes of decline in graft function is making the distinction between acute rejection and CMV infection, which generally cannot be made without a biopsy.[50] In the past, antirejection therapy was given based upon clinical findings alone, allowing the diagnosis of acute rejection to be made in retrospect if there was an improvement in the clinical picture. Currently, an aggressive approach for documenting acute rejection with FOB and TBLB is used. The TBLB specimens allow histologic detection of the presence and grade of acute rejection, cytomegalic inclusions, and OB. The key histologic finding in acute lung allograft rejection is that of perivascular mononuclear cell infiltrates. A working formulation has been developed and is shown in Table 12.6.[53] Bronchoalveolar lavage is done concomitant to TBLB but is useful mainly in identifying the presence of infection.

Acute rejection can be effectively controlled in most patients. Trulock has summarized the approach to treatment of acute lung transplant rejection and is similar to that followed by most lung transplant programs.[49] The basic components include:

1. High dose corticosteroids. This is usually given in the form of an intravenous bolus of methylprednisolone, 500-1000 mg daily for 3 days. Most patients respond to the first course of methylprednisolone.
2. An increase in the maintenance prednisone dose to 1mg/kg/day, then tapering back to the previous dose over 2 to 3 weeks.
   This approach has been found to be most useful in treating severe acute rejection episodes, especially if the oral prednisone dose has been drastically diminished or discontinued.
3. Persistent rejection despite the previous intervention is distinctly unusual but requires cytolytic therapy. Options include the use of OKT3 monoclonal antibody (5 mg/day for 10 to 14 days), or anti-thymocyte globulin (ATGAM, 10-20 mg/kg/day for 10 to 14 days) or rescue use of tacrolimus.[42-43]

CHRONIC REJECTION/BRONCHIOLITIS OBLITERANS SYNDROME

Obliterative bronchiolitis or bronchiolitis obliterans syndrome (BOS) is now regarded as a manifestation of chronic rejection.[47] The prevalence of OB in patients surviving more than 3 months after lung transplantation is 68%.[52] The actuarial 1, 3, and 5-year freedom from OB were 82%, 42% and 25% respectively.

Several risk factors for the development of OB/BOS have been identified, the predominant being late or recurrent/refractory acute rejections, lymphocytic bronchitis/bronchiolitis, noncompliance with the immunosuppressive medication, HLA mismatches at the A locus, total human leukocyte antigen mismatches and CMV pneumonitis.[54,55]

This chronic allograft dysfunction is associated with characteristic clinical, functional, and histologic changes. Obliterative bronchiolitis is consistently characterized by a reduction in pulmonary function parameters, most specifically in FEV_1, attributed to progressive airways obstruction. There is no significant reversibility after inhalation of short-acting beta-2 agonists. The onset of symptoms is mostly insidious, with progressive exertional dyspnea and cough. The characteristic histologic findings are obliterative bronchiolitis, which consists of dense fibrosis and scar tissue that obliterates the bronchiolar wall and lumen. In the later course of the disease, airway superinfections are frequently seen and colonization with *Pseudomonas aeruginosa* and *Aspergillus fumigatus* is common. At this time, high resolution CT scan often reveals bronchiectasis, diminution of peripheral vascular markings, hyperlucency due to air trapping, thickening of septal lines and mosaic phenomena. Once established, BOS may progress to severe airways obstruction with respiratory insufficiency and death. In other patients, progression may be arrested, either spontaneously or in response to treatment.

Although OB is a diagnosis that has to be made by the pathologist, it is often impossible to obtain adequate material on a TBB. The sensitivity of TBB for OB is only 28%, whereas the specificity is 75%. Because of this, in 1993, a committee sponsored by the International Society for Heart and Lung Transplantation proposed a clinical definition, called bronchiolitis obliterans syndrome (BOS). Based on pulmonary function criteria, rather than histology, BOS has been divided in four stages. Within each BOS category, there is a subtype a and b, based on no pathological evidence of OB or no pathological material for evaluation (a) or pathological evidence of OB (b). Recently, a modification to this staging system was proposed which includes a potential-BOS stage (BOS 0-p) that refers to patients with a 10% to 19% fall in FEV_1 or a 25% or greater decrease from baseline in midexpiratory flow rate (FEF_{25-75}).[56] The purpose of this additional stage is to signal the possible onset of BOS and to prompt closer surveillance, Table 12.7.

Unfortunately, therapeutic options for established BOS are limited.[48] The current treatment of chronic rejection has been summarized by Trulock and involves augmentation of the immunosuppressive drug regimen with high dose corticosteroids (methylprednisolone 1000 mg/day for 3 days, followed by prednisone 1 mg/kg/d tapering to the pretreatment dose over 2 to 3 weeks). If there is no stabilization in pulmonary function, cytolytic therapy with antilymphocyte agents such

12

Table 12.7. Bronchiolitis obliterans syndrome scoring system

BOS Stage	FEV_1; FEF_{25-75} (percent of baseline)*
0	FEV_1 >90% and FEF_{25-75} >75%
Potential BOS	FEV_1 81 to 90% and/or FEF_{25-75} <75%
1 Mild BOS	FEV_1 66 to 80%
2 Moderate BOS	FEV_1 51 to 65%
3 Severe BOS	FEV_1 <50%

*Best postoperative FEV_1 is defined as the average of two postoperative best FEV_1 measurements, 3 to 6 weeks apart.

•Grades 0 to 3 are subdivided into category A, without pathologic evidence of obliterative bronchiolitis, or category B, with pathologic evidence of obliterative bronchiolitis.

as ATGAM (10-20 mg/kg/d for 10 to 14 days) or OKT3 (5 mg/d for 10 to 14 days) may be utilized.[57] Options for more recalcitrant disease include switching from cyclosporine to tacrolimus based immunosuppression and total lymphoid irradiation. Tacrolimus rescue therapy has been shown to be effective in stabilizing pulmonary function in patients with BOS.[58-60] Although many patients show some improvement with augmented immunosuppression, the relapse rate remains high and patients remain at risk of steady progression of the disease and death. Retransplantation has been performed in selected patients with varying degrees of success.[46]

RESULTS OF LUNG TRANSPLANTATION

The success of lung transplantation demands attention to detail. Early results have improved secondary to refinements in donor and recipient selection, operative technique, and post-operative management.[15] Operative mortality rates have decreased as transplant centers acquire expertise in the management of these patients.[30]

LONG TERM SURVIVAL

As of the January 2002 update, the Registry of the International Society for heart and Lung Transplant contained survival data on 14,588 lung transplants from programs worldwide with up to 7 years of follow-up. Actuarial survival data for the reported international experience at 1,3, 5, and 7 years was 71%, 57%, 46%, and 36%, respectively. The half-life and the conditional half-life are longer after BLT than after SLT. Survival is significantly better after BLT than SLT for patients with COPD and alpha-1-antitrypsin deficiency emphysema. However, this does not apply for recipients with PPH or IPF. Actuarial survival by diagnosis does not differ substantially between the different groups.

FUNCTIONAL RESULTS

In general, the functional results after lung transplantation have been excellent for all of the diagnostic groups. In patients with emphysema, BLT consistently results in significantly greater post-operative improvements in FEV_1, arterial oxygen tension, and exercise tolerance compared to SLT.

In patients undergoing SLT for pulmonary vascular disease, there is dramatic improvement in pulmonary vascular resistance and right ventricular function. Although there is a resultant ventilation-perfusion imbalance, these patients experience a markedly improved functional result. There is no subjective dyspnea or limitation of exercise tolerance related to dead space ventilation of the native lung. However, development of BOS after SLT can lead to severe and precipitous decline in function. Although the data is inconclusive, many believe that BLT provides a longer period of improved functional result in patients with pulmonary vascular disease. Although pulmonary fibrosis represents a minority condition in most lung transplant programs, the functional results of SLT in these patients are excellent. The functional results of BLT for CF are also excellent.

CAUSES OF DEATH

During the first 30 days after transplantation, graft failure (16.4%) and non-cytomegalovirus infection (24.6%) are the principal fatal complications. After the first year OB is the principal cause of death, contributing to approximately one third of all deaths. Non-cytomegalovirus infections remain a significant cause of late mortality, contributing to approximately 25% and 16% of deaths at 3 and 5 years posttransplant respectively.[1,61,62] It is believed that the majority of fatal septic episodes and malignancies such as lymphoma arise as a consequence of the heightened immunosuppression to treat OB. It is therefore quite clear that BOS in lung transplantation is the main factor limiting long-term survival and that a better understanding of the immunologic and molecular mechanisms of this process is one of the main challenges facing clinical lung transplantation today.

REFERENCES

1. Hertz M, Taylor D, Trulock E et al. The Registry of the International Society for Heart and Lung Transplantation: nineteenth official report. J Heart Lung Transplant 2002; 21:950-970.
2. Maurer JR, Frost AE, Estenne M et al. International guidelines for the selection of lung transplant candidates. J Heart Lung transplant 1998; 17:1703-1706.
3. Zorn G. Lung transplantation in patients with thoracic malignant disease. In: Franco KL, Putnam J, eds. Advanced therapy in Thoracic Surgery. Hamilton, Ontario: BC Decker, 1998.
4. Sundaresan S, Semenkovich J, Ochoa L et al. Successful outcome of lung transplant is not compromised by the use of marginal donor lungs. J Thorac Cardiovasc Surg 1995;109:1075-1079.
5. Ware LB, Wang Y, Fang X et al. Assessment of lungs rejected for transplantation and implications for donor selection. Lancet 2002; 360:619-620.
6. Fiser SM, Kron IL, McLendon LS et al. Early intervention after severe oxygenation elevation improves survival following lung transplantation. J Heart Lung Transplant 2001; 20:631-636.
7. Bhorade SM, Vigneswaran W, McCabe MA et al. Liberalization of donor criteria may expand the donor pool without adverse consequence in lung transplantation. J Heart Lung Transplant 2000; 19:1199-1204.
8. Gabbay E, Williams TJ, Griffiths AP et al. Maximizing the utilization of donor organs offered for lung transplantation. Am J Resp Crit Care Med 1999; 160:265-271.

12

9. Pierre AF, Sekine Y, Hutcheon MA et al. Marginal donor lungs: a reassessment. J Thorac Cardiovasc Surg 2002; 123:421-428.

10. Follette DM, Rudich SM, Babcock WD. Improved oxygenation and increased lung donor recovery with high-dose steroid administration after brain death. J Heart Lung Transplant 1998; 17:423-429.

11. Cohen RG, Barr ML, Schenkel FA et al. Living-related donor lobectomy for bilateral lung transplantation in patients with cystic fibrosis. Ann Thorac Surg 1994; 57:1423-1427.

12. Couetil JA, Tolan MJ, Loulmet DF et al. Pulmonary bipartitioning and lobar transplant: A new approach to donor organ shortage. J Thorac Cardiovas Surg 1997; 113:529-533.

13. Sundaresan S, Shiraishi Y, Trulock EP et al. Single or bilateral lung transplantation for emphysema? J Thorac Cardiovasc Surg 1996; 112:1485-1494.

14. Pochettino A, Kotloff RM, Rosengard BR et al. Bilateral versus single lung transplant for chronic obstructive pulmonary disease: Intermediate-term results. Ann Thorac Surg 2000; 70:1814-1819.

15. de Hoyos A, Patterson GA, Ramirez JC et al. Lung transplantation: Early and late results. J Thorac Cardiovasc Surg 1992; 103:295-306.

16. Ramirez JC, Patterson GA, de Hoyos A et al. Bilateral lung transplantation for cystic fibrosis. J Thorac Cardiovasc Surg 1992; 103:287-294.

17. Meyers BF, Patterson GA, Cooper JD et al. Lung transplantation for pulmonary fibrosis. J Thorac Cardiovas Surg 2000; 120:99-108.

18. Waddell TK, Bennett L, Kennedy R et al. Heart-lung transplantation for Eisenmenger syndrome. J Heart Lung Transplant 2002; 21:731-737.

19. Sundaresan S, Trachiotis GD, Aoe M et al. Donor lung procurement: Assessment and operative technique. Ann Thorac Surg 1993; 56:1409-1417.

20. De Perot M, Mingyao L, Waddell TK et al. Ischemia-reperfusion-induced lung injury. Am J Resp Crit Care Med 2003; 167:490-511.

21. King RC, Binns OA, Rodriguez F et al. Reperfusion injury significantly impacts clinical outcome after pulmonary transplantation. Ann Thorac Surg 2000; 69:1681-1685.

22. Fiser SM, Tribble CG, Long SM et al. Ischemia-reperfusion injury after lung transplantation increases risk of bronchiolitis obliterans syndrome. Ann Thorac Surg 2002; 73:1041-1047.

23. Kesten S, de Hoyos A, Chaparro C et al. Aprotinin reduces blood loss in lung transplant recipients. Ann Thorac Surg 1995; 59:877-879.

24. Mojcik CF, Levy JH. Aprotinin and the systemic inflammatory response after cardiopulmonary bypass. Ann Thorac Surg 2001; 71: 745-754.

25. Pochettino A, Bavaria JE. Anterior axillary muscle sparing thoracotomy for lung transplantation. Ann Thorac Surg 1997; 64:1846-1851.

26. Meyers BF, Sundaresan SR, Cooper JD et al. Bilateral sequential lung transplantation without sternal division eliminates post-transplant sternal complications. J Thorac Cardiovasc Surg 1998; 117:358-365.

27. de Hoyos A, DeMajo W, Snell G et al. Preoperative prediction for the requirement of cardiopulmonary bypass in lung transplantation. J Thorac Cardiovasc Surg 1993; 106:787-796.

28. Triantafillou AN. Lung transplantation: Anesthetic considerations. Curr Top Gen Thorac Surg 1995; 3:171-178.

29. Miller J, de Hoyos A. An evaluation of the role of omentopexy and the early perioperative corticosteroid administration in clinical lung transplantation. J Thorac Cardiovasc Surg 1993; 105:247-252.

12

30. Meyers BF, Patterson GA. Lung transplantation. In Pearson GF, Cooper JD, eds. Thoracic Surgery. Philadelphia, Pennsylvania: Churchill Livingstone, 2002.

31. Date H, Triantafillou AN, Trulock EP et al. Inhaled nitric oxide reduces human lung allograft dysfunction. J Thorac Cardiovasc Surg 1996 May; 111(5):913-919.

32. Meyers BF, Sundt TM, Henry S et al. Selective use of extracorporeal membrane oxygenation is warranted after lung transplantation. J Thorac Cardiovasc Surg 2000; 120:20-30.

33. Herridge MS, de Hoyos A, Chaparro C et al. Pleural complications in lung transplant recipients. J Thorac Cardiovasc Surg 1995; 110:22-26.

34. Madill J, Maurer J, de Hoyos A. Comparison of pre-operative and post-operative nutritional states of lung transplant recipients. Transplantation1993; 56:347-350.

35. Snell G, de Hoyos A, Winton T et al. Pseudomonas cepacia infections in cystic fibrosis patients after lung transplantation. Chest 1993; 103:466-471.

36. Fishman JA, Rubin RH. Infections in organ-transplant recipients. N Engl J Med 1998; 338:1741-1751.

37. Wreghitt T. Cytomegalovirus infections in heart and heart-lung transplant recipients. J Antimicrob Chemother 1898; 23(Suppl E):49-60.

38. Zamora MR. Controversies in lung transplantation: Management of cytomegalovirus infection. J Heart Lung transplant 2002; 21:841-849.

39. Maurer JR, Snell G, de Hoyos A et al. Outcomes of lung transplantation using three different cytomegalovirus prophylactic regimens. Transplant Proc 1993; 25:1434-1435.

40. Limaye AP, Corey L, Koelle DM et al. Emergence of ganciclovir-resistant cytomegalovirus disease among recipients of solid-organ transplants. Lancet 2000; 356:645-649.

41. Reichenspurner H, Kur F, Treede H. Optimization of the immunosuppressive protocol after lung transplantation. Transplantation 1999; 68:67-71.

42. Keenan RJ, Konishi H, Kawai A et al. Clinical trial of tacrolimus versus cyclosporine in lung transplantation. Ann Thorac Surg 1995; 60:580-584.

43. Kur F, Reichenspurner H, Meiser BM et al. Tacrolimus (FK-506) as primary immunosuppressant after lung transplantation. Thorac Cardiovasc Surg 1999:47:174-178.

44. de Hoyos A. Maurer J. Complications after lung transplantation. Sem Thorac Cardiovasc Surg 1992; 4:132-146.

45. Date H, Trulock EP, Arcidi JM et al. Improved airway healing after lung transplantation: AN analysis of 348 bronchial anastomoses. J Thorac Cardiovasc Surg 1995; 110:1424-1435.

46. Cooper JD, Pearson FG, Patterson GA et al. Use of silicone stents in the management of airway problems. Ann Thorac Surg 1989; 47:371-382.

47. Novick RJ, Stitt L. Pulmonary retransplantation. Sem Thorac Cardiovasc Surg 1998; 10:227-231.

48. Cooper JD, Billingham M, Eagan T. A working formulation for the standardization of nomenclature and for clinical staging of chronic dysfunction in lung allogrfafts. J Heart Lung Transplant 1993; 12:713-720.

49. Trulock EP. Management of lung transplant rejection. Chest 1993 May; 103(5):1566-1576.

50. Trulock EP, Ettinger NA, Brunt EM et al. The role of transbronchial biopsy in the treatment of lung transplant recipients. An analysis of 2000 consecutive procedures. Chest 1992; 102:1049-1054.

51. de Hoyos A, Chamberlain D, Shvarztman R et al. Assessment of a standardized pathological grading system of acute rejection in isolated lung transplantation. Chest 1993; 103:1813:1818.

12

52. Hopkins PM, Aboyoun CL, Chajed PN et al. Prospective analysis of 1235 transbronchial lung biopsies in lung transplant recipients. J Heart Lung Transplant 2002; 21:1062-1067.

53. Yousem SA, Berry GJ, Cagle PT et al. Revision of the 1990 working formulation for the classification of pulmonary allograft rejection: Lung rejection study group. J Heart Lung Transplant 1996; 15(1Pt 1):1-5.

54. Sharples LD, McNeil K, Stewart S et al. Risk factors for bronchiolitis obliterans: a systematic review of recent publications. J Heart Lung transplant 2002; 21:271-281.

55. Husain AN, Siddique MT, Holmes EW et al. Analysis of risk factors for the development of bronchiolitis obliterans syndrome. Am J Resp Crit Care Med 1999; 159:829-833.

56. Estenne M, Maurer JR, Boehler A et al. Bronchiolitis obliterans syndrome 2001: an update of the diagnostic criteria. J Heart Lung Transplant 2002; 21:297-310.

57. Snell GI, Esmore DS, Williams TJ. Cytolytic therapy for the bronchiolitis obliterans syndrome complicating lung transplantation. Chest 1996; 109:874-878.

58. Kesten S, Chaparro C, Scavuzzo M et al. Tacrolimus as rescue therapy for bronchiolitis obliterans syndrome. J Heart lung Transplant 1997; 16:905-912.

59. Cairn J, Yek T, Banner NR et al. Time-related changes in pulmonary function after conversion to tacrolimus in bronchiolitis obliterans syndrome. J Heart Lung Transplant 2003; 22:50-57.

60. Fieguth HG, Krueger S, Wiedenmann DE et al. Tacrolimus for treatment of bronchiolitis obliterans syndrome after unilateral and bilateral lung transplantation: Transplant Proc 2002; 34:1884.

61. Chaparro C, Maurer J, de Hoyos A et al. Causes of death in lung transplant patients J Heart Lung Transplant 1994; 13:758-766.

62. Chaparro C, Chamberlain D, de Hoyos A et al. Acute lung injury in lung allografts. J Heart Lung Transplant 1995; 14:267-273.

12

Pediatric Transplantation
Part A: Heart Transplantation

Carl L. Backer, Elfriede Pahl, Constantine Mavroudis

There are significant differences in many aspects of heart transplantation for infants and children as compared to adults. A few examples are unique indications for transplantation such as hypoplastic left heart syndrome, technical issues of the implant for patients that need arch reconstruction or have complex anomalies such as situs inversus, postoperative immunosuppression protocols that must allow for growth and development, and rejection surveillance where central venous access is limited for biopsies.

HISTORY

The first successful series of pediatric heart transplants were reported out of Pittsburgh and Stanford. Fricker reported 14 transplants performed between 1981 and 1986 in children aged 2-16 years, with 8 long-term survivors.[1] Starnes reported 17 children aged 5 months to 14 years transplanted between 1981 and 1988 with 13 long-term survivors.[2] It was the introduction of Cyclosporine for immunosuppression that made success in pediatric heart transplantation possible. Neonatal cardiac transplantation was first performed by Leonard Bailey at Loma Linda University in 1985.[3] Pediatric heart transplants currently account for approximately 10% of all cardiac transplants.[4] Worldwide, there are approximately 200 centers performing pediatric heart transplants and since 1990 the number of pediatric heart transplants performed per year has been relatively constant, between 350 and 400.[5]

INDICATIONS FOR HEART TRANSPLANTATION IN CHILDREN

In neonates (1-30 days), the primary indication for cardiac transplantation is hypoplastic left heart syndrome. These babies have essentially no left ventricle, only a right ventricle. The other surgical option for these babies is an initial "Norwood" procedure followed by a bidirectional Glenn at 6 months of age and a Fontan at 2 years of age. There is currently much debate at different centers as to which approach is better, neonatal orthotopic cardiac transplantation (OCT) with arch reconstruction or staged palliation with an initial Norwood. In a recent review Starnes and associates concluded that given the existing data, these infants should be managed selectively on the basis of donor availability and family wishes.[6] In older children the two primary indications are cardiomyopathy and cardiac failure after conventional cardiac repairs. Indications for the 99 children transplanted at Children's Memorial Hospital from 1988 through 2003 are shown in

13

Organ Transplantation, 2nd edition, edited by Frank P. Stuart, Michael M. Abecassis and Dixon B. Kaufman. ©2003 Landes Bioscience.

Table 13A.1. Indications for cardiac transplantation at Children's Memorial Hospital 1988-2003

Diagnosis	# Patients
Infants	
Hypoplastic Left Heart Syndrome	19
Children	
Cardiomyopathy	51
Congenital structural disease, s/p failed surgical repair	26
Graft coronary disease	3
TOTAL	99

Table 13A.1. Table 13A.2 shows the broad indications by age group from the International Society for Heart and Lung Transplantation (ISHLT) Pediatric Registry.[5]

PRETRANSPLANT EVALUATION

The diagnosis of hypoplastic left heart syndrome (HLHS) is made by echocardiography (ECHO), these babies do not usually require cardiac catheterization. The initial medical support of these infants is quite specific and not intuitively obvious. Prostaglandin (PGE_1) is infused at 0.05 mcg/kg/min to maintain patency of the ductus arteriosus. The key to successful hemodynamic management is to maintain a delicate balance between the systemic and pulmonary vascular resistance. The FiO_2 is kept low (often at 0.21) to maintain the systemic arterial oxygen saturation at 75-80%. Higher O_2 concentrations will lead to a decrease in pulmonary vascular resistance (PVR), with too much pulmonary blood flow at the expense of decreased systemic perfusion. This may, in some cases, require a decrease in the inspired FiO_2 to 18%. The ventilation is controlled to maintain a pCO_2 of 45-55 mm Hg, again to keep the PVR elevated and maintain systemic perfusion.

For older children with cardiomyopathy or status post failed conventional cardiac procedures, cardiac catheterization is critical to determine the child's pulmonary vascular resistance (PVR) index and transpulmonary artery gradient (TPG).

$$PVRI(units/m^2) = \frac{Mean\ PAP\ (mm\ Hg)-PAWP\ (mm\ Hg)}{CI(L/min/m^2)}$$

$$TPG\ (mm\ Hg) = Mean\ PAP\ (mm\ Hg)-PAWP\ (mm\ Hg)$$

PAP = Pulmonary artery pressure, PAWP = Pulmonary artery wedge pressure

CI = Cardiac Index

After a prolonged period of time with congestive heart failure, many pediatric patients will have an elevated PVR. The pharmacologic reduction of the pretransplant PVR with vasodilator therapy (nitric oxide, milrinone, nitroglycerine, etc.) accurately predicts what the PVR will be after cardiac transplant.[7] If the PVR is above 6 units/m² and/or the TPG > 15 mm Hg, and the PVR is unresponsive to vasodilator therapy, heart transplantation may result in donor right ventricular failure with recipient death. For children with cardiomyopathy, many will

Table 13A.2. Indications for heart transplant by age group–ISHLT, 2002

Indication	Age		
	< 1 yr	1-10 yrs	11-17 yrs
Congenital Anomaly	74%	37%	24%
Cardiomyopathy	17%	50%	62%
Retransplantation	15%	5%	4%

Table 13A.3. Relative contraindications for pediatric cardiac transplantation

- "Fixed" PVRI ≥ 6 units/m^2
- "Fixed" TPG ≥ 15 mm Hg
- Active Infection
- Severe Metabolic Disease
- Multiple Congenital Anomalies
- Advanced Multiple Organ Failure
- Active Malignancy

respond with a reduction in their PVR prior to transplant by the administration of pulmonary vasodilators such as IV milrinone on a chronic basis pretransplant. The complete comprehensive evaluation of a child for heart transplant also includes a multidisciplinary evaluation by the renal, neurologic, infectious disease, and psychology services. Other contraindications for transplant are shown in Table 13A.3. For infants that have severe congestive heart failure not responsive to intubation, ventilation, and inotropic support, ECMO is currently the only available mechanical support. For older children over 25 kg, the Thoratec left ventricular assist device is currently available and has been used in a few centers with good results.

DONOR SELECTION AND PROCUREMENT

Like adult recipients, children are wait-listed through the regional organ bank and the United Network for Organ Sharing (UNOS). Each listing is specific for ABO compatibility and a certain weight range. In a few rare instances of fetal diagnosis of irreparable cardiac anomaly, babies are listed while still in utero with a plan to emergently deliver the child by C-section if a heart becomes available. If the strict size-matching criteria based on the adult experience were used, the supply of donor organs for children would be prohibitively small. We and other centers have, therefore, accepted organs from donors that are 0.8-2.5 times the recipient weight. Fullerton and colleagues reported success using hearts from donors over 3 times the recipient weight.[8] Evaluation of the donor function is made by assessing the donor hemodynamic status and degree of inotropic support along with the echocardiographic analysis. Final judgement of donor suitability is reserved for the visual inspection at the time of harvest. Diagnosis of cardiac donors that have been used at Children's Memorial Hospital included motor vehicle accident with head trauma, gunshot wound to the head, other head trauma, intracranial event, sudden infant death, and birth asphyxia. The cardiac harvest is performed through

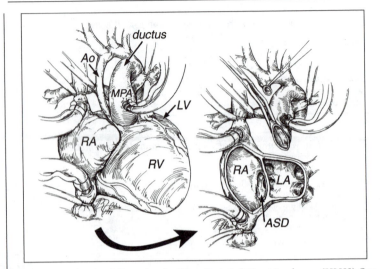

Fig. 13A.1. Cardiac Transplantation for Hypoplastic Left Heart Syndrome, (HLHS) Cannulation and Cardiectomy. The main pulmonary artery (MPA) is cannulated for arterial inflow, with the branch pulmonary arteries controlled with tourniquets. Systemic perfusion is through the ductus arteriosus and blood flow in the ascending aorta (Ao) is retrograde. The superior vena cava (SVC) and inferior vena cava (IVC) are cannulated for the venous return. Note the diminutive left ventricle (LV) and dominant right ventricle (RV). The second panel shows the recipient heart excised leaving the posterior right atrium (RA), atrial septal defect (ASD), and left atrium (LA). Systemic perfusion is being maintained.

a median sternotomy, usually as part of a multi-organ harvest. The donor is anticoagulated with 300 units/kg of heparin intravenously. The donor heart is arrested and preserved with cold crystalloid cardioplegia (30 cc/kg), infused under careful low pressure injection. For patients with HLHS, the entire donor aortic arch is harvested. For recipients with complex caval or pulmonary artery anatomy the innominate vein and/or right and left pulmonary arteries are also harvested with the donor. We attempt to keep the total donor ischemia time under four hours.

OPERATIVE TECHNIQUES OF IMPLANT

In pediatric patients, the original transplants were performed using the "right atrial" technique first described by Lower and Shumway.[9] For transplant of the infant with HLHS, the technique requires reconstruction of the hypoplastic aortic arch, which necessitates a period of circulatory arrest. For older children we and others have changed from the right atrial technique to using a "bicaval" anastomosis. The bicaval technique preserves the donor right atrial anatomy, which leads to significantly less tricuspid valve insufficiency and better preservation of sino-atrial node function. We have had to replace the tricuspid valve in 2 patients that had the right atrial technique of implant. Our technique of transplant for patients with HLHS is illustrated in Figure 13A.1 and 13A.2, and emphasizes mini-

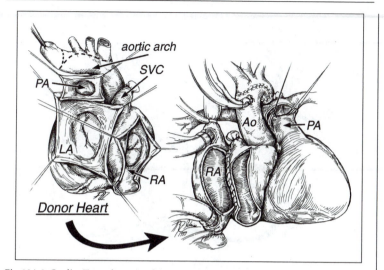

Fig. 13A.2. Cardiac Transplantation for HLHS: Donor Heart and Implant. The donor heart has been procured along with the aortic arch. The right atrium (RA) has been opened with an incision toward the SVC. Also shown are the left atrium (LA) and pulmonary artery (PA). The second panel shows the left atrial and aortic (Ao) anastomosis completed. Note the aortic cannulation is now in the donor aorta, and the heart and body are being perfused. The pulmonary artery (PA) anastomosis is being performed and the right atrial (RA) anastomosis will be done last.

mizing circulatory arrest time.[10] The recipient heart is excised and the left atrial anastomoses performed while systemic perfusion is maintained. Circulatory arrest is only used for the arch reconstruction. Systemic and cardiac perfusion are resumed for the right atrial and pulmonary artery anastomosis. Using this technique, the mean length of circulatory arrest was 26 minutes. The technique of OCT with a "right atrial" anastomosis is shown in Figure 13A.3, and OCT with "bicaval" anastomosis is shown in Figure 13A.4. With both techniques the aortic cross clamp is removed after the left atrial and aortic anastomosis to minimize ischemia time of the donor heart. The mean ischemia time in our series is 2.4 hours. The right atrial (or bicaval) and pulmonary artery anastomoses are performed with the donor heart perfused. The bicaval technique lends itself particularly well for patients that have had a previous Fontan operation, a previous Glenn shunt, or a prior atrial repair of transposition of the great arteries. Innovative use of donor vessels is required for patients with situs inversus or other complex anatomic anomalies.[11] For the SVC anastomosis we have used interrupted PDS (absorbable) suture to allow for growth.

POSTOPERATIVE MANAGEMENT

Because most pediatric patients will have had an elevated PVR preoperatively, pulmonary vasodilators are an integral part of the postoperative pharmacologic regime. These include nitric oxide, Milrinone, nitroglycerine, Nipride, and

13

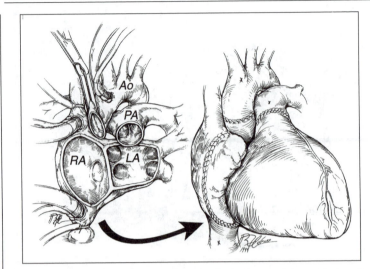

Fig. 13A.3. "Right Atrial" Cardiac Transplantation. In the first panel the recipient cardiectomy has been completed, leaving 4 anastomoses to be performed in the following sequence: 1) left atrial (LA), 2) aortic (Ao) 3) right atrial (RA), 4) pulmonary artery (PA). The donor heart is reperfused after the aortic anastomosis.

dobutamine. Most patients are paralyzed with Pancuronium and mildly hyperventilated for at least the first 24 hours postop. All patients are kept in strict reverse isolation. In our series of transplants at Children's Memorial Hospital, mean ICU stay was 19 ± 22 days, mean hospital stay was 37 ± 24 days.

IMMUNOSUPPRESSION

When infant heart transplants were first being performed there was some hope that there would be "neonatal tolerance" of the graft. This theory held that because the heart was implanted at such an early age it would come to be recognized as "self." This hypothesis has not proved true, and even babies transplanted as newborns must be kept on lifetime immunosuppression. Most centers use standard "triple therapy"—cyclosporine, azathioprine (or Cellcept), and prednisone. Approximately 40% of centers now use some form of induction therapy (OKT$_3$, Simulect, ATG, etc.). Some centers are now using Tacrolimus (FK-506) instead of cyclosporine (CSA) and most centers attempt to wean infants off of prednisone by 6 months posttransplant to allow for patient growth. Table 13A.4 shows the current ISHLT Pediatric Registry data for postoperative immunosuppression.

At Children's Memorial Hospital patients are given 5 mg/kg of cyclosporine (CSA) and 20 mg/kg of Cellcept p.o. preoperatively. Solu-Medrol is given at a dose of 10 mg/kg IV before cardiopulmonary bypass and a second dose of Solu-Medrol is given just before reperfusion of the donor heart. Simulect (Basiliximab-10 mg I.V. if < 35 kg, 20 I.V. mg if > 35 kg) has been used at our center as induction therapy since 2000, the first dose is given in the operating room after the bypass run, the second dose is given on day 4. Postoperatively, cyclosporine (CSA) is started

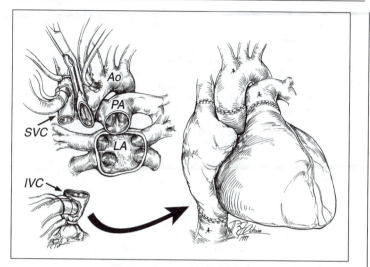

Fig. 13A.4. "Bicaval" Cardiac Transplantation. In the first panel the recipient cardiectomy has been completed. Note the entire recipient right atrium has been removed. The second panel shows the completed implant. The sequence of anastomoses are: 1) left atrial (LA), 2) Aortic (Ao), 3) inferior vena cava (IVC), 4) pulmonary artery (PA), 5) superior vena cava (SVC).

as an IV drip at 0.05-0.1 mg/kg/hr and continued for 2-3 days until gastrointestinal function is normal. At that time the patient is converted to oral cyclosporine dosing: 5-15 mg/kg/day given by mouth; t.i.d. dosing in newborns, b.i.d. in older children. We have used the enzyme multiplied immunoassay technique (EMIT) for monitoring the level and have kept that level at 250-300 ng/ml for the first 6 months, 200 ng/ml after 6 months, and at 150 ng/ml after 1 year. After a short (3 day) course of IV Solu-Medrol (10 mg/kg/day, day 1; 5 mg/kg/day, day 2; 3 mg/kg/day, day 3), prednisone is initiated at 1 mg/kg/day p.o. and tapered every two weeks until 0.2 mg/kg is reached by 8 weeks posttransplant. In neonates, prednisone is tapered off 6 months posttransplant. We previously used Azathioprine (1-2 mg/kg/day) but have converted to the use of Cellcept (mycophenolate mofetil). Cellcept is given at an initial dose of 50 mg/kg/day and titrated to a WBC count of 5,000-10,000. Some centers use OK-T$_3$ (murine monoclonal antibody) for induction therapy. Dose is 2.5 to 10 mg daily IV for 10-14 days. Other centers have reported the use of anti-thymocyte globulin (ATG) and antilymphocyte globulin

13

Table 13A.4. Pediatric heart transplant immunosuppression—ISHLT, 2002[5]

	Posttransplant % of Children on Various Immunosuppression Drugs				
	CSA	Tacrolimus	Azathioprine	MMF	Prednisone
Discharge	55%	45%	30%	50%	65%
Year 1	65%	35%	40%	30%	45%

Table 13A.5. ISHLT categories of acute cellular rejection[14]

- Grade 0–no evidence of cellular rejection
- Grade 1A–focal perivascular or interstitial infiltrate without myocyte injury
- Grade 1B–multifocal or diffuse sparse infiltrate without myocyte injury
- Grade 2–single focus of dense infiltrate with myocyte injury
- Grade 3A–multifocal dense infiltrates with myocyte injury
- Grade 3B–diffuse, dense infiltrates with myocyte injury
- Grade 4–diffuse and extensive polymorphous infiltrate with myocyte injury; may have
         hemorrhage, edema, and microvascular injury

(ALG) for induction therapy. We have only used OKT$_3$ and ATG for rescue therapy of severe rejection.

REJECTION SURVEILLANCE AND TREATMENT

Acute cardiac rejection is the most common cause of death in pediatric heart transplant recipients after 31 days and prior to 3 years posttransplant. After 3 years, chronic rejection (graft arteriopathy) is the most common cause of death. Clearly, rejection surveillance is extraordinarily important. In children, surveillance is done by clinical assessment, echocardiography, and endomyocardial biopsy. Still investigational are certain EKG parameters as assessed by an implanted pacemaker, new radionuclide imaging studies, and blood levels such as serum vascular endothelial growth factor (VEGF). Clinical parameters observed include change in activity or appetite, atrial or ventricular ectopy, and resting tachycardia (15-20 bpm over baseline). Because endomyocardial biopsy is invasive and not readily available to neonates with relatively small central veins and a thin right ventricle (risk of perforation), echocardiography is the key component of rejection surveillance in neonates and infants. The smallest child in our series to have a biopsy weighed 4 kg. By ECHO, a decrease in the peak rate of posterior wall diastolic thinning may be the most sensitive indicator of acute mild rejection in the infant.[12] Some patients develop a new pericardial effusion with rejection. Severe rejection is evidenced by an increase in LV posterior wall thickness and decrease in LV shortening fraction. Our patients undergo ECHO twice weekly in the first month, weekly in the second month, and monthly throughout the first year. In children over 6 months of age we have used endomyocardial biopsy—more frequently the older the child. It is our belief that any time there is a question regarding the ECHO findings, endomyocardial biopsy remains the "gold standard" for diagnosis of rejection.[13] Allograft rejection is graded according to ISHLT criterion (see Table 13A.5).[14]

We have a protocol that calls for yearly biopsies in older children. In these patients, coronary angiography is also performed on a yearly basis to look for transplant coronary artery disease. Our center has recently shown that intravascular ultrasound (IVUS) is more sensitive than angiography for the detection of TCAD.[15] We have also successfully used dobutamine stress echocardiogram to screen for TCAD.[16] Endomyocardial biopsy also provides tissue for the diagnosis of humoral/vascular rejection.[17] We have treated grade 2 or higher rejection with a 3 day pulse of IV methylprednisolone, 10 mg/kg. For more severe cellular rejection associated with hemodynamic compromise we have used OKT$_3$, and for severe humoral rejection plasmapheresis and cytoxan.[18] A significant problem at pediatric centers is the adolescent that decides he/she is "well enough" to stop taking their immunosuppression. This has been the major case of late death in our series.[19]

CHILDHOOD DISEASES

One of the great fears when pediatric transplantation began was that because of immunosuppression, the child would be at risk for severe complications from common childhood diseases. In actual experience, with careful "triple therapy" immunosuppression, most common childhood illnesses are well-tolerated. Routine immunizations (except for live virus vaccines) should resume at 12 weeks posttransplant. Transplant recipients have a normal response to routine immunization with diptheria, pertussis, typhoid, and Hepatitis B. Pediatric transplant recipients and their siblings should receive only inactivated polio. Measles, mumps and rubella vaccines are not given to these patients. Varicella-Zoster immune globulin should be given to pediatric heart transplant recipients within 72 hours after exposure to chicken pox. Currently, varicella vaccine is given to siblings, but not to the transplant recipient.

OUTCOMES AND LATE COMPLICATIONS

RESULTS

Surgical techniques of cardiac transplantation in children have been refined to a point now where, in most series, operative survival is at least 90%.[20] At Children's Memorial Hospital, we have had one operative mortality in the past 50 transplants (2% mortality). ISHLT Pediatric Registry data shows a current 5 year survival of 65% and a 10 year survival of 55%.[5] In the neonatal age group, the largest series (n=84) has been reported by Chiavarelli and colleagues with an operative mortality of 13% and a 5 year actuarial survival of 81%.[21] We have serially evaluated the cardiac index, left ventricular volume, and ejection fraction following heart transplants in both infants and children.[22] Cardiac output remains normal as indexed to the patient's body surface area and there is a steady appropriate rise in left ventricular end diastolic volume. The donor heart grows appropriately with the child!

GROWTH AND DEVELOPMENT

Longitudinal studies of linear growth of pediatric heart transplant recipients would indicate that growth is not normal. However, there are many variables such as preoperative growth failure caused by chronic disease, prolonged hospitalization, and other factors that make such studies difficult. Most pediatric heart transplant recipients are somewhat short for their age. Chronic prednisone may play a role. Developmental delay measured by the Denver Developmental Screening has not been documented. Most importantly, ISHLT Pediatric Registry data indicates that at 3 years postop over 95% of recipients have no activity limitations.[5]

GRAFT CORONARY ARTERY DISEASE

Accelerated graft coronary artery disease (CAD) is a severe limitation of both adult and pediatric heart transplant recipients. Selective coronary angiography most likely underestimates the true incidence, but it is probably present in 20-40% of patients at 3-4 years posttransplant.[23] Proposed etiologies include repeated cellular rejection episodes, hyperlipidemia, cytomegalovirus infection, and vascular rejection. Autopsies show a classic lesion of concentric intimal proliferation with an intact internal elastic lamina. The lesion is different than naturally occurring atherosclerosis. Surveillance is with coronary angiography and the recently introduced intraluminal ultrasound (IVUS).[15] Our center has also used Dobutamine stress echocardiography as a screening tool for significant graft CAD.[16] There is some evidence that captopril (an ACE inhibitor) may help prevent graft CAD.[24] For patients with graft failure secondary to severe graft CAD, retransplantation may become necessary. We have retransplanted three patients for graft CAD.

POSTTRANSPLANT LYMPHOPROLIFERATIVE DISEASE (PTLD)

Chronic immunosuppression may lead to the development of malignancy following heart transplantation. At Children's Memorial Hospital 10 patients have developed PTLD, with successful treatment in seven.[25] In some cases this is related to infection with Epstein-Barr virus.[26] PTLD presents with fever, malaise, leukopenia, and adenopathy. Early diagnosis with biopsy is important to determine whether the diagnosis is polyclonal or monoclonal, and to rule out infection. Most patients will respond to a reduction of immunosuppressive therapy, administration of acyclovir, or chemotherapy.

CONCLUSIONS

Heart transplantation can be performed in both neonates and children with a low operative mortality and excellent long-term survival. Currently this is the only therapy available for children with end-stage cardiomyopathy and structural heart disease not amenable to conventional surgical therapy. For neonates with HLHS, the results of transplant are still being compared to the results of staged reconstruction. Improving results with staged reconstruction and poor availability of donors for infants have swung this pendulum toward staged reconstruction. We have transplanted only 1 baby with HLHS in the past 5 years. In older children, proper recipient selection, particularly with regard to pulmonary vascular resistance and its response to vasodilators, may decrease perioperative mortality. Challenges still remaining include improved immunosuppression, graft CAD, PTLD, and donor shortages. Out on the horizon are the use of xenografts, permanent chimerism/tolerance, genetic modulation of the donor, and permanent mechanical support.

REFERENCES

1. Fricker RJ, Griffith BP, Hardesty RL et al. Experience with heart transplantation in children. Pediatrics 1987; 79:138-46.
2. Starnes VA, Bernstein D, Oyer PE et al. Heart transplantation in children. J Heart Transplant 1989; 8:20-6.

3. Bailey LL, Nehlsen-Cannarella SL, Doroshow RW et al. Cardiac allotransplantation in newborns as therapy for hypoplastic left heart syndrome. N Engl Med 1986; 315:949-51.

4. Hertz MI, Taylor DO, Trulock EP et al. The Registry of the International Society for Heart and Lung Transplantation: Nineteenth Official Report—2002. J Heart Lung Transplant 2002; 21:950-970.

5. Boucek MM, Edwards LB, Keck BM et al. The Registry of the International Society for Heart and Lung Transplantation: Fifth Official Pediatric Report—2001 to 2002. J Heart Lung Transplant 2002; 21:827-840.

6. Starnes VA, Griffin ML, Pitlick PT et al. Current approach to hypoplastic left heart syndrome. Palliation, transplantation, or both? J Thorac Cardiovasc Surg 1992; 104:189-94.

7. Zales VR, Pahl E, Backer CL et al. Pharmacologic reduction of pretransplantation pulmonary vascular resistance predicts outcome after pediatric heart transplantation. J Heart Lung Transplant 1993; 12:965-73.

8. Fullerton DA, Gundry SR, Alonso de Begona J et al. The effects of donor-recipient size disparity in infant and pediatric heart transplantation. J Thorac Cardiovasc Surg 1992; 104:1314-9.

9. Lower RR, Stofer RC, Hurley EJ et al. Successful homotransplantation of the canine heart after anoxic preservation for seven hours. Am J Surg 1962;104:302-6.

10. Backer CL, Idriss FS, Zales VR et al. Cardiac transplantation for hypoplastic left heart syndrome: A modified technique. Ann Thorac Surg 1990; 50:894-8.

11. Menkis AH, McKenzie N, Novick RJ et al. Expanding applicability of transplantation after multiple prior palliative procedures. Ann Thorac Surg 1991; 52:722-6.

12. Loker J, Darragh R, Ensing G et al. Echocardiographic analysis of rejection in the infant heart transplant recipient. J Heart Lung Transplant 1994; 13:1014-8.

13. Zales VR, Crawford S, Backer CL et al. The role of endomyocardial biopsy rejection surveillance in pediatric cardiac transplantation. J Am Coll Cardiol 1994; 23:766-71.

14. Rodriguez ER. The pathology of heart transplant biopsy specimens: revisiting the 1990 ISHLT working formulation. J Heart Lung Transplant 2003; 22:3-15.

15. Costello JM, Wax DF, Binns HJ et al. A comparison of intravascular ultrasound with coronary angiography for evaluation of transplant coronary disease in pediatric heart transplant recipients. J Heart Lung Transplant 2003; 22: 44-49.

16. Pahl E, Crawford SE, Swenson JM et al. Dobutamine stress echocardiography: experience in pediatric heart transplant recipients. J Heart Lung Transplant 1999; 18:725-32.

17. Zales VR, Crawford S, Backer CL et al. Spectrum of humoral rejection following pediatric cardiac transplantation. I Heart Lung Transplant 1993; 12:563-79.

18. Pahl E, Crawford SE, Cohn RA. Rodgers S. Wax D, Backer CL, Mavroudis C, Gidding SS. Reversal of severe late left ventricular failure after pediatric heart transplantation and possible role of plasmapheresis. Am J Cardiol 2000; 85:735-9.

19. Ringewald JM, Gidding SS, Crawford SE et al. Nonadherence is associated with late rejection in pediatric heart transplant recipients. J Pediatr 2001; 139:75-8.

20. Backer CL, Zales VR, Idriss FS et al. Heart transplantation in infants and children. Heart Lung Transplant 1992; 11:311-19.

21. Chiavarelli M, Gundry SR, Razzouk AJ et al. Cardiac transplantation for infants with hypoplastic left-heart syndrome. JAMA 1993; 270:2944-47

22. ZalesVR, Wright KL, Pahl E et al. Normal left ventricular muscle mass and mass/volume ratio after pediatric cardiac transplantation. Circulation 1994; 90(Part II):1161-1165.

13

23. Pahl E, Fricker FJ, Armitage J et al. Coronary arteriosclerosis in pediatric heart transplant survivors: limitation of long-term survival. J Pediatr 1990; 116:177-83.

24. Kobayashi J, Crawford SE, Backer CL et al. Captopril reduces graft coronary artery disease in a rat heterotopic transplant model. Circulation 1993; 88(11)286-90.

25. Pahl E, Crawford SE, Katz BZ et al. PTLD in a pediatric heart transplant cohort is not associated with traditional risk factors. J Heart Lung Transplant 2003; 22:S134.

26. Zangwill SD, Hsu DT, Kichuk MR et al. Incidence and outcome of primary Epstein-Barr virus infection and lymphoproliferative disease in pediatric heart transplant recipients. J Heart Lung Transplant 1998; 17:1161-6

13

Pediatric Transplantation
Part B: Kidney Transplantation

P. Stephen Almond

Transplantation is the optimal renal replacement therapy for infants and children with end-stage renal disease (ESRD). Compared to dialysis, a successful transplant at any age improves survival, allows for more normal growth and development, and provides an excellent quality of life. In large measure, these data are from living related (LR) recipients on cyclosporine (CsA) based immunosuppression. This chapter will review pediatric renal transplantation and the impact of these factors on outcome.

Pediatric renal transplant data are available from three sources. In 1987, North American centers agreed to pool pediatric data in the North American Pediatric Renal Transplant Cooperative Study (NAPRTCS). Currently, there are 135 participating centers. These data are published periodically and found by searching the title word "NAPRTCS". In May 1988, the United States Renal Data System (USRDS) was created to "collect and analyze information on the incidence, prevalence, morbidity, and mortality of ESRD in the United States". Since then the USRDS has published annual reports that are available by request or on line at www.med.umich.edu/usrds/. Finally, single center reports from large pediatric transplant centers are in the medical literature.

CAUSES OF KIDNEY FAILURE

According to the 1998 USRDS, the annual incidence of pediatric ESRD is 1,087, or 13 per million.[1] Many causes of pediatric ESRD are unique (Table 13B.1). Alport's syndrome is the association of nephropathy, deafness, and cataracts. Infantile polycystic kidney disease (ARPKD) is autosomal recessive, occurs one in 10,000 to one in 40,000 live births, and involves both kidneys and liver. Adult or autosomal dominant polycystic kidney disease (AKPDK) occurs one in 250 live births, but only 10% present in the first two decades of life. Hemolytic uremic syndrome (HUS) is a disease of infancy characterized by microangiopathic hemolytic anemia, renal cortical necrosis, renal failure, and thrombocytopenia. IgA nephropathy (Berger's disease) presents as a respiratory illness, hematuria, and glomerular mesangial IgA deposits. Henoch Schonlein syndrome (HSP) is characterized by purpuric skin lesions, abdominal pain, arthralgia, and renal dysfunction. Cystinosis is a rare, autosomal recessive, lysosomal storage disease that leads to the accumulation of intracellular cystine, cell death, and eventual end organ failure. Renal transplantation is appropriate for all diseases listed in Table 12B.1.

13

Organ Transplantation, 2nd edition, edited by Frank P. Stuart, Michael M. Abecassis and Dixon B. Kaufman. ©2003 Landes Bioscience.

Table 13B.1. Common causes of pediatric ESRD according to 1998 USRDS

Primary Disease	Incidence of Disease
Unspecified GN	13%
Focal glomerulosclerosis, focal glomerulonephritis	10%
Renal hypoplasia, dysplasia	9.2%
Congenital obstructive uropathy	5.7%
Lupus	5%
Hypertension	4.6%
Nephrolitiasis	4.5%
Alport's	2.9%
Membranoproliferative glomerulonephritis	2.5%
Polycystic kidneys, adult	2.5%
Chronic interstitial nephritis	2.4%
Chronic pyelonephritis	2.3%
Rapidly progressive glomerulonepritis	2%
Hemolytic uremic syndrome	1.9%
IgA nephropathy	1.2%
Henoch-Schonlein syndrome	1%
Medullary cystic disease	0.8%
Cystinosis	0.8%
Congenital nephrotic syndrome	0.8%
Prune belly syndrome	0.8%
Goodpastures synndrome	0.7%
Polycystic kidneys, infantile	0.6%
Wegner's granulomatosis	0.5%

RECIPIENT EVALUATION

The recipient evaluation is shown in Table 12B.2.[2] Potential recipients are evaluated by a team composed of pediatric nephrologists, surgeons, neurologists, nutritionists, psychologists, and social workers. The history and physical examination are aimed at identifying associated congenital anomalies, documenting height, weight, and head circumference, and determining neurologic development. Blood is obtained for CBC, platelet count, electrolytes, calcium, phosphorus, albumin, type and crossmatch, HLA typing, and viral serology (i.e., cytomegalovirus, Epstein-Barr virus, herpes simplex virus, varicella zoster, measles, mumps, and rubella). A urinalysis, chest x-ray, electrocardiogram, electroencephalogram, abdominal ultrasound, and vesicourethrogram (if indicated by history or prior renal ultrasound) or voiding cytourethrogram are also obtained. Pneumococcal and hepatitis B vaccines are given 6 weeks pretransplant.

The need and timing of native nephrectomy is also determined. Indications for nephrectomy include renal infection, severe hypertension, congenital nephrotic syndrome, reflux nephropathy, and polycystic kidneys.[2] Recipients < 16 kgs will have an intraabdominal transplant so nephrectomy is performed at transplant. Recipients > 16 kgs will have an extraperitoneal transplant so their native kidneys are removed pretransplant.

The selection criteria for pediatric recipients are liberal. Absolute contraindications include active infection, uncontrolled malignancy, and positive T-cell crossmatch.

13

Table 13B.2. Standard evaluation of pediatric kidney transplant candidates

---

History and physical examination
Laboratory tests
    Hematology (complete blood count with platelets and differential)
    Coagulation (prothrombin time, partial thromboplastin time, thrombin time)
    Chemistry (serum electrolytes, blood urea nitrogen, creatinine, liver function tests,
        lipid panel, serum electrophoresis, parathyroid hormone.
    Urine (urinalysis, urine culture, 24 hour urine for protein)
    Blood bank/immunology (ABO type, hepatitis profile, human immunodeficiency
        virus, HLA type, antileukocyte antibody screening.
Virology
    Cytomegalovirus
    Epstein-Barr virus
    Herpes simplex virus
    Varicella zoster virus
    Measles, mumps, rubella titers
Radiology
    Vesicourethrogram or (VCUG) voiding cystourethrogram
    Chest x-ray
    Bone age
Electroencephalogram
Consultations
    Social worker
    Dentists
    Neurologist
    Psychologist
    Nutritionist
    Cardiologist
Vaccines
    Pneumococcal
    Hepatitis B
Purified protein derivative

Reprinted with permission from Matas AJ, Najarian J. Pediatric Surgery 1998; 563-580. © Mosby-Year Book Inc.

13

Relative contraindications include active autoimmune disease, ABO incompatibility, and mental retardation. Certain diseases (i.e., membranoproliferative glomerulonephritis, focal sclerosing glomerulonephritis, HUS, HSP, and oxalosis) are known to recur in the allograft and may influence the timing of transplant, posttransplant management, and donor source. These issues should be discussed with the family pretransplant. Recipient weight is also a consideration. The smallest infant successfully transplanted was 5 kilograms.

Once the child has successfully completed the evaluation process, living donation is offered to the family. Potential donors have blood drawn for ABO typing, HLA typing, and crossmatch. Like recipients, potential donors are evaluated by a multidiciplinary team of internists, nephrologists, surgeons, psychologists, and social workers. The evaluation is aimed at uncovering hypertension, infection, malignancy, renal disease, and motivation for donation. Recipients without a living donor (LD) are placed on the cadaveric waiting list. USRDS data show that overall, 37% of children are transplanted within the first year of ESRD.[1]

PRETRANSPLANT MANAGEMENT

If dialysis is necessary, the method of dialysis (hemodialysis vs peritoneal) must be discussed and an access catheter placed. Peritoneal dialysis (PD), either continuous cycling or continuous ambulatory, is the preferred method and requires a peritoneal dialysis catheter.[3] The catheter of choice is a coiled, swan-neck, tunneled, double cuffed catheter. At operation, the coil is placed in the pelvis and an omentectomy preformed. Postoperative complications include tunnel infections, peritonitis, and catheter malfunction. The coil (vs. straight) catheter decreases discomfort associated with dialysate infusion. The swan neck, a permanent, 180° bend in the catheter, prevents outward catheter migration and assists in maintaining the catheter in the pelvis. The tunneling and double cuff (vs. nontunneled, single cuffed) decreases the incidence of peritonitis and tract infections. The omentectomy decreases catheter occlusion.

Hemodialysis (HD) is an option for older children. The catheter of choice is a double lumen, tunneled, cuffed catheter.[3] At operation, the catheter is placed in the right internal jugular vein (IJ) and the tip positioned at the superior vena cava/right atrial junction. Postoperative complications include infection, thrombosis, catheter malfunction, and venous stenosis. Tunneling and cuff(s) decrease infection. Right IJ catheters have the lowest incidence of vascular stenosis. Catheter malfunction due to thrombosis or a fibrin sheath is treated with intraluminal urokinase (74% successful) or fibrin sheath removal (95% successful), respectively.[3] All PD and HD patients are screened for nasal *Stapholcoccus aureus* and treated with intranasal Mupirocin if positive.

Medical management of ESRD children is aimed at preventing growth retardation, malnutrition, anemia, and renal osteodystrophy.[4] The etiology of growth retardation is multifactorial and related to age of ESRD onset, duration of ESRD, and glomerulofiltration rate. Other important factors include anorexia, malnutrition, anemia, chronic acidosis, and uremia. Malnourished or anorectic children are treated with enteral feedings by either a nasoenteric feeding tube or gastrostomy. Unlike adults, ESRD children are not protein restricted but rather receive additional protein due to their greater lean body mass/weight ratio and poor protein utilization. Growth hormone (0.05 mg/kg/day) is a safe and effective treatment for growth retardation. Anemia is due to lack of erythropoietin and iron deficiency. Potential recipients with long waiting time should receive erythropoietin (50-100 μ/kg 3 times per week), oral iron, and be followed to insure resolution of anemia. Renal osteodystrophy is the result of elevations in parathyroid hormone levels, hypocalcemia, hyperphosphatemia, and altered vitamin D metabolism. The result is an increase in bone resorption and poor bone mineralization. Treatment consist of a low phosphate diet, the use of phosphate binders (calcium carbonate, 10-20 mg calcium/kg/day) and vitamin D supplementation (calcitriol, 20-60 ng/kg/day).

ANESTHETIC CONSIDERATIONS

The anesthesiologist must have a good understanding of pediatric anesthesia and the operation. Compared to adults, children have increased surface area to mass ratio and lose heat rapidly when anesthetized and unclothed. To avoid hypothermia, the room, inhalational anesthetic, IV fluids, blood products, and child should be actively warmed. In small children, the graft requires a large portion of the circulating blood volume. To prevent hypotension and ensure immediate graft perfusion, the CVP is raised to 15 before unclamping. Unclamping washes preservation solution from the graft and acidotic, hyperkalemic blood from the lower body into the circulation. To prevent systemic acidosis and arrhythmias, 1 mEq/kg of sodium bicarbonate is given before unclamping and the anesthesiologist stands ready to treat arrhythmias. Finally, immediate graft function leads to large urinary losses of fluid and electrolytes. These losses are measured and replaced intraoperatively.

TRANSPLANT PROCEDURE

The transplant procedure has previously been described.[2] Recipients are brought to the operating room normothermic and euvolemic. Monitoring devices are placed and include ECG, pulse oximeter, rectal or oral temperature probe, and arterial line. Using flouroscopic guidance, a large, double lumen central line is placed. This is used for monitoring and administration of fluids and blood products intraoperatively and, if necessry, dialysis postoperatively. The abdomen and pelvis are prepped and draped and the Foley inserted on the field. The bladder is filled by gravity with antibiotic solution and the Foley clamped. Children < 16 kg are explored through a midline incision. The right colon is mobilized and control of the aorta, vena cava, and iliac arteries and veins obtained. If indicated, the native kidneys and ureters are removed. The aorta and vena cava are occluded with a single vascular clamp and the iliac arteries and veins controlled with vessel loops. The renal artery and vein are anastomosed to the distal aorta and vena cava using 7-0 and 6-0 prolene, respectively. Before reperfusion, the child is given 250 mg/kg of mannitol, 1mg/kg of furosemide, 1 mg/kg of sodium bicarbonate, and the CVP raised to 15 cms of water. The venous clamp is removed first followed by the arterial. The graft is bathed in warm saline and bleeding controlled with cautery and suture ligatures. The ureteral anastomosis is done using the Leadbetter-Politano technique. A double or single J ureteral stent is placed in the pelvis and the proximal end either in the bladder (double J) or brought out through the urethra and secured to the Foley. Children > 16 kilograms undergo an extraperitoneal approach.

RECOVERY

Most children are extubated at the end of the procedure and then transferred to the pediatric intensive care unit for one to one nursing care. Standard postoperative orders are shown in Table 13B.3.[2] Urine output, serum electrolytes, and body weight are monitored closely. Urine output is replaced hourly cc/cc with D5.45 NS with 10 mEq of sodium bicarbonate per liter. If urine output is > 8 cc/kg/hour, the dextrose concentration is decreased to 1% to prevent hyper-

Table 13B.3. Postoperative orders for pediatric kidney transplant recipients

Nursing
      Vital signs every 30 min for 4 hr, then every hour for 24 hr
      Central venous pressure and urine otput every hour for 24 hr
      Urine for glucose; abdominal girth; peripheral pulses; and nasogastric pH every
      4 hr
      Guaiac stool daily; turn, cough, and suction (as needed)
Laboratory
      Serum electrolytes, hematology, coagulation, and arterial blood gas immediately
      postoperative and at 4 hr postoperative
      Serum electrolytes, calcium, and phosphorus every 4 hr for 24 hr
      Daily electrolytes and hematoloty; coagulation and liver function tests twice
      weekly; cyclosporine levels 3 days week
Fluids
      1 ml IV for 1 ml urine each hour (5% dextrose in water/45% normal saline with
      10 mEq sodium bicarbonate/L if urine output < 8 ml/kg/hr; the glucose is changed
      to 1% dextrose in water if > 8 ml/kg/hr)
Medications
      Trimethoprim sulfa (2-4 mg/kg/day)
      Antacid (1 ml/dose for < 2 yr old, 2 ml for 2-5 yr, and 3 ml for 5-10 yr)
      Nystatin 4 times/day, docusate, pain medication, and immunosuppression
Radiology
      Chest radiograph immediately postoperative and on postoperative day 1
      Stent study on postoperative day 6.

Reprinted with permission from: Matas AJ, Najarian J. Pediatric Surgery 1998; 563-580.
© 1998 Mosby-Year Book Inc.

glycemia. Electrolytes (sodium, potassium, bicarbonate, calcium, phosphorus, and magnesium) are checked every 2 hours and replaced. Trimethoprim is given to prevent Pneumocystis carinii infection and Legionaries disease. Antacids and nystatin are used to prevent gastrointestinal bleeding and thrush, respectively.

    The child is transferred to the ward when urine output and electrolytes have stabilized. Infants and malnourished children are started on parenteral nutrition. Children with a single J stent have a stentogram on postoperative day 5. If there is no leak, the stent and Foley are removed. Those with a double J stent have it removed endoscopically at 6 weeks.

IMMUNOSUPPRESSION

    Immunosuppression is based on either cyclosporine (CsA) or FK506. Children received either polyclonal (ATGAM 10 – 20 mg/kg/day for up to 14 days post transplant) or monoclonal [OKT3 (for less than <30 kg 2.5 mg/day and for >30 kg 5 mg/day give for 7 to14 days) or daclizumab (12 mg per meter squared with maximum of 20 mg/dose given on the day of transplant and on postoperative day 4)] antibody preparations as induction-type therapy. Rapamycin (9 mg per metered squared on day one and 2 mg per metered squared maintenance dose with maximum of 6 mg per dose given once per day) mycophenolate mofetil (600 mg/meter squared/dose twice a day), prednisone (2 mg/kg/day tapered to 0.5 mg/kg/

day at 1 month and then to a maintenance dosage of 0.25 mg/kg/day at 1 year post transplant), and azathioprine (5 mg/kg/day tapered to 2.5 mg/kg/day by post transplant day 7) are started on the day of transplant. For maintenance immunosuppression, most centers use either CsA or FK506 (with or without rapamycin), MMF or AZA, and prednisone. Cyclosporine (2.5 mg/kg twice a day) or FK506 (.10 to .30 mg/kg twice a day) is added when serum creatinine is <3 mg/dl and levels are measured periodically.

REJECTION

Although most pediatric renal allograft recipients are rejection free (56%), rejection is still the most common reason for readmission. Acute rejection usually occurs in the first 6 months posttransplant. The incidence is higher in cadaver (vs. living donor) recipients, recipients not given antibody induction (vs. those given antibody induction), and retransplant (vs. primary) recipients.[6] The usual symptoms of rejection include malaise, weight gain, fever, abdominal pain, allograft tenderness, decreased urine output, and an elevation of serum creatinine. In small children, however, rejection is more subtle. Compared to adults, small children have less muscle mass. Therefore, serum creatinine, a byproduct of muscle metabolism, is lower. In addition, the change in serum creatinine seen with a decrease in renal function is much smaller in young children (vs. adults). Small recipients of adult kidneys also have significant renal reserve. Therefore, a large portion of the kidney can be lost (to rejection) before classic signs of rejection are obvious. For these reasons, the indications for renal biopsy in pediatric recipients include a 25% rise in serum creatinine over baseline, unexplained fever for five days, and new onset or worsening hypertension.[5] In addition to the usual risks associated with renal biopsy (bleeding, hematuria, and a nondiagnostic biopsy), the small but real risk of allograft loss must be discussed. The biopsy can be performed percutaneously or open. Percutaneous biopsy can be done under general or local anesthesia with sedation. Children with intraabdominal grafts may require ultrasound or CT guidance for percutaneous biopsy.

The treatment of rejection depends of the severity of the rejection episode, the time between the transplant and rejection, and prior antirejection treatments. In general, recipients with early (< 1 month) posttransplant, biopsy proven rejection are treated with antibody and prednisone. Thereafter, rejection episodes are initially treated with prednisone; antibody is added if the rejection is prednisone resistant.

RADIOLOGIC CONSIDERATIONS

Radiologic studies are an important part of the preoperative, perioperative, and postoperative evaluation. Preoperative studies include a chest x-ray, hand films, and a voiding cytourethrogram (VCUG). The chest x-ray is used as a screening tool to rule out pulmonary infections, congenital heart disease, and metastatic lesions. The hand xrays are taken to determine bone age. The relationship of bone age to chronological age is important in determining growth potential and treating growth retardation. The voiding cytourethrogram (VCUG) evaluates the size of the bladder, bladder outlet obstruction, reflux, and bladder emptying. In the perioperative period, the stentogram rules out a urine leak. In the early postoperative period, radiologic studies determine causes of graft dysfunction. A renogram is used to show blood flow to the graft. Ultrasound will show (1) blood flow to the graft, (2) the presence of perinephric masses, (3) ascites, and (4) hydronephrosis.

Invasive, pediatric, radiologic procedures are similar to adults. Percutaneous nephrostogram is used to determine the level of obstruction in recipients with hydronephrosis. If the obstruction is due to a ureteral stricture, it can be dilated with a ballon catheter and a ureteral stent placed across the stricture. A nephrostomy tube is placed to drain the upper urinary tract. Arterial strictures can be dilated with ballon angioplasty.

EARLY SURGICAL COMPLICATIONS

In the early postoperative period, the pediatric recipient can experience any complication associated with operation (bleeding, infection, wound complications, etc.). Those unique to renal transplant recipients usually revolve around graft dysfunction. Signs and symptoms of graft dysfunction include anorexia, malaise, weight gain, fever, abdominal pain, allograft tenderness, decreased urine output (< 1 cc/kg/hour), and elevated creatinine. The differential diagnosis includes vascular occlusion, urinary obstruction, urine leak, prerenal azotemia, rejection, CsA toxicity, and acute tubular necrosis. The diagnostic approach to this common scenario is systematic.[2] The recipient is examined and the Foley irrigated to remove clots and demonstrate patency. A duplex ultrasound is obtained to rule out vascular thrombosis, external compression (hematoma, lymphocele), and hydronephrosis. If duplex ultrasound is not available or cannot exclude thrombosis, a renogram is done. Vascular thrombosis requires immediate exploration, but usually results in graft loss. Ureteral obstruction, usually a late complication due to stricture, requires percutaneous nephrostogram to determine the level of obstruction and a nephrostomy tube to drain the upper urinary tract. If possible, the stenosis is dilated and an internal stent placed across the obstruction into the bladder. If urine flows across the obstruction via the stent, the nephrostomy tube is elevated for 24 hours and then clamped for 24 hours. If there are no signs or symptoms of obstruction and no hydronephrosis on repeat ultrasound, the nephrostomy tube is removed. Urine leaks occur at the ureteroneocystomy. The child presents with ascites and a rising creatinine. Ultrasound shows a well vascularized kidney and ascitic fluid high in creatinine. Renal scan will show extravasation into the peritoneal cavity. Treatment requires re-exploration and ureteral reimplantation. In cases of ureteral necrosis, pyeloureterostomy with the native ureter or pyelocystostomy may be necessary. As previously discussed, acute rejection requires a biopsy. Prerenal azotemia is suspected based on clinical grounds (i.e., unbalanced in's and out's, a decrease in body weight, and low CVP) and responds promptly to volume. Acute tubular necrosis can be seen on a biopsy obtained to rule out rejection. CsA toxicity responds to lowering the CsA dose. (See Chapter on Immunosupression).

EARLY OUTPATIENT CARE

The majority of recipients are discharged 2-3 weeks posttransplant. Initially, labs are drawn three times per week. The child is seen every 2 weeks for the first 2 months and then monthly for 6 months.

RECURRENCE OF DISEASE

Several diseases are known to recur in the allograft and adversely affect long-term outcome.[2] Children with oxalosis have a high rate of recurrence leading some to believe they should not be transplanted. There are, however, two other alternatives. The first is performing a LD kidney transplant followed by aggressive medical management aimed at increasing the solubility of urinary calcium oxalate. The second is a combined kidney liver transplant, wherein the liver transplant cures the metabolic disease preventing recurrence in the kidney. The latter option is the treatment of choice for infants where the liver enzymatic defect is likely to be complete. Children with membranoproliferative glomerulo-nephritis have a high rate of recurrence which accounts for 20% of graft losses. FSGS has two variants, both of which can recur. The first leads to ESRD within 2 years of diagnosis and is assoicated with a high incidence of posttransplant recurrence. These children should have a cadaver transplant. The second variant has a slower progression to ESRD and a lower incidence of recurrence. These children should have a LD transplant. In children with HUS, those with diarrhea (vs. those without diarrhea) have a higher incidence of recurrence.

LONG-TERM OUTCOME

Long-term results are measured in many ways. Patient and graft survival rates are the most obvious, but allograft function, quality of life, and growth and development are also important. Compared to dialysis, a successful transplant improves survival. Pediatric death rates for patients on dialysis are 3.8 per 1000 patient years vs 0.4 per 1000 patient years for recipients of a successful trans-plant.[1] Improvements in transplant outcome are largely due to the use of CsA (Fig. 13B.1).[6] One and 5 year patient survival in recipients on CsA (vs. those not on CsA) are 100% (vs. 94% and 86%, respectively). Corresponding 1 and 5 year graft survival for recipients on CsA (vs. those not on CsA) are 93% and 77% (vs. 82% and 68%, respectively). In recipients on CsA, donor source has no impact on patient or graft survival. Importantly, USRDS and single center studies show no difference in survival based on transplant age.[1,4,5]

In decreasing order of frequency, the causes of graft loss include chronic rejection, acute rejection, recurrence of primary disease, death with function, and technical failure. Risk factors for graft loss have been identified.[7] For cadaver recipients, risk factors include age of recipient < 2 years, cold storage time > 24 hours, no antibody induction, donor age < 6 years, previous transplant, African American race, and no HLA DR matches. For LD recipients, the only risk factor is delayed graft function in recipients < 1 year old. The majority of recipients (80%) require at least one readmission.[6] The mean number of readmissions is 3 and the reasons for readmission are rejection, infection, urologic complications, and dehydration.

13

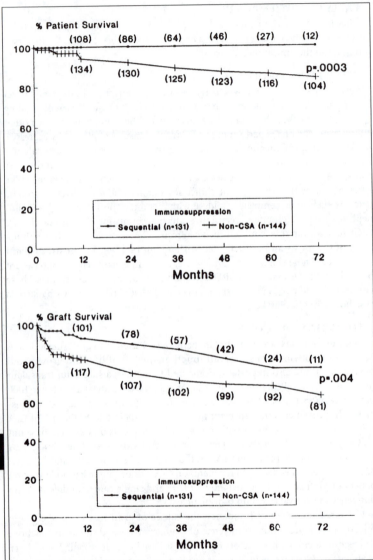

Fig. 13B.1. (A) Patient survival for recipients on sequential (ñ CsA) vs. immunsuppression. Number of patients at risk is indicated in parentheses. (B) Graft survival recipients on sequential vs. on CsA immunosuppression. Number of patients at risk is indicated in parentheses. Reprinted with permission from Almond PS, Matas AJ, Gillingham K et al. Transplantation 1992; 53:46-51. © 1992 Williams & Wilkins.

Table 13B.4. *Serum creatinine (mg%) with sequential immunosuppression, living donor vs. cadaver*

Follow-up	Living donor	Cadaver	P value
1 week	0.6 ± 0.6 (n=57)	1.7 ± 1.9 (n=26)	0.008
1 month	0.6 ± 0.5 (n=79)	1.5 ± 1.3 (n=46)	< 0.001
1 year	0.6 ± 0.5 (n=64)	1.2 ± 0.7 (n=34)	0.009
2 years	1.2 ± 1.1 (n=49)	1.1 ± 0.5 (n=29)	—
3 years	1.1 ± 0.7 (n=34)	1.1 ± 0.5 (n=20)	—
4 years	1.0 ± 0.4 (n=20)	1.1 ± 0.6 (n=13)	—
5 years	1.1 ± 0.7 (n=11)	1.0 ± 0.4 (n=8)	—
6 years	1.4 ± 1.3 (n=5)	1.3 ± 0.7 (n=2)	—

Reprinted with permission from Almond PS, Matas AJ, Gillingham K et al. Transplantation 1992; 53:46-51. © 1992 Williams & Wilkins.

Renal allograft function for CsA recipients is shown in Table 13B.4.[6] For the first posttransplant year, serum creatinine is significantly lower in LD (vs. cadaver) recipients and in primary (vs. retransplant) recipients. Overall, there is no significant difference in serum creatinine levels in CsA (vs. non-CsA) recipients suggesting no long-term affect of CsA-nephrotoxicity on graft function.

The impact of dialysis and transplantation on growth and development is a major focus. Growth data are reported as z scores which are defined as the patients measurement (i.e., height, weight, or head circumference) minus the 50th percentile for age and sex, divided by the standard deviation (for that measurement) for age and sex. NAPRATCS data show potential recipients are 2.2 standard deviations (SD) below the mean for height.[7] Two factors that impact on posttransplant growth are initial height deficit and recipient age.[8] Children < 1 year at transplant and those with the lowest z score are the only children that demonstrate catch up growth (i.e., growth ≥ 1 SD). Children older than 6 years show neither accelerated growth nor an adolescent growth spurt. Head circumference is a direct reflection of brain growth. Infants on dialysis have decreased head circumference and cognitive deficits. A successful transplant significantly improves head circumference and decreases developmental delay.

13

Acknowledgment
Special thanks to Dr. Oscar Salvatierra for reviewing and editing this chapter.

References
1. United States Renal Data System, USRDS 1998 Annual Data Report. Bethesda, MD: The National Institutes of Health, National Institutes of Diabetes and Digestive and Kidney Disease, 1998.
2. Matas AJ, Najarian JS. Kidney Transplantation. In: Neill JA, Rowe MI, Grosfeld JL, Fonkalsrud EW, Coran AG, ed. Pediatric Surgery. 5th ed. Mosby-Year Book Inc., 1998; 563-580.
3. Warady B, Watkins S. Current advances in the therapy of chronic renal failure and end stage renal disease. Seminars in Nephrology 1998;18:341-354.

4. Hanna JD, Foreman JW, Chan JCM. Chronic renal insufficiency in infants and children. Clinical Pediatrics 1991;30:365-384.

5. Matas AJ, Chavers BM, Nevins TE et al. Recipient evaluation, preparation, and care in pediatric transplantation: The University of Minnesota protocols. Kidney Int 1996;49:S99.

6. Almond PS, Matas AJ, Gillingham K, et al. Pediatric renal transplants results with sequential immunosuppression. Transplant 1992;53:46-51.

7. Tejani A, Cortes L, Sullivan EK. A longitudinal study of the natural history of growth posttransplant. Kid Int 1996; 49:S103-108.

8. McEnery PT, Stablein DM, Arbus G, et al. Renal transplantation in children. A report of the North American pediatric renal transplant cooperative study. N Engl J Med 1992;326;1727-32.

13

Pediatric Transplantation
Part C: Liver Transplantation

Estella M. Alonso, Riccardo A. Superina

Liver transplantation in children shares many similarities with that in adults. There are, however, important differences. The diseases which lead to liver insufficiency in children are dissimilar, and the interval between diagnosis of liver dysfunction and rapid deterioration in health due to liver disease is often more compressed in the child as compared to the adult. Liver disease in infancy interferes with the critical period of growth and development in the early years of life and adds to the urgency of transplantation so that losses can be regained. In the following chapter, we have concentrated on the areas in which the management of children before and after transplantation differs most from that in older patients.

CAUSES OF LIVER FAILURE UNIQUE TO CHILDREN

The most common indication for liver transplantation in children is Biliary Atresia.[1] Biliary Atresia is a progressive fibro-inflammatory destruction of the extrahepatic biliary tree, which develops in approximately 1 in 15,000 newborns. The etiology of the disorder remains unknown, but as many as 10% of affected children will have associated developmental abnormalities such as polysplenia, malrotation, and intra-abdominal vascular anomalies. Some infants will benefit from the Kasai procedure but the majority will have progressive biliary cirrhosis despite surgical intervention. Metabolic liver diseases that result in cirrhosis, such as alpha-1-antitrypsin deficiency and Wilson's disease are also common indications for liver transplantation in children. Approximately 5% of the children receiving liver transplants have fulminant hepatic failure. Inborn errors of metabolism without cirrhosis such as Crigler-Najjar syndrome or Ornithine transcarbamylase deficiency are uncommon but important indications as well. Liver transplantation can be performed in children with malignancy, which is limited to the liver such as hepatocellular carcinoma or hepatoblastoma.

13

RECIPIENT EVALUATION, SELECTION CRITERIA, LISTING PROCESS

Patients with chronic liver disease are not actively listed for transplant unless they are judged to have less than six months life expectancy. Predicting life expectancy is dependent upon the form of liver disease. Biliary atresia, for example, has a very predictable progression. Patients who do not have successful biliary drainage following the Kasai procedure invariably reach endstage liver disease with hepatic insufficiency by two years of age. It is best to proceed with transplantation

Organ Transplantation, 2nd edition, edited by Frank P. Stuart, Michael M. Abecassis and Dixon B. Kaufman. ©2003 Landes Bioscience.

when they begin to have linear growth failure and the first complications of portal hypertension. Patients with familial cholestatic syndromes, which ultimately lead to cirrhosis, may have a less predictable course. Growth failure is characteristic of these syndromes even when liver function is preserved. Signs of advancing portal hypertension and liver synthetic failure are the earliest indications for transplant in this group. Children with metabolic defects, which are corrected by transplantation, are approached with a different strategy. In this setting, the goal should be to perform the transplant before the patient develops significant complications from the metabolic defect. The child with fulminant hepatic failure should undergo transplant as soon as a suitable organ is available, since fewer than 25% of these patients will survive without transplant. Table 12C.1 summarizes the medical complications that indicate the need to proceed with transplant.

The preoperative evaluation of a child awaiting liver transplantation includes, establishing the etiology, predicting the timing of the need for transplant, and identifying anatomic abnormalities or other organ system impairment, which would complicate the surgical procedure. Children with cirrhosis should show signs of hepatic insufficiency, such as growth failure or coagulopathy, or have significant complications of portal hypertension, such as ascites or variceal bleeding before liver transplant is performed. A child who has not developed these complications may have many years of good quality of life prior to the need for liver transplant.

Table 13C.1. Medical complications indicating the need for liver transplantation in pediatric patients

---

1) **Biliary Atresia**
   a) Status post failed Kasai procedure
   b) Recurrent ascending cholangitis
   c) Complications of cirrhosis as listed below
2) **Cirrhosis of any etiology with the following complications**
   a) Growth failure
   b) Ascites which is refractory to medical management
   c) Episodes of variceal bleeding which are refractory to sclerotherapy and/or TIPS
   d) Hypersplenism causing thrombocytopenia
   e) Liver synthetic failure
   f) Other major systemic complications
3) **Fulminant hepatic failure**
4) **Neonatal liver failure**
5) **Inborn errors of metabolism**
   a) Tyrosinemia
   b) Glycogen storage disease
   c) Crigler-Nijjar Syndrome
   d) Ornithine transcarbamylase deficiency
   e) Other defects with the potential to cause neurologic or other major systemic complications
6) **Unresectable hepatic tumors without extension**

ANESTHETIC CONSIDERATIONS

Children present anesthetic challenges different from those in adults, but those under 10 kg in size present unique anesthetic considerations. Intravenous access is best secured with a large single lumen tunneled catheter directly into the internal jugular or subclavian vein. Large bore catheters are essential for adequate fluid and blood replacement since babies have small circulating volumes which are rapidly depleted in the face of ongoing hemorrhage. In addition to a fluid replacement catheter, a double lumen central catheter is also desirable for drug administration. An arterial line is also essential. Tunneled catheters provide long term IV access in infants who may otherwise present great difficulties in drawing blood or starting IVs as outpatients.

Heat preservation may be a problem in small children who are often hypothermic during a long transplant. Devices to direct warm air along the body of an infant are very helpful in maintaining core temperature above 35° during the operation. Rapid infusion systems should be available if blood loss is anticipated to be heavy. Cell saver devices are impractical in children less than 10 kg in size because of the relatively small volume of blood lost in these children.

It is useful to have at least two anesthetists who are experienced in the care of children during a transplant operation. One set of hands, however knowledgeable, may not be sufficient to deal with the many tasks that are necessary for a successful anesthetic. Blood monitoring for blood gases, serum electrolytes, calcium, magnesium and blood counts should be done at least once per hour, even in stable situations. Monitoring of coagulation parameters is essential for the management of the bleeding complications often seen during a liver transplant. A thromboelastogram device may be helpful in pinpointing a coagulation defect in the face of ongoing bleeding with abnormal bleeding times.

TRANSPLANT PROCEDURE AND INTRAOPERATIVE CONSIDERATIONS

Operative techniques in pediatric liver transplantation have evolved significantly in the last decade. The most important advances include the techniques to reduce the size of a large donor's liver in order to transplant a portion into a small child, the procurement of grafts from living donors and the division of a liver to transplant into a child and an older recipient.[2]

The transplant operation itself can be divided into three phases: the removal of the native liver, the anhepatic phase during which the new organ is implanted, and the reperfusion phase.

EXPLANT OF DISEASED LIVER

The removal of a diseased liver in the child is often the most hazardous part of the liver transplant procedure. Much of the dissection is done with cautery to minimize bleeding or with sharp dissection with ligation of all structures that will be divided. Patients with previous Kasai operations for biliary atresia have extensive scarring in the right upper quadrant increasing the difficulty of dissection.[3] After opening the abdomen, adhesions between the liver, colon, stomach and abdominal wall obscure clear entry into the abdominal cavity. The portoenterostomy

13

must be traced into the hilum and divided close to the liver in order to access the hepatic artery and portal vein. The hepatic artery and portal vein must be divided close to the liver beyond their bifurcation in order to preserve length. Patients with biliary atresia will often have unusual anatomical features such as preduodenal portal vein, retroperitoneal continuation of the inferior vena cava, left sided vena cava and situs inversus abdominis. Anatomical anomalies are common and are not always diagnosed with pretransplant imaging.

Long standing portal hypertension may contribute to progressive shunting of mesenteric blood away from the liver leading to hypoplasia of the portal vein. In extreme cases the portal vein size is less than 2 mm in diameter and flow is minimal and makes the portal vein unsuitable for use as a conduit of mesenteric blood to the new liver. One must trace the portal vein to the confluence of the splenic and superior mesenteric veins where the flow of blood is acceptable. Vein grafts from the donor may be used to bridge the distance between the native vein and the donor liver.[4] In pediatric transplantation, the native inferior vena cava is often left in place.[5] The vena cava is mobilized completely, but the adrenal vein is preserved. Retrohepatic branches may be ligated and divided individually. Alternately, clamps can be placed above and below the liver on the IVC and the liver dissected sharply off the vein. Vein branches can then be oversewn under direct vision.

IMPLANTATION OF THE NEW ORGAN

Over 50% of infant transplants are done with organs that have been reduced in size from older donors.[6,7] These include size reduced cadaveric organs, split livers where the right lobe is preserved for transplantation into another recipient, and living donor procured segments.

Cadaveric organs that are reduced in size for a single recipient only, are reduced in size on the back table. When the donor's weight is less than four times that of the recipient, the entire left lobe can usually be transplanted. All structures to the right lobe are divided and ligated. The parenchyma is then divided in the plane of the inferior vena cava and gall bladder. Particular attention must be paid to the area of the confluence of the hepatic veins to insure that all major veins are secured. If the donor is more than four times the weight of the recipient, the size of the liver will only permit implantation of segments 2 and 3. Parenchymal division takes place to the right of the falciform ligament.

For organs that are considered for transplantation into two recipients, the division of the liver can be done either in or ex vivo depending on the logistics of the transplant, cold ischemic time, and preference of the teams.[8,9] Important technical features in splitting a liver are preservation of segment 4 arterial supply, division of the bile duct preferably beyond the confluence of the segments 2 and 3 ducts, and preservation of as much of the length of the left portal vein. Hepatic venous anomalies are uncommon. The left hepatic vein is usually sufficient to provide adequate venous drainage of segments 2 and 3 or even the entire left lobe.

When the operative field is ready for implantation, the graft is positioned in such a manner as to facilitate proper portal vein orientation. If the liver is piggybacked onto the cava, the venous orifices of all 3 hepatic veins may be used as a

13

single confluence to ensure proper venous drainage of the graft. Alternately, a new caval orifice can be created more inferiorly on the vena cava orienting the liver more medially, and allowing the portal vein ends to come together with less tension. Fine absorbable monofilament sutures are used for the venous anastomosis. The artery is anastomosed using fine interrupted nonabsorbable sutures and an operating microscope in infants[10] or when the donor vessels are small such as in living donor transplants.

REPERFUSION PHASE

After reperfusion, a number of metabolic changes occur. Calcium requirements decrease or stop, the serum bicarbonate level rises and potassium may fall. Additionally, PTT may go up. The coagulopathy may require additional fresh frozen plasma, but more aggressive coagulation factor replacement with cryoprecipitate is not advisable unless bleeding is life threatening. Over correction of the coagulopathy seen after reperfusion may lead to hepatic arterial thrombosis.

During this last phase, the Roux loop is constructed if necessary, and the bile duct anastomosis is completed.

Children usually require biliary reconstruction using a Roux loop of bowel. The choledochojejunostomy is done over a stent in cases of very small ducts in order to ensure that the back wall is not accidentally included in the front wall reconstruction. In cases of grafts from living donors or where the liver is split, separate ducts from segments 2 and 3 may be encountered which require individual attachment to the bowel.

RECOVERY AND INTENSIVE CARE

The postoperative phase of care is characterized by careful monitoring for graft function, fluid replacement and electrolyte monitoring. Portal vein and hepatic arterial flow is also monitored by ultrasound. Some bleeding is not infrequent in the postoperative period but does not often require urgent re-operation. Progressive abdominal distention resulting in difficulty in ventilation or decreasing urine output may indicate that a return to the OR is indicated for evacuation of blood and clots. Some form of antithrombosis prophylaxis therapy is desirable. We use a heparin infusion at 10 units/kg/hour as prophylaxis against arterial thrombosis and replace drain output with fresh frozen plasma in infants to correct the hypercoagulable state that may exist following transplantation.[11]

Ultrasound examination is carried out on the first postoperative day to examine arterial and portal venous flow. In the absence of satisfactory flow in either vessel, an immediate return to the operating room is indicated for revision of the vascular anastomoses. Arteriography is rarely necessary to confirm ultrasound findings. Vessels, particularly the portal vein can frequently be revised to restore satisfactory blood flow.[12] Although the success of intervention in cases of arterial occlusion is less predictable, a substantial proportion of these vessels can be opened up with a combination of thrombolytic therapy in the OR, postoperative anticoagulation and careful revision of the anastomoses to exclude technical factors.[13]

Arterial hypertension in the ICU is common and can usually be managed by careful fluid management, diuretic therapy and calcium channel blockers. Pro-

13

phylactic antibiotics are continued for 48 hours usually, and anti CMV prophylaxis is implemented in cases where the donor is CMV positive.[14]

Patients may be ventilator dependent for a few days following transplantation. Failure to wean from ventilator support is usually secondary to abdominal distention, but intra thoracic pathology should also be suspected. Diaphragmatic dysfunction on the right side from phrenic nerve clamp injury should be suspected when there are no other obvious causes for ongoing ventilator dependence.

IN-PATIENT CARE AFTER ICU

FEEDINGS

Feedings are generally restarted between the third and fifth post-operative day depending on whether bowel surgery has been performed as part of the transplant. Babies may require feeding tube supplementation particularly when oral intake was poor prior to the transplant. Unless there are complications, the diet can be quickly advanced to age appropriate food supplemented in some cases with special liquid enteral feeds.

IMMUNOSUPPRESSION

The resolution of postoperative ileus and implementation of oral feeding calls for vigilance in the monitoring of immunosuppressive medication levels as absorption of these drugs may increase significantly in the days following transplantation. Usually, the transition between intravenous and oral administration of medication takes place during this week unless other complications arise.

INFECTIONS

If infections occur in the early postoperative period, they are usually bacterial. Sources of infection include intra abdominal sites such as the bowel or biliary tract, particularly in children who have been transplanted for biliary atresia. Other common sites are intravenous catheter and arterial infusion sites.

Early and aggressive investigation and treatment are essential for the successful resolution of these infections. Careful physical examination, chest X-ray and abdominal ultrasound along with culture of blood, urine and wounds will usually uncover the source. Empirical treatment with antibiotics is often recommended when the source of the fever is undetermined. Antibiotics, which cover enterobacter and enterococci, are important in the early post transplant period. Yeast infections should always be suspected if a septic picture continues and there is no improvement with antibiotics. Children, who have been on pre transplant steroids, those with bile leaks or bowel perforations and those with arterial thrombosis have been shown to be at higher risk for the development of *Candida* infections.

SPECIAL CONSIDERATION FOR IMMUNOSUPPRESSION

The choice of immunosuppressive therapy for a child must balance the need to prevent rejection against the desire to allow normal growth and development and to avoid infectious complications. The approach to therapy in pediatric liver transplantation has largely been adapted from adult experience. Most centers use a triple drug regimen, which includes cyclosporine, azathioprine and corticosteroids. A new oral formulation of cyclosporin has improved intestinal absorption even in the setting of poor bile flow.[15,16] This is particularly valuable for small infants who typically require large doses of cyclosporine secondary to poor intestinal absorption. Cyclosporine is usually administered intravenously during the initial postoperative period, but the oral preparation can be absorbed adequately even on the first postoperative day. The cyclosporine dose is adjusted to yield a 12-hour trough level of 250-350 ng/ml as measured by TDX. Azathioprine is administered at a constant dose of 1 mg/kg/day and methylprednisolone or prednisone doses are gradually tapered from 2mg/kg/day to .3 mg/kg/day over the first month. Tacrolimus is gaining acceptance as an alternative to cyclosporine in pediatric liver transplant recipients. Because it is more potent than cyclosporine, patients treated with tacrolimus are less dependent on steroid administration, and may avoid steroid related complications, such as growth failure and hypertension. Unfortunately, tacrolimus can be difficult to administer to smaller children since it is not available in a liquid formulation. It does not cause the cosmetic side effects associated with cyclosporine, but it may contribute to anorexia and chronic gastrointestinal symptoms, which are not common in children treated with cyclosporine. There is also a growing concern that post transplant lymphoproliferative disease (PTLD) is more common in children who have received tacrolimus.

The use of monoclonal antilymphocyte antibodies, such as OKT3, either for induction or for treatment of steroid resistant rejection is becoming less common in pediatrics since this therapy has also been identified as a risk factor for PTLD.[17] Chimeric antibody preparations for specific T-cell markers are being tested in pediatric solid organ recipients and may prove to be safe and effective alternatives to current antilymphocyte therapy.

ASSESSMENT OF GRAFT FUNCTION, DIAGNOSIS AND TREATMENT OF REJECTION

Allograft rejection occurs in approximately two thirds of children following liver transplantation. The peak incidence of acute rejection is within the first 2-6 weeks following the transplant. Fever, jaundice and abdominal pain are typical symptoms at presentation. Frequent monitoring of biochemical indicators of cholestasis and hepatocellular injury may allow the clinician to suspect rejection prior to the onset of typical physical signs. Since the laboratory and physical signs are not specific for rejection, the diagnosis of rejection must always be confirmed by histology. Rejection episodes are treated in a step-wise fashion. The first step is an intensified steroid regimen, which includes intravenous boluses of methylprednisolone (10-20 mg/kg/day for 3 days) occasionally, followed by tapering doses of oral steroids. If there is no improvement in the biochemical parameters or the liver histology, the next step in treatment might be an antilymphocyte preparation such as OKT3, or conversion from cyclosporine to tacrolimus. Chronic rejection can occur either following an episode of refractory acute rejection or denovo weeks to months after transplant. Chronic rejection is characterized by a slow progression of the clinical signs of cholestasis without many constitutional symptoms. Tacrolimus may be effective in reversing this progression. Although rejection is commonplace, less than 10% of children lose their liver to chronic or ongoing acute rejection.

DIAGNOSTIC RADIOLOGY

PLAIN FILMS
A preoperative chest film is always done to exclude significant pulmonary disease that may be a contraindication to proceeding with transplantation.

In the postoperative period, evaluation of the abdomen and chest with plain radiographs is done to exclude pleural effusions, pneumonia, pulmonary edema and intraabdominal free air. Bowel perforations may show few overt clinical signs, and careful evaluation of abdominal plain films may provide the first evidence of unusual air or fluid collections that may indicate intestinal or biliary anastomotic dehiscences.

ULTRASOUND
The single most useful radiological test in pediatric liver transplantation is the abdominal ultrasound.

Posttransplant day 1 ultrasound is essential for determining the early patency of all vascular anastomoses. The quality of ultrasound studies has improved dramatically over the last several years and obviates the need for angiograms in most cases to confirm arterial or portal venous occlusion. Doppler studies now can provide information about the velocity of flow in vessels as well as the resistive index in the hepatic artery. This information, while not always helpful, may provide the basis for future advances in the evaluation of vascular anastomoses.

Ultrasound also provides information regarding the presence of fluid collections. It may serve as the definitive study to determine the size and nature of a fluid collection, but CT scanning usually is more helpful in determining the need for further interventions or treatment.

COMPUTERIZED TOMOGRAPHY
CT scanning gives a wealth of information on the anatomic state of the liver. Periportal edema is a common finding after transplantation and can persist for weeks. Fluid between the left lobe and the stomach or behind the cut surface of the liver is also common and does not necessarily signify infection.

CT scanning is the most definitive study in the investigation of fever. Fluid collections detected by ultrasound can be studied in greater detail with CT. Liver parenchymal lesions are seen in great detail on CT, including perfusion defects, intrahepatic fluid collections and bile duct dilatation.

INTERVENTIONAL RADIOLOGY CONSIDERATIONS
Interventional radiology (IR) has become a full partner in the management of many post transplant issues.[18-20]

Aspiration of intraabdominal fluid collections to rule out infections is the most common reason for consultation with IR. The decision to go on to surgical drainage can then be taken depending on the outcome of the aspiration and the nature of the fluid.

Investigation of the biliary tract is an area where IR has had a very positive impact. The biliary tree is relatively inaccessible in many children because of the biliary enteric anastomosis usually present. Percutaneous transhepatic cholangiography (PTC) can provide information regarding biliary strictures, leaks, bile cultures and most importantly provides an opportunity for possible corrective measures.

Transhepatic insertion of biliary stents has provided long term and in many cases permanent correction of postoperative biliary strictures. Balloon dilatation of strictures and passage of indwelling stents provide both short and long-term palliation and even permanent solutions to biliary strictures. Stents may be left in place for 8-12 weeks after which they may be removed. Repeated dilatations may be necessary. Placement of permanent stents in the biliary tree has been attempted but long term results have not been gratifying because of the build up of sludge in these stents with resulting obstruction and sepsis.

Vascular interventions are less common. Diagnostic arteriography and venography following transplantation have become less often used as the diagnostic accuracy of ultrasound has improved.

Arterial infusion of thrombolytic agents has been used to treat arterial thrombosis with little to support further use of this technique. Venous obstructions however have been successfully approached through interventional radiology.

Posttransplant portal vein and hepatic vein stenoses have been successfully treated by using transhepatic or transvenous introduction of balloon dilators followed in some instances by placement of permanent indwelling vascular stents. Long term patency of portal veins treated for stenotic areas has been well documented. Hepatic vein lesions may be more difficult to treat because of the confluence of the hepatic veins with the inferior vena cava making stent placement more difficult.

EARLY SURGICAL COMPLICATIONS

BLEEDING

Posttransplant bleeding, when it occurs, is usually slow but persistent. Bleeding into the abdomen occurs to some degree in most transplant patients and postoperative transfusion is often required. Blood is evacuated through the drains left in place at the time of the transplant, but blood and blood clots can occlude the drains and still accumulate in the abdomen resulting in progressive distension. A return to the OR is indicated when abdominal distension interferes with ventilation, renal perfusion and lower limb perfusion. Laparotomy results in the immediate improvement in renal perfusion and ventilation. The bleeding point, if found, will usually be a small arterial branch in the hilum of the liver or along the course of the hepatic artery. Bleeding is more frequent in reduced size or split liver transplants, but requires laparotomy in less than 10% of cases.

NONFUNCTION

Primary nonfunction occurs in less than 5% of cases, but requires urgent retransplantation when it is diagnosed. Reduced size and split liver transplantation may result in a higher incidence of nonfunction. Therefore, more stringent criteria are used for selection of livers to be reduced in size or split than for those used as whole organs. In general, organs intended for splitting or size reduction must be from donors less than 40 years of age, with near normal enzymes and less than 10% fat.

VASCULAR THROMBOSES

Early re-exploration of clotted arteries has not been reported to be as successful in children as in adults. Takedown of the arterial anastomosis, infusion of urokinase into the graft and trimming back both the donor and recipient ends of the artery can result in restoration of arterial flow on occasion and is probably worth doing in all cases when the thrombosis is detected early. Microvascular surgical techniques have resulted in a decrease in thrombotic complications in small children. When the artery does clot, however, the clinical course can be unpredictable. Retransplantation is always necessary if significant biliary damage has occurred. Some children, however, have acceptable liver function and heal any ischemic damage with few or no serious sequelae.

PORTAL VEIN OCCLUSION

Unlike arterial thrombosis, portal vein occlusion can almost always be reversed when diagnosed early. Localized thrombus, kinking of the vein, or extrinsic compression can usually be reversed. Even if liver function is acceptable, the portal vein should be declotted to prevent the long-term problems of portal hypertension and cavernous transformation.

BILIARY PROBLEMS

Choledochojejunostomies are the most common form of biliary hook up in children. Serious biliary leaks may signify arterial thrombosis, and in that setting, retransplantation may be the most practical option. Operation and evacuation of infected collections, attempted repair of biliary dehiscences and insertion of drains must be done in all cases of bile leaks associated with fever and a septic clinical state.[21] Small leaks, which appear to be adequately drained, may be safely observed even if the artery is not open.

Bile leaks may also originate from secondary ducts that may not be visible at the time of the transplant. Although this may occur with whole liver transplantation, it is more common after living related or split liver transplantation when small segmental ducts at the liver parenchymal transection plane either at the cut surface or within the liver plate secrete bile into the abdominal cavity. Small ducts at the cut surface may safely be oversewn. Those at the plate may signify segmental ducts and should be anastomosed to the bowel with a separate choledochojejunostomy.

INTESTINAL PERFORATIONS

Bowel injuries from cautery burns may become evident during the first postoperative week. Leukocytosis and abdominal distention should always raise the possibility of an occult perforation.[22] Leaks from bowel anastomotic suture lines or from de novo perforations in areas not subjected to extensive dissection may also occur. Early diagnosis is essential to prevent recurring infected collections. When the abdominal cavity is heavily contaminated with bowel contents and peritonitis is well established, planned returns to the OR are useful for regular peritoneal washing and prevention of the persistence of infected collections.

EARLY OUTPATIENT CARE

The routine length of hospitalization after liver transplantation is approximately three weeks. After discharge the patients are monitored frequently to allow the clinician to recognize the early signs and symptoms of rejection and infection. After the first month to six weeks, follow-up is weekly and then ultimately monthly.

LONG-TERM FUNCTION, OUTCOME, GROWTH AND DEVELOPMENTAL RESULTS, FOLLOW-UP AND REHABILITATION

Following the first year after transplant, laboratories are obtained quarterly and children are examined twice yearly to monitor growth and look for signs of chronic graft dysfunction. Cyclosporine and tacrolimus doses are weaned to approximately 50% of the initial values after the first year following transplant. At

---

Table 13C1.2. Recommended immunization schedule for liver transplant recipients

Begin the following schedule **6 months after** the transplant	
Hepatitis B	Month 7, Month 9, Month 12
DTP	Resume standard schedule
H. influenza type b	Resume standard schedule
Polio	Resume standard schedule Patient and sibs must receive **IPV***
Measles, Mumps, Rubella	Month 7 if not previously protected, confirm vaccine response with titers
Varicella	Month 7 if not previously protected, confirm vaccine response with titers **
Pneumovax	Required for patients with splenectomy or asplenia***
Hepatitis A	Newly recommended vaccine for immunocompromised patients including organ transplant recipients
Influenza	Yearly

* Inactivated Polio Vaccine
** Patients may experience low grade fever and vesicles at injection site.
*** Penicillin prophylaxis is also recommended for these patients.

13

18 months after transplant most children can be switched to alternate day steroids. Children receiving primary therapy with tacrolimus may tolerate steroid withdrawal six months after transplant. Unfortunately, it is nearly impossible to predict which children will tolerate complete withdrawal of all immunosuppression. We have treated children who were taken off all immunosuppression for serious complications who did not develop graft rejection, even in long-term follow-up. These children are a minority. Once the immunosuppressive regimen has been decreased somewhat, children can resume a routine schedule of immunizations, see Table 12C.2. The intramuscular polio vaccine preparation should be substituted for the live attenuated vaccine. The approach to immunization with measles vaccine and varicella vaccine can be more liberal. Even though a poor response rate to these two vaccines has been noted in this population, serious consequences of immunization even in children on standard levels of immunosuppression have not been reported. In addition, most liver transplant recipients receive hepatitis B vaccine and yearly influenza vaccine as determined by their local physician. Obviously, children who have a history of asplenia or splenectomy are also immunized with Pneumovax.

One of the most important aspects of long-term follow-up care is monitoring growth and development. Poor linear growth is not uncommon in the first six months after transplant. The onset of catch-up growth is usually between 6 and 24 months after the transplant and can be improved with early withdrawal of corticosteroids. Developmental delay is common in infants in the first year following liver transplant, but steadily improves as children reach school age. Most pediatric liver transplant recipients do have normal school performance once they have rehabilitated from the transplant.

Outpatient management also focuses on patient education, and monitoring compliance. Addressing patient concerns about the cosmetic side-affects of their medication are important issues as well. The final objective of the outpatient visit is to evaluate chronic medical disabilities secondary to the transplant. Most children have minimal medical complaints. A few children are plagued with chronic minor infections. Occasionally, persistence of these infections will warrant a decrease in their immunosuppression to clear the pathogen naturally.

OVERALL RESULTS

The results of liver transplantation in children have steadily improved over the last decade.[23] In recent results, 1 and 5 year survival for children with chronic liver disease exceeds 90%. Transplantation in the setting of fulminant failure is excellent before the onset of severe neurological complications or the progressive failure of other organs such as the kidneys and lungs.

The goal of transplantation is to restore children with liver disease to normal life. Improved immunosuppression and more accurate techniques for the early detection of viral diseases have helped to make this goal an achievable reality.

Organ availability remains one of the largest stumbling blocks on the road to timely transplantation. It is hoped that techniques such as split liver transplantation and living donation in the acute setting, if used on a more regular basis, may substantially reduce waiting times and waiting list mortality for children in need of a liver transplant.

REFERENCES

1. Whitington PF, Balistreri WF. Liver transplantation in pediatrics: indications, contraindications, and pretransplant management. J Pediatr 1991; 118:169-77.
2. Malago M, Rogiers X, Broelsch CE. Reduced-size hepatic allografts. Ann Rev Med 1995; 46:507-12.
3. Sandler AD, Azarow KS, Superina RA. The impact of a previous Kasai procedure on liver transplantation for biliary atresia. J Pediatr Surg 1997; 32:416-9.
4. Shaw BW, Jr., Iwatsuki S, Bron K, Starzl TE. Portal vein grafts in hepatic transplantation. Surg, Gynecol Obstet 1985; 161:66-8.
5. Nery J, Jacque J, Weppler D et al. Routine use of the piggyback technique in pediatric orthotopic liver transplantation. J Pediatr Surg 1996; 31:1644-7.
6. Bilik R, Greig P, Langer B, Superina RA. Survival after reduced-size liver transplantation is dependent on pretransplant status. J Pediatr Surg 1993; 28:1307-11.
7. Otte JB, de Ville de Goyet J, Reding R et al. Pediatric liver transplantation: From the full-size liver graft to reduced, split, and living related liver transplantation. Pediatr Surg Intern 1998; 13:308-18.
8. Pichlmayr R, Ringe B, Gubernatis G, Hauss J, Bunzendahl H. [Transplantation of a donor liver to 2 recipients (splitting transplantation)—A new method in the further development of segmental liver transplantation]. Langenbecks Archiv fur Chirurgie 1988; 373:127-30.
9. Rogiers X, Malago M, Habib N et al. In situ splitting of the liver in the heart-beating cadaveric organ donor for transplantation in two recipients [see comments]. Transplantation 1995; 59:1081-3.
10. Mori K, Nagata I, Yamagata S et al. The introduction of microvascular surgery to hepatic artery reconstruction in living-donor liver transplantation—Its surgical advantages compared with conventional procedures. Transplantation 1992; 54:263-8.
11. Leaker MT, Brooker LA, Mitchell LG, Weitz JI, Superina R, Andrew ME. Fibrin clot lysis by tissue plasminogen activator (tPA) is impaired in plasma from pediatric patients undergoing orthotopic liver transplantation. Transplantation 1995; 60:144-7.
12. Bilik R, Yellen M, Superina RA. Surgical complications in children after liver transplantation. J Pediatr Surg 1992; 27:1371-5.
13. Langnas AN, Marujo W, Stratta RJ, Wood RP, Li SJ, Shaw BW. Hepatic allograft rescue following arterial thrombosis. Role of urgent revascularization. Transplantation 1991; 51:86-90.
14. King SM, Superina R, Andrews W et al. Randomized comparison of ganciclovir plus intravenous immune globulin (IVIG) with IVIG alone for prevention of primary cytomegalovirus disease in children receiving liver transplants. Clin Infec Dis 1997; 25:1173-9.
15. Whitington PF, Alonso EM, Millis JM. Potential role of Neoral in pediatric liver transplantation. Transpl Proc 1996; 28:2267-9.
16. Superina RA, Strong DK, Acal LA, DeLuca E. Relative bioavailability of Sandimmune and Sandimmune Neoral in pediatric liver recipients. Transpl Proc 1994; 26:2979-80.
17. Morgan G, Superina RA. Lymphoproliferative disease after pediatric liver transplantation. J Pediatr Surg 1994; 29:1192-6.

13

18. Zajko AB, Sheng R, Zetti GM, Madariaga JR, Bron KM. Transhepatic balloon dilation of biliary strictures in liver transplant patients: A 10-year experience. J Vasc Intervent Radiol 1995; 6:79-83.

19. Orons PD, Zajko AB, Bron KM, Trecha GT, Selby RR, Fung JJ. Hepatic artery angioplasty after liver transplantation: Experience in 21 allografts. J Vasc Intervent Radiol 1995; 6:523-9.

20. Adetiloye VA, John PR. Intervention for pleural effusions and ascites following liver transplantation. Pediatr Radiol 1998; 28:539-43.

21. Heffron TG, Emond JC, Whitington PF et al. Biliary complications in pediatric liver transplantation. A comparison of reduced-size and whole grafts. Transplantation 1992; 53:391-5.

22. Beierle EA, Nicolette LA, Billmire DF, Vinocur CD, Weintraub WH, Dunn SP. Gastrointestinal perforation after pediatric orthotopic liver transplantation. J Pediat Surg 1998; 33:240-2.

23. Goss JA, Shackleton CR, McDiarmid SV et al. Long-term results of pediatric liver transplantation: an analysis of 569 transplants. Ann Surg 1998; 228:411-20.

13

Pediatric Transplantation:
Part D: Pediatric Lung Transplantation

David A. Fullerton

INTRODUCTION

Pediatric lung transplantation has been performed in a limited number of centers worldwide since 1990, and the number of pediatric lung transplants has decreased since the mid-1990's. According to the International Society of Heart and Lung Transplantation (ISHLT) registry, approximately 60 lung transplants are performed annually in patients under the age of 18 years. The majority of these patients are adolescents (Fig. 13D.1). Most children undergo double lung transplantation using the bilateral sequential technique. Cardiopulmonary bypass is routinely used for pediatric lung transplantation.

INDICATIONS FOR LUNG TRANSPLANTATION

As in adults, the indications for lung transplantation are: 1) end-stage disease of the pulmonary parenchyma and 2) pulmonary hypertension (PH). The indications in children vary by age. In infants, pulmonary vascular disease is the most

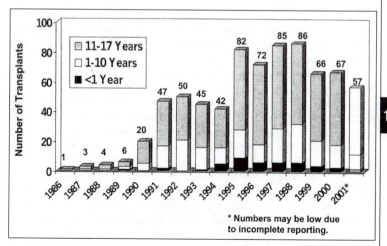

13

Fig. 13D.1. Age distribution of pediatric lung recipients by year of transplant. (From Boucek MM et al. The Registry of the International Society of Heart and Lung Transplantation: Fifth Official Pediatric Report-2001 to 2002. J Heart Lung Transplant 2002; 21:827-840.)

Organ Transplantation, 2nd edition, edited by Frank P. Stuart, Michael M. Abecassis and Dixon B. Kaufman. ©2003 Landes Bioscience.

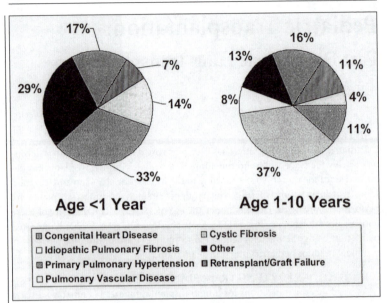

Fig. 13D.2. Indications for pediatric lung transplant beween 1991 and 2001, by age of recipient. (From Boucek MM et al. The Registry of the International Society of Heart and Lung Transplantation: Fifth Official Pediatric Report-2001 to 2002. J Heart Lung Transplant 2002; 21:827-840.)

common indication. Among adolescents, the principal indication for lung transplantation is cystic fibrosis (CF). It is important to note that diagnoses leading to end-stage parenchymal disease in children differ from adults (Fig. 13D.2). Leading indications for transplantation in adults include pulmonary fibrosis and emphysema, both of which are uncommon in children.

Most patients with CF survive well into adulthood with current medical regimens. However, a relatively small percentage of patients with CF become refractory to medical therapy and require lung transplantation as children and adolescents. Indications for pediatric lung transplantation secondary to CF include increasingly frequent hospitalizations for pulmonary infections and pulmonary distress, progressive weight loss despite nutritional supplementation, and worsening gas exchange evidenced by progressive hypercapnia and hypoxemia.

Pulmonary hypertension (PH) may be primary or secondary to congenital heart disease. Primary PH is uncommon in children and is seldom life threatening during childhood. Therefore, children undergoing lung transplantation for PH usually have secondary PH arising from congenital heart disease. Pulmonary hypertension ultimately progresses to loss of exercise tolerance, cyanosis, polycythemia, and right heart failure. If lung transplantation is indicated because of a correctable cardiac anomaly, the cardiac defect should be repaired at the time of lung transplantation. The presence of an uncorrectable cardiac defect requires consideration of heart-lung transplantation.

CONTRAINDICATIONS

Patients should not have irreversible organ damage (other than their pulmonary disease). Creatinine clearance should be near normal. Left ventricular function should be normal. Relative contraindications include severe malnutrition, presence of significant neurologic defects, poorly controlled diabetes mellitus, active collagen vascular disease, a skeletal abnormality which significantly impairs pulmonary mechanics. The presence of panresistant organisms in the patient's sputum may prelude transplantation.

The patient's support system is of paramount importance. Inadequate support from the child's immediate or extended family may be a relative contraindication for transplantation.

PRETRANSPLANT MANAGEMENT

The waiting time for pediatric lung transplantation is commonly longer that a year, and the mortality rate on the waiting list is at least 20-25%. Therefore, once the estimated life expectancy of the child reaches approximately 2 years, the child should be listed for transplantation. While on the waiting list, the child's nutritional status must be optimized and often requires supplemental enteric feedings. If on steroids, the dosage must be weaned to the lowest possible dose consistent with acceptable pulmonary function (preferably less than 0.1mg/kg/day prednisone). Patients should be enrolled in monitored physical therapy and pulmonary rehabilitation activities to optimize functional status. Sputum should be cultured regularly to exclude the presence of resistant organisms. All pulmonary infections must be recognized immediately and aggressively treated. Hemoglobin levels should be regularly checked in cyanotic patients; phlebotomy may be required for hemoglobin levels above 16 gm/dl.

TRANSPLANT PROCEDURE

DONOR ORGAN SELECTION

Donor lungs are scarce for pediatric lung recipients. No more that approximately 15% of cardiac donors are suitable lung donors, reflecting the pulmonary injury associated with brain death and resuscitation of an organ donor. A diagnosis of asthma or other pulmonary disease in the donor is a strong relative contraindication go lung donation. In addition to a chest x-ray without infiltrates, the suitable lung donor should have an arterial paO_2 greater than 300 mmHg on an FIO_2 of 1.0 plus PEEP $5cmH_2$). All prospective lung donors should undergo flexible bronchoscopy as part of the evaluation process. Bronchoscopic findings, which contraindicate lung donation, include evidence of gastric aspiration and significant amount of purulent secretions.

Appropriate donor-recipient size matching is very important. Using the chest x-ray, the vertical and transverse dimensions of the donor are matched as closely as possible to the recipient's. Donor lungs, which are a bit small, may be expected to expand to fill the hemithorax. Donor lungs which are a bit too large may be trimmed with a stapling device at the implantation procedure. Donor-recipient bronchial size matching seems more closely associated with height than weight, so most programs list recipients with a range of 3-4 inches above and below the height of the recipient.

The lung procurement operation is conducted as for adults. The volume of pulmoplegic solution administered is adjusted for the size of the donor in a dose of 40 ml/kg.

RECIPIENT PROCEDURE

Most pediatric lung transplant procedures are performed using cardiopulmonary bypass. Its use obviates the need for single lung ventilation which may be very difficult to achieve in children; small airways may make bronchial blockers and double lumen endotracheal tubes difficult or dangerous to use. In children with cystic fibrosis, thick and copious secretions may be impossible for the anesthesiologist to control. Therefore, the implantation procedure is usually more safely performed using cardiopulmonary bypass. Further, use of cardiopulmonary bypass permits irrigation of the open bronchus with antibiotic irrigation without cross-contamination of the contralateral lung.

A "clamshell" incision through the fourth or fifth intercostal space provides excellent surgical exposure to both pleural spaces, the mediastinal structures, and the heart should an intracardiac defect require correction. The aorta and right atrium are cannulated for initiation of cardiopulmonary bypass. Aprotinin should be used to decrease bleeding. As the pulmonary hilar structures are dissected, extreme care must be taken to avoid injury to the phrenic nerves. Once the diseased lungs are removed, the recipient's mainstem bronchi are irrigated with antibiotic irrigation containing an aminoglycoside.

The implantation procedure is performed as in adults. The left lung is usually transplanted first. Implantation of the left lung requires retraction of the heart toward the patient's right. Implantation of the left lung first permits the heart to recover from this while the right lung is implanted.

Three anastomoses are required to implant a lung: pulmonary vein, pulmonary artery, and bronchus. It is easiest to perform the pulmonary venous anastomosis first using a running 4-0 polypropylene suture. The bronchial anastomosis is performed next. Great care must be taken to cut the donor bronchus as short as possible: two cartilaginous rings above the bronchial bifurcation into upper and lower lobes. The membranous trachea is sewn with a running 4-0 polypropylene suture and the cartilaginous portion is sewn with interrupted figure-of-eight 4-0 sutures. Most pediatric bronchial anastomoses are performed in an end-to-end fashion. Should significant bronchial size disparity be found between the donor and recipient bronchi, a telescoping anastomosis may be performed. The pulmonary artery anastomosis is then performed, again with a running 4-0 suture. The right lung is then implanted in the same fashion.

POSTOPERATIVE MANAGEMENT

Because reperfusion of the transplanted lungs rapidly leads to pulmonary edema, perioperative fluid restriction is essential. Following lung implantation but prior to cessation of cardiopulmonary bypass, the bypass circuit should be utilized to remove as much fluid as possible via ultrafiltration. Postoperatively, administration of intravenous fluids must be minimized and the patient should be aggressively diuresed with furosemide. If intravascular volume is needed, colloid rather than crystalloid should be administered. Such diuresis typically leads to a contraction alkalosis, and a base excess of 8-10 is acceptable.

To control the mechanics of ventilation and to optimize efforts to eliminate pulmonary edema, it is advisable to keep the child pharmacologically paralyzed, sedated and mechanically ventilated for 24-48 hours postoperatively. The F_iO_2 should be weaned to achieve a p_aO_2 of at least 70 mm Hg. Prior to extubation, fiberoptic bronchoscopy should be performed to suction retained secretions.

Prophylactic antibiotics (vancomycin and aztreonam) should be given for 48 hours, by which time the culture results of the donor's sputum culture should be known. Posttransplant, patients should continue to receive all antibiotics which were being administered preoperatively, especially in patients with cystic fibrosis. Bacteria grown from the donor's sputum may cause an infection in the recipient, no matter how innocuous the organism may seem; the recipient should receive antibiotics which will specifically cover such bacteria. Lung transplant recipients should receive a 6-week course of intravenous ganciclovir as prophylaxis against CMV infection. For most children, an indwelling silastic venous catheter is necessary for blood sampling and intravenous drug administration.

IMMUNOSUPPRESSION

Most lung transplant programs use an immunosuppression protocol similar to that employed in their pediatric heart transplant program. Traditionally, a "triple drug" protocol based on steroids, azathioprine and cyclosporine is used.

Intravenous cyclosporine may be administered by continuous infusion to achieve a whole blood level of cyclosporine of 300-350 ng/ml in the early postoperative period. To achieve this blood level typically requires a cyclosporine dosage of 0.5-1.0 mg/hr, but the dosage must be determined empirically and adjustments determined by blood level. During the first 24-72 hours posttransplant, a cyclosporine level between 200 and 250 ng/ml is acceptable and will minimize the risk of early renal impairment. This is especially important since diuresis is vital in the early postoperative period. Azathioprine 2 mg/kg (i.v. or p.o.) is given daily and the dosage decreased or held if the white blood cell count falls below 5000. Methylprednisolone 1 mg/kg/day (i.v. or p.o.) in divided doses is started immediately posttransplant and slowly tapered (over weeks) to approximately 0.2 mg/kg/day.

Some lung transplant programs have employed immunosuppression protocols using tacrolimus and mycophenolate mofetil instead of cyclosporine and azathioprine. One immunosuppression protocol has not been shown to be better than another, but tacrolimus may offer advantage in terms of ease of administration in children.

REJECTION

Clinically, rejection is suggested by tachypnea, fever and malaise. Suspicion is greater if patients are found to have: peripheral oxygen desaturation (pulse oximetry), pulmonary infiltrates noted on chest x-ray and decrement in pulmonary function studies, particularly the FEV$_{1.0}$. Bronchoscopy and transbronchial biopsy should be performed for diagnostic confirmation. Rarely, open lung biopsy may be required for diagnosis.

Pulse steroid therapy is the first line treatment for rejection. Rejection which is refractory to steroids should be treated with OKT3.

SURGICAL COMPLICATIONS

Two complications are unique to lung transplantation: 1) lung dysfunction secondary to reperfusion injury and 2) bronchial anastomotic problems.

The surgeon must be prepared to deal with reperfusion injury of the transplanted lung which is typically manifest soon following lung reperfusion. Minor cases are manifested by relatively mild pulmonary edema and are easily treated with aggressive diuresis and mechanical ventilation using positive end-expiratory pressure (PEEP) of 5-10 cm H$_2$O. Severe cases are notable for marked hypoxemia, severe pulmonary edema and poor pulmonary compliance. In addition to diuresis and mechanical ventilation, inhaled nitric oxide (NO) (2-20 ppm) is valuable in improving oxygenation. Extreme cases may fail to respond to inhaled NO and life-threatening hypoxemia may require extracorporeal membrane oxygenation (ECMO) for survival.

Bronchial anastomotic complications occur in fewer than 10% of cases and result from ischemia of the donor bronchus. The anastomosis may dehisce or more commonly may become stenotic; the diagnosis may be confirmed with a chest CT scan or bronchoscopy. Anastomotic dehiscence is marked by pulmonary distress and sepsis and may be life-threatening; it may be avoided by cutting the donor bronchus as short as possible at the time of implantation. Anastomotic stenosis should be suspected because of wheezing or dyspnea. Intrabronchial stent placement across the anastomosis and frequent follow-up bronchoscopy procedures should be performed. Frequent bronchoscopic procedure and balloon dilation may be required.

OUTCOMES

For the first several months following transplantation, children and their families are enrolled in a multispecialty rehabilitation program. Cardiopulmonary reconditioning is achieved in a monitored setting under the supervision of physical and pulmonary rehabilitation specialists. Patients and their families are instructed in the use of spirometers and pulse oximetry; patient logs are kept at home recording these data. Families are educated in the significance of changes in these data. Because the risk of infection is highest during the first 3 months posttransplant, patients are secluded from crowds until their immunosuppression regimen is down to lowest levels; in the interim, patients wear surgical masks when in public.

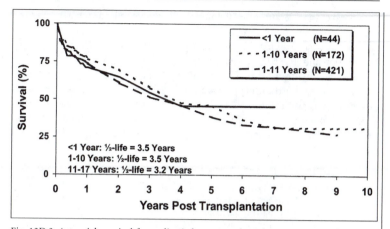

Fig. 13D.3. Actuarial survival for pediatric lung transplants between 1990 and 2001, by age group. (From Boucek MM et al. The Registry of the International Society of Heart and Lung Transplantation: Fifth Official Pediatric Report-2001 to 2002. J Heart Lung Transplant 2002; 21:827-840.)

Hospital readmission is relatively common (54%) during the first year posttransplant, principally for infection. But by 2 years follow-up only 13% of patients require readmission, again for treatment of infection (Fig. 12D.3). Functional status among lung transplant recipients is excellent. According to the ISHLT Registry, 96% of patients have no activity limitations at 2 year follow-up (Fig. 12D.4).

Regardless of preoperative diagnosis, the overall survival following lung transplant is 63% at 3 years and 54% at 5 years (Fig. 13D.3). Data from the ISHLT Registry suggest that recipient age has little impact on survival following pediatric lung transplant. The half-life following lung transplantation is approximately 3.5 years. Patients with pulmonary hypertension have a higher perioperative mortality rate, as do children who are mechanically ventilated pretransplant or undergoing lung retransplantation. It is noteworthy that among pediatric patients undergoing lung transplantation for cystic fibrosis, the one-and three-year actuarial survival rates were 90% and 73%, respectively. These data in patients with cystic fibrosis compare favorably with lung transplantation in adults.

The mortality rate posttransplant is highest during the first year following the transplant procedure. The principal cause of perioperative death is primary graft failure. During the first year posttransplant death may occur from infection, hemorrhage, and cardiovascular death. Beyond the first year posttransplant, bronchiolitis obliterans (OB) (63%) and infection (22%) are the 2 leading causes of late death. The third leading cause of late death is posttransplant lymphoproliferative disorder (15%).

As in adults, OB is the principal limitation of all long-term survivors of pediatric lung transplantation, affecting up to 40% of three-year survivors. In the absence of infection or rejection, OB is heralded by progressive dyspnea and peripheral oxygen desaturation. Pulmonary function studies are characterized by progressive decline in $FEV_{1.0}$ and FEF_{25-75} to levels less than 70% of posttransplant baseline. The diagnosis may be suspected by failure to resolve with empiric therapy for rejection and may be confirmed by transbronchial biopsy.

Following lung transplantation, pediatric patients are subject to several morbidities. By 1 year posttransplant 33% are hypertensive and by 5 years, 71%. Renal insufficiency is also common: 27% by 5 years posttransplant. Further, diabetes mellitus is found in 25% of patients overall posttransplant. Of interest, 30% patients with cystic fibrosis develop diabetes posttransplant as compared to only 5% of patients without cystic fibrosis.

SELECTED READINGS

1.   Boucek MM, Edwards LB, Berkeley MKE et al. The Registry of the International Society of Heart and Lung Transplantation: Fifth Official Pediatric Report-2001 to 2002. J Heart Lung Transplant 2002; 21:827-840.
2.   Mendeloff EN, Huddleston CB, Mallory GB et al. Pediatric and adult lung transplantation for cystic fibrosis. J Thorac Cardiovasc Surg 1998; 115:404-414.
3.   Spray TL. Lung transplantation in children with pulmonary hypertension and congenital heart disease. Semin Thorac Cardiovasc Surg 1996; 8:286-296.
4.   Spray TL, Mallory GB, Canter CE et al. Pediatric lung transplantation: Indications, technique, and early results. J Thorac Cardiovasc Surg 1994; 107:990-1000.
5.   Huddleston CB, Bloch JB, Sweet SC et al. Lung transplantation in Children. Ann Surg 2002; 236:270-276.

13

Anesthesia for Organ Transplantation

Andre DeWolf, Yoogoo Kang and Laurence Sherman

After the major advances in effective immunosuppression and organ preservation in the early 1980s, there was an exponential increase in the number of
transplant procedures and transplant centers. Anesthesiologists had to learn quickly
how to care for these patients with organ failure and allow these complex procedures to be performed without further insults to other organs of the recipient.
After these hectic years, organ transplants became more common procedures in
these institutions, with anesthesiologists better prepared. Thus, after the initial
development of anesthetic protocols, we have seen more delicate refinements in
these anesthetic techniques. Anesthetic improvements are based on better
understanding of the pathophysiology of organ failure and surgical procedures, but success can only be attained by paying attention to countless details.
This chapter summarizes the anesthetic management for liver, heart, lung, kidney,
and pancreas transplantation.

LIVER TRANSPLANTATION

Providing anesthetic care to a patient undergoing liver transplantation is one
of the most challenging tasks for an anesthesiologist. Liver transplantation is a
complex and formidable procedure, frequently involving major hemodynamic
changes due to preload and afterload changes, massive blood loss, coagulation
changes, acid-base changes, and electrolyte changes (hyperkalemia, hypocalcemia)
of magnitudes that are unseen in any other procedure. Furthermore,
patients with severe liver disease may have significant dysfunction of several other
organ systems. Proper management starts with an appropriate and complete
preoperative work-up; however, only the issues that are of significant importance
for the anesthetic management of patients undergoing liver transplantation will
be discussed here.

PREOPERATIVE EVALUATION

In patients with acute or subacute liver failure requiring urgent transplantation
there is little time to do an extensive preoperative evaluation. Most patients
however have to wait a significant amount of time for their transplant, and
therefore evaluation should be complete. Anesthesiologists should be part of the
multidisciplinary team determining appropriate candidacy for transplantation for
patients with severe liver disease.

14

Cardiovascular System

Patients with severe liver disease have a hyperdynamic circulation, with a high
cardiac output and a low systemic vascular resistance. The systemic vasodilation

Organ Transplantation, 2nd edition, edited by Frank P. Stuart, Michael M. Abecassis
and Dixon B. Kaufman. ©2003 Landes Bioscience.

is reflected by dilated capillaries and peripheral arteriovenous shunting, probably the result of abnormal nitric oxide and endothelin metabolism, although other factors may also play a role. The high cardiac output is achieved by an increase in both resting heart rate and stroke volume, and leads to an increased mixed venous oxygen concentration. Echocardiography typically shows mild four-chamber enlargement; this should not be interpreted as congestive heart failure. Similarly, because of the systemic vasodilation, mild hypotension (systolic blood pressure 90-100 mm Hg) is frequently seen, and again is not necessarily an indication of left ventricular dysfunction.

Coronary artery disease (CAD) should be excluded in patients with risk factors (diabetes mellitus, positive family history). In patients with limited mobility due to severe ascites or encephalopathy, dobutamine stress echocardiography may be the preferred initial preoperative test. Whether CAD is treated surgically or medically, liver transplantation in these patients carries an increased perioperative mortality rate (approximately 31%) with a 3-year mortality rate of 50%. This high mortality rate has to be considered when the decision is made to accept a patient with CAD for liver transplantation. Echocardiography has the additional advantage that overall cardiac function can be assessed and other problems can be diagnosed (pericardial effusion, valvular disease). Overall, patients with cardiac dysfunction may not tolerate the intraoperative hemodynamic changes. In addition, the increase in systemic vascular resistance after successful transplantation represents an increase in afterload for the left ventricle, and may lead to overt cardiac failure. Abnormal cardiac function may be seen in patients with hemochromatosis and alcoholic liver disease, and while this may not be apparent preoperatively at rest, a dobutamine stress echocardiography may elicit reduced cardiac reserve. Other appropriate tests in selected patients include stress electrocardiography, resting echocardiography, or stress echocardiography; cardiac catheterization may be necessary to make the final diagnosis.

Pulmonary hypertension is seen more frequently in patients with portal hypertension than in the general population for unknown reasons. Because of high perioperative mortality, liver transplantation is probably contraindicated in patients with severe pulmonary hypertension (systolic pulmonary artery [PA] pressure > 60 mm Hg, mean PA pressure > 40 mm Hg) and in patients with moderate pulmonary hypertension (systolic PA pressure 45-60 mm Hg, mean PA pressure 35-40 mm Hg) when right ventricular dysfunction is present. Screening for pulmonary hypertension is best accomplished by electrocardiography (right axis deviation, right ventricular hypertrophy, right ventricular strain), chest radiography (prominent PA), and questioning the patient for symptoms like fatigue, dyspnea on exertion, substernal chest pain, and hemoptysis. Evaluation is done by transthoracic echocardiography, and the diagnosis is confirmed by right heart catheterization.

Pulmonary System

Routine preoperative evaluation should include chest radiograph. Pulmonary function tests and blood gas analysis should be done when indicated. Lung dysfunction can be independent from liver disease or can be the direct result of severe liver disease. Intrinsic pulmonary dysfunction in patients with severe liver disease, such as emphysema or asthma has the same incidence as in patients without liver disease. The severity of these irreversible diseases should be considered when accepting a patient for transplantation. However, some lung dysfunction is the direct result of liver disease. Restrictive pulmonary disease can be the result of tense ascites or pleural effusion, which is especially common on the right side.

Hepatopulmonary syndrome is a condition that is unique to severe liver disease: it is the result of abnormally dilated precapillaries and capillaries in the pulmonary circulation, leading to significant ventilation/perfusion mismatch and hypoxia. Dilated capillaries are more common in the bases of the lungs, leading to orthodeoxia (lower arterial pO_2 in the upright position). Oxygen administration improves oxygenation, which should not occur in situations with right-to-left shunting. However, injection of agitated saline (echogenic contrast) during echocardiography shows the contrast in the left atrium and ventricle about 3-4 cardiac cycles after their presence in the right atrium or ventricle, suggesting the presence of an intrapulmonary right-to-left shunt. This combination is unique to hepatopulmonary syndrome. Thus, the condition is confirmed by contrast-enhanced echocardiography, and pulmonary angiography or radionuclide scanning is rarely necessary to confirm the diagnosis. The presence of the hepatopulmonary syndrome is not a contraindication to liver transplantation, because the syndrome is reversible after successful transplantation, although many have a prolonged recovery in the intensive care unit.

Central Nervous System

Cerebral function can be affected because of an excess of metabolites that are normally metabolized by the liver or because of abnormal metabolites. Hepatic encephalopathy, common in acute fulminant failure, has been attributed to abnormal ammonia metabolites. However, other factors contribute to hepatic encephalopathy: cerebral edema, changes in neurotransmitter concentrations and blood-brain barrier function, decreased cerebral metabolic rate, uncoupling of cerebral blood flow, and increased intracranial pressure. By itself, encephalopathy is not a contraindication to liver transplantation, but if severe it may require tracheal intubation for airway protection. Seizures and subarachnoidal bleeding also can affect consciousness. Evaluation includes the use of computed tomography scans of the head, electroencephalography, transcranial Doppler blood flow determination, and epidural intracranial pressure determination.

14

Renal System

Renal dysfunction may be the result of hypovolemia, acute tubular necrosis, terminal renal disease, or hepatorenal syndrome. The hepatorenal syndrome is caused by abnormal distribution of renal blood flow, due to hormonal imbalances, resulting in low urine output, a low urinary sodium concentration (< 5 mmol/L) and a high urine/plasma creatinine ratio. The hepatorenal syndrome is reversible after successful transplantation. Patients with non-reversible renal failure should be considered for combined liver-kidney transplantation.

Acid-Base Balance and Electrolytes

Diuretic therapy can result in intravascular volume depletion, hyponatremia, and hypokalemia. Excessive antidiuretic hormone activity can contribute to the hyponatremia, which should be corrected very slowly to avoid central pontine myelinolysis. Hypokalemia rarely requires potassium administration. Hyperkalemia can be present in patients with renal failure, and usually requires dialysis. Metabolic alkalosis is due to hypokalemia and drainage of gastric secretions, and metabolic acidosis can be the result of compromised tissue perfusion in severely ill patients. Correction of all these problems is difficult preoperatively.

Carbohydrate Metabolism

Hypoglycemia can occur with fulminant hepatic failure. Chronic liver disease may lead to insulin resistance and high glucagon levels, although hyperglycemia is rarely seen.

Hematology and Coagulation

Almost all candidates for liver transplantation have synthetic liver dysfunction, and therefore have coagulopathy as a result of decreased production of coagulation factors (especially factors II, V, VII, IX, and X). Fibrinogen levels may be high, normal, or low. In addition, portal hypertension leads to sequestration of platelets in the spleen, and therefore thrombocytopenia contributes to the coagulopathy. However, coagulation changes are complex because the liver also produces inhibitors of coagulation and fibrinolytic proteins (plasminogen and alpha2-antiplasmin), and because activated coagulation factors are normally cleared by the liver. This may lead to varying degrees of disseminated intravascular coagulation. Correction of the coagulopathy is best done intraoperatively, except when patients are bleeding acutely or when coagulopathy is extreme (prothrombin time [PT] > 20 s, platelet count < 20,000,000/mL). Anemia may be the result of continuing gastrointestinal bleeding, erythrocyte destruction in the spleen, and reduced production in the bone marrow. Preoperatively patients need routine hemostatic evaluation, with special analyses for items such as preexisting red cell alloantibodies. If HLA antibodies are present they will not only affect graft survival but also reduce in vivo yields of transfused platelets, unless special products are selected. Patients with broadly reactive red cell or HLA antibodies require careful preoperative planning between surgeons, anesthesiologists, and coagulation/transfusion specialists. Rare circulating anticoagulants can sometimes be managed by preoperative plasmapheresis.

ANESTHETIC MANAGEMENT

Pharmacokinetic and Pharmacodynamic Changes

The pharmacology of many anesthetics is changed in the presence of liver disease and during liver transplantation. This is the result not just of altered metabolism by the liver, changes in liver blood flow, and drug protein binding, but also by changes in the volume of distribution of the drugs. However, this usually does not interfere significantly with the use of anesthetics intraoperatively, because although the duration of action of many of the anesthetics may be prolonged, most patients after liver transplantation are not extubated immediately but require a postoperative ventilation period of at least a few hours. Similarly, the pharmacodynamic changes that may occur are handled by titrating the drugs to effect. Therefore, unless massive drug overdosing occurs, the altered pharmacology is only relevant if the anesthesiologist wants to extubate the patient at the end of the procedure.

PREOPERATIVE PREPARATION

The anesthesia team should be experienced; at least two anesthesia providers and an experienced anesthesia technician should be available. The blood bank should be prepared to supply packed red blood cells, fresh frozen plasma, cryoprecipitate, and platelets in large quantities. Special equipment should be available, and is presented in Table 13.1. The operating room table and the arm boards should be padded to avoid nerve or skin injury.

INDUCTION AND MAINTENANCE OF ANESTHESIA

One of the main concerns is the possibility of aspiration of gastric contents after induction of anesthesia but before tracheal intubation. Patients with liver disease, severe ascites, and/or recent gastrointestinal bleeding may not have an empty stomach. Therefore, cricoid pressure is routinely applied during induction of anesthesia. Most commonly thiopental, propofol, or etomidate are used to induce anesthesia, and succinylcholine provides the most rapid paralysis permitting fast intubation. Nondepolarizing muscle relaxants have been used in patients with hyperkalemia.

Maintenance of anesthesia is accomplished with a combination of intravenous narcotics (e.g., fentanyl), benzodiazepines (e.g., midazolam, lorazepam), muscle relaxants (e.g., pancuronium, cisatracurium), and inhaled anesthetics (e.g., isoflurane, desflurane). Cardiovascular drugs such as lidocaine, atropine, dopamine, epinephrine (10 µg/mL and 100 µg/mL) should be available. Other drugs that should be available include epsilon-aminocaproic acid, protamine sulfate, calcium

Table 14.1. Equipment required for liver transplantation anesthesia

Anesthesia machine with air supply
Multichannel patient monitor with pulse oximeter
Multigas analyzer
Cardiac output monitor
Cardiac defibrillator
Drug infusion pumps
Warming blanket
Forced air warmer
Heated humidifier
Rapid infusion system
Autotransfusion system
Thromboelastographs (TEG)
Transesophageal echocardiography (TEE)

14

chloride, sodium bicarbonate, tromethamine (THAM), dextrose, and insulin. Positive end-expiratory pressure (5 cm H_2O) is frequently applied to improve oxygenation in the presence of tense ascites and upper abdominal retractors and to prevent atelectasis. The humidification of inspired gases, a forced air warmer, increasing the room temperature, and appropriate draping by the surgeon in order to prevent the patient to become wet may all aid in the prevention of hypothermia.

POSTINDUCTION PREPARATION

Arms are extended at a 90° angle in an attempt to avoid brachial plexus injury. An orogastric tube is placed to drain gastric secretions; nasogastric tubes are avoided to reduce the chance of nasal bleeding. Two large-bore intravenous catheters (7-8.5 Fr) are placed after induction of anesthesia to allow blood transfusion. The choice of veins for these catheters depends on whether venovenous bypass is used; in general the antecubital vein on the side where the axillary vein is cannulated for venovenous bypass is avoided. Also, subclavian veins are only used as a last resort because accidental subclavian arterial puncture may lead to intrathoracic bleeding in patients with significant coagulopathy. The most commonly used veins are the right antecubital vein and the internal jugular veins, with the external jugular veins as acceptable alternatives.

The radial arterial catheter is usually inserted after induction of anesthesia unless the patient is hemodynamically unstable. An additional femoral arterial catheter is placed because it gives more accurate information regarding central aortic pressure, especially during the anhepatic state and immediately after graft reperfusion. A pulmonary artery catheter is placed, most frequently through an internal jugular vein; commonly the pulmonary artery catheter has been modified to determine mixed-venous oxygen saturation, while another modification allows continuous cardiac output, or right ventricular ejection fraction and end-diastolic volume determination (RVEDV). Intermittent determination of arterial blood gas tension, acid-base status, electrolytes (including ionized calcium) and hematocrit or hemoglobin is obligatory.

INTRAOPERATIVE LABORATORY TESTS

The tests presented in Table 14.2 should be performed every hour, or more frequently when indicated. Tests should be performed at the following times: baseline, every hour thereafter, 5 min after onset of anhepatic state, every 30 min during anhepatic state, 15 min before graft reperfusion, 5 and 30 min after graft reperfusion, and then every hour. Many institutions use thromboelastography (TEG) instead of or in addition to more standard coagulation tests.

INTRAOPERATIVE CARE

The liver transplantation procedure is divided into three stages: the preanhepatic, anhepatic, and neohepatic stage. Surgeons may use venovenous bypass to decompress the inferior vena cava and portal vein during the anhepatic stage of the procedure when the inferior vena cava and portal vein are clamped. Some surgeons use it never; others use it only if the patient does not readily tolerate a trial clamping

---

Table 14.2. Laboratory tests during liver transplantation

Arterial blood gas analysis and acid-base state
Electrolyte levels (Na^+, K^+, Ca^{++}, Cl^-)
Blood glucose level
Thromboelastography (TEG)
Platelet count, prothromin time (PT), partial thromboplastin time (PTT), fibrinogen

---

of the inferior vena cava; and yet others use it in virtually all patients. Another technique is side-clamping of the inferior vena cava to allow end-to-side anastomosis of donor cava to recipient cava (piggy-back technique). Although this technique was designed to maintain flow in the inferior vena cava, the side-clamping usually results in a significant reduction in blood flow. Venovenous bypass is infrequently used in pediatric liver transplantation.

HEMODYNAMIC MANAGEMENT

Most anesthesiologists feel that the circulation has to be maintained hyperdynamic perioperatively in order to maintain tissue perfusion. However, this may not be possible during the anhepatic stage, because venous return is significantly reduced when the inferior vena cava is clamped, even if venovenous bypass is used. During the preanhepatic stage, hypotension is most commonly due to hypovolemia related to bleeding and insensible fluid losses, and is treated by fluid administration. Ionized calcium concentrations should be normalized by the administration of calcium chloride. Determination of RVEDV and transesophageal echocardiography (TEE) may help when the interpretation of more routine hemodynamic monitoring is difficult. Small amounts of vasoconstrictors/inotropic agents (dopamine, epinephrine) are rarely necessary to maintain an adequate perfusion pressure.

During the anhepatic stage, when cardiac output is lower, there is a compensatory increase in systemic vascular resistance, usually resulting in preserved blood pressure. There are several different surgical techniques of handling the inferior vena cava during this stage: simple cross-clamping, side-clamping of the inferior vena cava (piggy-back), and the use of venovenous bypass. The latter techniques result in a smaller decrease in the heart's preload. On graft reperfusion, there is more of an increase in venous return with the simple cross-clamping technique, and therefore fluid management before unclamping of the vessels and graft reperfusion has to take this into account to prevent hypervolemia after graft reperfusion.

The neohepatic stage starts with unclamping of the portal vein and inferior vena cava. Graft reperfusion is usually associated with a severe reduction in systemic vascular resistance and an increase in venous return, leading to arterial hypotension in about 30% of the patients. This post-reperfusion syndrome is probably the result of the sudden release of cold, acid, and hyperkalemic solution from the graft, but probably other released substances play a role as well. Usually myocardial contractility seems to be preserved, but some patients may develop

14

short-lived myocardial depression. The post-reperfusion syndrome responds readily to small amounts of epinephrine (10-50 μg). However, severe bradycardia or even sinus arrest within the first few minutes after graft reperfusion has been seen in a few patients, but fortunately is quickly reversed by boluses of epinephrine (100-250 μg) and chest compressions.

During the remainder of the neohepatic stage, cardiac output returns to the values seen during the preanhepatic stage. Sometimes excessive vasodilation associated with graft reperfusion persists for 1-2 h, requiring low-dose epinephrine or dopamine infusion. Normally, the graft starts to be metabolically active very early after reperfusion, and therefore $CaCl_2$ and $NaHCO_3$ are rarely necessary to maintain a normal metabolic state.

FLUID AND TRANSFUSION MANAGEMENT

Maintaining normovolemia intraoperatively is probably the most important task for the anesthesiologist. At the same time, it is probably also the most difficult task. The main reason for maintaining the patient's volume status is that this is the only way for the cardiovascular system to remain hyperdynamic during the procedure. Cardiac filling pressures may not accurately reflect the volume status of the patient, because the compliance of the heart and thoracic cage changes significantly during the transplant as the result of the use of retractors and varying pressure on the diaphragm. TEE and determination of RVEDV probably allow a more accurate estimation of the patient's volume status.

Although the average blood loss has gradually decreased over the last 15 years, it is impossible to predict blood loss in individual patients. Indeed, blood loss can still be substantial (more than 10 times blood volume). Intraoperative autologous transfusion (cell saver) may reduce the need for packed red blood cells from the blood bank. However, virtually all coagulation factors and platelets are removed during the process. Most anesthesiologists aim for a hematocrit of 25-30%, which should be sufficient to provide adequate oxygen transport. Because most patients have coagulopathy, one unit of fresh frozen plasma is usually administered for each unit of transfused packed cells. More fresh frozen plasma may be necessary to correct coagulopathy. Cryoprecipitate and platelets are given based on coagulation tests. Most liver transplant anesthesiologists use a type of rapid infusion device to allow transfusion of large amounts of fluids and blood products, allowing for adequate warming of the solution. Transfusion devices ideally should have air detectors to avoid intravenous infusion of air. A commonly used transfusion system is the Rapid Infusion System® (Haemonetics, Inc. Braintree, Mass.) (Fig. 14.1). This device allows transfusion of up to 1.5 L/min of a warmed blood mixture.

COAGULATION MANAGEMENT

Intraoperative coagulopathy is the result of preoperative coagulopathy, thrombocytopenia, and platelet dysfunction; intraoperative dilution of coagulation factors and platelets; excessive fibrinolysis; and hypothermia. Although the coagulopathy that occurs during liver transplantation usually can be corrected by transfusion of fresh frozen plasma, cryoprecipitate, and platelets, pharmacologic

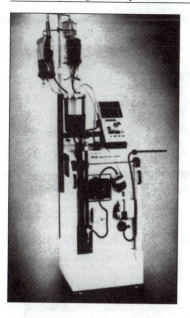

Fig. 14.1. The rapid infusion system (Haemonetics, Inc., Braintree, Mass).

intervention may allow normalization of the coagulation with less blood products and reduced time for complete surgical hemostasis. Blood coagulability is determined by TEG, platelet count, and routine coagulation tests (PT, PTT, fibrinogen level), but observation of the surgical field reveals invaluable information. TEG measures global hemostasis of plasma proteins and platelet interactions. Only TEG shows fibrinolysis in a timely manner. In addition, TEG allows the in-vitro use of pharmacologic intervention, greatly expanding the value of this monitoring technique (Figs. 14.2, 14.3). However, some feel that TEG lacks the specificity for guiding blood component replacement that is found with measuring platelets, fibrinogen, PT, and PTT. Thus in problem cases both systems may have a role. The frequency of hemostatis testing is determined by the degree of coagulation dysfunction, the degree of surgical difficulty, and the surgical field.

Blood component therapy is based on hemostatic testing, TEG, and adequacy of surgical hemostasis. When component replacement is indicated, it should be done to a level of adequacy, not normality. Patients may have successful transplants with only moderate blood loss with platelet counts of 40-50,000,000/mL or fibrinogen concentrations of 100-125 mg/dL. When fibrinolysis is present, epsilon-aminocaproic acid is frequently used for reversal, and some programs use continuous administration for prophylaxis. In either approach very low doses are usually effective (single dose 250-500 mg). Aprotinin has been advocated, with some data suggesting decreased blood loss. Others are concerned about the potential for thrombosis with the prophylactic use of antifibrinolytic agents, and additional research is required. After graft reperfusion, heparinoid effect can be seen even after flushing of the donor organ, but protamine reversal is rarely necessary.

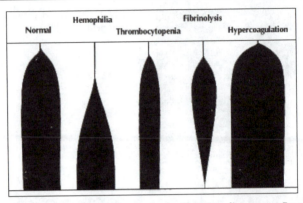

Fig. 14.2. Thromboelastographic patterns of patients with various disease states. Reproduced with permission from Kang YG. Monitoring and treatment of coagulation. In: Winter PM, Kang YG eds. Hepatic Transplantation: Anesthetic and Perioperative Management. New York: Praeger 1986:155. By permission of Praeger Publishers, 1986.

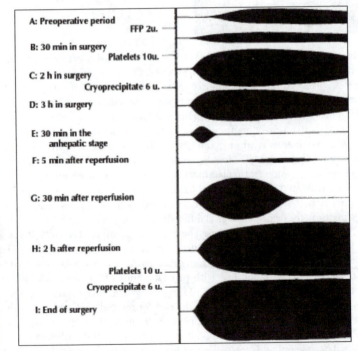

Fig. 14.3. Thromboelastographic patterns of patient undergoing liver transplantation. Reproduced with permission from Kang YG. Monitoring and treatment of coagulation. In: Winter PM, Kang YG eds. Hepatic Transplantation: Anesthetic and Perioperative Management. New York: Praeger 1986:155. By permission of Praeger Publishers, 1986.

Unfortunately, the coagulation changes during liver transplantation are incompletely understood: although activation of the fibrinolytic system has been well documented, especially during the anhepatic state and immediately following graft reperfusion, it is possible that there is also activation of the coagulation system in some patients, possibly leading to intravascular coagulation. Pulmonary thromboembolism has been reported in some patients.

METABOLIC MANAGEMENT

Ionized hypocalcemia is a recognized complication of blood transfusion in patients with liver failure because of their decreased ability to metabolize citrate. Patients undergoing liver transplantation may develop hypocalcemia when transfused during the preanhepatic and anhepatic stage, requiring calcium chloride administration. Hyperkalemia may develop in patients undergoing massive transfusion or having renal failure. This is best treated with glucose and insulin, which forces potassium into the cells, or washing of the red blood cells to reduce the potassium load. Metabolic acidosis is more common in patients with liver disease, especially when there is tissue hypoperfusion or massive transfusion. Sodium bicarbonate may increase sodium levels too quickly, possibly resulting in central pontine myelinolysis. However, tromethamine (THAM) also corrects metabolic acidosis but does not contain sodium, and therefore the use of tromethamine (THAM) contributes to the correction of metabolic acidosis but at the same time can ameliorate the hypernatremic effects of sodium bicarbonate. Ionized hypomagnesemia has been documented, but its clinical consequences are unknown. Therefore, the administration of magnesium sulfate is still controversial.

SELECTED READINGS
1. Marquez J, Martin D, Virji MA et al. Cardiovascular depression secondary to ionic hypocalcemia during hepatic transplantation in humans. Anesthesiology 1986; 65:457-461.
2. Kang YG, Lewis JH, Navalgund A et al. Epsilon-aminocaproic acid for treatment of fibrinolysis during liver transplantation. Anesthesiology 1987; 66:766-773.
3. Krowka MJ, Cortese DA. Hepatopulmonary syndrome: an evolving perspective in the era of liver transplantation (editorial). Hepatology 1990; 11:138-142.
4. De Wolf AM, Begliomini B, Gasior T et al. Right ventricular function during liver transplantation. Anesth Analg 1993; 76:562-568.

14

HEART TRANSPLANTATION

The number of heart transplant procedures has exploded in the mid-1980s, making heart transplantation a well-established and standardized procedure today. Because of the shortage of organ donors, this increase has now reached a plateau. The one-year survival rate is currently > 80%, with most survivors returning to a fairly normal active lifestyle. One of the main concerns for the anesthesiologist is to get the patient safely onto cardiopulmonary bypass without further damage to any of the other essential organs.

PREOPERATIVE EVALUATION

Most candidates for heart transplantation have end-stage cardiac failure as a consequence of ischemic or idiopathic cardiomyopathy. Other indications for heart transplantation include terminal valvular lesions and congenital anomalies that are not amenable to other surgical or medical therapy. Symptoms include severely limited physical activity and shortness of breath with limited activity or at rest. Recipients are usually < 60 years old, should be healthy otherwise or have organ dysfunctions that are reversible after heart transplantation, and should be compliant with medical instructions.

Cardiovascular System

Patients with end-stage cardiac disease have low cardiac outputs and high filling pressures despite optimal medical management. The initial response to left ventricular dysfunction is an increase in left ventricular end-diastolic volume; this may temporarily restore stroke volume at the cost of an increase in left atrial pressure. Eventually right atrial pressure will increase also, leading to the classic signs and symptoms of congestive heart failure. In addition, the systemic vascular resistance may increase in an attempt to maintain blood pressure. Atrial arrhythmias are common, and some patients with ventricular arrhythmias may have received an automatic implantable cardiac defibrillator (AICD).

Preoperative evaluation should include right and left heart catheterization to evaluate the pulmonary circulation and accurately determine intracardiac pressures. A pulmonary vascular resistance > 6 Wood units is a contraindication for heart transplantation because it frequently leads to failing of the right ventricle of the newly transplanted heart; these patients may be candidates for combined heart-lung transplantation. Moderate increases in pulmonary vascular resistance are usually tolerated if the transplanted heart functions well.

The dilated cardiomyopathy can lead to mural thrombi, which may be treated with chronic anticoagulation. Coronary angiography may determine the presence of treatable coronary lesions. Patients may receive pharmacologic support, frequently consisting of a combination of vasodilators and inotropic agents. Alternatively, a circulatory assist device may be in place (intraaortic balloon counterpulsation, ventricular assist device, artificial heart). Importantly, the prolonged low cardiac output and venous congestion will affect the function of other organs, and therefore all organ systems have to be evaluated preoperatively.

Pulmonary System

Prolonged left ventricular dysfunction results in increased left atrial and pulmonary venous pressure, which leads to an increase in pulmonary vascular resistance caused by hypertrophy of the musculature of the pulmonary arteries. The increase in left atrial pressure also results in an increase in lung water, causing ventilation/perfusion mismatch, increased airway resistance, decreased lung compliance, and increased work of breathing. Pleural effusions will reduce functional residual capacity and possibly further impair oxygenation.

Hepatic System

Chronic passive congestion of the liver may result in a reduced production of coagulation factors and other proteins. In addition, drug metabolism may be altered.

Gastrointestinal System

Like most other patients undergoing transplant procedures, these patients are considered to have full stomachs, not just because of the possible recent food intake, but also because some patients ingested immunosuppressants just before arriving in the operating room. Also, the high catecholamine concentrations may lead to reduced gastric emptying.

Renal System

Renal dysfunction is common as a result of low cardiac output and the aggressive use of diuretics to treat congestive heart failure. The use of diuretics may also result in hyponatremia and hypokalemic metabolic alkalosis, which may require potassium replacement. Similarly, hypomagnesemia can lead to arrhythmias and may require supplemental therapy with magnesium sulfate.

ANESTHETIC MANAGEMENT

The single most important aspect of anesthetic management is to avoid further deterioration of the circulatory system, attempting to preserve the function of other organ function until cardiopulmonary bypass is started. Therefore, myocardial depressant drugs are avoided and increasing inotropic support may be necessary.

Pharmacokinetic and Pharmacodynamic Changes

Although the liver dysfunction and the changes in volume of distribution may be altered, this usually does not require a significant change in drug dosing intraoperatively.

Immediate Preinduction Preparation and Monitoring

Premedication is frequently avoided, not just because these patients are more sensitive to sedatives, but also because these patients are usually well-informed regarding their procedure, and therefore tolerate the transfer to the operating room very well. The anesthesia equipment that is used for routine cardiac procedures should be sufficient for heart transplants. Monitoring includes electrocardiography (leads II and V5), pulse oximetry, and multigas analysis. Urine output is followed. Bladder/rectal, esophageal, and pulmonary arterial temperature are measured. A femoral arterial catheter is placed before induction of anesthesia to allow determination of central aortic pressure, because there may be a discrepancy between radial arterial and central aortic pressure, especially immediately after cardiopulmonary bypass. An oximetric pulmonary artery catheter is inserted through the right internal jugular vein, allowing continuous mixed-venous oxygen saturation (SvO_2) measurement. Cardiac output is determined using thermodilution technique. A long sheet is used to cover the pulmonary artery catheter, which allows it to be pulled back into the superior vena cava during cardiopulmonary bypass and readvanced into the pulmonary artery after cardiopulmonary bypass. A transesophageal echocardiography (TEE) probe is placed after induction of anesthesia, and allows for additional monitoring of cardiac function and volume status.

14

Induction and Maintenance of Anesthesia

Communication with the surgical team is essential before starting induction of anesthesia in order to decrease ischemic time of the organ. The timing should take into consideration the longer surgical time needed in patients undergoing reoperation.

Frequently, the induction and maintenance of anesthesia does not differ much from the techniques used in other patients with poor cardiac function undergoing cardiac anesthesia. However, cricoid pressure is usually applied to prevent aspiration of gastric contents. A combination of narcotics (e.g., fentanyl, sufentanil) and muscle relaxants (e.g., pancuronium, rocuronium, cisatracurium) is administered, and their hemodynamic effects are followed carefully. Narcotics are administered until loss of consciousness. The choice of muscle relaxants depends mainly on the desired change in heart rate. Alternatively, a hypnotic agent that is devoid of myocardial depressant effects (etomidate) has been used with succinylcholine to provide rapid intubating conditions.

Although the anesthetic agents that are used have no significant direct hemo-dynamic effects, the loss of consciousness by itself may lead to a reduction in sympathetic output from the brain stem and circulating catecholamines, leading to hypotension as a result of systemic vasodilation and possibly a further decrease in myocardial contractility. In addition, venous dilatation may exacerbate the mildly hypovolemia that is frequently present as a result of aggressive diuretic therapy. Therapy of hypotension during induction of anesthesia is usually guided by the hemodynamic monitoring: vasodilation may be treated with volume administra-tion and small amounts of vasoconstrictors (e.g., phenylephrine 50-100 µg boluses), while reductions in myocardial contractility can be treated with inotropic agents (e.g., dopamine or dobutamine 3-5 µg/kg/min). More accurate monitoring and therefore also more directed treatment can be initiated once the TEE probe has been placed after tracheal intubation. It has to be recognized that volume therapy can exacerbate congestive heart failure, and that vasoconstrictors can increase left ventricular afterload and therefore decrease cardiac output. Also, catecholamines can lead to arrhythmias which are frequently poorly tolerated in these patients. Reaction to intubation or surgical stimulation can be different from that seen in other patients: light anesthesia may not be reflected as hypertension and tachycardia, but merely as a decrease in cardiac output, an increase in pulmonary capillary wedge pressure and systemic vascular resistance, and a decrease in SvO_2. This requires additional anesthetics or the use of vasodilators (nitroglycerine or nitroprusside).

Anesthesia is maintained with a combination of narcotics (e.g., fentanyl 50-100 µg/kg followed by 5-10 µg/kg/h, or sufentanil 10-15 µg/kg followed by 1-2 µg/kg/h), muscle relaxants, and amnestic agents (e.g., midazolam 5-10 mg, lorazepam 2-4 mg). However, the addition of benzodiazepines to high-dose narcotics may be associated with mild reductions in cardiac output and systemic vascular resistance, resulting in hypotension. Low-dose dopamine, mannitol and diuretics (furosemide) are frequently used intraoperatively, especially in patients with preoperative renal dysfunction. However, there is little evidence that this management improves postoperative renal function.

Cardiopulmonary Bypass

Anticoagulation is achieved with normal doses of heparin (300 U/kg) prior to cardiopulmonary bypass. Some patients with hepatic congestion or prolonged heparin administration preoperatively may have reduced plasma concentrations of antithrombin III and therefore may be resistant towards the effects of heparin. Heparin resistance is promptly corrected by administration of antithrombin III (1000-1500 U), although giving 1-2 units of fresh frozen plasma is an acceptable alternative. The aorta is cannulated in the normal fashion. The pulmonary artery catheter is withdrawn into the superior vena cava, and the superior and inferior vena cava are cannulated separately to allow excision of the heart. After initiation of cardiopulmonary bypass, the aorta, pulmonary artery, and atria are transected and the heart is excised, leaving a cuff of the right and left atrium to allow anastomosis to the donor right and left atrium. This is followed by anastomosis of the aorta and pulmonary artery, and after rewarming the patient is weaned from cardiopulmonary bypass.

Weaning from Cardiopulmonary Bypass

The patient is weaned from cardiopulmonary bypass using the same principles as those for any cardiac procedure. Thus, heart rate and rhythm, volume status, contractility, and afterload are optimized. This frequently requires the administration of isoproterenol, dopamine, or dobutamine. The choice of agent is mainly determined by the systemic vascular resistance. The volume management is guided by the filling pressures and TEE. Direct observation of the heart in the surgical field reveals right ventricular function. After weaning, the pulmonary artery catheter is readvanced into the pulmonary artery. Appropriate monitoring will allow the anesthesiologist to determine whether weaning from cardiopulmonary bypass is successful. Adequate circulation results in a cardiac index of > 2 L/min/m^2 and a SvO$_2$ of $> 70\%$.

Management after Cardiopulmonary Bypass

Most transplanted hearts depend on the exogenous administration of catecholamines for the first few days after transplantation. Therefore, their administration should not be interrupted at any point, especially during and immediately after the transfer of the patient to the intensive care unit.

Biventricular failure immediately after transplantation can be the result of inadequate preservation or hyperacute rejection, and may require inotropic support, biventricular assist device, or artificial heart. Right ventricular failure can be seen in patients with mild-moderate pulmonary hypertension. Excessive volume loading should be avoided; maintaining perfusion of the right ventricle by optimizing blood pressure and inotropic support is essential. A right ventricular assist device may be required.

14

SELECTED READINGS

1. Ghignone M, Girling L, Prewitt RM. Volume expansion versus norepinephrine in treatment of a low cardiac output complicating an acute increase in right ventricular afterload in dogs. Anesthesiology 1984; 60:132-135.

2. Stern DH, Gerson JI, Allen FB et al. Can we trust the direct radial artery pressure immediately following cardiopulmonary bypass? Anesthesiology 1985; 62:557-561.

3. Demas K, Wyner J, Mihm FG et al. Anaesthesia for heart transplantation. A retrospective study and review. Br J Anaesth 1986; 58:1357-1364.

4. Berberich JJ. Anesthesia for heart and heart-lung transplantation. In: Fabian JA, ed. Anesthesia for organ transplantation. Philadelphia: JB Lippincott Company, 1992:1-19.

LUNG TRANSPLANTATION

Candidates for lung transplantation include patients with end-stage pulmonary disease but preserved right and left ventricular function. Pulmonary diseases include restrictive, obstructive, infective, and pulmonary vascular. Most candidates for single lung transplantation have pulmonary fibrosis, emphysema, chronic obstructive pulmonary disease, and alpha1-antitrypsin deficiency. Patients with pulmonary hypertension who do not yet have right ventricular failure may also be acceptable candidates for single lung transplants. A double lung transplant is performed when leaving one diseased lung is place would lead to complications (chronic bilateral infection such as cystic fibrosis and bronchiectasis, or severe air trapping).

Heart-lung transplantation is indicated in patients with end-stage pulmonary vascular disease that can be the result of cardiac lesions (e.g., Eisenmengers' syndrome) or pulmonary disorders (e.g., pulmonary hypertension), resulting in irreversible failure of both heart (right or left ventricle) and lungs.

The main intraoperative problems include hypoxemia, hypercarbia, and right ventricular failure.

PREOPERATIVE EVALUATION

Pulmonary System

Pulmonary function tests and ventilation/perfusion scans will help in the determination which lung is diseased the most. Most patients require supplemental oxygen to improve oxygenation, although most are still mildly hypoxic. Hypercarbia is common. Patients with little or no functional reserve may be more prone to hemodynamic instability during induction of anesthesia.

Cardiovascular System

Right ventricular function has to be determined, especially in patients with pulmonary hypertension. The degree of right ventricular dysfunction and its potential reversibility will determine what type of procedure is indicated and whether the patient is likely to require cardiopulmonary bypass. This is usually accomplished by echocardiography or radionuclide scans. In addition, left ventricular failure or coronary artery disease should be excluded.

Hepatic System
As in candidates for heart transplantation, passive congestion of the liver may result in decreased synthesis of proteins and drug metabolism.

SURGICAL PROCEDURE

Single vs Double Lung Transplant
The preferred procedure in most patients is single lung transplantation. A posterolateral thoracotomy is performed with the patient in the lateral position, but with one groin exposed to allow cannulation of femoral vessels for cardiopulmonary bypass. If double lung transplantation is indicated, it is usually accomplished using sequential single lung transplants using the clamshell thoracosternotomy in the supine position. Again, at least one groin is exposed. The native lung with the most severe pathology should be transplanted first.

Need for Cardiopulmonary Bypass
Cardiopulmonary bypass is avoided if possible because it prolongs the procedure, and results in more perioperative blood loss and an increased need for platelets and coagulation factors after cardiopulmonary bypass. The decision to use cardiopulmonary bypass is based on the right ventricular function after clamping of the pulmonary artery of the lung that is excised. If cardiac output after clamping of the pulmonary artery decreases significantly and/or if worsening right ventricular dysfunction is observed on transesophageal echocardiography (TEE), it is very likely that cardiopulmonary bypass will be necessary. Most patients with pulmonary hypertension will require cardiopulmonary bypass because the right ventricle does not tolerate a further increase in afterload. Also, the patients with pulmonary hypertension who will need correction of cardiac anomalies (e.g., atrial septum defect, ventricular septum defect) will require cardiopulmonary bypass. Finally, cardiopulmonary bypass is used when there is intractable hypoxia during one-lung ventilation, despite appropriate interventions.

Use of femoral cannulas to establish partial cardiopulmonary bypass may be adequate treatment for right ventricular failure. However, there is still blood flow through the native lung and ejection by the left ventricle, and therefore oxygenation of arterial blood in the upper half of the body is sometimes inadequate. Therefore, determination of PaO_2 from a blood sample from a radial artery has to be done to assure adequate oxygenation of blood perfusing the heart and brain. Measures should be instituted to improve oxygenation of blood flowing through the native lung. If necessary, ventricular fibrillation is induced during cardiopulmonary bypass in order to interrupt flow through the native lung.

ANESTHETIC MANAGEMENT
Preoperative sedation is minimal because these patients have limited cardiopulmonary reserve. Intravenous catheters are usually placed in upper extremities. However, in patients undergoing sequential double lung transplantation using the clamshell thoracosternotomy who will have both arms bent at the elbows and suspended from the ether screen, catheters in antecubital veins

14

are avoided. After the recipient is transferred to the operating room, a right radial arterial catheter is placed. In patients who are likely to require cardiopulmonary bypass a femoral arterial catheter is also inserted. An oximetric pulmonary artery catheter is usually placed after induction of anesthesia. Standard non-invasive monitoring includes electrocardiography (lead II and V5), blood pressure cuff, pulse oximetry, and multigas analysis. TEE is very helpful in the evaluation of volume status and right ventricular function, and may help in the evaluation of the vascular anastomoses.

Anesthetic induction should keep the patient's recent oral intake into consideration. Because these procedures are done with little notice, cricoid pressure usually has to be applied. In patients with compromised right ventricular function, anesthetic agents that do not depress cardiac function are used (e.g., fentanyl, sufentanil, etomidate, muscle relaxants, benzodiazepines). In patients with preserved right ventricular function, low concentrations of inhaled anesthetics are usually well tolerated.

Single Lumen Endotracheal Tube vs. Double-Lumen Tube

Double lung transplants performed on cardiopulmonary bypass usually receive a single lumen endotracheal tube. All other patients will require a double-lumen tube, and its correct position is verified using fiberoptic bronchoscopy.

Management of Hypoxemia

Hypoxemia can occur at any time during lung transplantation, but is most common during one-lung ventilation. Hypoxemia is most severe about 20 min after initiation of one-lung ventilation, and initial treatment should include judicious use of positive end-expiratory pressure (PEEP) to the ventilated lung, oxygen insufflation to the non-ventilated lung, and clamping of the pulmonary artery of the nonventilated lung. Pneumothorax on the ventilated side should always be considered. If hypoxia is refractory to all interventions, cardiopulmonary bypass should be initiated.

Management of Mechanical Ventilation

Mechanical ventilation in patients undergoing lung transplantation can result in air trapping due to incomplete exhalation, especially in patients with obstructive pulmonary disease. Significant air trapping may lead to increases in intrathoracic pressure and hemodynamic compromise because of a reduction in venous return. This problem can be diagnosed by disconnecting the patient from the breathing circuit for about 30 seconds; the blood pressure will returns to baseline if the cause of hemodynamic instability was hyperinflation of the lungs and air trapping. This is best treated with increasing expiratory time, resulting in hypoventilation and permissive hypercapnia, which is usually well tolerated as long as oxygenation can be maintained.

Mechanical ventilation with or without PEEP in any patient may lead to an increase in pulmonary vascular resistance. This may not be tolerated well in patients with right ventricular dysfunction. Increasing the volume status may not be the best intervention; inotropic support may have better results. The use of pulmonary vasodilators in these circumstances is frequently not very effective. It is important to maintain coronary perfusion pressure in order to preserve the oxygen supply to the right ventricle.

Sudden reductions in oxygen saturation and hypotension can be the result of tension pneumothorax, especially in patients with bullous disease or fibrotic lungs.

After reperfusion, the transplanted lung may dysfunction, or the native lung may not tolerate PEEP or develop air trapping; this situation requires differential lung ventilation. However, if the transplanted lung functions well, the double lumen endotracheal tube is replaced by a single lumen tube at the end of the procedure.

Coagulopathy

Coagulopathy may be induced by cardiopulmonary bypass, although lung transplantation by itself may be associated with activation of the coagulation and fibrinolytic systems. Thus, double lung transplantation and use of cardiopulmonary bypass is associated with more significant bleeding, and frequently requires platelet administration. Aprotinin, epsilon-aminocaproic acid, tranexamic acid, and DDAVP have all been used in lung transplantation.

Lung Reperfusion

Some degree of pulmonary edema is common in the transplanted lung after reperfusion. Significant edema requires the use of high levels of PEEP, diuresis, and volume restriction. Severe pulmonary edema requires differential lung ventilation, or in the case of double lung transplantation, the use of extracorporeal membrane oxygenation.

Postoperative Analgesia

A thoracic epidural catheter may be placed preoperatively in patients with a very low chance for cardiopulmonary bypass. However, more frequently, the epidural catheter is placed early postoperatively after correction of any persisting coagulopathy.

SELECTED READINGS
1. Haluszka J, Chartrand DA, Grassino AE et al. Intrinsic PEEP and arterial PCO_2 in stable patients with chronic obstructive pulmonary disease. Am Rev Respir Dis 1990; 141:1194-1197.
2. Quinlan JJ, Buffington CW. Deliberate hypoventilation in a patient with air trapping during lung transplantation. Anesthesiology 1993; 78:1177-1181.

14

KIDNEY TRANSPLANTATION

Although chronic dialysis improves live expectancy, kidney transplantation improves the quality of life. Therefore, kidney transplantation has now become a commonly performed and standardized surgical procedure. Patients with end-stage renal disease but with otherwise normal life expectancy are good candidates for this procedure. The tolerable ischemic time for kidneys is up to 48 h, and therefore cadaveric kidney transplants are semi-elective procedures, while living-related kidney transplants are elective.

PREOPERATIVE EVALUATION

Renal failure ultimately results in the uremic syndrome: these patients are unable to regulate their volume status and composition of body fluids, leading to fluid overload, metabolic acidosis, and hyperkalemia. In addition, there is secondary organ dysfunction with neuropathy, anemia, platelet dysfunction, hypertension, congestive heart failure, pericardial or pleural effusions, muscle weakness, osteodystrophy, nausea, vomiting, and impaired cellular immunity.

Renal System

Candidates for renal transplantation usually have end-stage renal failure. However, with the advent of living-related kidney transplantation, some of the recipients may have pre-terminal renal disease that does not yet require dialysis. When the patient is treated with dialysis, it is important to determine the volume status and electrolyte concentrations immediately preoperatively.

Electrolyte Changes

Hyperkalemia above 5.5 mmol/L should be corrected by dialysis before the patient is transferred to the operating room. Hypocalcemia may result from decreased intestinal absorption of calcium, resulting in secondary hyperparathyroidism, leading to bone decalcification.

Acid-Base Status

Chronic metabolic acidosis is the result of impaired excretion of hydrogen ions and impaired reabsorption of bicarbonate by the kidneys, and is associated with a compensatory respiratory alkalosis. Severe metabolic acidosis requires dialysis.

Cardiovascular System

Most patients with end-stage renal disease have a hyperdynamic cardiovascular system to compensate for chronic anemia, which is the result of inadequate production of erythropoietin, uremic depression of bone marrow, and erythrocyte membrane fragility. Chronic anemia (hemoglobin levels of 6-8 g/dL) is usually well tolerated and should not require preoperative transfusion. Chronic arterial hypertension is common due to elevated concentrations of renin and angiotensin, and is usually treated with antihypertensive agents such as angiotensin-converting enzyme inhibitors, vasodilators, beta-antagonists, and calcium channel blockers. Uncontrolled hypertension can result in hypertensive cardiomyopathy, which may be aggravated by hyperlipidemia. Left ventricular hypertrophy and coronary artery disease are not uncommon. Some of these patients may have silent ischemia as a result of uremic neuropathy. Significant uremic pericarditis is uncommon. Some patients have diabetes mellitus, and therefore the degree of coronary artery disease has to be determined in these patients.

Nervous System

For the anesthesiologist, uremic neuropathy of the sympathetic nervous system may lead to hemodynamic instability and unexpected changes in heart rate and rhythm.

Coagulation System

Coagulopathy may result from thrombocytopenia, abnormal platelet function, and residual heparin effect from hemodialysis.

ANESTHETIC MANAGEMENT

Pharmacokinetic and Pharmacodynamic Changes

The pharmacology of some anesthetics is changed in the presence of renal failure due to reduced excretion (e.g., pancuronium) or altered protein binding and volume of distribution (e.g., midazolam, diazepam, thiopental). However, the pharmacology of many other anesthetics is not affected to a clinically significant degree, while most drugs are tolerated well if titrated to effect.

Monitoring

Routine monitoring includes electrocardiography, noninvasive blood pressure determination, pulse oximetry, multigas analysis, and peripheral nerve stimulation. For patients with coronary artery disease or poorly controlled hypertension an arterial catheter should be placed to allow continuous blood pressure monitoring and better control of blood pressure. A central venous catheter may be placed after induction of anesthesia to assist in the perioperative volume management, but is only helpful in sicker patients. The central venous catheter is commonly avoided in living-related kidney transplants. A pulmonary artery catheter is rarely indicated. Intraoperative laboratory tests should include determination of hemoglobin or hematocrit, serum electrolytes, blood glucose, and acid-base state.

Induction and Maintenance of Anesthesia

Patients with diabetes may have decreased gastric emptying, and therefore rapid sequence induction should be used. Most commonly used induction agents include thiopental, propofol, and etomidate. Succinylcholine is acceptable if hyperkalemia is not present. Other muscle relaxants that are commonly used are rocuronium and cisatracurium. Anesthesia is maintained with a combination of inhaled anesthetics (e.g., isoflurane, desflurane), narcotics (e.g., fentanyl, sufentanil) and nitrous oxide.

Fluid Management

Most anesthesiologist assure adequate perfusion of the graft by inducing mild hypervolemia and hypertension after revascularization. Fluids are administered to obtain a central venous pressure of 10-15 mm Hg. Achieving a systolic blood pressure of 120-140 mm Hg sometimes requires the use of dopamine infusion (3-10 µg/kg/min). In addition, most anesthesiologists also administer mannitol (12.5-25 g) and furosemide (10-40 mg) after release of the vascular clamps to promote urine production.

SELECTED READINGS

1. Kirvela M, Scheinin M, Lindgren L. Haemodynamic and catecholamine responses to induction of anaesthesia and tracheal intubation in diabetic and non-diabetic uraemic patients. Br J Anaesth 1995; 74:60-65.

2. Heino A, Orko R, Rosenberg PH. Anaesthesiological complications in renal transplantation: A retrospective study of 500 transplantations. Acta Anaesth Scand 1986; 30:574-580.

PANCREAS TRANSPLANTATION

The incidence of insulin-dependent diabetes mellitus (IDDM) is approximately 0.5%, indicating that between 1 and 1.5 million people in the US have IDDM. Although subcutaneous injection of insulin is a well-established therapy, it does not result in normal glucose metabolism. Pancreas transplantation is indicated in patients with extremely labile IDDM despite complex insulin regimens and in patients with hypoglycemia unawareness, resulting in poor quality of life. In addition, patients with severe neuropathy, especially autonomic neuropathy, may benefit from pancreas transplantation. Pancreas transplantation should also be considered in patients who already require immunosuppression, most frequently because of a kidney transplant. After successful pancreas transplantation, there is total independence from exogenous insulin administration, and glucose levels should be normal although responses to oral and intravenous glucose administration may be slightly abnormal. Secondary complications will gradually improve, although this may take several years. Importantly, quality of life is significantly improved after pancreas transplantation.

PREOPERATIVE EVALUATION

IDDM results in secondary complications which are the result of microvascular disease and nonenzymatic glycosylation of proteins. The secondary complications are more severe in patients with poor control of IDDM, but eventually all patients will develop secondary complications. Secondary complications include nephropathy, retinopathy, neuropathy, and cardiovascular disease.

Cardiovascular System

The cardiovascular complications of IDDM include ischemic heart disease, idiopathic cardiomyopathy, peripheral vascular disease, and hypertension. Because of the autonomic neuropathy myocardial ischemia and infarction may be silent. Therefore, even in patients without angina, preoperative evaluation should include dobutamine stress echocardiography or adenosine thallium scintigraphy. Coronary angiography may be required in selected patients.

Neuropathy

Abnormal nerve conduction will affect motor, sensory, and autonomic nerves. This results in abnormal cardiovascular and cardiorespiratory reflexes (e.g., orthostatic hypotension), cardiovascular lability (e.g., resting tachycardia), esophageal dysfunction, gastroparesis with delayed gastric emptying, and sudden death.

Renal System

Nephropathy develops in about half the patients with IDDM, and is caused by microvascular changes in the glomeruli and peritubular capillaries.

ANESTHETIC MANAGEMENT

Monitoring

Hemodynamic monitoring includes electrocardiography (leads II and V5), direct invasive arterial pressure monitoring, and central venous pressure monitoring. Pulse oximetry, multigas analysis, temperature measurement, and peripheral nerve stimulation are routinely performed.

Induction and Maintenance of Anesthesia

Because autonomic dysfunction may result in delayed gastric emptying, rapid frequency induction is commonly used. Many anesthetic agents are acceptable for induction and maintenance of anesthesia, as long as hemodynamic stability is maintained. Autonomic neuropathy of sympathetic nerves may require the use of vasoactive substances.

Metabolic Control

Although there is some evidence that tight glucose control (70-100 mg/dL) may result in better allograft function, commonly the glucose level is kept in the 100-150 mg/dL range. This may require continuous infusion of insulin (0.5-2 U/h) guided by frequent blood glucose determinations. Usually glucose is infused as well (75-100 mL/h of 5 % glucose solution) to prevent hypoglycemia. The transplanted pancreas very quickly becomes metabolically active, and insulin should not be required to maintain normoglycemia postoperatively.

SELECTED READING

1. Burgos LG, Ebert TJ, Asiddao C et al. Increased intraoperative cardiovascular morbidity in diabetic with autonomic neuropathy. Anesthesiology 1989; 70:591-597.

14

Psychiatric Issues in Organ Transplantation

John E. Franklin and Roslyn M. Paine

The myriad of technical advances in solid organ transplant over the past 20 years has challenged the field in expected and unexpected ways. The increased survival rates and quality of life has increased referrals and demand for organ transplantation. This increased demand however has highlighted the shortage of available donor organs. This dilemma serves as a backdrop to many of the psychosocial issues discussed in this Chapter. These issues include selection criteria, dealing with long waits for transplant, the anxiety of where to list and rule changes in allocation of organs. Hopefully with advances in areas such as xeno-transplantation, artificial organs, islet cell transplants, split livers and increased donor registration some of the issues discussed here will become relatively mute, much as risk/ benefit data has made transplantation decisions relatively easy for patients and physicians in recent years. This Chapter will be divided into 1) what transplant personnel should know about the general psychosocial care of transplantation patients; 2) the role of mental health specialists in transplant; 3) specific issues regarding liver, small bowel, kidney, pancreas, heart and lung transplants.

GENERAL PSYCHOSOCIAL ISSUES IN TRANSPLANTATION

PATIENT, FAMILY EVALUATIONS

Why do some patients come to transplant and many others who meet medical criteria do not? The selection process includes such factors as primary physician awareness, patient knowledge base and motivation, health status, family and friends, managed care companies and geographical location. When we encounter patients on the transplant service, they, on some level, have cleared some of these hurdles. Much of the selection process has happened before patients come for their first transplant clinic evaluation. This is important to realize, as much of our job is to create a supportive and therapeutic environment to get patients through not only the medical challenges, but also psychosocial challenges of transplant. What often makes the challenge of supporting patients more difficult is the geographic distances to transplant centers, managed care concerns that have the potential of fracturing care and the need for communication between multiple medical services. It must be kept in mind the maze of tests, medical people and often frustrating systems that patients undergoing transplant are interfacing. Although this is burdensome to patients, this process has the potential for building good therapeutic alliances between patient/institutions and it provides an observation

15

Organ Transplantation, 2nd edition, edited by Frank P. Stuart, Michael M. Abecassis and Dixon B. Kaufman. ©2003 Landes Bioscience.

period of the patient's strengths and weaknesses and their coping mechanisms. A patient may fail this test, such as a patient who relapses to alcohol or drugs or patients who become grossly noncompliant with medical care. Although these situations may make transplant doubtful, we need to determine what is going on and respond in a therapeutic way. For the relapsing alcoholic, it may be referral for rehabilitation with the possibility of being active on transplant list post rehabilitation. For patients who are noncompliant, a referral to another transplant center might be appropriate. However, for the vast majority of "transgressions" by patients we want to understand issues from the patient's perspective, maintain clear expectations and help identify and support patient positive coping mechanisms.

Table 15.1. outlines the general psychosocial evaluation format for all solid organ transplants. Several visits may be necessary before a thorough evaluation can be completed. The patient's energy level, mental status and the presence or lack of presence of family may play an important factor. In general, it is always preferable to have patients bring family members with them for psychosocial evaluations. Family members often provide information that the patient cannot and can collaborate important information such as recent substance use. Their presence at the meeting also provides an important opportunity to observe family dynamics. It is equally important in certain instances to interview patients and families separately to allow them to express any concerns or questions they may have. Family members may be reluctant to say anything that jeopardizes transplant candidacy. Potential donors may have ambivalence about donation or organ recipients may express a feeling of pressure to undergo a procedure they do not want. Even the timeliness of scheduling appointments can be instructive as to motivation or ability to follow through. Some patients and families are fairly sophisticated regarding medical issues. This apparent sophistication may belie underlying psychosocial difficulties in the family. On the other hand, some patients and families present so chaotically initially that it seems they will not make it through the procedure, but they do. Medical personal should be careful not to make unalterable judgments of people based on initial presentation, personnel bias or third party information. It is the practice of many transplant centers to

---

Table 15.1. Essentials of psychosocial evaluation for tranplant

— Clear introduction
— Review of medical history and patient's perspective of illness
— General medical history
— General psychiatric history
— Family history
— Drug and alcohol history
— Social history
— Current living situation and support
— Insurance and pertinent financial concerns
— Meet with family together or separately
— Assess patient motivation and sophistication
— Listen for patient concerns
— Urine toxicology as necessary

15

have a multidisciplinary conference where patient candidacies are discussed. The benefit of this approach includes the advantage of assessing people who have interacted with the team in different settings. Surgeons, other physicians, nurses, social workers, etc., in addition to the mental health specialists, all have observational skills important to understanding a patient situation. An opportunity for all staff to air support and concerns about a patient's candidacy results in consensus decisions that are usually, but not always correct. In addition, it creates a more cohesive staff and provides clearer messages to patients and families. It is important to realize that it is not only physicians caring for and taking responsibility for these patients. Health care providers such as nurse coordinators, social workers, ICU nurses and floor nurses feel acutely the stress of caring so intensely for transplant patients. Making everyone feel some part of the decision making and policy of selection is important. A number of prognostic rating scales for psychosocial functioning in transplantation have been devised. The Psychosocial Assessment of Candidates for Transplant (PACT) is a general rating scale that can be used for all solid organs. However, these rating scales need further validation and the fact is there will always be exceptions to the rule. Sometimes the experienced evaluator will have hunches or are able to draw on similar experiences with similar patients to help with patient selection.

The issue of patient selection is important to all transplants; however, as we will discuss later, areas such as liver transplantation in recovering alcoholics, still create controversy and misunderstanding. Sometimes there are situations where the patient clearly contributed to the rejection by noncompliance with medications. The experience for most patients who experience organ rejection is often one of shock, disappointment, sadness and unexpressed guilt. The decision to retransplant has to be carefully weighted. The reasons for non-compliance have to be fully understood and the transplant team has to be convinced that it will not happen again. Unfortunately, at this time many patients die awaiting transplant, which is particularly difficult for families and medical staff to accept. Patient and family support groups can be very helpful to help with the process, give practical advice and emotional support. Many patients and families contribute time in organizations such as The Transplant Recipient International Organization (TRIO), which provides patient information and promotes donor registration.

ABSOLUTE V. RELATIVE CONTRAINDICATIONS FOR TRANSPLANT

A survey was conducted by UNOS regarding the transplantation selection processes of the vast majority of U.S. transplantation centers. At most centers active substance abuse and major psychiatric disorders such as schizophrenia were judged to be close to absolute contraindications for transplant. In the real world, however, exceptions are granted based on other important factors. In addition, the maturity of the transplant center and the expertise of its staff may play a crucial role in patient selection of difficult cases. A high volume, more established transplant center may afford to be able to transplant riskier cases. There are stated universal guidelines; however, there are few hard and fast rules to selection based on psychosocial criteria. The survey also highlighted that most transplant centers stated that they had some mental health personnel associated with the program to help with psychosocial evaluations: psychiatrists, psychologists, social workers or psychiatric nurses.

As mentioned, active substance abuse and major psychiatric disorders such as schizophrenia and dementia may be absolute contraindications. Other strong relative contraindications include active suicidal ideation, unstable bipolar disorder, gross noncompliance with prior treatment, antisocial personality disorder and no social support. Even the risk of some of these contraindications can be acceptable if there is evidence of long-term stability or a good chance for recovery. For example, a schizophrenic patient who has been stable over a number of years may do perfectly fine with the whole procedure. Approximately one of five people with alcoholic cirrhosis may not fit the criteria for alcohol dependence as will be discussed later. These individuals may have been chronic heavy drinkers who present to transplant centers with cirrhosis being the first major stigmata of their drinking pattern. Individuals who have led stable lives and have good family collaboration may be reasonable risks for transplant without long observational periods. The transplant team itself often functions in crucial roles in psychosocial recovery. For example, an isolated, unemployed man may use the transplant team as a major support system. More difficult to resolve may be the relative contraindications to transplant such as less serious noncompliance, poor social support, problematic personality disorders and the vast array of mood disorders. In addition, you find individuals who aren't clinically depressed, but seem to have given up the will to live. These patients can be particularly difficult to deal with post-transplant, when effort and rapid participation in rehabilitation is needed to prevent complications. The general goal is short inpatient stays to prevent hospital complications.

DONOR EVALUATIONS

Increasingly, transplant programs are exploring the new ways to use live human donors when possible. Obviously this has been a practice with renal transplants for some time due to the fact that most individuals do reasonably well with one remaining kidney. The first successful kidney transplant was done in 1954 when an identical twin brother was used as the source of the kidney. Currently, in addition to kidneys, transplant surgeons now use portions of the liver, lung, heart, and pancreas from living donors.

Several different types of living organ donors have been identified, which include genetically related donors, emotionally related donors, "Good Samaritan donors," and donors-at-large. Genetically related donors can include first-degree relatives or more distant relatives, while emotionally related donors include spouses, partners, and friends. "Good Samaritan donors" generally refer to people that have no relationship or a distant relationship to the recipient. Finally, donors-at-large have been identified as those who wish to donate an organ in the absence of any direct or indirect relationship to the recipient. The number of living donations in recent years has increased due to increasing numbers of individuals needing transplantation and an inadequate supply of cadaver organs. In many Asian cultures, there have traditionally been many more living donor organs used than cadaver organs. This is due to cultural beliefs that include the Confucian tenet of respect for bodily integrity, as well as controversy over the acceptance of a definition for brain death, which limits the availability of cadaver organs.

There have been several factors identified with regards to what motivates people to donate organs. The two most commonly cited reasons for donation include helping to save a family member's life and improving the donor's own quality of life. Other factors that have been identified include guilt for past behaviors and concerns about family disapproval or forms of family pressure. It is important to note that all donations should be altruistic and free of any coercion either by family members or the medical team. One ethical dilemma that is currently being debated is whether or not to provide financial incentives for organ donation to help alleviate the organ shortage. Those in favor of financial incentives argue that an increased supply of organs would result in many more lives being saved, which would mitigate any potentially questionable ethical issues. Another potential consequence of financial incentives that has been argued is that the money from the sale of an organ might have a positive impact on the economic well being of donors. The arguments against financial incentives include a departure from American and western society's ethical standards, the fear of the human body and its parts being treated as commodities, and the potential consequence of exploiting the financial distress of the poor.

Many potential donors are in favor of donating immediately after they learn about the possibility of a living-related transplantation. This is especially true in cases where parents donate to their children. In the situation where parents are donating a part of their liver to a child there is a small risk of complications. In addition, there may also be unrealistic expectations or fantasies that may set parents up for extreme guilt or self-reproach if the donor liver fails. Determining if there are unrealistic expectations and forming an alliance if further support is needed may be the goals of psychiatric referral. However, psychiatric evaluation and follow-up support is essential for all living donors. Important elements that should be included in any donor evaluation include: donor motives and decision-making processes; description of the relationship with the recipient; prior and current beliefs about organ donation; ability to provide informed consent; attitudes of significant others toward decision to donate; availability of support from family and/or friends; past and current psychological problems and treatments; past and current substance use history; any current life stresses; and financial preparations for time off work. The majority of donors do well in terms of psychological adjustment. However, throughout the evaluation process and in follow-up it is critical to keep in mind the well being of the donor and to make sure that physically, emotionally, and financially they will not be at risk for complications.

EMERGENCY EVALUATIONS

There are situations when a decision to transplant is made on an emergency basis. What most likely presents as an emergency situation today is acute liver failure due to toxins such as acetaminophen; however, there are numerous medical conditions that present as acute liver failure. A typical situation might be a young, troubled person who is drinking and decides to take an overdose of acetaminophen. Acetaminophen toxicity itself could cause liver failure; however, alcohol increases the risk due to its increase of toxic metabolites. Treatment is usually effective; however, when it is not, acute liver failure may be fatal without transplantation. More often these patients are already confused, comatose and may rapidly develop increased intracranial pressure. Because of their mental status, it may be impossible to interview them. Often from family you can get a history of depression, personality problems or substance abuse. Sometimes the problems appear out of the blue to the family. In many cases a decision will be made to transplant even in a situations where suicide was a clear intent. Depression is potentially a treatable condition and suicidality might be a transient state. Conversely, in situations where you have young adults who have clear histories of intractable substance abuse or severe antisocial personalities you may decide it is not a reasonable risk. In situations where you decide not to transplant, it is important to clearly inform the family regarding the rationale for the decision and refer to other institutions if possible.

THE DIFFERENTIAL OF COMMON PSYCHIATRIC DISORDERS IN TRANSPLANTATION

The prevalence of psychiatric issues in transplant patients is probably comparable to the prevalence of other serious medical illnesses. However, as we will discuss later differences in end organ damage and their etiologies may make for differences. Baseline psychological factors surrounding suffering from early onset type 1 diabetes mellitus may raise the possibility of different problems in these patients than older cardiac transplant patients. The most common disorders you will encounter in all transplants are delirium, depression, anxiety, adjustment disorders, organic mood and personality disorders, substance abuse and lifelong personality disorders. What often makes diagnosis challenging is the great overlap of psychiatric conditions, personality, culture, and physical symptoms of disease which masquerade as psychological symptoms, medication side effects, staff anxiety and presence of cognitive disorders. Often only good detective work or time clarifies some of these issues. The detective work includes getting good pre-morbid histories and systematically ruling out medical causes and medication side effects. Often no definitive etiology can be ascertained and the cause is unknown or multi-factorial. Even when it is clear what is causing the psychiatric symptoms, some symptoms such as delirium can only be controlled after the medical cause abates: steroid-related mood changes or insomnia secondary to high dose steroid administration.

MOOD DISORDERS

The most common affective symptom that we encounter in the transplant population is depression. Depression is a subjective symptom that can range from mild, transient sadness secondary to a known stressor such as illness, to more serious clinical syndromes such as major depression or bipolar disorder. The most obvious serious medical emergency in depression is suicide. Depression as a symptom

15

is probably not any more common in transplant patients than any other serious life threatening illness. Most transient, sad feelings are appropriate and dealt with in ways that are characteristic for individuals. Depressive symptoms can be the result of the many stressors surrounding the primary disease or other life complications. When they are clearly beyond the severity of what one would reasonably expect and these symptoms do not meet criteria for major depression then the depression is labeled an adjustment disorder, with depressed mood. 30% of post liver donors can experience depression and 2-17% of post-op liver and heart transplant patients meet criteria for major depression. In transplant patients, quiet delirium and transient anxiety can easily be misinterpreted as adjustment disorder with depression. There is a great overlap in prevalence between depression and anxiety disorders. When people have major illness many of the physical symptoms mimic symptoms of depression such as insomnia due to pain or metabolic abnormalities, fatigue due to the primary illness or medications, loss of appetite/weight loss and concentration/memory problems. In many cases we have to look at the cognitive features of depression such as hopelessness, helplessness, guilt, and poor self-esteem to help make the distinction between illness and depression.

The DSM-IV signs and symptoms of major depression are listed in Table 15.2. Major depression, as discussed earlier, is only a relative contraindication for transplant. In most cases major depression is a treatable condition, even in the medi-

---

Table 15.2. Major depressive episode

A. Five (or more) of the following symptoms have been present during the same 2-week period and represent a change from previous functioning, at least one of the symptoms is either (1) depressed mood or (2) loss of interest or pleasure.

Note: Do not include symptoms that are clearly due to a general condition, or mood-incongruent delusions or hallucination.

1. Depressed mood most of the day, nearly every day, as indicated by either subjective report (e.g., feels sad or empty) or observation made by others (e.g., appears tearful).
   Note: In children and adolescents, can be irritable mood.
2. Markedly diminished interest or pleasure in all, or almost all, activities most of the day, nearly every day (as indicated by either subjective account or observation made by others).
3. Significant weight loss when not dieting or weight gain (e.g., a change of more than 5% of body weight in a month), or decrease or increase in appetite nearly every day.
   Note: In children, consider failure to make expected weight gains.
4. Insomnia or hypersomnia nearly every day.
5. Psychomotor agitation or retardation nearly every day (observable by others, not merely subjective feelings of restlessness or being slowed down).
6. Fatigue or loss of energy nearly every day.
7. Feelings of worthlessness or excessive or inappropriate guilt (which may be delusional) nearly every day (not merely self-reproach or guilt about being sick).
8. Diminished ability to think or concentrate, or indecisiveness, nearly every day (either by subjective account or as observed by others).
9. Recurrent thoughts of death (not just fear of dying), recurrent suicidal ideation without a specific plan, or a suicide attempt or a specific plan for committing suicide.

cally ill. We will discuss pharmacological approaches to treatment in a following section. It must be noted however that large scale studies in relatively healthy individuals show that some forms of talking therapies such as cognitive-behavioral or interpersonal therapy can be equally as effective in treating non-melancholic major depressions. Melancholic major depression may be a more psychotherapy resistant, genetic form of the illness. It must be noted, however, that there is no contraindication to using both medications and therapy. This is the most common practice in medical settings. Psychotherapy in transplant patients that are feeling up to it can proceed much like therapy in healthy individuals, especially in post-transplant patients. Transplant issues include guilt, shame, denial, the stress of long waiting periods, unrealistic expectations, complications such as rejection, re-entering the work world and relinquishing family responsibilities. The general mode of therapy is here and now, supports patient's strengths, and entails active listening, but also involves giving advice and connecting people to resources.

The most commonly used antidepressants used in transplantation medicine are listed in the pharmacology section. SSRIs are indicated for initial treatment. It can take 4-8 weeks to show some signs of improvement. If there is partial improvement the dose can generally be pushed to two to three times the initial dose at weekly intervals. Elderly, the seriously ill and patients with compromised liver metabolism may have to start at half the dose with close monitoring of side effects. Nefazadone and fluvozamine are relatively contraindicated due to their P450 3A4 interaction with prograf and cyclosporine. Ritalin, a psychostimulant, can be helpful in de-energized, depressed, medically ill patients. Stimulants, if effective, work more rapidly, and increase energy and appetite.

DELIRIUM

Delirium is extremely common in the post-operative period of transplant. The DSM IV signs and symptoms are outlined in Table 15.3. An assessment should be made of associated features such as sleep disturbance, increased or decreased psychomotor activity and labile emotions. Understanding the patient's pre-morbid baseline is also important. The cardinal features of delirium that help distinguish

---

Table 15.3. Delirium

A. Disturbance of consciousness (i.e., reduced clarity of awareness of the environment) with reduced ability to focus, sustain, or shift attention.

B. A change in cognition (such as memory deficit, disorientation, language disturbance) or the development of a perceptual disturbance that is not better accounted for by a preexisting, established, or evolving dementia.

C. The disturbance develops over a short period of time (usually hours to days) and tends to fluctuate during the course of the day.

D. There is evidence from the history, physical examination, or laboratory findings that the disturbance is caused by the direct physiological consequences of a general medical condition.

15

it from other cognitive disorders include the waxing and waning of symptoms over a 24 hour period, characteristic, but not definitive diffuse slowing on EEG and the fact that it is usually reversible. Common causes of delirium in transplant patients include in no particular order: metabolic disturbances, vascular disturbances, hypoxia from cardiac or pulmonary causes, infection, hepatic or renal failure, drug toxicity, drug or alcohol withdrawal, and endocrinopathies. Hepatic failure produces false CNS neurotransmitters that may play a direct role in hepatic confusion. There is however, no direct correlation between blood ammonia levels and degree of encephalopathy. The immediate clinical implications of delirium include the patient being a danger to him/ herself or others and the pain and suffering that can result from hallucinations and delusions. Many times patients do not remember their perceptional disturbances. At other times patients can remember in excruciating detail, frightening delusions post-operatively that are remembered as traumatic experiences. Delirium should be addressed in an appropriate way to minimize suffering. In addition, delirium can be a heralding event of something seriously going wrong in the body. Having a good understanding of current or pre-morbid psychiatric conditions will help with diagnosis. Delirium is a common cause for the patient to be declared incompetent to sign for medical procedures. Management includes coordinated care with all physicians, reviewing history and pertinent laboratory and radiological examinations. Basic tests include CBC with differential, chemistry panels including electrolytes, liver and renal function, arterial gases, EKG, blood cultures V/Q scans, CT and MRI of the brain and other organs and lumbar puncture. Cognitive tests such as trail making tests or digit span can highlight mild cases.

Treatment entails identifying and treating reversible causes of delirium, monitoring safety (patient may often need a sitter or soft restraints), educating patient and family regarding the diagnosis and reassuring that in most cases the delirium is reversible. It is helpful to provide regular reorientation and provide as much familiarity in the environment as possible. One preventive strategy is to have patients bring in pictures and other familiar items from home for long hospital stays. The environment should not be over or under stimulating as either can promote perceptual disturbances and confusion. The ICU, either pre or postoperatively, can be particularly disturbing to patients due to high activity in the unit or the feeling of relative isolation. Somatic interventions can include antipsychotics, droperidal, and benzodiazepines (however they should be avoided in liver failure patients) and cholinergics for people whose delirium is caused by anticholinergics. Haloperidol (Haldol) is a standard antipsychotic medication used in delirium. It can be administrated PO, IM or IV. IV administration has to be pushed slowly (torsades de pointes EKG change is a rare complication), however this administration has the advantage of decreasing the chance of extrapyramidal side effects. The initial dose is 1-2 mg Q 2-4 hr (0.25-0.5 mg Q4 for elderly patients). Olanzepine (Zyprexa) 2.5 to 20 mg, has increasingly been found to be an effective antipsychotic in delirium. Neuroleptic malignant syndrome, sometimes caused by high dose neuroleptics, presents with the clinical triad of confusion, rigidity with increased CPK, and hyperthermia and is treated by cessation of neuroleptics and dopamine agonists. Inadequately treated pain can also increase anxiety and cause agitation.

SUBSTANCE ABUSE AND DEPENDENCE

Substance abuse disorders have the second highest prevalence of any of the mental health disorders, second only to anxiety disorders. Approximately one out of every ten adults in the general population suffers from consequences of alcohol abuse and one year prevalence of substance abuse is 6.7%. These percentages are even higher among men. Screening for substance abuse is an essential component in the care of any patient. In transplantation, it is even more crucial to fully characterize any substance abuse problems in patients. In this section we will define the core features of substance abuse, discuss the use of screening instruments and cover recent advances in treatment. The special issue of transplanting alcoholics with liver disease will be discussed in the liver transplant section. It is important to realize that although addiction can be characterized as a relapsing disease, it is treatable. Abstinence rates with alcohol dependence over a lifetime approach two thirds of patients. Thus, it is important to have an informed degree of optimism to work successfully with this population. For many patients substance abuse may be in the remote past and it should not greatly bias our decisions. There is strong evidence from longitudinal studies that five years of sobriety corresponds with extremely low rates of return to drinking In addition to alcohol and illegal drugs, prescription drug abuse and nicotine addiction can be major issues in this population. Chronic pain syndromes can also be a byproduct of years of illness. Concerns regarding pain control in patients that have been on chronic opiates are not an infrequent concern. Obviously, nicotine addiction is a major risk factor for heart and lung disease. Nicotine addiction is also a vexing addiction to address. Nicotine relapse rates exceed cocaine and heroin relapse rates.

The DSM IV diagnosis of psychoactive substance dependence is included in Table 15.4. The signs and symptoms include symptoms of physical tolerance and withdrawal. The psychosocial sequelae may include impairment of interpersonal relations, employment or self-care. Any individual not in acute pain, exposed to high dose opiates over an extended period of time will develop physical dependence. They will experience withdrawal on cessation. There are two major neurobiological systems involved in addiction. One is the dopaminergic reinforcement system, which is crucial for craving and for the habitual pattern to develop and the other are the systems that cause symptoms of withdrawal. Fear of withdrawal is a major motivation for continued use. The core feature of the psychological addiction is the loss of control over use and the chronic obsessive craving for the substance which can be chronic, episodic and intense. Understanding the process as a disease may help individuals overcome the addictive process and focus on the task of breaking through denial and accepting help.

The gold standard for diagnosis is a clinical interview, where the goal is determining substance abuse patterns, the severity of use and the consequences of use. In addition, co-morbid disorder should be ruled out. There are however, several screens that have been used in the general and medical populations. The best known

15

Table 15.4. Substance dependence

A maladaptive pattern of substance use, leading to clinically significant impairment or distress, as manifested by three (or more) of the following, occurring at any time in the same 12-month period:

1. tolerance, as defined by either of the following:
    a. a need for markedly increased amounts of the substance to achieve intoxication or desired effect
    b. markedly diminished effect with continued use of the same amount of the substance

2. withdrawal, as manifested by either of the following:
    a. the characteristic withdrawal syndrome for the substance (refer to Criteria A and B of the criteria sets for withdrawal from the specific substances)
    b. the same (or a closely related) substance is taken to relieve or avoid withdrawal symptoms

3. the substance is often taken in large amounts or over a longer period than was intended

4. there is a persistent desire or unsuccessful efforts to cut down or control substance use

5. a great deal of time is spent in activities necessary to obtain the substance (e.g., visiting multiple doctors or driving long distances), use the substance (e.g., chain-smoking), or recover from its effects

6. important social, occupational, or recreational activities are given up or reduced because of substance

7. the substance use is continued despite knowledge of having a persistent or recurrent physical or psychological problem that is likely to have been caused or exacerbated by the substance (e.g., current cocaine use despite recognition of cocaine-induced depression, or continued drinking despite recognition that an ulcer was made worse by alcohol consumption)

are the CAGE and the MAST. The CAGE is composed of four questions 1) have you ever thought you should cut back on your drinking, 2) felt annoyed by people criticizing your drinking, 3) felt guilty or bad about your drinking, 4) had a morning eye opener to relieve a hangover or nerves. For gross screening, a positive response to two out the four would warrant further evaluation. In transplant centers a positive response to any of the questions warrants further assessment by a substance abuse expert. The second major screening devise is the Michigan Alcoholism Screening Test (MAST) which is a 25 item self-screening test. There are similar brief screening instruments for drug abuse. In a transplant center, endorsements of any illegal drug at all warrants evaluation by a substance abuse expert. Determining the amount of use is certainly important as a guide to the severity of the problem, the treatment course and the possible need for detoxification.

The recovery process starts with recognition of the problem and often results in physician referral. The actual recovery process uses a variety of support and educational systems such as drug treatment programs and self help groups. Physicians in their role of trusted healers can often make a difference in getting people to break through denial and to seek help. In nicotine addiction all physicians should be familiar with basic counseling techniques and understand the use of the various pharmacological aids such as nicotine gum and patches, buproprion and nicotine inhalers. All physicians should understand the basic principles of detoxification. The general principle is substituting another drug which is cross tolerant to the drug of abuse and slowly weaning it from the system so neuroreceptors have time to readapt. The gold standard treatment of alcohol withdrawal is benzodiazepines, which are relatively safe and prevent the most serious neuropsychiatric complications of withdrawal such as seizures and delirium tremens. The most important steps are recognizing the need for coverage, following signs and symptoms of withdrawal such as vital sign changes, tremors and agitation and adjusting benozdiazepine doses as needed. One excellent tool for monitoring withdrawal symptoms is the CIWA-AR; however, its use in medical patients may be more difficult to interpret due to other conditions that may be mimicking symptoms. Medically ill individuals, however, are at higher risk for severe complications of withdrawal. Long and shorter acting benzodiazepines can be used for these purposes. Shorter acting benzodiazepines are preferred in patients with hepatic dysfunction or in patients that are elderly. A standard detoxification order in liver patients is Lorazepam .5 to 2 mg P.O or I.V. Q.I.D. on day one adjusted up or down based on response and tapered over 3-5 days. Once a patient is in delirium tremons (D.T.'s) the care is supportive with benzodiazepines, neuroleptics, droperidal, opiates, proprofol and paralysis all being options to decrease agitation. When D.T.'s seem intractable a switch to a long acting barbiturate like Phenobarbital may be helpful. Newer pharmacological approaches to aid in long term abstinence with drugs include 1) naltrexone for alcohol and opiate addiction, 2) methadone and buprenorphine for opiate addiction, and 3) pending FDA approval approaches such as acamposate for alcohol use. Treatment of the other psychiatric co-morbidity is also crucial for long-term success.

Adjustment Disorders

Adjustment disorders defined as a maladaptive reaction to a known stressor are common psychiatric diagnoses in transplant patients. Often the illness itself or complications present as the most likely stressor. The patients also have life problems that don't necessarily go away when they need a transplant. Maladaptive reactions come in the form of symptoms such as anxiety, depression, behavioral problems, mixed emotions, physical complaints, social withdrawal or general poor functioning. An adjustment disorder may or may not signal the underlying possibility of a more serious psychiatric problem. Usually the symptoms are time limited and cease when the stressor stops. This is not to say that patients do not feel very symptomatic during this time and need direct intervention. Usually the approach is supportive therapy where stress and feelings are acknowledged and worked through as necessary. Often medications may be helpful for anxiety or sleep.

15

ANXIETY

Anxiety can be thought of as having both physical and mental components. The physical components are often experienced as fight/flight symptoms of motor tension and autonomic hyperactivity such as shortness of breath, palpitations, sweating, nausea and diarrhea. The mental components can be excessive vigilance, worry or frank panic. The differential of anxiety in the medical setting is extensive. The most common medical disorders that present with anxiety are pulmonary processes such as embolus, hyperthyroidism, complex partial seizure, vascular events, hypoglycemia, drug side effects such as steroid psychosis and theophyline for asthma, adrenal disorders and pheochromocytoma. The psychological manifestations of anxiety range from adjustment disorder as stated previously to anxiety disorders such as panic disorder, posttraumatic stress syndrome, generalized anxiety disorder and anxious personality disorders. The most useful approach for mild anxiety symptoms is reassurance when possible and addressing the underlying cause. The anxiety of waiting for transplant is very common but usually patients have the ability to minimize the severity by psychological defenses such as healthy denial, rationalization and intellectualization. These anxieties are usually episodic and transient. Sustained anxiety with extreme worry, panic or chronic insomnia needs psychiatric referral and usually some medications. The most common medications are benzodiazepines for acute anxious mood; SSRI's for panic attacks and a trial of busperone (buspar) for mild, chronic anxiety. It is generally advisable to limit the use of benzodiazepines to acute periods of anxiety to prevent iatrogenic addiction. A typical approach involves an initial dose of Lorazepam 0.5-1 mg PO T.I.D. or Clonazepam 0.5-1mg PO B.I.D. The incidence of addiction in individuals without previous addictions or high risk factors is uncommon. There are patients who will do better in long term use on low dose benzodiazepines.

Insomnia is a frequent complication of anxiety, depression and the pain and discomfort of medical illness. Several commonly used medications including stimulants can cause insomnia. Patients with hepatic cirrhosis very commonly complain of sleep disturbance as a core feature of their disease. Pharmacological approaches to liver-related insomnia may not be effective in liver patients and benzodiazepines may not be advisable in liver patients. In other patients, sedative-hynotics such as benzodiazepines and Zolipidem are useful short-term sleep aids. Sedating antidepressants such Trazadone and Mirtazapine in low doses can also be very helpful. Sometimes the sleep disorder may be a part of a primary sleep disorder such as sleep apnea or restless leg syndrome. In all cases you want to teach good sleep hygiene such as avoiding stimulants and exercise at night, going to bed and waking up at a regular schedule and using the bed for and associating the bed with sleep.

PERSONALITY DISORDERS

Personality is a combination of innate temperament and learned character. Hopefully, everyone has a personality or at least personality traits. When personality traits are maladaptive across time and most situations, exist over a lifetime and have significant effects on functioning, then the personality can be considered as a disorder. The personalities of patients can sometimes be very noncontributory to the quality of their care and how we go about treating them. In other situations our patient's and our own personalities can be crucial to outcome. Recognizing personality traits, being able to minimize miscommunications and distortions, maximizing coping skills of patients and building therapeutic alliances are the tools of a good physician. Patients may be histrionic (dramatic, attention seeking), obsessive (rigid, perfectionists), narcissistic (self-involved, controlling), dependent (demanding, clinging), masochistic (long suffering), schizoid (unsociable) or paranoid (mistrustful, blaming). In most cases these personalities do not get in the way of patient care. There are incidences where you may have a strong reaction to a patient's personality and feel it is affecting care. The important thing is to be able to take a step back and sort out your own feelings and thoughts. Often patients react to illness in the way that they have reacted to all problems in their lives and often physicians are seen much like other primary caretakers in their lives such as parents or other authority figures. Examples might include a dependent patient who fears abandonment, a histrionic man who is seductive with nurses, a narcissist who devalues the medical staff, or paranoid patient who appears angry and confused. It must be noted that many apparent personality disorders are in reality transient regressive behaviors in response to stress and they are not indicative of long-term problems. What makes some of these issues more pertinent in the transplant population is the seriousness of many situations and the need for good doctor/patient communication and alliance. When you find yourself having very strong emotional reactions to patients or acting out in ways that are not typical for you, then you may be dealing with a patient with a personality disorder. In these cases it is crucial to discuss your feelings with colleagues or get formal psychiatric consultation. Borderline personality disorders sometime only become clearly evident when the staff as a whole realizes that the patient is splitting the staff into all good and all bad caregivers. A true DSM IV antisocial personality is a relative contraindication for transplant. Antisocial personalities have a serious disregard for the feelings of other people. True antisocial personality must be distinguished from people who may interface with the criminal system for other reasons.

NEUROPSYCHIATRIC SIDE EFFECTS OF COMMON TRANSPLANT MEDICATIONS

Table 15.5 is a list of common transplant medications that have neuropsychiatric effects. It often can be difficult to know the offending agent because drugs frequently are prescribed concurrently. Cyclosporine is a lipophilic polypeptide that is derived from a fungus, Tolypocladium, and has been a mainstay of immunosuppression since 1978. The microemulsion form, Neoral, allows for greater bioavailability of cyclosporine. Neuropsychiatric side effects include anxiety, headache tremor, white matter changes, central pontine myolisis, cortical bleeding, ataxia, seizures, disorientation and visual hallucinations. Corticosteriods, such as Prednisone, are associated with anxiety; depression, delirium and mania are generally a dose-related manner. Prednisone doses above 40 mg are associated with

15

Table 15.5. Common transplant medications with neuropsychiatric side effects

Cyclosporine
Corticosteroids
OKT3
Prograf
Acyclovir
Ganciclovir
Alpha interferon
Ciprofloxacin
Cephalosporins
Sulfonamides
Gentamicin
Amphotericin B
Penicillin
Metronidazole
Ketoconazole
Any medications with anti-cholergic effect

---

higher incidence of steroid psychosis. OKT3 is a monoclonal antibody used for immunosuppression. Delirium, seizures, tremor, cerebral edema, aseptic meningitis and encephalopathy have been reported, even on the first dose. The side effects of Tacrolimus (Prograf), include prominent neurotoxicity, renal dysfunction, increased blood sugars, headache, anxiety, tremor, restlessness, insomnia, psychoses and parasthesis. The antiviral agent, acyclovir, can cause tremor, confusion, lethargy, depression, seizures, agitation and delirium. Gancyclovir for CMV can cause headache, delirium, seizures and hallucinations. Alpha interferon used for hepatitis C infection can cause anxiety, irritability and most notably serious depression especially in individuals prone to depression. Common antibiotics such as ciprofloxacin, sulfonamides, gentamicin and celpalosporins can cause delirium and hallucinations. Many antibiotics and antifungal agents have been associated with depression. Ampotericin B an antifungal agent is more clearly associated with delirium. There are numerous medications that have some degree of anticholinergic effect. Anticholinergic delirium has the classic picture of increased temperature, red skin and delirium. Many medications that have minimal anticholinergic effects can have a culminative effect when added together.

PSYCHOTROPIC MEDICATIONS

Table 15.6 is a listing of useful psychotropic medications in the transplant setting. Nortriptyline and despiramine are secondary amine tricyclic antidepressants. Tricyclics continue to be as effective as other antidepressants however side effects and high suicide potential have limited their use. Tricyclics are particularly effective with concurrent neuropathic pain, as second line antidepressants and in situations where blood levels are helpful. Amitriptyline is a tertiary tricyclic that is highly sedating and has higher anticholinergic effects. It is useful as a pain adjuvant and sleep aid. Selective serotonin reuptake inhibitors (SSRIs) have taken on the role of first line medications for depression due to their favorable side effect profile and safety in overdose. Although they are all equally effective there are

Table 15.6. Psychotrophic medications useful in transplantation (starting doses)

Antidepressants	Antipsychotics
Tricylcics	Haloperidol .5-5 mg BID
	Resperidone 0.5-1 mg BID
Nortriptyline 10 mg BID	Olanzepine 10 mg QD
Desipramine 25 mg BID	Seroquel 25-50 mg BID
Amitriptyline 25-50 mg BID	
	Antianxiety/Hypnotics
SSRIs	
	Lorezepam 0.5 P.O. TID
Fluoxetine 10-20 mg AM	Alprazolam .25 mg TID
Sertratine 25-50 mg AM	Clonazepam .5-1 mg BID
Paroxetine 10 mg HS	Buspirone 5 mg TID
Citalopram 10 mg QD	Zolidem 5-10 mg HS
	Diazepam 2-5 mg QD
	Chlordiazepoxide 25-50 mg QD
Others	Oxazepam 15 mg TID
Ritalin 2.5-5 mg 8am/12noon	
Buspropron 100 mg BID	**Mood Stabilizers**
Trazodone 50 mg HS	
Venlafaxine 37.5 mg BID	Lithum 600-900 mg QD Divided dose
Mirtazapine 15 mg HS	Carbamazepine 200 mg BID
	Dialproex Sodium 250 mg P.O. TID

differences in the degree of selectivity and interaction with other neurotransmitter systems. Fluoxetine was the first SSRI; it has a fairly long half-life. A long half-life gives you fairly steady state blood levels; however in poor metabolizers the drug could be difficult to clear. Prozac is a fairly potent 2D6 inhibitor and may have clinically significant interactions with other drugs metabolized at this isozyme. Sertraline is a good middle of the road SSRI. Sertraline's most troublesome clinical side effect is G.I. distress. Paroxetine is a strong 2D6 inhibitor but the half-life is not as long as fluoxetine. Paroxetine has mild anticholinergic effects so it can be initially sedating and it can induce its own metabolism. Citalopram was the newest SSRI to reach the U.S. market although it had been used in Europe for years. Citalopram has essentially no clinically significant drug-drug interactions. Many of the SSRI's come in sustained release formulations.

Nefazodone and Fluoxamine are not useful medications in a transplant population because of their strong inhibition of 3A4 isozymes. Many of the transplant medications including cyclosporine are metabolized by this isozyme. Ritalin as mentioned earlier can be helpful in de-energized, depressed medical patients who need appetite stimulation. Cautions to its use are situations where tachycardia or lowered seizure threshold is to be avoided. Bupropion, a dopamine agonist, may be particularly helpful in patients who need a more stimulant effect or with patients who are trying not to smoke cigarettes. Trazadone is probably not an effective antidepressant, but it is very useful in the medical setting for sedation to help with insomnia or anxiety. Venlafaxine is both a serotonin reuptake inhibitor

15

and a norepinephrine reuptake inhibitor. At low dose it is essentially a serotonin reuptake inhibitor. Venlafaxine may be useful as a medication to try when a SSRI fails. Mitazapine is a designer drug that selectively blocks some serotonin subtype receptors that cause nausea, agitation and sexual dysfunction. Mitazapine at lower doses can be sedating and help with quick relief of insomnia.

The antipsychotics used in delirium have been highlighted previously. Haloperidol is still the most widely used antipsychotic in medical settings. The newer atypical antipsychotics such as risperidone, olanzepine and seroquel have shown promise in decreasing extrapyramidal symptoms and treating the negative/blunted affect of psychotic disorders. Risperidone does have extrapyramidal effects at higher doses and reduced clearance with liver disease, olanzepine can cause greater weight gain and sedation with chronic use and seroquel requires multiple dosing and is associated with sedation and postural hypotension. Seroquel is also metabolized by 3A4 and may interact with transplant medications.

Lorazepam and oxzepam are relatively short acting benzodiazepines and alprazolam is intermediate. Clonazepam, diazepam and chlordiazepoxide are useful, longer acting benzodiazepines that should be avoided in patients with poor hepatic function or poor perfusion. Busipirone is a sometimes useful anti-anxiety medication for people with generalized anxiety. It is not useful in the short term as it can take up to 3-4 weeks to show effects. Zolidem is a very useful sedative/hypnotic that has some associated abuse potential.

Mood stabilizers have been found to be useful in a variety of psychiatric conditions. The classic indication is for bipolar or manic-depressive illness. Lithium may be particularly useful for long term stability. Levels must be watched closely in patients with body fluid fluctuations. Dialproex Sodium (Depakote) has become a first line treatment for acute mania and mixed bipolar disorders. Depakote may increase ammonia levels in liver cirrhosis. Neurotonin and lamictal are better tolerated in liver patients. Carbamazepine is a mood stabilizer where liver function tests must be watched closely.

SPECIAL ISSUES IN ORGAN TRANSPLANTATION

LIVER TRANSPLANTATION

The most salient psychosocial issue and controversy in liver transplantation is the transplantation of patients with alcoholic liver cirrhosis. There is slowly emerging data on the most important questions initially asked about liver transplantation in alcoholics. These scientific questions have always existed in the context of the larger questions of ethics, moralization and the reality of public perception. The scientific questions have been about survival, return to drinking and quality of life. The ethical question goes as follows: if we have a scarce resource like donor livers should we allow alcoholics who presumably shoulder a greater responsibility for their illness to receive equal consideration for transplantation? The fact is that alcohol cirrhosis and recently hepatitis C, which in some cases is the consequence of IV drug abuse, are the most common causes for cirrhosis. As the waiting lists rise, it will be important to have data to support the rationale in transplantation individuals who suffer from alcoholism.

First, is the survival of the alcoholic equal to other transplant patients? The answer in both the short and long term is yes. Second, does the quality of life in post-transplant alcoholics look similar to other transplants? Again, the answer is yes. In fact, an otherwise healthy alcoholic recipient can have much less post-transplant morbidity than older or sicker patients. The third question involves the issue of a return to drinking. The data suggests that carefully screened alcoholics have a low return to drinking. Recent data suggest that approximately 13% of alcoholics return to some level of drinking in one year. It is important to note that this is any drinking at all. Most centers can only cite less than a handful of transplanted alcoholics that have run into serious medical consequences from return to drinking. The return to drinking in alcoholic transplants is considerably lower than in non-alcohol related transplanted patients. Three year post transplant data suggests that as time goes on a greater percentage, approximately a third of alcoholics will pick up a drink again. The long term follow up of this cohort is important. The attention now has shifted to how to best select suitable candidates for transplant. The initial approach was to use length of sobriety as a predictor of return to drinking. As mentioned previously, five years of sobriety in general alcoholic populations is correlated with extremely low return to drinking. In the transplant population, most patients have been sober less than five years. One initial study suggested that six months of sobriety might be a reasonable predictor for return to drinking. Although there has been some data that supported that conclusion, the bulk of research has not been able to demonstrate six months as a gold standard. What is probably more important is the severity of alcoholism, the number of failed treatment attempts in the past, other psychiatric co-morbidity including drug use, willingness to enter treatment, follow through and having a sober, supportive home environment. Refer to Figure 15.1 as a suggested algorithm in assessing alcoholics for transplant. One exception may be the poly-substance abuser, which has been shown in some series to be associated with poorer outcomes. With long wait times most programs have a chance to put patients through the test of compliance. There are obviously no guarantees as people can and do return to drinking when they feel well again. However, the reality is that liver transplantation is associated with a low incidence of relapse compared to the 60% relapse in three months in natural circumstances.

KIDNEY

Renal failure can be very devastating to a person's quality of life, even if the renal failure can be managed reasonably well medically with dialysis. For many patients who have dialysis complications or whose quality of life or life span is compromised by dialysis, a renal transplant is a reasonable option. The pre-selection process should screen patients like any other transplant and select individuals who are going to take care of their graft, whether they receive it from a live or cadaver donor. Due to the high numbers of kidney patients on transplant lists, it is often not feasible to do a psychiatric screening of all of them. However, if a patient has been on dialysis, psychosocial assessments can often be obtained from a patient's dialysis unit. Kidney recipients can wait long periods, with little direct contact

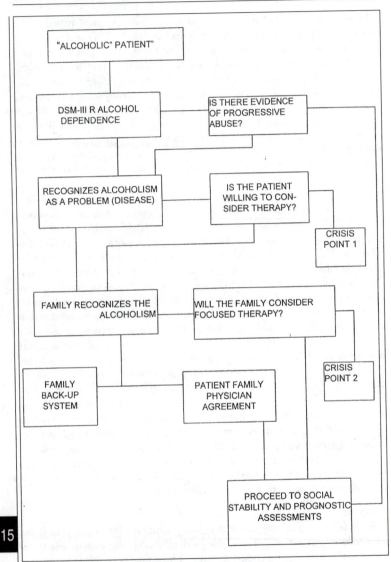

Fig. 15.1A. Pretransplant alcohol-dependence assessment I.

with the transplant team and suddenly get "beeped" for a transplant. This uncertainty of "when" can generate anxiety. The most common problem seen by mental health personnel is post-transplant patients who stop taking their medications and reject their organ. Many times these patients want to be re-listed for another transplant. Reasons for noncompliance range from denial, lack of insight, finances

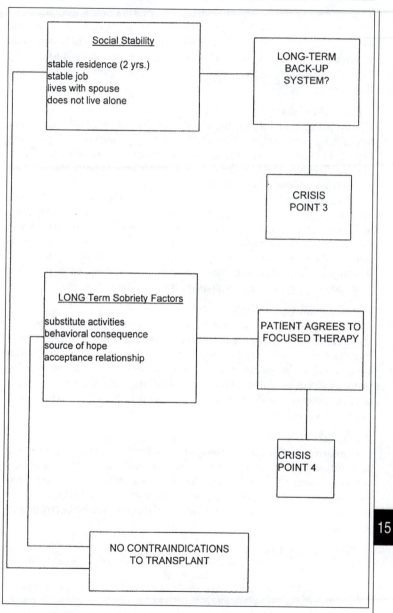

Fig. 15.1B. Pretransplant alcohol-dependence assessment II. From Beresford et al. A Rational Approach to Liver Transplantation Psychosomatics. 31(3):241-254. Reproduced with permission.

(patients need to stay on top of funding sources for transplant medications by contacting their social worker), chaotic lifestyles, personality disorders and other mental problems. Also, these patients may be suffering from medical consequences of their primary disease. Only with very careful evaluation and demonstration by the patient that poor self-care is clearly behind them would one consider re-transplantation.

KIDNEY/PANCREAS

These patients are distinguishable by the fact that they most often have a history of being brittle diabetics from childhood. The dynamics of having an early, unpredictable and often dramatic chronic disease can interfere with normal separation/individuation and emotional development in some patients. Many of these patients have underlying low self-esteem and issues of control with authority figures like parents and physicians. Some are and will continue to have to deal with other end organ damage like retinal hemorrhage and neuropathy. Quality of life series suggest that this group may have a more protracted recovery period before their quality of life changes. Frequently it may be up to a year before patients really see the positive risk/benefit to the surgery. It is important to express hope that the procedure will arrest the disease process, but not to oversell the procedure to the point that patients are not prepared for complications.

HEART, LUNG TRANSPLANTS

Many of these patients have history of depressive disorders, anxiety and substance abuse. Heart and lung transplants also can raise ethical concerns of transplanting patients who may have had knowledge that their alcohol or smoking could contribute to disease. Heart and lung failure patients can have such impaired quality of life and low survival rates that transplant is the only option. The selection process with these patients is conducted with the same general principles as described above.

CONCLUSION

Consideration of psychosocial issues in transplant is crucial to the good care of these patients. There continues to be areas of needed research. The care of these patients requires a true multidisciplinary approach with nurses, doctors, social workers, ethicists, mental health and substance abuse experts. Understanding of the basic psychosocial issues is important for all staff involved in the care of these patients.

SELECTED READINGS

1. Rodrigue, J. R. 2001. Biopsychosocial Perspectives on Transplantation (1st edition). New York: Plenum Publishing Corporation.
2. deVilla VH, Lo C, Chen C. Ethics and rationale of living-donor liver transplantation in Asia. Transplantation, 2003; 75(3):S2-S5.
3. American Psychiatric Association. 1994. Diagnostic and statistical manual of mental disorders, 4th ed. Washington, DC.

Infections in Transplant Recipients

Valentina Stosor

INTRODUCTION

Over the past two decades, significant advances were made in the management of infections occurring after transplantation. Even so, infection remains a leading complication of organ transplantation, and the prevention and management of such infections are an important element of care in transplant recipients. Infections are associated with allograft rejection, and therefore, a key to a successful transplantation is the prevention, diagnosis, and treatment of infectious complications.

In this chapter, the most important infectious disease issues that affect different organ-transplant populations are reviewed, including the prophylaxis of infection after transplantation (Table 16.1).

PRETRANSPLANTATION INFECTIOUS DISEASES EVALUATION

Prevention is, above all, the most important approach to infection in transplant recipients. This begins with a rigorous evaluation to identify previous infections and potential active infectious processes in all candidates before transplantation.

A complete history and physical examination is performed during the pretransplantation screening evaluation. Information regarding past infections is sought, such as childhood illnesses (chickenpox, rubella, measles, and infectious mononucleosis), recurrent sino-pulmonary infections, viral hepatitis, and sexually transmitted diseases. Allergies to antimicrobial agents are documented. Past immunization records are reviewed, and immunizations are administered or updated, if necessary. These vaccines include the inactivated polio, tetanus-diphtheria, influenza, pneumococcal, varicella (if nonimmune), hepatitis B, and hepatitis A (if nonimmune), *Haemophilus influenzae* type B (pediatric patients), and measles-mumps-rubella (pediatric patients). Dietary habits are obtained, including drinking water source and consumption of raw or undercooked meat, unpasteurized milk products, and seafood.

Epidemiological exposures are identified through social, sexual, recreational, occupational, and pet and wild animal exposure histories. Certain workplace settings, such as healthcare facilities, prisons, and homeless shelters, increase the risk for exposure to infectious agents, especially tuberculosis. Residence or travel exposure to certain agents can identify candidates that are at risk for reactivation of infection after transplantation. Examples of these pathogens include: *Histoplasma capsulatum* (Ohio and Mississippi river valleys), *Coccidiodes immitis* (Southwest-

16

Organ Transplantation, 2nd edition, edited by Frank P. Stuart, Michael M. Abecassis and Dixon B. Kaufman. ©2003 Landes Bioscience.

Table 16.1. Prophylaxis of infection in transplant recipients

Type of Anti-Infective Prophylaxis	Allograft Type				
	Liver	Kidney	Pancreas	Heart	Lung
Bacteria	Oral selective bowel decontamination	TMP/SMX or quinolones for UTI prophylaxis	Peri-operative antibiotics		
Fungi	Oral selective bowel decontamination High risk only: consider fluconazole	Fluconazole for candiduria fluconazole	High risk only: consider fluconazole	------ amphotericin B	Consider aerosolized Consider itraconazole if *Aspergillus* isolated from sputum
PCP			TMP/SMX for six months		
Toxoplasma			TMP/SMX for six months		
CMV	For CMV D+/R-: oral GCV for three months				For CMV D+/R: IV GCV for 4 weeks, then oral GCV for 3 months
HSV			Acyclovir for 30 days		
M. tuberculosis		Consider INH for + PPD with epidemiological exposures			

16

Table 16.2. Routine infection evaluation of organ transplant candidates

All Candidates
- Serologic screening for syphilis (VDRL or RPR)
- Serologic testing for CMV
- Serologic testing for HSV

- Serologic testing for VZV
- Serologic testing for EBV
- Serologic testing for *Toxoplasma*
- Serologic testing for hepatitis A, B, and C
- Tuberculin skin testing (PPD)
- Chest radiograph
- Urinalysis and urine culture

Candidates with Specific Exposure History
- Serologic screening for *Histoplasma capsulatum*
- Serologic screening for *Coccidioides immitis*
- Serologic screening for *Strongyloides stercoralis* (or 3 fecal specimens)

---

ern United States, Northern Mexico), and *Strongyloides stercoralis* (Southeast Asia, Puerto Rico, certain areas of rural Kentucky, Tennessee and Louisiana).

Pre-transplantation infection testing (Table 16.2) includes a complete blood count (CBC) with differential, blood chemistries, urinalysis and urine culture, a tuberculin skin test (PPD), and a chest radiograph (CXR). Serologic evaluation for toxoplasmosis (particularly for heart transplant), syphilis (RPR or VDRL), cytomegalovirus (CMV), herpes simplex viruses (HSV), and varicella zoster virus (VZV) are routinely obtained. Additional screening for viruses includes serologic testing for human immunodeficiency virus (HIV), hepatitis A virus (HAV), hepatitis B virus (HBV), and hepatitis C virus (HCV). All candidates with the appropriate endemic exposures should have serologies for *S. stercoralis* and the endemic mycoses, *H. capsulatum* and *C. immitis*. In lung transplant candidates, especially those with cystic fibrosis, sputum cultures are important for the detection of colonization of the respiratory tract by pathogens such as *Aspergillus* species and *Burkholderia cepacia*. These organisms pose a risk for the development of significant infections after transplantation as the recipient becomes immunosuppressed. CT imaging of the paranasal sinuses should be considered for patients with cystic fibrosis and others with a history of recurrent or symptomatic sinus infections.

TIMING OF POST-TRANSPLANTATION INFECTIONS

Susceptibility of a host to the development of an infection after transplantation varies according to specific environmental circumstances, surgical factors, and the level of immunosuppression at any given time. Transplant-related infections are categorized according to the relative risk periods that correspond to the evolution of immune deficiencies and technical factors that render the recipient vulnerable to infection. While it is important to entertain infectious etiologies within this context, the transplant physician should consider a broader differential diagnosis of infection in the transplant recipient, as atypical pathogens and presentations of disease can occur in these hosts throughout the post-transplant course.

16

First Month Post-Transplantation

At this time the degree of immunosuppression is not high enough to render patients susceptible to opportunistic pathogens; therefore, the most frequent infections are related to surgical and nosocomial complications such as bacterial and candidal wound infections, urinary tract infections, nosocomial pneumonias, and central venous catheter-associated bacteremias and fungemias. When dealing with infected recipients in this time period, consideration is given to the type of organ transplanted and the anatomical details of the surgical intervention, as infection is often related to such factors. Patients may have infection related to prolonged stays in the intensive care unit including *Clostridium difficile*-associated diarrhea or infections with antimicrobial-resistant bacteria such as vancomycin-resistant enterococci. HSV is the only common viral pathogen of this period, manifesting as stomatitis in HSV-seropositive patients.

The Early Post-Transplant Period

This period extends from the second to the sixth month after transplantation. During this time, the level of immunosuppression is most intense. Patients are at high risk of developing serious opportunistic infections including CMV infection and disease, *Pneumocystis carinii* pneumonitis (PCP), invasive aspergillosis, disseminated toxoplasmosis, dermatomal and disseminated VZV infection, and certain bacterial infections such as listeriosis. Pre-existing infectious agents that can also reactivate during this period include *Mycobacterium tuberculosis* and the endemic mycoses.

The Late Post-Transplant Period

Beyond six months after transplantation, recipients that have not had allograft rejection episodes requiring increase immunosuppression are at risk for the usual community-acquired infections; however, certain opportunistic infections can still manifest at this late time. Examples include tuberculosis, cryptococcosis, nocardiosis, and herpes zoster. Recurrence of HBV and HCV infections is also seen.

The time frame outlined above is "set back" whenever allograft rejection is treated and immunosuppression is augmented with anti-CD3 monoclonal antibodies (OKT3), anti-lymphocyte globulin (ALG), steroid boluses, or radiation. Recipients with chronic allograft rejection also remain at higher risk for the development of opportunistic infections.

BACTERIAL INFECTIONS

Bacterial infections occur in 33-68% of liver transplants, 21-30% of heart transplants, 54% lung transplants, 35% of pancreas transplants, and 47% of kidney transplants. The types of infections encountered and the bacterial etiologies differ depending on the transplanted organ and the surgical technique used. Below, bacterial infections are reviewed in the context of the transplanted organ.

16

LIVER TRANSPLANTATION

Overview

The most common bacterial infections following liver transplantation are intra-abdominal and surgical wound infections. Cholangitis, abscesses, and intra-vascular device-related bacteremias are other frequent complications. The main risk factors associated with intra-abdominal infections are prolonged surgical time, high transfusion requirements during surgery, re-operations, early rejection, re-transplantation, and CMV infection.

The type of biliary anastomosis influences the risk of post-transplant infections. The choledochostomy is associated with less infection risk because the native sphincter of Oddi is maintained. The presence of a Roux-en-y choledocojejunostomy, required in patients with abnormalities of the extra-hepatic biliary system, is associated with a higher incidence of post-operative cholangitis as well as infection following liver biopsy and cholangiography procedures after transplant. This is likely caused by reflux of intestinal contents and microbial flora into the biliary system.

The specific bacterial etiologies of infection after liver transplantation depend on the gastrointestinal flora of the recipients. In the absence of exposure to anti-microbial agents, the enteric gram-negative bacilli such as coliforms, the *Enterobacteriaceae*, and occasionally, *Pseudomonas* species, are the predominant organisms causing infection, along with enterococci, staphylococci, and the anaerobes. Infections with vancomycin-resistant enterococci (VRE) are a significant problem in some transplant centers.

Clinical Presentation

Recipients with wound infections or intra-abdominal abscesses may present with any of the following: fever, abdominal pain, wound dehiscence, and purulent wound drainage. On physical examination, pain may be elicited by palpation, and guarding and rebound may be present. Laboratory abnormalities may include leukocytosis with a left shift. However, the absence of leukocytosis or fever does not exclude the presence of infection in the transplant recipient. The definitive diagnosis is achieved by imaging studies such as computerized tomography (CT), ultrasound (US), or magnetic resonance imaging (MRI).

Cholangitis usually manifests with fever and right upper quadrant pain, with tenderness and rebound on palpation of the abdomen. Hyperbilirubinemia, elevated transaminases and alkaline phosphatase levels, as well as leukocytosis, may be found on laboratory testing.

Treatment

Fluid collections are aspirated and cultured. Optimal treatment requires specific pathogen identification by culture of purulent secretions, fluid collections, and blood and antimicrobial susceptibility testing. If an abscess is confirmed, drainage, whether surgical or CT/US-guided, is necessary to achieve an appropriate therapeutic response. Antimicrobial regimens are tailored to the specific pathogen isolated, and empiric coverage for known colonizing flora is appropriate in

16

seriously ill patients while awaiting final culture results. The use of cephalosporins, fluoroquinolones, carbapenems and beta-lactam+beta-lactamase inhibitor combinations are appropriate choices for complicated intra-abdominal and surgical wound infections.

Cholangitis is managed medically with intravenous antibiotics if there is adequate biliary flow. When obstruction of the biliary tree is present, a therapeutic procedure, such as ERCP with dilatation, is performed.

Again, third- and sometimes second-generation cephalosporins, as well as the other antibiotics mentioned above, are reasonable antimicrobial choices depending on the culture results.

Prevention

Adequate surgical technique is one of the most important factors that prevent infectious complications. Because of the high morbidity and mortality of gram-negative infections, the eradication of oral and gut flora may be desirable. This is accomplished with the use of oral selective bowel decontamination (OSBD), consisting of nonabsorbable antibiotics that eliminate gram-negative aerobic bacteria and fungi. These regimens spare gram-positive and anaerobic organisms that have an antagonistic effect on the growth of gram-negative pathogens. OSBD is most effective when initiated one week prior to transplant surgery and continued post-operatively for one to three weeks. Whereas the overall incidence of infection with OSBD is not different than without it, there is a substantial decrease in the incidence of gram-negative bacteremias, which carry a high mortality as mentioned above. After OSBD, most infections in patients are secondary to gram-positive organisms. Some centers have noted the emergence of infections caused by resistant gram-positive organisms such as VRE; however, the exact relationship to OSBD is unclear.

Peri-operative prophylaxis with an extended-spectrum cephalosporin is routinely administered to prevent surgical wound infections. In general, antibiotics are continued for 24 to 48 hours post-operatively. It is recommended that liver recipients also receive antimicrobial prophylaxis prior to post-transplant cholangiograms, liver biopsies, and any other manipulations of the biliary tract.

RENAL TRANSPLANTATION

Overview

The most frequent bacterial complications of renal transplantation arise in the urinary tract. Predisposing factors include renal insufficiency, decreased urine flow through the urinary epithelium, prolonged bladder catheterization, and underlying medical conditions such as diabetes mellitus. The bacteria implicated in such infections are the same as for the general population and include enteric gram-negative bacilli, enterococci, staphylococci, and *P. aeruginosa*.

Surgical wound infections caused by gram-positive cocci and gram-negative bacilli also occur after renal transplantation. Finally, line-related bacteremias can complicate the post-operative course.

16

Clinical Presentation

Urinary tract infections (UTI) may manifest as acute pyelonephritis and systemic illness with high fever, pain around the graft site, and laboratory data indicative of leukocytosis and active urinary sediment. Alternatively, renal allograft recipients with UTIs can be asymptomatic and present without pyuria. Because of this, a high index of suspicion is required, and routine surveillance urine cultures are often performed after transplantation. In febrile patients, blood cultures are obtained.

Treatment

Choice of therapy for UTIs depends on specific antimicrobial susceptibilities of the bacteria isolated from urine and blood cultures. Fluoroquinolones are widely used in this population; cephalosporins are alternative agents. Anaerobic organisms are rarely involved in these infections and are not routinely covered. For infections caused by coagulase-negative staphylococci or by ampicillin-resistant enterococci, vancomycin is the antimicrobial agent of choice. Length of treatment depends upon the severity of the infection, with two weeks or longer duration of therapy for pyelonephritis. Recurrent infections of the urinary tract prompt further investigation with imaging studies to exclude anatomic abnormalities and obstruction.

Surgical wound infections require appropriate debridement and adjunctive antimicrobial therapy. Empiric coverage is directed at gram-positive cocci and gram-negative bacilli until deep culture data is available to guide antimicrobial therapy.

Prevention

The use of trimethoprim/sulfamethoxazole (TMP/SMX) reduces the incidence of UTIs and blood stream infections after renal transplantation. Such an approach offers additional protection against opportunistic pathogens such as *P. carinii*, *Listeria monocytogenes*, and *Nocardia* species. For sulfa-sensitive recipients, fluoroquinolones are alternative prophylaxis agents. Typically, prophylaxis is continued for six months after transplantation.

To prevent surgical wound infections, a cephalosporin antibiotic is administered peri-operatively and continued for less than 24 hours.

HEART TRANSPLANTATION

Overview

The main bacterial infection complicating heart transplantation is nosocomially-acquired ventilator-associated pneumonia caused by gram-negative bacteria, including *P. aeruginosa*, *Klebsiella pneumoniae*, and the other *Enterobacteriaceae*. Of special interest are wound infections, especially mid-line sternotomy infections caused by *Staphylococcus aureus* and coagulase-negative staphylococci. Mediastinitis and line-related bacteremias also occur after heart transplantation. Endocarditis, related to valvular lesions caused by repeated endomyocardial biopsies and colonization by circulating bacteria, is occasionally reported after transplantation.

16

Clinical Presentation

Pneumonia complicating heart transplantation presents as persistent respiratory failure and inability to wean from mechanical ventilation. In addition, fever may be present, and respiratory secretions are purulent. Radiographic findings that demonstrate consolidative changes aide in the diagnosis and follow-up of patients. Respiratory tract cultures are collected routinely in all mechanically-ventilated patients to determine optimal antimicrobial therapy in the event of an established infection.

Mediastinitis is a serious complication of heart transplantation, presenting as fever, leukocytosis, and signs of systemic toxicity. CT imaging establishes the diagnosis and extent of the infection.

Sternal wound infections may present early in the post-transplantation course as poor healing or dehiscence of the wound or later in the transplant course with sinus tract formation and purulent discharge. CT or nuclear medicine imaging studies may help in defining this infectious process.

Treatment

Antimicrobial therapy for ventilator-associated pneumonia is guided by the results of respiratory tract cultures. Sternal wound infections and mediastinitis require surgical debridement and adjunctive antimicrobial therapy that provides coverage against gram-positive pathogens.

Prevention

Efforts to prevent nosocomial pneumonia include aggressive attempts to wean patients from mechanical ventilation and pulmonary hygiene measures in extubated patients. Peri-operative prophylaxis targeted at gram-positive bacteria may prevent sternal wound infection. Heart transplant recipients are given standard antimicrobial prophylaxis for endocarditis when undergoing any high-risk procedures.

LUNG TRANSPLANTATION

Overview

Pneumonia is a very common complication of lung transplantation, with an overall prevalence of 60% in lung recipients. The heightened susceptibility to lung infection stems from several factors related to the allograft such as impaired cough reflex of the lung allograft, poor mucociliary clearance, ischemia to the explanted lung, abnormal lymphatic drainage, diffuse reperfusion injury, and airway inflammation caused by rejection and resulting in bacterial colonization. After single lung transplantation, the allograft may become infected from the remaining native lung. Gram-negative bacteria, including the *Enterobacteriaceae* and *P. aeruginosa*, account for the majority of post-transplant pneumonias. Other important pathogens are *S. aureus*, *H. influenzae*, and *Streptococcus pneumoniae*. *B. cepacia* colonization is associated with high morbidity and mortality after transplantation for cystic fibrosis.

Mediastinitis and sternal wound infections are other important post-operative infections in lung transplant recipients. The most serious complications after lung transplantation are leakage or dehiscence of the bronchial or tracheal anastomosis.

Clinical Presentation

As with heart transplantation, nosocomial pneumonia often presents as persistent respiratory failure requiring ongoing mechanical ventilatory support and the radiographic finding of consolidation. Fever and leukocytosis may be absent. Cultures of lower respiratory tract secretions before and after transplantation are essential for managing these infections. Post-surgical wound infections and mediastinitis present in a manner as described for heart transplant recipients.

Treatment

The treatment of pneumonia is guided by the antimicrobial susceptibilities of the pathogens isolated from respiratory tract specimens. Aggressive therapy is warranted, and double antimicrobial regimens are administered to patients infected with *P. aeruginosa*, *B. cepacia*, or multi-drug resistant *Enterobacteriaceae*. Post-surgical wound infections and mediastinitis require debridement and antimicrobial therapy.

Prevention

Because of the high risk of infection, lung transplant recipients are given antimicrobial prophylaxis determined by cultures of respiratory tract secretions of both donors and recipients. Very aggressive prophylaxis regimens including two or three antibiotics and inhaled aminoglycosides are considered for patients colonized with *B. cepacia* or other multiply-resistant gram-negative bacteria. Peri-operative prophylaxis is directed at gram-positive bacteria to prevent post-surgical wound infections. After transplantation for cystic fibrosis, antibiotic prophylaxis is continued for 14 days post-transplantation or until purulent respiratory secretions resolve. In addition, some authorities advocate routine sinus surgery in cystic fibrosis patients prior to transplant in order to minimize respiratory infections after transplant.

PANCREATIC TRANSPLANTATION

The most common infectious complications after pancreatic transplantation are surgical wound infections and intra-abdominal abscesses. UTIs may occur in recipients who have urinary drainage of exocrine secretions of the allograft because of bacterial overgrowth in the bladder. Other infections have been described after transplant, including abdominal wall cellulitis, peri-pancreatic abscesses, and peritonitis. Important pathogens include gram-positive cocci, followed by gram-negative and anaerobic bacteria.

16

These infections are managed in a similar manner as described for liver and renal transplant recipients. There are no additional antimicrobial prophylaxis guidelines specific to pancreas transplantation, Trimethaprim-sulfamethyodeis (TMP/SMX) is used to prevent UTI, and standard peri-operative antibiotics are employed.

OTHER IMPORTANT BACTERIAL INFECTIONS

Nocardia species

The incidence of *Nocardia* infections after transplant ranges from 0.7 to 3%, and *N. asteroides* is the most commonly implicated species in most reported series. This infection can present years after organ transplantation. Lung involvement, as evidenced by pneumonia, pulmonary nodules, and abscesses, is common. Less commonly, patients may have brain abscesses, meningitis, and skin involvement. Beaded, branching gram-positive bacilli can be detected on gram-stain of lower respiratory tract specimens and abscess material. The diagnosis is confirmed by the isolation of *Nocardia* species from culture. All patients require imaging of the central nervous system to exclude the possibility of brain abscesses.

The sulfonamides, such as TMP/SMX, are the preferred treatment for nocardiosis. Alternative effective agents include minocycline, the carbapenems, ceftriaxone, cefotaxime, ciprofloxacin, and most recently, linezolid. TMP/SMX, in doses typical for PCP prophylaxis, offers some protection against *Nocardia*.

Listeria monocytogenes

Infections caused by *L. monocytogenes* tend to complicate the early post-transplant course, from weeks to the first two months after transplant; however, listeriosis can also present years after transplant. Infection results from the ingestion of contaminated food products. Listeriosis most commonly manifests as central nervous system involvement with meningitis, meningo-encephalitis, or encephalitis, followed by primary bacteremia. More unusual manifestations include pneumonia, arthritis, endophthalmitis, endocarditis, peritonitis, myocarditis, and hepatitis. Patients may present with fever, headache, meningismus, and altered mental status; focal neurological findings and seizures can occur. The cerebral spinal fluid (CSF) analysis reveals polymorphonuclear pleocytosis and hypoglycorrhachia, and the gram stain of the fluid may or may not reveal gram-positive bacilli. The diagnosis is made by the isolation of *L. monocytogenes* from culture of blood and CSF or other sterile site. The treatment of choice is ampicillin in combination with an aminoglycoside, and for patients with penicillin hypersensitivity, TMP/SMX is an alternative agent. TMP/SMX prophylaxis has a protective effect against *Listeria* infections.

Legionella Species

Legionellosis, most commonly presenting as pneumonia, can occur at any time after transplantation as a nosocomial or community-acquired infection. It often occurs early in the post-transplant course or co-incidentally with steroid therapy or allograft rejection. The presenting symptoms include fever, malaise, chills, dyspnea, chest pain, nonproductive cough, and diarrhea. Radiographic findings include unilateral or bilateral dense pulmonary infiltrates that can progress to cavitation. The diagnosis is made by detecting *Legionella* antigen in urine, direct fluorescent antibody staining of respiratory secretions or tissue specimens, and culturing lower respiratory secretions on supplemented media. Treatment options include the fluoroquinolones, erythromycin, and the newer macrolides. Rifampin

16

can be added to quinolone or macrolide therapy; however, there are significant drug interactions with the calcineurin inhibitors and prednisone.

MYCOBACTERIAL INFECTIONS

MYCOBACTERIUM TUBERCULOSIS

Overview

Tuberculosis is considered a serious complication of organ transplantation, with associated mortality approaching 30%. The worldwide incidence of *M. tuberculosis* infections in recipients of organ transplantation is 0.35% to 15%. Transplant recipients are at risk for both primary and reactivation infection, and disseminated infection occurs more frequently than in immunocompetent hosts. The diagnosis and treatment of tuberculosis in this population is often delayed because of atypical and extra-pulmonary presentations of this disease. This infection is rarely transmitted by the allograft.

Clinical Presentation

Transplant recipients may present with cavitary pulmonary, genitourinary, intestinal, cutaneous, central nervous system, bone, or disseminated disease. Because these many different presentations, symptoms and signs depend on the site(s) of involvement. Usually, patients have fever accompanied by malaise, night sweats, and weight loss.

The diagnosis of tuberculosis requires a high index of suspicion. The tuberculin skin test is positive in approximately 25% to 33% of infected transplant recipients. Tuberculosis needs to be excluded as a diagnosis in any transplant recipient with pulmonary infiltrates. In cases of suspected tuberculosis, acid fast smears and special cultures are performed on the appropriate clinical specimens. In the event that the initial expectorated sputum and gastric washings are unrevealing, bronchoalveolar lavage or lung biopsy are required to make the diagnosis. Histopathology of biopsy specimens can demonstrate granulomas. All *M. tuberculosis* isolates require susceptibility testing to exclude resistance to anti-tuberculosis agents.

Treatment

Transplant patients with active tuberculosis require nine to twelve months of therapy with at least two bactericidal agents to which the isolate is susceptible, preferably with combinations of isoniazid, rifampin, and pyrazinamide. Several regimens are available and have been extensively reviewed in the literature. The treatment of multi-drug resistant tuberculosis poses a particular challenge, and second-line agents must be used. However, outcomes in other immunosuppressed populations have not been very satisfactory, and surgical approaches to eliminate disease are undertaken frequently.

Anti-tuberculous agents have potential toxicities and interactions with immunosuppressive agents: isoniazid is hepatotoxic, rifampin is hepatotoxic, streptomycin is ototoxic and nephrotoxic, and ethambutol may cause optic neuritis.

16

Rifampin is a potent inducer of hepatic metabolic enzymes. During rifampin therapy, careful monitoring of immunosuppressant drug levels (such as cyclosporine and tacrolimus) is recommended because rifampin therapy is associated with subtherapeutic immunosuppressive drug levels and allograft rejection. Rifabutin may be an alternative agent to rifampin as it has a lesser effect on drug levels.

Prevention

Although the majority of transplant candidates are anergic, tuberculin skin testing (PPD) may identify candidates at risk for reactivation tuberculosis and, therefore, is performed during the pre-transplantation evaluation. Chemoprophylaxis with isoniazid for nine to twelve months is recommended in the following situations: recipients with a positive PPD prior to transplantation, radiographic evidence of old active tuberculosis and no prior prophylaxis, a history of inadequately treated tuberculosis, or close contact to a patient with active tuberculosis; recipients of allografts from a known infected or inadequately treated donor; or recipients with a newly positive PPD (recent converters). Chemoprophylaxis may be considered for recipients from highly endemic areas.

NONTUBERCULOUS MYCOBACTERIA

Overview

The mycobacteria other than tuberculosis are occasional causes of post-transplant infections. Such infections typically occur late (years) after transplantation. These organisms are ubiquitous in the environment and include the following pathogens: *M. kansasii*, *M. avium* complex, *M. fortuitum*, *M. chelonae*, *M. marinum*, and *M. abscessus*. *M. marinum* infections are associated with exposure to fresh- or salt-water aquariums or swimming pools.

Clinical Presentation

The clinical presentations can vary depending on the site(s) of infection. Frequently, patients present with chronic skin lesions or joint and tendon involvement in the absence of systemic manifestations. Pulmonary and disseminated infections also occur including fever, adenopathy, and intestinal compromise.

The diagnosis relies on biopsy and culture of suspicious lesions. Stains for acid-fast bacilli and special cultures are required on all pathology specimens. Unlike *M. tuberculosis*, granuloma formation is not common.

Treatment

Specific treatment recommendations for atypical mycobacterial infections are not available because of their infrequent nature. Treatment is guided by the results of special susceptibility testing of any culture isolates.

Agents effective against *M. avium* complex include clarithromycin or azithromycin, rifabutin, and the fluoroquinolones. *M. fortuitum* infections are typically treated by surgical debridement and adjunctive antimicrobial therapy with one or more of the following agents: cefoxitin, amikacin, imipenem, TMP/

16

SMX, fluoroquinolones, tetracyclines, and azithromycin or clarithromycin. Agents active against *M. kansasii* include isoniazid, rifampin, and ethambutol. No specific recommendations are available for the prevention of atypical mycobacterial infections in the transplant recipient.

FUNGAL INFECTIONS
Fungal infections are a major cause of morbidity and mortality in transplant recipients. The reported incidence ranges from 5% in renal allograft recipients to almost 50% in liver recipients. While most of these infections occur in the first six months after transplantation, fungal infections are occasionally seen several years post-transplantation. The reported mortality for such infections exceeds 30%. Recent advances have been made in the area of antifungal therapy, and the impact of these new therapies on the morbidity and mortality of fungal infections in transplant recipients is currently unknown.

CANDIDA SPECIES

Overview
The most common source of *Candida* infection is gut translocation or, alternatively, intravascular catheters. The risk factors associated with invasive fungal infection include: the use of high-dose corticosteroids, the administration of broad-spectrum antimicrobials, episodes of allograft rejection requiring increased immunosuppression, and allograft dysfunction. In the case of liver transplantation, the presence of a Roux-en-Y choledocojejunostomy, CMV infection, the administration of OKT3, and re-transplantation are additional risk factors (Table 16.3).

Recipients of renal allografts are at risk for UTIs with these organisms because of underlying medical conditions, such as diabetes mellitus, and the use of indwelling urinary drainage catheters. Pancreas-kidney recipients also are at additional risk because urinary pH changes associated with exocrine secretion drainage favor bladder colonization with *Candida*.

Clinical Presentation
Candida infections can present in multiple ways including intravascular catheter infections with sepsis and fever, intra-abdominal abscesses, urinary tract infections. Mediastinitis can complicate heart and lung transplantation.

The diagnosis is made by isolating *Candida* species from culture of appropriate clinical specimens. In the setting of documented or suspected candidemia, fundo-

Table 16.3. Risk factors associated with invasive fungal infections

- Receipt of high-dose corticosteroids
- Receipt of broad-spectrum antimicrobial agents
- Multiple rejection episodes requiring heightened immunosuppression
- Allograft dysfunction
- Concomitant infection with immunosuppressive viruses (namely, CMV)

16

scopic eye examination may reveal endophthalmitis or lesions suggestive of septic emboli. The role of surveillance cultures for the diagnosis of fungal infections is unknown, as many patients colonized with *Candida* never demonstrate clinical infection.

Treatment

Most *Candida* species are susceptible to amphotericin B and its lipid-based preparations. Azoles, such as fluconazole and itraconazole, are alternatives for infections caused by *C. albicans*; however, *C. krusei* and *C. glabrata* demonstrate significant resistance to these agents. Lipid-based amphotericin B preparations are less nephrotoxic as compared to conventional amphotericin B. This benefit is offset by greater expense and less clinical experience with their use. New antifungal agents such as caspofungin, an echinocandin, and voriconazole, a triazole, have activity against *Candida* species, but extensive clinical experience is lacking. Potential drug interactions exist for both the azoles and the echinocandins with the immunosuppressant agents.

Prevention

For liver transplant recipients, oral bowel decontamination with nystatin decreases the rate of gut colonization with *Candida* species, but its benefit in reducing post-operative fungal infections is unclear. Although fluconazole has demonstrated efficacy in the prevention of *Candida* infection after liver transplantation, its universal use is discouraged because of the potential for emergence of azole-resistant fungi. Fluconazole prophylaxis is acceptable for high risk liver allograft recipients, especially in the setting of repeated surgeries, prolonged operation time, renal failure, and high intra-operative transfusion requirement. Fluconazole courses beyond four weeks are not warranted. In kidney and pancreas-kidney recipients, pre-emptive treatment of asymptomatic candiduria may be indicated. There is anecdotal efficacy reported with fluconazole prophylaxis after pancreas transplantation, and it may be considered for high risk recipients such as those with an enteric drainage procedure, pancreas after kidney transplantation, peritoneal dialysis before transplantation, reperfusion pancreatitis, and re-transplantation.

ASPERGILLUS SPECIES

Overview

Invasive aspergillosis is one of the most devastating infectious complications of organ transplantation. The incidence of this infection ranges between 1-2% in liver and kidney allograft recipients, 5% heart recipients, and up to 15% of lung recipients. The portal of entry in the vast majority of cases is the respiratory tract through environmental exposures. Once in lung tissue, *Aspergillus* causes ulceration, necrosis, and tissue and blood vessel invasion. Once angioinvasive, dissemination to distant sites occurs. While, historically, most infections present in the first three months following transplantation, recent trends suggest that late onset infections are occurring more frequently than in the past. Risk factors for invasive disease include prolonged operation time, renal failure, neutropenia, CMV infection, and heavy immunosuppression, especially high dose corticosteroids and OKT3. Additional risk factors for liver recipients include allograft dysfunction, fulminant hepatic failure before transplantation, and re-transplantation. For lung recipients, airway specimen cultures positive for *Aspergillus* and obliterative bronchiolitis are risk factors.

Clinical Presentation

The clinical presentation depends on the site(s) of infection. Respiratory symptoms such as dyspnea, cough, pleuritic chest pain, and fever predominate. Hemoptysis is a sign of invasive disease. *Aspergillus* disseminates to the brain, liver, spleen, kidneys, heart, blood vessels, bone, joints, and gastrointestinal tract.

The diagnosis of invasive infection with *Aspergillus* relies on the isolation of the organism in the appropriate setting; histopathology of biopsy specimens is required to determine invasiveness when this mold is recovered from respiratory specimens. CT imaging is helpful in evaluating lung disease, and CNS imaging by CT and MR will delineate brain abscesses. The isolation of *Aspergillus* species from the respiratory secretions of any transplant recipient requires a prompt and thorough investigation to exclude invasive infection; this is especially true for lung allograft recipients.

Treatment

Recent studies have demonstrated that voriconazole has greater efficacy than amphotericin B for the treatment of invasive aspergillosis. This offers some promise given that invasive aspergillosis carries an 80% to 100% mortality rate, even with the use of amphotericin B. Caspofungin is also approved for the treatment of refractory aspergillosis. Extensive clinical experience with these new agents is lacking, and the role of combination therapy for invasive mold infections has not been studied.

Prevention

There are no firm recommendations for prophylaxis against *Aspergillus infections* in transplant recipients. Some advocate the use liposomal amphotericin B in select high risk liver recipients; however, this approach is not without expense or potential significant toxicity. In lung and heart-lung recipients, inhaled amphotericin B may reduce the frequency of *Aspergillus* infections. Some centers advocate the use of oral itraconazole for lung recipients with demonstrated airway colonization by *Aspergillus* species.

ZYGOMYCOSIS

Infections caused by molds, such as *Rhizopus* species, *Mucor* species, and *Absidia* species, are reported in up to 9% of transplant recipients. Exposure occurs as a result of inhalation or cutaneous inoculation of spores. Risk factors for such infections include corticosteroid therapy, metabolic abnormalities such as diabetic ketoacidosis, or deferoxamine therapy. Typical clinical presentations include rhinocerebral infection, nodular or cavitary pulmonary disease, gastrointestinal

16

involvement, skin and soft tissue infection, and disseminated disease. The diagnosis is made by biopsy of involved areas, both by histopathology and culture. Amphotericin B remains the therapy of choice, although surgical debridement is the optimal therapy to control these, often fatal, infections.

CRYPTOCOCCUS NEOFORMANS

Overview
Cryptococcus neoformans, a fungus present in soil and bird-droppings, is acquired by inhalation. Infections caused by *C. neoformans* occur throughout the post-transplant course, and only approximately one half of cases occur within the first year after transplantation.

Clinical Presentation
This fungus has a strong predilection for the central nervous system, causing meningitis and, occasionally, brain abscesses. Patients present subacutely with low-grade temperature, headache, and altered mental status. Signs of frank meningismus are usually absent. Alternate presentations include pulmonary disease with pneumonitis, nodules, and lung abscess; soft tissue, skin or joint infection; and fever of unknown origin.

The diagnosis of cryptococcosis is made by the detection of antigen in serum and CSF. Examination of the CSF is required in any transplant recipient who presents with cryptococcal infection from any site. With meningitis, the CSF analysis demonstrates lymphocytic pleocytosis, elevated protein, and low glucose; the opening pressure is elevated. Fungal cultures of CSF and blood should be performed.

Treatment
Amphotericin B is the treatment of choice for cryptococcal infection. 5-flucytosine may be added to amphotericin B for the first phase of treatment. Fluconazole has been used to treat meningitis in other patient populations but is not well studied in transplant recipients.

The total duration of treatment for cryptococcosis is a minimum of eight to ten weeks. Meningitis is initially treated with amphotericin B for the first two weeks followed by fluconazole for the remainder of therapy. Important adjunctive measures include repeated lumbar punctures to relieve elevated intracranial pressure, repeat analysis of CSF to monitor response to therapy, and periodic monitoring of the serum cryptococcal antigen to detect relapses.

Prevention
Primary prophylaxis directed against *C. neoformans* is not advocated. It remains unclear whether secondary prophylaxis or suppression with fluconazole is necessary in transplant recipients who have completed treatment for cryptococcal infection.

16

PNEUMOCYSTIS CARINII

Overview

Infections caused by *P. carinii* occur during the first six months after transplantation in organ recipients who are not receiving prophylaxis. With appropriate prophylaxis, this infection has been effectively eliminated in transplant recipients.

Clinical Presentation

Patients present subacutely with fever, dyspnea, and nonproductive cough. The classic radiographic finding is bilateral diffuse interstitial and alveolar infiltrates. The laboratory evaluation reveals elevated lactate dehydrogenase (LDH) levels and hypoxemia. The definitive diagnosis is made by direct staining of induced sputum, bronchoalveolar lavage, or lung tissue specimens with fluorescent monoclonal antibodies, methenamine silver, calcofluor white, or Wright-Giemsa stain. Co-infection with CMV can occur and requires investigation.

Treatment

Pneumonitis caused by *P. carinii* is treated with high-dose TMP/SMX, 15 to 20 mg/kg/day, for a duration of 21 days. In cases associated with severe hypoxemia, adjunctive corticosteroid therapy is warranted. For patients intolerant of TMP/SMX, pentamidine, clindamycin plus primaquine, and atovaquone are alternative treatment options.

Prevention

Low-dose TMP/SMX, one single or double strength tablet daily, administered for six months after transplantation has effectively reduced the incidence of post-transplant PCP. Alternative regimens are dapsone, aerosolized pentamidine, and atovaquone. Patients with ongoing significant immunosuppression, allograft rejection, and allograft dysfunction require a longer period of prophylaxis.

ENDEMIC MYCOSES

The endemic mycotic pathogens include *H. capsulatum*, *C. immitis*, *Blastomyces dermatitidis*, and *Paracoccidiodes brasiliensis*. These infections occur at any point after transplantation as a result of primary acquisition or reactivation of disease. It is unclear whether blastomycosis or paracoccidiomycosis occur at any greater frequency in transplant recipients than in the general population.

Organ recipients who reside or who have traveled to the southwestern United States or northern Mexico, and certain areas of Latin America are at risk for coccidioidomycosis. Common clinical presentations include disseminated and pulmonary infection. Patients present with nonspecific symptoms such as fever, night sweats, malaise, cough, and dyspnea. Chest radiographic findings include interstitial, alveolar, nodular, or lobar infiltrates; a miliary pattern; cavitary lesions; and hilar adenopathy. Other potential sites of infection include the blood, skin, brain, liver, urine, bone and joints, and muscle, including myocardium. The diagnosis is made by serology, histopathology, and culture. The treatment of choice is amphotericin B, and an alternative agent is fluconazole.

16

Histoplasmosis is endemic in the central United States, along the Ohio and Mississippi River Valleys, as well as many other countries. This infection usually occurs as a result of inhalation, and transmission by the allograft has been documented. Disseminated infection is the most common presentation in transplant recipients, whose symptoms are nonspecific, including fever, night sweats, nonproductive cough, headache, myalgias, and muco-cutaneous lesions. Hepatosplenomegaly and pancytopenia may be present. Chest radiographic findings include interstitial or miliary infiltrates, focal consolidation, and hilar adenopathy. The diagnosis is made by serologic testing, direct detection of antigen is urine, histopathology, and culture of appropriate specimens such as respiratory tract secretions and tissue, blood, bone marrow, and other affected tissues. The treatment of choice is amphotericin B; alternative agents include the lipid-based amphotericins and itraconazole.

There are no firm recommendations for prevention of the endemic mycotic infections after transplant; some centers advocate pre-transplantation screening of at-risk candidates by serologic testing and chest radiography and azole prophylaxis for candidates with evidence of prior infection.

VIRAL INFECTIONS

Viruses are important pathogens associated with significant post-transplant morbidity and mortality. Viral infections result from acquisition of new infection or reactivation of latent viruses. The herpesviruses, in particular, are responsible for common and, sometimes, severe infection syndromes in the transplanted host.

CYTOMEGALOVIRUS

Overview

Cytomegalovirus is the most common infection in transplant recipients after the first month post-transplantation. For the transplant recipient, CMV has three major implications: it causes disease associated with substantial morbidity and mortality, it augments immunosuppression that is associated with increased risk of PCP and other infections, and it is associated with allograft rejection. After transplantation, CMV disease is defined as symptomatic viremia or end-organ infection.

Most individuals develop CMV infection at some point during their lifetime, and after the acute phase of the illness, the virus persists in a latent stage within the host. Depending on the recipient's and donor's previous exposure to the virus, transplant recipients are at different risk of developing CMV disease post-transplantation. Thus, three patterns of infection are observed:

1. Primary infection occurs when a CMV-seronegative allograft recipient receives cells latently infected with CMV from a seropositive donor, resulting in viral reactivation (CMV donor+, recipient-).
2. Reactivation infection occurs when endogenous latent virus reactivates in a seropositive recipient (CMV D+ or -, R+).
3. Superinfection (reinfection) occurs when a seropositive recipient receives an organ from a seropositive donor and the virus that reactivates is that of donor origin (CMV D+, R+).

The D+R- patients represent the highest risk group for the development of post-transplant CMV disease, with up to 60% of recipients in this category manifesting this infection. Lower risk groups are the D+R+ and D-R+ in which the incidence of CMV disease ranges from 20-40%. The use of anti-lymphocyte therapies such as OKT3 for induction or rejection increases the risk of any seropositive recipient (R+) for CMV disease, such that these recipients are treated as another high risk group. The D-R- group represents the lowest risk group and may rarely develop primary infection after receipt of unscreened blood products or community exposure. Based on historical data, depending on the organ transplanted, lung and gut recipients are at high risk for CMV disease, whereas liver, pancreas, and heart recipients fall into an intermediate category, and renal transplant recipients represent the lowest risk group.

Clinical Presentation

Cytomegalovirus disease is found most commonly one to four months after transplantation. Manifestations of CMV range from asymptomatic viremia to lethal disseminated disease. Mild to moderate disease presents with fever, malaise, headache, arthralgias and myalgias. Laboratory abnormalities include leucopenia and thrombocytopenia. End organ involvement usually correlates with the type of transplant, thus, hepatitis occurs in liver recipients, glomerulopathy in renal transplants, pancreatitis in pancreatic transplants, and pneumonitis in lung and heart-lung recipients. Other organs that can be affected are the gut (gastritis, esophagitis, colitis), central nervous system (encephalitis, polyradiculopathy), and retina (retinitis). Colitis usually presents with diarrhea that is occasionally bloody, and it may be complicated by the formation of ulcers and perforation. Retinitis is significantly less common in transplant recipients than in patients with HIV infection.

Laboratory abnormalities with CMV end-organ involvement depend on the affected organ system. CMV pneumonitis appears radiographically as bilateral interstitial, unilobar, and nodular infiltrates. Hepatitis manifests with elevated transaminases, alkaline phosphatase, and gamma-glutamyltransferase with minimal increases in bilirubin levels. The definitive diagnosis of CMV organ involvement often requires biopsy of the affected organ and pathologic demonstration of inclusion bodies or detection of CMV in tissue by immunohistochemistry or in-situ hybridization techniques.

To detect replicating virus in fluids or tissue, several techniques have been used including tube cell culture that demonstrates the cytopathic effect of the virus after 7-14 days of incubation and rapid shell-vial culture that uses fluorescent labeled antibodies and yields results in a shorter period of time, usually after 24-48 hours. Serologic testing is not reliable for the diagnosis of acute CMV infection.

Rapid tests that detect early disease are blood CMV antigenemia and PCR techniques. These two methods are used to determine which patients may benefit from preemptive administration of antiviral drugs in an effort to avoid the development of overt CMV disease.

16

Treatment

The currently available antivirals for the treatment and prevention of CMV disease are acyclovir, valacyclovir, ganciclovir, valganciclovir, foscarnet, and cidofovir. Acyclovir, valacyclovir, ganciclovir, and valganciclovir are available in oral formulations. CMV hyperimmune globulin is available in different preparations as well.

Treatment of established CMV disease currently relies on the intravenous administration of ganciclovir. Induction treatment at doses of 5 mg/kg twice daily are used, and maintenance doses are 5 mg/kg/daily. Duration of treatment varies depending on the severity of the disease; viremia may be treated with a regimen of 14 days at full doses whereas end-organ involvement usually requires longer courses. Oral ganciclovir has been used as maintenance therapy to prevent relapses of CMV. Valganciclovir offers improved oral bioavailability over GCV; however, limited clinical data are available to support its routine use for the treatment of transplant-related CMV infection. Side effects of GCV include bone marrow suppression, hemolysis, renal toxicity, rash, liver function abnormalities, and infusion site reactions.

D+R- patients with CMV infection who have received multiple courses of ganciclovir are at risk for the development of antiviral drug resistance. Ganciclovir resistance, usually caused by viral mutations in the UL97 gene, must be considered in patients with poor clinical response or persistent viral shedding during treatment.

The use of foscarnet in transplant recipients is less well studied. Side effects of this drug include nephrotoxicity, hyper- and hypophosphatemia, hyper- and hypocalcemia, nausea, vomiting, and seizures. Its use is reserved for patients who are intolerant of or have failed to respond to GCV.

Some centers advocate CMV hyperimmune globulin as an adjunctive therapy for CMV disease but other studies fail to confirm its effectiveness. Combination of this agent with GCV may be useful, especially in patients with severe life threatening disease, such as CMV pneumonitis.

Prevention

There are two main strategies for the prevention of CMV disease after transplantation. The first is the administration of antiviral prophylaxis to prevent the occurrence of CMV disease. The second approach is pre-emptive therapy, whereby patients at risk are monitored for laboratory evidence of subclinical CMV infection, usually by the CMV antigenemia or quantitative CMV PCR assay, and initiated on antiviral therapy if subclinical infection is detected. There is significant debate in the literature in regards to the effectiveness of these different approaches for the prevention of CMV disease, although both strategies are acceptable practices.

In general, the pre-emptive therapy strategy is effective in CMV-seropositive recipient groups. However, D+R- patients require oral GCV prophylactically. Because of improved oral availability, many centers now use valganciclovir instead of GCV, although clinical data is lacking. Data on renal transplant recipients sug-

16

gests that valacyclovir is an effective alternative in this population. Patients in any group, except D-R-, receiving OKT3 or anti-lymphocyte antibodies are considered candidates for GCV prophylaxis since their risk of CMV disease is substantially increased. Prophylaxis is continued for the first three months after transplantation. One exception to this approach is lung and heart-lung recipients that fall into the D+R- group; since they are at the highest risk, intravenous full dose GCV for two weeks and followed by maintenance doses to complete three months of prophylaxis after transplantation seem warranted to prevent severe disease.

For individuals in the D-R- group, CMV-negative blood products are administered if needed to prevent primary CMV infection and subsequent risk of disease.

EPSTEIN-BARR VIRUS

Overview

Epstein-Barr virus infection plays an important role in the development of post-transplant lymphoproliferative disorder (PTLD). Most transplant recipients have been infected with EBV at some point during their lifetime, and the virus persists within the host in a latent state. The pathogenesis of PTLD involves the heightened replication of EBV-infected lymphocytes triggered by agents such as OKT3 and the polyclonal anti-lymphocyte globulins. In addition, immunosuppression impairs the ability of virus-specific cytotoxic T-lymphocytes to control the expression of EBV-infected transformed B-cells, leading to the polyclonal and monoclonal proliferation of lymphocytes that constitutes PTLD. EBV may be of donor origin in EBV-seronegative recipients who receive an organ from a seropositive individual.

The incidence of PTLD ranges from 1% to more than 20% depending on the organ transplanted: 1% in renal transplant recipients, 2% in liver recipients, 2-4% in heart recipients, 2-8% in lung recipients, 11% in kidney-pancreas recipients, 7-11% in intestine recipients, and 13-33% in multi-organ recipients. Risk factors associated with the development of PTLD include EBV-seronegativity prior to transplantation, OKT3 or polyclonal anti-lymphocyte antibody therapy, CMV seromismatch, and CMV disease. Primary EBV infection is a strong predictor of PTLD.

Clinical Presentation

Primary infection with EBV may cause a syndrome characterized by malaise, fever, headaches, and sore throat. PTLD may develop at any time post-transplantation manifesting as a mononucleosis syndrome with fever, adenopathy, and sore throat; fever of unknown origin; allograft dysfunction; respiratory symptoms with pulmonary infiltrates; and weight loss. The definitive diagnosis of PTLD relies on histopathologic examination of biopsy specimens. Quantitative PCR techniques may help determine patients at high risk for the development of this disease before overt signs and symptoms manifest; however, this approach remains experimental.

16

Treatment

Once PTLD is established, the effectiveness of antiviral treatment has been disappointing. The main therapeutic approach in these cases is reduction of immunosuppression to the extent possible. Multifocal disease with organ involvement carries the worst prognosis, and in this setting, chemotherapy or radiotherapy is usually indicated. An alternative option is immunotherapy based on sensitized T-lymphocytes obtained from patients before transplant and stimulated ex vivo; such therapy is under investigation and is showing promising results.

Prevention

In EBV-seronegative patients, the use of EBV-seronegative organs constitutes the best option to prevent PTLD, especially in high-risk groups such as lung, gut, and pancreas recipients. It is unclear whether GCV is effective for the prevention of EBV-associated PTLD. Some data suggest that intravenous GCV followed by oral acyclovir during the first three months following transplantation decreases the incidence of PTLD in liver, kidney, and kidney-pancreas EBV-seronegative recipients, but there is not enough data to recommend this approach.

HERPES SIMPLEX VIRUS

Overview

Infection with HSV in transplant recipients most commonly represents reactivation of latent virus. Up to 80% of adult transplant recipients are seropositive for HSV, indicating prior infection. Following primary infection, HSV remains latent in sensory nerve ganglia, and reactivation often occurs during the first month post-transplantation in up to 40% of organ recipients. The use of OKT3 is associated with higher frequency of reactivation.

Clinical Presentation

HSV reactivation most often manifests as oral or genital mucocutaneous lesions. Because of depressed cell-mediated immunity, organ recipients are at risk for more severe disease, delayed healing of skin lesions, and occasionally, visceral or disseminated involvement such as pneumonitis, tracheobronchitis, esophagitis, or hepatitis.

Mucocutaneous involvement presents as painful vesicular and ulcerative lesions; the appearance may be different than that observed in immunocompetent individuals. Pneumonitis is rare and usually seen in conjunction with other pulmonary infections. Ulcerative esophagitis manifests as dysphagia and odynophagia and can resemble or occur concomitantly with candidiasis. Hepatitis may occur and be rapidly progressive and fatal. Disseminated HSV is occasionally reported. Encephalitis is not seen with greater frequency than in immunocompetent individuals.

The diagnosis of HSV is made by direct immunofluorescent antibody staining, Tzanck smear, or culture of tissue and body fluids. Serodiagnosis is possible if IgM is detected or a four-fold rise in IgG titers is noted. In the case of HSV pneumonitis, the definitive diagnosis relies on histopathologic examination of biopsy specimens, since the recovery of HSV from tracheal secretions may represent reactivation of virus in the oropharyngeal cavity.

Treatment

Acyclovir is considered the treatment of choice for HSV infections. Oral administration of 200 mg five times daily is effective in mild disease. Alternative oral preparations with better bioavailability are valacyclovir and famciclovir. Serious cases with disseminated infection or organ involvement requires treatment with intravenous acyclovir. Acyclovir may cause nephrotoxicity due to the precipitation of drug crystals in the renal tubules; other serious side effects include confusion, delirium, and seizures. The dose of acyclovir is adjusted according to the creatinine clearance. Acyclovir resistance has occurred in organ transplant recipients with HSV infection; foscarnet is considered the drug of choice in this situation.

Prevention

The use of low-dose acyclovir (200 mg every six to eight hours) for the first month after transplantation is an effective prophylactic regimen in all seropositive transplant recipients. Other potential options are valacyclovir and famciclovir.

VARICELLA ZOSTER VIRUS

Overview

Post-transplantation, VZV causes herpes zoster in seropositive individuals (90% of the adult population). The remaining 10% of patients are at a risk for primary infection. Up to 13% of transplant recipients develop herpes zoster during the first six months post-transplantation.

Clinical Presentation

Typical dermatomal skin lesions are the usual presentation of herpes zoster. Disseminated disease occurs as well, with multiple dermatome involvement. Dermatomal pain without skin eruption has been described.

Primary VZV infection is transmitted via contact with infected individuals; it may occur at any time after transplant and is potentially serious with pneumonia, skin lesions, hepatitis, encephalitis, pancreatitis, and disseminated intravascular coagulation.

The diagnosis is most often made clinically, but culture of VZV and direct immunofluorescent antibody staining or Tzanck smear of appropriate clinical specimens are used for confirmatory purposes.

Treatment

The treatment of localized dermatomal zoster is treated with oral acyclovir, valacyclovir, or famciclovir. In severe cases, disseminated disease, or primary infection, intravenous acyclovir is administered initially and patients are monitored carefully. The duration of treatment is usually ten days.

16

Prevention

Varicella immune-globulin is administered to VZV-seronegative recipients exposed to acutely infected individuals. Prophylactic low-dose oral acyclovir may prevent VZV infections although this has not been formally studied.

OTHER VIRUSES

Human herpesviruses 6 and 7 reactivate after transplantation. The role of these viruses in clinical disease is currently under active investigation, as well as their interactions with other pathogens such as CMV and role in allograft rejection. There are no standard assays available for the diagnosis of these infections.

Hepatitis B and C viruses are common causes of end-stage liver disease and important indications for liver transplantation. The risk of recurrent infection with either virus is more than 80% after transplantation and morbidity can be high, especially in the case of hepatitis B. Viral hepatitis may be transmitted to recipients by the organs of infected donors. There is considerable ongoing debate regarding the use of organs from infected donors (mainly hepatitis C) in emergent transplantation of live-saving organs, and consensus has not been reached.

The polyomaviruses, BK virus (BKV) and JC virus (JCV), are ubiquitous viruses that cause subclinical and latent infections in humans. 80% of the adult population is seropositive for each of these viruses. These viruses may reactivate in the setting of immunosuppression, resulting in distinctive clinical syndromes. BKV reactivation after renal transplantation may result in tubulointerstitial nephritis, leading to progressive allograft dysfunction and eventual graft loss. In addition, BKV reactivation may present as ureteral stenosis leading to obstructive nephropathy. The diagnosis of BK nephropathy is suggested by the presence of characteristic "decoy" cells on cytologic examination of urine. Because the histopathological findings may be confused with allograft rejection, the definitive diagnosis is established by the demonstration of polyomavirus inclusions in renal biopsy specimens. The use of PCR assays with blood and urine specimens for the diagnosis of BK nephropathy are being investigated. Progressive multifocal leukoencephalopathy, caused by the JCV, is less commonly encountered in organ recipients than in patients with acquired immunodeficiency syndrome. Treatment for polyomavirus infections has generally focused on supportive care and reduction of immunosuppression. A variety of antiviral agents are under investigation, with cidofovir demonstrating some promise.

Transplant recipients are also at heightened risk of developing infection with the related human papillomaviruses. These viruses are associated with neoplasms such as squamous carcinoma of the cervix. An association with skin carcinomas has been suggested but not definitively confirmed.

Parvovirus B19 infection may cause profound aplastic anemia after transplantation. Other respiratory viruses with potential to cause severe disease in transplant recipients include the respiratory syncytial virus, the adenoviruses, and the influenza viruses.

PARASITIC INFECTIONS

Strongyloides stercoralis and *Toxoplasma gondii* warrant special attention because of the distinctive features of these pathogens and the potential for severe infection in organ transplant recipients.

STRONGYLOIDES STERCORALES

Overview

Strongyloidiasis is a helminthic infection endemic to tropical and subtropical areas of the world, including Southeast Asia, the Caribbean, and West Africa. Rural areas of Kentucky, Tennessee, and Louisiana are the endemic foci within the United States. Infection results in a diarrheal illness with peripheral eosinophilia, but the organism can be maintained in the gastrointestinal tract asymptomatically for decades. The lifecycle of *S. stercoralis* is complex and unique in that autoinfection of the host occurs; larvae transform into an infectious form in the intestine and invade the intestinal mucosa and perianal skin. This process maintains the infection by constant re-introduction of infectious forms into the hosts. In immunocompromised patients, autoinfection may result in the hyperinfection syndrome, a form of disseminated strongyloidiasis in which an accelerated lifecycle and excessive helminth burden occur. The mortality rate of hyperinfection syndrome approaches 70%. A complication of this syndrome is gram-negative bacteremias and shock caused by the disruption of normal intestinal barriers during invasion of the gut mucosa by larvae.

Clinical Presentation

In patients harboring *S. stercoralis*, symptoms develop in the first six months after transplantation. The intestinal form presents as abdominal pain, diarrhea, nausea, and vomiting. Hyperinfection syndrome presents with tachypnea, respiratory distress, cough, hemoptysis, and enterocolitis. Fever and rash may also be present. The chest radiograph demonstrates alveolar or interstitial infiltrates. On laboratory data, eosinophilia, if present, is helpful.

The definitive diagnosis is made by examining stool specimens for the presence of larvae. The sensitivity of this method is improved by the examination of multiple specimens and concentration techniques. Sputum and duodenal aspirates may be examined for the presence of larvae. Serologic testing is also available.

Treatment

The treatment options for strongyloidiasis include ivermectin, 200 mcg/kg/day for two days; albendazole, 400 mg/day for three days; or thiabendazole, 25 mg/kg for two days. The hyperinfection syndrome requires seven to ten days of therapy. Secondary bacteremias are treated with antimicrobial agents.

Prevention

Prevention is aimed at candidates who have resided or traveled extensively to endemic areas. Prior to transplantation, candidates are screened for the presence of this infection, either serologically or by examination of multiple stool speci-

16

mens for larvae. Established infection is treated before transplant. Interestingly, cyclosporine A (CsA) has activity against *S. stercoralis*; recipients treated with CsA are less likely to develop the hyperinfection syndrome.

TOXOPLASMA GONDII

Overview
Toxoplasmosis after transplantation is most frequently caused by reactivation of latent disease. Seronegative heart transplant recipients are at the greatest risk for developing this disease when receiving an organ from a seropositive donor; in such cases, the risk of primary infection is 50% if prophylaxis is not administered. This infection is also potentially transmitted by other organs, such as liver, and blood products.

Clinical Presentation
Most infections occur during the first two months after transplantation. Primary infection manifests as fever, malaise, and generalized lymphadenopathy. Other presentations include meningo-encephalitis, pneumonitis, pericarditis, myocarditis, and retinitis.

The definitive diagnosis is made by the histologic demonstration of trophozoites surrounded by inflammatory reaction. Serologic testing is not very helpful for the diagnosis of infection.

Treatment
The options for therapy are one of the following: pyrimethamine and sulfadiazine or pyrimethamine plus clindamycin. Folinic acid is administered with either regimen to prevent myelotoxicity. Treatment is continued for two to three weeks after resolution of the acute infection.

Prevention
After transplantation, toxoplasmosis is prevented by the use of TMP/SMX in the same doses used for PCP prophylaxis for at least six months. Pyrimethamine is an alternative agent for sulfa-intolerant patients. All heart transplant candidates and donors require serologic screening to determine the post-transplant risk of toxoplasmosis.

VACCINATION IN TRANSPLANT RECIPIENTS
Two issues limit the overall effectiveness of vaccination strategies in transplant recipients. First, transplant recipients may have declining antibody levels and diminished antibody responses to previous vaccine antigens once they become severely immunosuppressed (loss of previous immunity). Secondly, available evidence suggests that transplant recipients have diminished, although not absent, responsiveness to immunization (reduced vaccine efficacy). This is best demonstrated in kidney, liver and heart recipients after immunization with the pneumococcal vaccine.

16

Solid organ transplant recipients require periodic assessment of immunization status for vaccine-preventable illnesses, beginning during the pre-transplantation evaluation. Routine immunizations are administered or updated as long as possible before the transplant to allow for the development of immunity; these vaccinations include the hepatitis B series, hepatitis A, pneumococcal, yearly influenza, and tetanus-diphtheria. For VZV-seronegative transplant candidates, immunization with the varicella vaccine should be considered. In general, live attenuated virus vaccines are contraindicated in severely immunosuppressed hosts because of the potential for viral reactivation. Also, household contacts of transplant recipients should not undergo immunization with live viruses because of the potential for secondary infections.

SELECTED READINGS

1. Chow JW, Yu VL. Legionella: a major opportunistic pathogen in transplant recipients. Sem Resp Infect 1998; 13:132-39.

2. Fishman JA, Rubin RH. Infection in organ-transplant recipients. New Eng J Med 1998; 338:1741-51.

3. Husain S, Wagener MM, Singh N. *Cryptococcus neoformans* infection in organ transplant recipients: variables influencing clinical characteristics and outcome. Emerg Infect Dis 2001; 7:375-381.

4. Ljungman P. β-Herpesvirus challenges in the transplant recipient. J Infect Dis 2002; 186(Suppl):S99-109.

5. Patel, R. Infections in recipients of kidney transplants. Infect Dis Clin N Am 2001; 15:901-952.

6. Patel R, Paya C. Infections in solid organ transplant recipients. Clin Microbiol Rev 1997; 10:86-124.

7. Patterson JE. Epidemiology of fungal infections in solid organ transplant recipients. Transpl Infect Dis 1999; 1:229-36.

8. Rubin R, Tolkoff-Rubin N. Antimicrobial strategies in the care of organ transplant recipients. Antimicrob Agents Chemother 1993; 37:619-24.

9. Sia IG, Patel R. New strategies for prevention and therapy of cytomegalovirus infections and disease in solid-organ transplant recipients. Clin Microbiol Rev 2000; 13:83-121.

10. Singh N. Antifungal prophylaxis for solid organ transplant recipients: seeking clarity amidst controversy. Clin Infect Dis 2000; 31:545-53.

11. Singh N, Paterson DL. *Mycobacterium tuberculosis* infection in solid-organ transplant recipients: impact and implications for management. Clin Infect Dis 1998; 27:1266-77.

12. Snydman DR. Infection in solid organ transplantation. Transplant Infect Dis 1999; 1:21-28.

13. Soave R. Prophylaxis strategies for solid-organ transplantation. Clin Infect Dis 2001; 33(Suppl 1):S26-31.

14. Villacian JS, Paya CV. Prevention of infections in solid organ transplant recipients. Transpl Infect Dis 1999; 1:50-64.

16

Early Medical Problems Common to Many Recipients

Joseph P. Leventhal and William A. Schlueter

EARLY MEDICAL PROBLEMS COMMON TO MANY RECIPIENTS

Complications following organ transplantation can be categorized as early or late, depending upon whether they occur within six months or later in the posttransplantation period. However, some of these complications can overlap in their time of occurrence. Posttransplantation complications can also be divided into organ specific complications, and those common to all transplantation procedures. Global complications following transplant are often related to the immunosuppressive drugs used to control allograft rejection, and include hypertension, cardiovascular disease, hyperlipidemia, infection, drug-induced nephrotoxicity, metabolic bone disorders, and malignancy. The reader is referred to Chapters 2,16 and Appendix 1 (in this text) for a review of immunosuppressive drugs and their long-term effects.

Many problems in the early posttransplant period are related to the surgical procedure itself (mechanical), while others are strictly medical problems. These early complications are best managed at a transplantation center to avoid the consequences of misdiagnosis and inadequate evaluation. Table 17.1 summarizes the early complications associated with different forms of organ transplants. The reader is referred to appropriate chapters in this text for more detailed discussion of many of the organ-specific mechanical complications, as well as for a comprehensive review of posttransplant infections. The remainder of this Chapter will focus on problems encountered in patients undergoing renal, pancreatic, and hepatic transplantation.

RENAL TRANSPLANTATION

GRAFT DYSFUNCTION

Causes of early renal allograft dysfunction include rejection, vascular complications such as renal arterial or venous thrombosis, urinary tract obstruction, urine leak secondary to ureteral necrosis/anastomotic disruption, and acute tubular necrosis (Table 17.2). The spectrum of graft dysfunction may range from frank anuria (< 50cc of urine/day) to oliguria (< 500cc/day), to moderate dysfunction manifested through sluggish fall in serum creatinine. It is essential to make the correct diagnosis of in cases of posttransplant graft dysfunction, since the management of each complication is very different. Vascular complications such as

Organ Transplantation, 2nd edition, edited by Frank P. Stuart, Michael M. Abecassis and Dixon B. Kaufman. ©2003 Landes Bioscience.

Table 17.1. Early complications of organ transplantation (adapted from references)

Transplant	Surgical/Mechanical	Medical
Renal	Lymphocele	Acute rejection
	Urine leak	Ureteral obstruction
	Renal artery stenosis	Delayed graft function/ATN
	Vascular thrombosis	CsA/FK506 nephrotoxicity
	Bleeding	Prerenal/hypovolemia
	Drug toxicity	
	Infection	
	Recurrent disease	
Liver	Hepatic artery/portal vein thrombosis	Acute rejection
	Biliary obstruction/leak	Preservation injury
	Vena cava obstruction	Recurrent disease
	Liver infarction	Infection (cholangitis)
	Bleeding	Drug toxicity
Pancreas	Vascular thrombosis	Rejection
	Bleeding (GI vs. bladder)	Pancreatitis
	Pancreatic leak	Infection
	Peripancreatic abscess	Bleeding (GI vs. bladder)
Heart	Primary nonfunction	Rejection
	Arrhythmias	Infection
	Poor function	Poor function
	Bleeding	
	Infection	
	Hypotension	
	Pulmonary artery hypertension	
Lung	Pulmonary venous obstruction	Reperfusion injury
	Dehiscence of airway anastomosis	Acute rejection
	Bronchomalacia	Infection
	Phrenic nerve injury	Pulmonary embolism
	Bleeding	
	Pleural effusion	

renal arterial or venous thrombosis require immediate surgical intervention. Hyperacute rejection of an allograft is a very rare event that invariably leads to loss of the organ immediately posttransplant; confirmation of blood group compatibility and proper crossmatching should be performed if suspected. Acute rejection can occur in the first week after transplant and should always be considered in the evaluation of early graft dysfunction; allograft biopsy is an important diagnostic tool that should be used if the diagnosis is unclear. The reader is referred to Chapters 2 and 6 for management of allograft rejection.

Urinary tract obstruction leading to oliguria or anuria in the immediate posttransplant period is commonly caused by clot obstruction of the bladder catheter. Catheter irrigation and/or replacement should be performed to ensure patency. Obstruction may also be caused through external compression by a lymphocele or by development of a stricture at the ureteroneocystostomy. Lymphoceles represent fluid collections which form as a consequence of the retroperitoneal vascular dissection performed during the transplant procedure.

17

Table 17.2. Differential diagnosis of acute renal allograft dysfunction in the immediate posttransplant period

	Rejection	Renal artery thrombosis	Ureteral necrosis/ anastomotic disruption	Obstruction	ATN	Hypovolemia
Urine output	Oliguria/ anuria (HAR)*	anuria	anuria	Anuria/oliguria	Anuria/oliguria	oliguria
Hypertension	+	+	+/-	+/-	+/-	-**
Renal scintigraphy	Reduced uptake/ no uptake(HAR)*	No uptake	Good uptake; but poor/no excretion into bladder. Isotope can be seen in perirenal space from leak.	Good uptake, delayed or no excretion into bladder	Good uptake, delayed to good excretion into bladder	Good uptake, excretion
Ultrasound	Enlarged kidney; loss of cortico-medullary differentiation	Normal size kidney	Normal size kidney; perinephric fluid collection with leak	Dilated collecting system	Normal size or large kidney	Normal size kidney
Arteriogram	Patent renal artery	Occluded renal artery with filling defects	Patent renal artery	Patent renal artery	Patent renal artery	Patent renal artery
Management	IV steroid pulse vs. antibody therapy; graft nephrectomy if HAR	Surgical exploration, thrombectomy possible graft nephrectomy	Ureteral reconstruction	Relief of obstruction	Observation and supportive care; hemodialysis if indicated	Volume challenge and continued IV fluids

*-hyperacute rejection

**-prolonged hypotension can precipitate the development of ATN

Lymphoceles typically manifest weeks or months posttransplant, and are easily detected by ultrasonography. These collections may be drained percutaneously or surgically into the abdominal cavity. Anastomotic ureteral obstruction is often treated by nephrostomy tube placement, followed by balloon dilatation and stenting. Surgical revision is often required. Functional obstruction from bladder distension may also be observed in male patients with prostatic hypertrophy, or in diabetics with neurogenic bladders.

Posttransplant oliguria or anuria caused by ischemic acute tubular necrosis (ATN) occurs commonly after cadaveric renal transplantation, with incidence reported from 10 to 50%. ATN following live donor transplantation is very uncommon, likely related to the minimal warm ischemic time, short preservation time, and excellent health status of the donor. ATN has been associated with highly sensitized recipients, prolonged (> 24hrs) cold ischemia time, cadaver donor age > 50 years, perioperative hypotension, and high dosing of calcineurin inhibitors. The management of posttransplant delayed graft function consists of general supportive measures, intermittent dialysis, avoidance of nephrotoxic medications, and continued close observation. The duration of ATN is typically 1-3 weeks, but can last as long as 3 months. Intermittent noninvasive imaging of these allografts should be considered to ensure adequate blood supply and rule out urine leak or obstruction. Allograft biopsy should be performed in cases of prolonged ATN (i.e., more than 2-3 weeks) to rule out the development of rejection. Recovery from ATN is associated with steady increase in daily urine volumes and a reduction is serum creatinine rise between dialysis treatments. Most patients experiencing ATN go on to acquire an adequate GFR.

THROMBOEMBOLIC DISEASE

The risk of thromboembolic disease is increased following renal transplantation. Associated risk factors include older recipient age, postoperative immobilization, use of high dose steroids, treatment with calcineurin inhibitors such as cyclosporine, and increased blood viscosity from posttransplant erythrocytosis. All patients undergoing renal transplantation require some form of deep vein thrombosis prophylaxis, such as sequential leg compression devices. Diagnosis and management of venous thrombosis with appropriate tests and anticoagulation is no different than in other clinical situations.

HEMODIALYSIS

PRETRANSPLANT

Hemodialysis is often performed preoperatively to correct: elevated BUN and creatinine levels, hyperkalemia, metabolic acidosis and volume overload. Care is taken during the dialysis treatment to remove sufficient fluid to attain a euvolemic state and to avoid volume contraction since volume contraction would render the transplanted kidney ischemic. Correction of the uremic state improves the bleeding diathesis common to these patients. In addition, by dialyzing the patient pretransplant dialysis may be held for several days posttransplant if the kidney doesn't immediately function. By withholding dialysis at this juncture, you may

17

hasten recovery from acute tabular necrosis (ATN) by preventing fresh episodes of dialysis induced ischemia to the transplanted kidney.

POSTTRANSPLANT

Hemodialysis is occasionally required in patients posttransplant. The need for dialysis is used to define delayed graft function (DGF). The decision to dialyze a patient is made on a case-by-case basis made jointly by the transplant team and the nephrologist. The indication for dialysis (such as hyperkalemia, metabolic acidosis, volume overload, pericarditis, and uremic encephalopathy) must be weighed against the risks of the dialysis procedure (hypotension inducing fresh episodes of ATN and thereby contributing to a delay in kidney function). Hyperkalemia, volume overload and metabolic acidosis can often be managed medically without the need for dialysis. Pericarditis or uremic encephalopathy are indications for emergent dialysis. The transplant team can provide prognostic information concerning the expected time until recovery of renal function from the appearance of the kidney at the time that vascular flow is established and from the cold ischemic time (the time from harvesting of the kidney until transplantation). With this information, delaying dialysis a few days is appropriate if signs point toward an imminent restoration of renal function (such as a progressive increase in urine output, BUN and creatinine rate of rise is diminishing or plateauing and the kidney appeared very fresh when the circulation was reestablished). Finally, there is a psychological benefit to the patient by avoiding dialysis posttransplant. Some patients experience emotional upset when the kidney does not begin to function immediately after the transplant and their anxiety is heightened by the need for dialysis.

GOUT

Following a successful kidney transplant serum uric acid levels decrease from the levels seen in patients on dialysis. Hyperuricemia due to reduced uric acid clearance, is a common complication posttransplant. In the cyclosporine-era, the frequency of hyperuricemia has increased from 25-55% to 56-84%, and gout from 5-28%. Cyclosporine and tacrolimus damage the renal tubules decreasing uric acid secretion into the urine, and contribute to hyperuricemia. Volume contraction due to diuretics increase uric acid retention by the kidney and contribute to hyperuricemia Aspiration of the affected joint is imperative to make the diagnosis and to rule out septic arthritis and other conditions. A definitive diagnosis of gout is made by identification of polymorphonuclear leukocytes containing phagocystosed uric acid crystals in joint fluid examined using polarized light microscopy. Uric acid crystals are needle shaped and negatively birefringent, whereas calcium pyrophosphate crystals seen in pseudogout are positively birefringent. Hydroxyapatite complexes, diagnostic of apatite disease can only be identified by electron microscopy and mass spectroscopy.

Treatment with nonsteroidal anti-inflammatory drugs should be avoided. Colchicine may be administered orally or intravenously. Oral administration is effective but often causes severe gastrointestinal toxicity. The oral dose has a more

17

conservative protocol than in nontransplant patients (up to three hourly doses of 0.6 mg on the first day). Intravenous colchicine results in more rapid relief of pain with fewer gastrointestinal side effects but has the potential for severe myelosuppresion, myopathy and neuropathy and therefore is best avoided. Glucocorticoids are indicated for patients who are refractory to colchicine or in whom colchicine is contraindicated. Prednisone 40-60 mg/day is given until a response occurs, then tapered rapidly. Alternatively, parenteral adrenocorticotrophic hormone (ACTH) can be quite effective for treating acute attacks of gout. If colchicine is to be used on a continuing basis in the presence of impaired renal function, it should be given on an alternate day schedule in order to minimize neuromuscular toxicity.

Allopurinol is never given during an acute attack. It is used to prevent future attacks. Use of Allopurinol poses several problems: 1) Allergic reactions occur in 10-20% of patients who take the drug chronically and this should not be taken lightly, when there are hepatic, renal, and severe skin lesions associated with the allergy, the mortality rate is up to 25%, 2) Allopurinol potentiates Azathioprine myelosuppression. Dose reduction and careful monitoring of the WBC count is required, 3) Implementation of therapy with Allopurinol may precipitate an acute attack of gout. Therefore, to prevent gout attacks it would be more prudent to discontinue diuretics and follow a regimen of vigorous oral hydration. If this is not successful in preventing attacks of gout, Colchicine may be prescribed on an alternate day schedule to prevent attacks. If this is not successful, then one may consider changing from Azathioprine to Mycophenolate mofetil this is because Allopurinol inhibits the metabolism of Mercaptopurine (the active metabolite of Azathioprine) resulting in increased mercaptopurine toxicity. When coadministration with Azathioprine cannot be avoided, the dose of Azathioprine should be reduced by 75% and the dose of Allopurinol should be initiated at 50 mg/day. Allopurinol does not interfere with mycophenolate mofetil metabolism. Therefore, it may be given with mycophenolate mofetil with less of a risk of inducing leukopenia.

HYPERKALEMIA

Hyperkalemia is a common finding in transplant patients. It is usually due to a combination of physiologic derangements and medications that impair potassium excretion in the urine. The physiologic derangements that impair potassium excretion in the urine include: a reduction in the glomerular filtration rate (GFR), damage to the distal nephron (the major site of potassium secretion into the urine), and suppression of plasma aldosterone levels. Cyclosporine (and tacrolimus) may cause hyperkalemia by decreasing the GFR (by causing vasoconstriction of the afferent arteriole) and have a direct toxicity to the distal nephron where potassium is secreted into the urine. Trimethoprim/sulfamethoxazole (Bactrim) decreases potassium excretion in the urine by blocking the sodium for potassium exchange in the distal nephron. ACE inhibitors, angiotensin II receptor antagonists (such as Losartan) and nonsteroidal anti-inflammatory agents may cause hyperkalemia by suppressing plasma aldosterone levels. In addition, plasma aldosterone levels may be suppressed as part of the syndrome of hyporeninemic hypoaldosteroidism,

17

particularly in diabetic patients. Endogenous catecholamines result in enhanced cellular uptake of potassium. This mechanism helps attenuate hyperkalemia in patients with renal impairment. When nonselective beta-blockers used for treating hypertension or coronary artery disease, they may exacerbate hyperkalemia by blocking the beta 2 receptors thereby preventing the transcellular uptake of potassium.

For the treatment of hyperkalemic emergencies see Table 17.3. When hyperkalemia is mild and chronic, therapeutic measures should be as specific as possible. Reduction in dietary potassium intake and avoidance of drugs that provoke hyperkalemia (if possible). For example, in place of a nonselective beta-blocker, a Beta-1 selective beta-blocker (such as metoprolol) may be used. Or a calcium channel blocker may be used. Calcium channel blockers have been demonstrated to increase the transcellular uptake of potassium and thereby reduce serum potassium levels. Diuretics, which increase potassium excretion in the urine, may be used to reduce potassium levels in patients in whom dietary restriction of potassium and rearranging of other medications is not sufficient to correct hyperkalemia completely. In patients with aldosterone deficiency, Fludrocortisone (Florinef) replacement may be indicated. Florinef may cause salt retention. The incidence of cardiovascular disease and hypertension among transplant patients is very high. The administration of Florinef in this setting may cause edema and worsening of hypertension and may precipitate congestive heart failure. For these reasons, loop diuretics provide a safer alternative.

HYPOMAGNESEMIA
Renal magnesium wasting and hypomagnesemia are common in transplant patients. Hypomagnesemia may be caused by poor dietary intake (malnutrition, alcoholism), malabsorption from the GI tract (diarrhea, malabsorption syndrome) and increased magnesium losses in the urine (cyclosporine, tacrolimus, diuretics, aminoglycoside antibiotics, and amphotericin B).

Table 17.3. Therapeutic approach to severe hyperkalemia

Medication	Dosage	Peak Effect
Calcium Gluconate*	10-30 ml of 10% solution	< 5 min
Insulin and Glucose*	Insulin, 5 U IB bolus, followed by Bw/min in 50 ml of 20% glucose	30-60 min
Sodium Bicarbonate	100 ml 1.4% solution 2 mmol/min	Variable 1-6 h
Albuterol	20 mg in the nebulized form	60 min
Kayexalate	Enema (50-100 g) Oral (40g)	2-4 h
Hemodialysis		30-60 min

*Denotes recommended initial therapy.
Hemodialysis is not listed as immediate therapy since it is usually not available within the first hour or so of arrival to the hospital. If available, it could replace insulin therapy.
Adapted from Rombola G. Batlle DC; Hyperkalemia, P. 49. In Jacobson H. Stricker G. Hklar S et al (eds):
Principles and practice of Nephrology. BC Decker. Philadelphia. 1990. With permission.

Clinical manifestations of hypomagnesemia are generally not seen until serum magnesium level are less than 1mg/dl, and may include: lethargy, confusion, tremor, positive Trousseau's and Chvostek's signs, muscle fasciculations, ataxia, nystagmus, tetany and seizures. EKG abnormalities include prolonged PR and QT intervals. Atrial and ventricular arrhythmias may occur, especially in patients receiving Digoxin.

Treatment of hypomagnesemia must be given with extreme care in patients with impaired kidney function because of the risk of causing hypermagnesemia. For severe symptomatic hypomagnesemia serum magnesium level (< 1mg/dl) see Table 17.4. Correct mild or chronic hypomagnesemia with Mag-Ox 400 mg tablets (contains 240 mg of elemental magnesium) q.d.-b.i.d. The major side effect of Mag-Oxide tablets is diarrhea.

HYPOPHOSPHATEMIA

Renal phosphate wastage and hypophosphatemia is the most common divalent ion abnormality in transplant patients. The causes include: 1) secondary hyperparathyroidism which may persist for up to a year after kidney transplantation (PTH causes the kidney to waste phosphate in the urine; 2) glucocorticoids inhibit the tubular reabsorption of phosphate and contribute to phosphate wastage in the urine; 3) calcium containing antacids prescribed to correct hypocalemia bind dietary phosphate in the GI tract, and 4) the adequacy of renal function i.e., a well-functioning kidney excretes more phosphate while a poorly functioning kidney does not.

Clinical manifestations of hypophosphatemia are only seen if serum phosphorus is less than 1 mg/dl. Muscle weakness, rhabdomyolysis, respiratory failure from diaphragmatic muscle weakness and congestive heart failure may occur. Other manifestations include: paresthesias, confusion, seizures, coma, hemolysis, platelet dysfunction, and metabolic acidosis. Treatment of hypophosphatemia must be given with extreme care in renal failure to prevent causing hyperphosphatemia Moderate hypophosphatemia can be treated orally with skim milk (contains about 1000 mg of phosphorus per liter) or Neutra-phos tablets (250 mg of phosphorus q-day to q.i.d. Oral phosphorus can cause diarrhea. Treatment of severe hypophosphatemia (< 1.0 mg/dl) if the patient is asymptomatic may be by oral replacement 3 g/day for a week. If the patient is symptomatic, replacement should be intravenously 2 mg/kg body weight as the sodium salt infused over six hours, then phosphorus should be rechecked and continued until serum phosphorus exceeds 1 mg/dl. Intravenous phosphorus may produce hypocalcemia and metastatic calcifications (especially if the calcium—phosphorus product exceeds 60 mg/dl).

OCULAR DISEASE

Posterior subcapsular cataracts are a common complication of long-term treatment with glucocorticoids. Patients with cataracts complain of visual impairment on bright sunny days or in response to bright lights where the pupil is forced to constrict and focus the image to the path occupied by the cataract.

17

Table 17.4. *Therapeutic approach to severe hypocalcemia, hypomagnesemia, hypophos-*
phatemia

		Dosage
(1)	Hypocalcemia Calcium	10 ml of a 10% solution IV over 10 minutes followed by 50 ml of a 10% solution of 500 ml D5W over 8 hours then recheck.
(2)	Hypomagnesemia	2 gms of 50% solution IV over 8 hours then recheck (may need to be continued for 3 days to correct the intracellular deficits) Monitor deep tendon reflexer for a decrease In DTA suggest the development of hypermagnesemia.
(3)	Hypophosphatemia	2 mg/kg IV over 6 hours then recheck

Transplant patients with diabetes mellitus should have regular eye examinations to prevent the complications of diabetic retinopathy. Whether patients with diabetic retinopathy would benefit with stabilization or improvement in their vision following early pancreas transplant remains unanswered. Eye infections in transplant patients can be devastating and require emergent diagnosis and treatment. Herpetic keratoconjunctivitis, CMV retinitis, toxoplasma chorioretinitis and ophthalmic herpes zoster are among the eye infections which require timely diagnosis and treatment.

MUSCLE WEAKNESS
Muscle weakness in the transplant recipient may have several causes including 1) Electrolyte abnormalities (hypophosphatemia, hypomagnesemia, hypocalcemia, severe hypo- or hyperkalemia), 2) glucocorticoids therapy, 3) persistent hyperparathyroidism, 4) Other medications (cyclosporine or tacrolimus in combination with an HMG – CoA reductase inhibitor, HMG – CoA reductase inhibitor in combination with a fibrinic acid derivatives (gemfibrizol, fenofibrate, clofibrate) or niacin, or colchicine in patients with impaired renal function), 5) hypo- or hyperthyroidism, 6) infections with CMV. Diagnosis is made by discontinuing medications that may cause weakness or at least reducing the dose, blood tests to look for electrolyte abnormalities, check the PTH level, and check thyroid functions as well as specific biochemical abnormality, creatinine kinase enzymes (CPK), and aldolase level. An electromyogram and muscle biopsy may be required.

NONMALIGNANT SKIN DISEASE
The most common and debilitating skin problems arise from the long-term use of glucocorticoids. The manifestations include: striae, friable skin, ecchymosis, acneform rash and cushingoid features. In addition, cyclosporine may cause thickening of the skin, hypertrichosis, gingival hyperplasia, epidermal cysts, pilar keratosis, and folliculitis. Treatment includes reducing the dosage of these medications to the lowest level necessary to maintain immunosuppression and topical measures.

Fungal infections of the skin are common among transplant recipients include: tinea, molluscum contagiosum, malassezia furfur and Candida. Treatment with topical agents is usually adequate, and systemic spread is extremely rare.

Viral warts occur in more than 50% of transplant recipients, usually in sun-exposure areas. The causative agent of warts is the Human Papilloma Virus (HPV) of which there are 18 different subtypes. Type 5 may predispose to squamous cell carcinoma of the skin. Warts and skin cancer often appear in the same patient. Genital warts (condyloma acuminata) result from sexual transmission of a Human Papilloma Virus (HPV). The lesions are often resistant to therapy with topical preparation. Further, the lesions may recur after surgical removal, fulguration, cryotherapy, or laser therapy. Both condylomata and cervical neoplasias occur with an increased incidence following transplantation and often appear together in the same patient.

HYPER- AND HYPOCALCEMIA

Approximately 10% of transplant recipients develop hyperkalemia. The hyperkalemia is due to persistent hyperparathyroidism. Secondary hyperparathyroidism is improved following successful kidney transplant due to restoration to normal levels of the active form of vitamin D [1,25 (OH2 D3] and the reversal of hyperphosphatemia. In most cases, mild hyperkalemia resolves slowly over 6-12 months as the parathyroid glands involute. Severe or persistent hyperkalemia may require parathyroidectomy. The indications for parathyroidectomy: 1) Acute, severe hyperkalemia (calcium level > 14 mg/dl) elevation of the PTH levels and evidence of continuing or worsening bone disease of hyperparathyroidism (ostitis fibrosa cystica) and proximal myopathy. Phospage supplementation is usually all that is needed to keep plasma calcium and phosphate at acceptable levels while the parathyroid glands involute.

Patients who have undergone parathyroidectomy prior to transplantation may develop severe hypocalcemia posttransplant. This is caused by 1) PTH deficiency, 2) temporary deficiency of the active form of vitamin D [1,25 (OH) 2 VitD3] immediately following transplantation before new kidney begins to function and converts the inactive 25 (OH VitD3 to the active form, 3) calcium loss in the urine by the newly functioning kidney, and 4) steroid-induced reduction in intestinal calcium symptomatic.

Treatment of severe symptomatic hypocalcemia see Table 17.4. Treatment for asymptomatic hypocalcemia is the addition of oral calcium supplements as calcium carbonate to provide 1 gm of elemental calcium t.i.d. each day and 1,25 (OH)2 D3 in a starting dose of 0.25 mg p.o. each day.

HYPOCHLOREMIC METABOLIC ACIDOSIS

The finding of hyperchloremia and hypocarbonatremia may be due to hyperchloremic metabolic acidosis or primary respiratory alkalosis. The pH on an arterial blood gas will differentiate between these two possibilities. The differential diagnosis for a hyperchloremic metabolic acidosis includes: urinary diversions such as an ileal conduit, Addison's disease, diarrhea and renal tubular acidosis (RTA). One can differentiate among these possibilities with the patients

17

history, physical examination, and obtaining a spot urine sample and measuring sodium, potassium, and chloride. Ideally, direct measurement of urinary ammonium would help distinguish between renal and nonrenal metabolic acidosis. Urinary ammonium determinations, however, are technically difficult to perform. The urinary anion gap, defined as (Ung +UK)-Ucl is used to indirectly estimate the urinary ammonium excretion and normally ranges from −10 to +10. The urinary anion gap represents the amount of unmeasured anions in the urine. These unmeasured anions are accompanied by acid excreted in the form of ammonium. In patients with nonrenally induced metabolic acidosis (diarrhea), increased acid production is accompanied by increased renal acid excretion as the kidney compensates. As a result, the kidney excretes ammonium chloride (NH4CL) lending to a disproportionate increase in urine chloride in relation to urine sodium and potassium. Therefore, the urine anion gap is a large negative value (> -10). In contrast, when the kidney induces a metabolic acidosis by failing to excrete acid at a normal rate, urinary ammonium chloride is very low, leading to a positive urinary anion gap. For example, the urinary anion gap is positive (> +10) in RTA.

Renal tubular acidosis is most commonly due either to a defect in acid secretion by the distal tubule and/or aldosterone deficiency. Cyclosporine and tacrolimus commonly cause distal tubule damage to the sites where acid is secreted (into the urine) and may also render the distal tubule unresponsive to aldosterone. In the precyclosporine era, the development of RTA indicated immune-mediated impairment of hydrogen ion secretion as part of the rejection process. In addition, some patients are aldosterone deficient, primarily diabetics.

Treatment of metabolic acidosis attempts to correct the serum bicarbonate to near normal levels to improve growth in children and ameliorate the effects of acidosis on the bones. The distal tubule normally excretes 1 mEq of acid/kg body weight per day. Since we do not want to cause a metabolic alkalosis bicarbonate therapy may be initiated at 0.5 mEq per kg body weight and then adjusted by the serum bicarbonate level.

SELECTED READINGS

1. Rao Uk ed Renal Transplantation Surgery Clinic North American 1998
2. Danovitch GM ed Handbook of Kidney Transplantation 2nd edition 1996, Little, Brown, and Co.

Late Complications of Transplantation

Bruce Kaplan and Herwig-Ulf Meier-Kriesche

With improvements in surgical technique and the advent of more potent and selective immunosuppressive agents, early complications of whole organ transplantation have been reduced. This improvement in early outcome has led to a greater emphasis on the management of late complications of the transplant patient. The fact that improvements in long-term graft (and patient) survival have not kept pace with the improvement in short-term survival points out the necessity of concentrating on these late problems. Broadly, late complications can be classified into several areas: Those affecting the kidney, e.g., chronic rejection and recurrent renal disease, cardiovascular diseases, metabolic disease, bone disease, and infectious complications. Further, these complications can either be secondary to the medications used to control the acute rejection process, comorbid disease states, the aging process, or some combination of the three. This chapter will attempt to cover these late complications with concentration on pathophysiology as well as theoretical and practical treatment options.

CHRONIC ALLOGRAFT NEPHROPATHY

The past 15 years has seen a marked improvement in early graft survival. However, while later graft survival has not kept pace with the remarkable gains made in early (1 year) graft survival. The most common cause of late allograft loss is due to the process of chronic allograft nephropathy (formerly referred to as chronic rejection). Chronic allograft nephropathy can be defined as a chronic deterioration in renal function which cannot be explained by other known processes (e.g., acute rejection, recurrent disease) which is also accompanied by characteristic histologic lesions of interstitial fibrosis, vasculopathy, and glomerular sclerosis. The diagnosis of chronic allograft nephropathy (CAN) is therefore one of exclusion and it is important to search critically for treatable causes of the renal deterioration (particularly if the histologic lesions are nonspecific).

The etiology of chronic allograft nephropathy has to this date not been fully elucidated. Both antigen dependent and antigen independent processes may be important.

POSSIBLE IMMUNOLOGIC ETIOLOGIES

In rat cardiac allograft models, both class I and class II MHC mismatches are associated with a mononuclear cell infiltrate which is accompanied by vascular thickening and arteriosclerotic lesions common in clinical chronic rejection in hearts and CAN in kidneys. This effect may be accentuated by acute nonspecific injury as it has been demonstrated that ischemic injury may upregulate class II

18

Organ Transplantation, 2nd edition, edited by Frank P. Stuart, Michael M. Abecassis and Dixon B. Kaufman. ©2003 Landes Bioscience.

MHC molecules important in the allo/inflammatory response. Several animal models have demonstrated a humoral component to the lesion of chronic rejection. Rats tolerant to MHC class I allopeptides show little acute cellular rejection, however do develop antidonor antibodies. These antidonor antibodies, particularly IyGIb closely correlate to the development of vascular lesions in the allograft. Another line of evidence implicating humoral response is that of xenografts, where a lesion similar to chronic rejection develops, which seems to correlate to antidonor antibody and b-lymphocytes in the graft. Much recent interest has centered on the role of transforming growth factor β (TGF-β) and chronic allograft nephropathy. Suthanthiran and colleagues have demonstrated that intragraft TGF-β1 mRNA correlates with the presence of chronic allograft nephropathy in clinical specimens. It is also interesting to note that the fibrotic lesion produced by CsA in the salt depleted Sprague Dawley rat may be secondary to angiotension II mediated TGF-β production and that at least experimentally, maneuvers that decrease AII (e.g., ACE inhibitors) seem to decrease CsA mediated fibrosis in this animal model.

POSSIBLE NONIMMUNOLOGIC ETIOLOGIES

Ischemia and reperfusion injury seems to produce a multitude of changes conducive to end organ damage. It is widely appreciated that ischemia is capable of producing a number of cytotoxic leukocyte mediators which can damage endothelium. In addition, ischemic injury can upregulate a number of cell surface molecules, e.g., adhesion molecules ICAM, VCAM, LFA-1, MHC-1, and II molecules which may contribute to further tissue injury. Halloran has proposed an alternative view of the process whereby through a multiplicity of injuries a state of premature aging occurs and leads to the premature development of fibrosis and sclerosis.

RISK FACTORS FOR CHRONIC ALLOGRAFT NEPHROPATHY

A number of risk factors have been identified to be related to the development of CAN and a decrease in long-term graft survival. It appears that a decrease in the number of donor nephrons relative to recipient body mass may be a correlate to decreased long-term function. The pathophysiology of this relationship may be related to the hyperfiltration theory of renal injury as proposed originally by Brenner. Studies from Minnesota have demonstrated a relationship of acute rejection and the subsequent development of CAN and shortened long-term graft survival. A large retrospective study demonstrated a correlation of CAN and low CsA (Sandimmune formulation) dosage (< 4 mg/kg/day). A similar relationship of low Sandimmune dose and CAN was also demonstrated in a separate single-center study. A study from Houston demonstrated a correlation of variability of cyclosporine levels and chronic rejection in their population. Several studies have demonstrated a relationship of the severity of an acute rejection episode and the subsequent development of chronic allograft nephropathy. Reinsmoen has studied the effect of donor hyporesponsiveness in MLC to donor antigen and has shown a correlation of hyporesponsiveness and a decrease in chronic rejection.[15] In addition, Kerman has demonstrated an increase in donor anti-HLA antibodies in patients who develop CAN as opposed to those who follow a more benign clinical course.

As can be seen the issue of chronic allograft nephropathy is filled with much uncertainty and remains one of the great areas of future research interest.

HYPERTENSION

Hypertension commonly occurs post-renal transplantation. As many as 70% of renal allograft recipients receiving CsA have hypertension severe enough to require medical therapy. Hypertension in this setting is usually multifactorial in nature. Pre-existing disease, renal insufficiency, and immunosuppressant medication (CsA and Prednisone) all may contribute to the hypertension in patients post renal transplantation. While no randomized, controlled study has been performed, it would be reasonable to assume that hypertension in this setting incurs the same risk of end organ damage as it does in the population at large.

It appears that among the medications given posttransplantation, CsA appears to have the most negative effect on blood pressure control. It is interesting to note that while Tacrolimus shares the same immunosuppressant mechanism of action as CsA it does not seem to produce as much hypertension. CsA probably produces elevated pressure by both volume dependent and volume independent mechanisms of action. CsA will induce a predictable increase in tubular sodium resorption and thus increases plasma volume. In addition, CsA can produce vasoconstriction in several vascular beds, particularly in the afferent arteriole of the glomerulus. CsA-induced vasoconstriction appears to be related to an increase in endothelin levels, a decrease in nitric oxide, and an imbalance of vasodilating and vasoconstrictive prostanoids. Calcium channel blockers of all three classes appear to be the best agents in obviating the renal vasoconstriction induced by CsA.

Many patients with pre-existing hypertension often will have a worsening of their hypertension posttransplant and require more intensive medical therapy.

Stenosis of the transplanted renal artery can be the etiology of a resistant posttransplant hypertension. Patients with RAS of the transplanted artery will often have a deterioration of their renal function as blood pressure is controlled. In these patients, an interventional procedure (e.g., stenting, angioplasty or bypass) often is helpful in both controlling blood pressure and preserving renal function.

Treatment of posttransplant hypertension should broadly have the same goal as treatment in the nonrenal transplant recipient, that of prevention of end organ damage to the heart, kidney, and prevention of cerebral vascular events. In this regard, basic principles of hypertension control should be followed in this setting, e.g., β-blockers for patients with coronary artery disease. Table 18.1 outlines various antihypertensive medications and lists possible positive and negative effects of these agents.

There appears to be no consensus on the routine use of any particularly antihypertensive regimen in the posttransplant period. However, certain theoretic and practical issues should be kept in mind. In general, calcium channel blockers have been shown to ameliorate CsA-induced renal vasoconstriction. In addition to this possible advantage, there are certain caveats to their use. Both Verapamil and Diltiazem can significantly increase CsA and tacrolimus levels, probably by both inhibiting cytochrome P450 IIIA activity as well as interstitial p glycoprotein

18

Table 18.1. Various antihypertensive medications

Medication	Positives	Negatives
Specific β₁ blockers, e.g., Atenolol, Metoprolol	Agents of choice in patients with CAD and good LV function.	↓ cardiac output. May ↓ renal blood flow.
Alpha 1 blockers, e.g., Terazosin	May help in patients with BPH. May ↑ RBF.	Orthostatic hypotension.
Alpha 1 blocker, + nonspecific β-blocker, e.g., Labetelol	May be effective for patients who need both vasodilatation and heart rate control.	Bronchospasm. Avoid with PVD. Possible ↑ of CVA. Induce hyperkalemia. ↓ CO.
Calcium Channel Blockers: a. dihydropyridines, e.g., Nifedipine	Best agents to prevent CsA-induced vasoconstriction.	Edema (nonsodium retentive). Reflex tachycardia. Avoid with active CAD.
b. Diltiazem	Can prevent CsA-induced vasoconstriction.	↑ CsA, Tacrolimus levels.
c. Verapamil	Can prevent CsA-induced vasoconstriction.	↑ CsA, Tacrolimus levels. Constipation. Bradycardia. ↓ CO.
ACE inhibitors, e.g., enalapril, angiotensin II blockers, e.g., losartan	May be best agents for patient with ↓ LV function or LV dilatation. May be best agents to decrease proteinuria experimentally may decrease CsA medicated renal fibrosis.	Hyperkalemia. Deterioration in renal function (particularly in patients with RAS).
Alpha 2 agonists, e.g., Clonidine	Effective in many people. Decreases central sympathetic discharge.	Drowsiness, dry mouth.
Direct vasodilators, e.g., Hydralazine, Minoxidil	Hydralazine often used with patients intolerant to ACEI or AII blocker with CHF. Minoxidil effective in refractory patients.	Reflex tachycardia. Na⁺ retention.
Diuretics, e.g., Furosemide	Often necessary in volume extended patients.	Electrolyte disorder. Experimentally may exacerbate CsA-induced fibrosis.

activity. In general, the dyhydropyridines exhibit much less pharmacokinetic interaction with either of these agents. The dihydropyridine calcium channel blockers may exacerbate CsA-induced gingival hyperplasia as well. Nonspecific β-blockers may exacerbate CsA-induced hyperkalemia. ACE inhibitors can precipitate a decrease in GFR in the setting of CsA or tacrolimus afferent arteriola

18

vasoconstriction. Volume depletion induced by diuretics may activate the Renin-AII system and at least in experimental models may worsen CsA-induced chronic toxicity. Again, the goal is to individualize therapy for the particular patient, bearing in mind both long-term benefits and keeping in mind the specific issues involved with patients on current immunosuppression therapy.

HYPERLIPIDEMIA

Hyperlipidemia is common in the posttransplant period. As with many of the other complications seen in the transplant recipient, this hyperlipidemia is often multifactorial. It has been noted that the pattern of increased lipids changes from the period of time on dialysis to the period after transplantation. Patients on hemodialysis often show increased triglycerides along with elevated LDL levels, the posttransplant period is characterized by a more type IIA pattern. Post-transplantation HDL levels, as well as LDL levels, may become elevated. In addition, other lipoproteins, e.g., lipoprotein A, are also elevated.

A number of the immunosuppressive medications used contribute to this hyperlipidemia. CsA tends to increase both LDL cholesterol as well as lipoprotein A. Tacrolimus may have less effect in this regard than CsA (perhaps secondary to the different carrier protein used by these drugs, CsA-lipoproteins and tacrolimus-albumin). Corticosteroids increase both LDL and HDL proportionally. There is evidence, however, that this increased HDL is not a "protective form of HDL". Rapamycin can produce marked elevations in triglycerides and the mechanism of this effect is currently under investigation.

Comorbid disease also can adversely effect the lipid profile in the posttransplant recipient. Diabetes can produce the typical hyperglyceridemia and hypercholesterolemia seen in this patient population. Significant proteinuria can markedly elevate both LDL cholesterol and lipoprotein A. Finally, as hyperlipidemia is epidemic in the general population (possibly secondary to dietary factors) there is no reason to believe that the transplant recipient would be immune to these factors.

Treatment of hyperlipidemia in these patients broadly should follow the guidelines used for most patients with dyslipidemias. Special considerations in the transplant recipient are described below.

The serum levels of Hmg-CoA reductase inhibitors are increased by the concurrent use of cyclosporine, leading to an increased risk of rhabdomyolysis and liver dysfunction. It should be noted that this effect of CsA on Hmg CoA reductase inhibitors has marked drug variability, i.e., certain of the "Statin" levels are increased to a much greater extent than others. An increased incidence of rhabdomyolysis has also been observed with the use of fibrate-type agents. In patients with significant proteinuria, interventions that decrease proteinuria, e.g., institution of ACE inhibitors, also may improve the lipid profile. Finally, the utility of dietary intervention and weight loss, while difficult to institute in these patients, may also be of use.

18

POSTTRANSPLANT DIABETES MELLITUS

Anywhere from 5-40% of recipients of renal allografts will develop posttransplant diabetes mellitus (PTDM). This hyperglycemia is generally medication-related. Corticosteroids increase gluconeogenesis, and produce end organ resistance to insulin effect. Both CsA and Tacrolimus directly inhibit islet cell release of insulin and are additive in their effect on blood glucose with corticosteroids. It appears that increase in age, a positive family history for diabetes, and African American race all predispose to the development of PTDM. Clinically, Tacrolimus appears to produce more hyperglycemia than CsA.

Treatment of PTDM may respond to dietary counseling as well as judicial lowering of steroids or calcineurin inhibitors. Oral hypoglycemics are frequently used and one must remember that most of these have at least partial renal excretion and therefore must be used carefully in patients with renal impairment. Troglitazone (Rezulin) seems to induce the cytochrome IIIA system and can decrease CsA and probably Tacrolimus levels. Insulin therapy is frequently necessary and the same guidelines that are used in other settings should be utilized in the posttransplant patient.

RECURRENT RENAL DISEASE

The impact of recurrent renal disease on long-term graft survival has been difficult to quantify as the initial disease of the recipient is often unknown and changes on biopsy that are characterized as chronic rejection often overlap with those of recurrent renal disease.

Diabetic Nephropathy: Changes consistent with diabetic retinopathy will almost always occur in patients who lost their native kidney from diabetes mellitus. However, given the prolonged time from exposure to over disease (~10 years) this rarely is a cause of allograft loss. Pancreatic transplantation appears to be predictive and may even reverse changes associated with diabetic nephropathy.

Focal Segmental Glomerulosclerosis (FSGS): Primary FSGS recurs frequently (~30-40%) if a patient has had graft loss due to recurrent FSGS, subsequent graft loss due to FSGS may run as high as 90%. Recurrence can occur as early as one week posttransplant and is heralded by massive proteinuria. Graft loss is frequent in recurrent FSGS. Recurrence of FSGS has been reported to be associated with a 50kd protein in the serum. High serum levels of this factor have been associated with a high rate of recurrent disease. Lowering of these levels has been associated with remission of disease. At this point in time there is no optimal treatment for recurrent FSGS, plasmapheresis, increased doses of calcineurin inhibitors, and cytotoxic therapy have been used with varying degrees of success.

Membranoproliferative Glomerulonephritis (MPGN): Type I MPGN recurs in up to 30% of allografts and may be a cause of early graft loss in up to 40% of these patients.

Type II MPGN recurs in the majority of patients and may cause graft loss in almost 50% of these patients. To date, there is no definitive treatment for either type I or type II MPGN.

IgA Nephropathy (IgAN): Histologic recurrence occurs frequently in patients with IgA nephropathy. Early graft loss is very rare due to recurrence. It was thought that recurrent IgA disease had little impact on long-term graft survival, however, more recent data indicates that recurrent IgA may decrease graft survival in a small number of patients.

Hemolytic Uremic Syndrome (HUS): Patients who have lost native function secondary to HUS have up to a 50% chance of recurrent disease. This may be exacerbated by the use of calcineurin inhibitors. If a familial form of HUS occurs, native nephrectomy has been reported to decrease the incidence of recurrence in these patients.

Systemic Lupus Erythematosis (SLE): SLE may recur in the renal allograft, however is a relatively rare cause of allograft loss. Transplantation should be delayed if severe activity of disease (or serologic evidence) is present.

Anti-GBM Disease: Anti-GBM rarely recurs. Patients should not be transplanted until evidence for anti-GBM antibodies has disappeared.

Pauci-immune Glomerulonephritis: Anti neutrophilic antibody related glomerulonephritis rarely recurs, severe extrarenal manifestations may be present. Generally, a one year wait after quiescence of the renal disease is advised.

Sickle Cell Disease: Patients with sickle cell disease can be safely transplanted. However, a higher rate of sickle crisis may be observed. Graft survival appears to be poorer in patients with sickle cell disease.

Alports Syndrome: A genetic defect in type IV collagen synthesis often produces renal failure. Patients with alports syndrome generally do very well however, there are anecdotal reports of patients with alports disease may rarely develop anti-glomerular basement membrane disease.

PREGNANCY

Successful pregnancy is rare in patients on maintenance dialysis. While it is not common, successful pregnancy can be achieved in many patients posttransplantation. As with other renal disease, degree of renal function offers the best prognostic sign of any parameter. Patients with creatinines less than 2 mg/dl have the greatest likelihood of successful full term pregnancy. As renal function decreases, the chances of successful pregnancy fall.

The effect of pregnancy on allograft function can be variable. In patients with good baseline function, pregnancy is well tolerated. During pregnancy, GFR may increase as may proteinuria. In patients with compromised renal function, pregnancy may have an adverse effect on allograft function and survival.

Pregnancy-induced hypertension and pre-eclampsia occur approximately four times more frequently in transplant recipients as in normal pregnancies. The diagnosis of pre-eclampsia is difficult in these patients and may require close follow-up of the patient. Urinary tract infections are more frequent in pregnancy as is allograft pyelonephritis.

Numerous normal pregnancies have been reported for women who have been on cyclosporine, azathioprine, and prednisone. While there are theoretic concerns with the use of these agents in pregnancy, none of them are absolutely contrain-

18

dicated in pregnancy. Newer agents, e.g., mycophenolate mofetil and rapamycin, have unknown effects and at this writing, should be avoided if conception is contemplated.

POSTTRANSPLANT ERYTHROCYTOSIS (PTE)

Elevated hematocrits occur in up to 15% of all patients postrenal transplantation. The etiology of this erythrocytosis is not clear. Elevated erythropoietin levels have been found in some patients, but this is not a universal finding. Some investigators have hypothesized either altered feedback regulation of epo production, while others have postulated an increase in end organ epo sensitivity.

Treatment for PTE should be initiated in any patient with a hematocrit of >52 to avoid thromboembolic and central nervous system complications. Before initiation of therapy, a search should be made for other reasons for erythrocytosis such as hemoconcentration or hypoxemia. For patients with true PTE the treatment of choice is either ACE inhibitor therapy or phlebotomy. Recent reports indicate that the angiotensin II receptor blockers also have utility in this setting. Theophylline can also decrease HCTs in PTE, however, the patient often poorly tolerates the use of Theophylline.

POSTTRANSPLANT MALIGNANCIES

Transplantation and the use of immunosuppressive medications seem to increase the risk of some, but not all, types of malignancies. The incidence of colon, breast, prostate, and lung cancer does not seem to be increased in posttransplant recipients. The incidence of renal cell carcinoma is difficult to assess, however, does not appear to be greater in renal transplant recipients than what is found in the nontransplant end stage renal disease population.

The incidence of lymphomas, skin cancer, and Kaposi's sarcoma does appear to be significantly increased in posttransplant recipients.

SKIN CANCER

Skin cancer occurs at a frequency up to twenty times greater in posttransplant recipients than in the general population. While basal cell carcinoma is more common in the general population, squamous cell carcinoma seems to occur more frequently in transplant recipients. In addition, melanoma occurs more frequently in transplant recipients. In general, whatever the type of skin malignancy that occurs, it tends to take a more aggressive course in the transplant recipient than in the nonimmunocompromised individual. Early detection and aggressive management should be incorporated in all long-term follow-up of transplant recipients. Sun exposure should be dissuaded and regular complete skin evaluations should be employed.

KAPOSI'S SARCOMA (KS)

Kaposi's sarcoma is not infrequent in the posttransplant period. It has recently been related to the presence of Herpes virus VIII. However, as of this writing, the casual relationship of HSV VIII and KS has not been totally established. KS usually presents as a bluish and often macular skin lesion. However, it may affect

18

other sites including the oropharynx, lung, and other internal organs. KS may respond to decreased or cessation of immunosuppression, but at times may require either chemotherapy or radiotherapy.

LYMPHOMA

Most lymphomas posttransplant are B-cell lymphomas of the non-Hodgkin's type. The nomenclature of B-cell lymphoma in this setting is confusing and is often referred to as posttransplant lymphoproliferative disease (PTLD). Almost all cases of PTLD are related to infection with Epstein-Barr Virus (EBV). Patients initially seronegative for EBV who receive organs from donors serologically positive for EBV seem to be at highest risk for the development of PTLD.

As opposed to nontransplant lymphomas, PTLD is often extra-nodal and may first present as involvement of such disparate sites as the GI tract, lung, or kidney. The risk of PTLD also seems to be related to the degree and type of immunosuppression. In particular, repeated anti-lymphocyte antibodies seem to incur a higher risk for development of PTLD. However, PTLD also seems to have a tendency to increase when very large doses of either Cyclosporine or Tacrolimus were used compared to lower doses of these agents that are now in common use.

Therapy for PTLD varies, occasionally a decrease in immunosuppressive medications may reverse the process. This option is most appropriate in vital organ transplants. Cessation of immunosuppression completely is also an option. Chemotherapy and radiotherapy can also be employed. The use of Acyclovir has also been advocated. Regardless of therapy, the prognosis for this entity is generally poor. Much discussion has been made of the clonality and presence of gene rearrangements, particularly as to guide therapy. Whether this distinction truly benefits patients remains to be seen.

SKELETAL COMPLICATIONS

Renal transplantation is associated with several functional and structural abnormalities of the skeletal system. Some of these complications such as persistent hyperparathyroidism, aluminum-associated bone disease, B-2 microglobulin amyloidosis may be preexistent at the time of transplantation. Subsequent development of osteopenia, osteonecrosis and gout may cause additional skeletal problems. The most common posttransplant skeletal disorders are immunosuppression related bone disease, painful leg syndrome and avascular bone necrosis.

Osteoporosis is prevalent after renal transplantation and is characterized by decreased bone mineral density [BMD], fractures and skeletal pain. The most significant bone loss is noted during the early posttransplant period when corticosteroid dosage is high and seems to correlate directly with total amount of corticosteroid used. Additionally cyclosporine and persistent elevation of parathyroid hormone levels may exacerbate the bone loss after transplantation. The kidney-pancreas allograft recipients seem to be at an increased risk for fracture due to steroid induced osteopenia as compared to recipients of kidney alone. Treatment for - after transplantation is quite challenging since significant bone loss has already occurred. Current attempts to treat these patients include steroid withdrawal regimens, alendronate, calcium supplements and calcitonin.

18

Ten percent of renal transplant patients experience the painful leg syndrome during the posttransplant period. This is characterized by symmetrical pain in the knees and ankles during the first few months of the posttransplant period and the pain is relieved in a few weeks. Appearance of the syndrome coincides with the initiation of Cyclosporine therapy and may be relieved by addition of calcium channel blockers. Plain x-rays and bone scans do not help. MRI scans often show stress fractures. Pathogenesis of this syndrome is not known and there is no correlation with pretransplant hyperparathyroidism.

Prevalence of avascular necrosis of the hip in renal transplant patients is between 5% and 37%. This is a most disabling complication that develops between 6 months and 2 years in the posttransplant period. The patient often presents with pain on the affected side and is frequently severe enough to impede walking. Bilateral avascular necrosis of the hips is common and several other bones may also be affected in 30% of patients. This complication is directly due to steroids and there seems to be a relationship between the cumulative steroid dose and necrosis as well as other bone fractures. Diagnosis can be made with the currently available MRI technology. When patients develop severe symptoms surgical treatment often becomes necessary. There are many surgical approaches such as drilling of the femoral head and neck, free fibular grafting, cup arthroplasty, hemiarthroplasty and cemented or cementless total hip arthroplasty.

SELECTED READINGS

1. Cecka JM, Terasaki PI. The UNOS scientific renal transplant survey. In: Terasaki PI, Cecka JM, eds. Clinical Transplants 1994. Los Angeles: UCLA Tissue Typing Laboratory, 1995; 1-18.
2. Paul LC. Chronic renal transplant loss. Kidney Int 1995; 47:1491-1499.
3. Raisanen-Sokolowski A, Glysing-Jensen T, Mottram P et al. Sustained anti-CD4/CD8 blocks inflammatory activation and intimal thickening in mouse heart allografts. Arterioscl Thromb Vasc Biol 1997; 17:2115-2122.
4. Russell ME, Wallace AF, Hancock WW et al. Upregulation of cytokines associated with macrophage activation in the Lewis to F344 rat chronic cardiac rejection model. Transplantation 1995; 59(4):572-578.
5. Mhoyan A, Cramer DV, Baquerizo A et al. Induction of allograft nonresponsiveness following intrathymic inoculation with donor class I allopeptides. I. Correlation of graft survival with anti-donor IgG antibody subclasses. Transplantation 1997; 64:1665-1670.
6. Kerman RH, Susskind B, Kerman DH, Lam M et al. Anti-HLA antibodies detected in posttransplant renal allograft recipient sera correlate with chronic rejection. Transplant Proc 1997; 29:1515-1516.
7. Tullius SG, Tilney NL. Both alloantigen-dependent and-independent factors influence chronic allograft rejection. Transplantation 1995; 59:313.
8. Brenner BM, Cohen RA. Milford EL. In renal transplantation, one size may not fit all. J Am Soc Nephrol 1992; 3:162.
9. Almond PS, Matas A, Gillingham K et al. Risk factors for chronic rejection in renal allograft recipients. Transplantation 1993; 55:752-757.
10. Cosio FG, Pelletier RP, Falkenhain ME et al. Impact of acute rejection and early allograft function on renal allograft survival. Transplantation 1997; 63:1611-1615.

18

11. Reinsmoen NL, Matas AJ. Evidence that improved late renal transplant outcome correlates with the development of in vitro donor antigen-specific hyporeactivity. Transplantation 1993; 55:1017.

12. Curtis JJ. Hypertension following kidney transplantation. Am J Kidney Dis 1994; 23:471-475.

13. Massey AZ, Kasiske BL. Posttransplant hyperlipidemia: Mechanisms and management. JASN 1996; 7:971-977.

14. Markell MS, Armenti V, Danovitch GM et al. Hyperlipidemia and glucose intolerance in the postrenal transplant patient. J Am Soc Nephrol 4(Suppl):S37, 1994.

15. Sumrani NB, Delaney V, Ding Z et al. Transplantation 1991; 51:343.

16. Savin VJ, Sharma R, Sharma M et al. Circulating factor associated with increased glomerular permeability to albumin in recurrent focal segmental glomerulosclerosis. New Eng J Med 1996; 334:878, 1996.

17. Armenti VT, Ahlswede KM, Ahlswede BA et al. Variables affecting birthweight and graft survival in 197 pregnancies in cyclosporine-treated female kidney transplant recipients. Transplantation 1995; 59:472-476.

18. King GN, Healy CM, Glover MT et al. Increased prevalence of dysplastic and malignant lip lesion in renal transplant recipients. New Eng J Med 1995; 332:1052.

19. Kuo PC, Dafoe DC, Alfrey EJ et al. Posttransplant lymphoproliferative disorders and Epstein-Barr virus prophylaxis. Transplantation 1995; 59:135.

20. Churchill MA, Spencer JD. End stage avascular necrosis of bone in renal transplant patients: The natural history. J Bone Joint Surg Br 1991; 73:618-620.

21. Grotz WH, Mundlinger FA, Gugel B et al. Bone fracture and osteodenistometry with dual energy x-ray absorptiometry in kidney transplant recipients. Transplantation 1994; 58:912-915.

22. Chiu MY, Sprague SM, Bruce DS et al. Analysis of fracture prevalence in kidney pancreas allograft recipients. JASN 1998; 9:677-683.

23. Grotz WH, Mundlinger FA, Gugel B et al. Bone loss after kidney transplantation: A longitudinal study in 115 graft recipients. Nephrol Dial Transplant 1995; 10:2096.

24. Grotz WH, Rump LC, Niessew A et al. Treatment of osteopenia and osteoporosis after kidney transplantation. Transplantation 1998; 66:1004-1008.

18

Organ Transplantation Finance

Michael M. Abecassis, Dixon B. Kaufman and Frank P. Stuart

INTRODUCTION

Unusually complex and expensive, organ transplantation is perhaps the most intensely regulated of medical disciplines. The extensive system of financial tracking and reimbursement for organ transplants reflects this complexity. Unfortunately, the surgeons and physicians most directly involved in evaluating and caring for transplant patients and donors are often largely unaware of how—or even whether—compensation occurs for their services.

Although a hospital "owns" and is responsible for administration of a transplant program, it is the medical director appointed for each transplanted organ who plans, organizes, and leads the extensive staff in executing the transplant center's activities. Transplant surgeons and physicians must understand the rules and accounting practices peculiar to transplantation so they can work effectively with their medical directors and to optimize reimbursement for their services.

Without knowledge and advocacy of the fiscal issues surrounding organ transplantation, the fiscal health of a seemingly robust transplant program may be undermined. In contrast, when the hospital and its physicians and surgeons are equally concerned about the other's fiscal integrity, transplant centers can thrive and better provide high-quality organ transplant services.

MEDICARE REIMBURSEMENT BASICS

Reimbursement for chronic hemodialysis and kidney transplantation was scarce until the mid-1960s and haphazard until 1972, when Congress amended the Social Security Act to include coverage for end-stage renal disease for qualified patients. Because the amendment made Medicare the primary kidney transplantation insurer for more than 90% of the United States population, it was Congress and the U.S. Department of Health and Human Services, rather than the commercial health insurance industry, that wrote the rules and accounting practices for kidney transplantation. As transplantation of heart, lung, liver, pancreas, and intestine evolved, the rules and accounting practices applicable to the kidney were extended to other transplantable organs.

For a hospital to bill Medicare for transplant services, it must first apply for certification for each organ it proposes to transplant. If approved, the hospital then becomes a certified transplant center for those organs. Medicare reimbursement occurs through three main channels: diagnostic related groups (DRGs) paid to the hospital, organ acquisition cost centers (OACC) paid to the hospital, and physician reimbursement.

Organ Transplantation, 2nd edition, edited by Frank P. Stuart, Michael M. Abecassis and Dixon B. Kaufman. ©2003 Landes Bioscience.

DIAGNOSTIC RELATED GROUPS

Diagnostic related groups (DRGs) define the payment a hospital receives for the organ transplant itself. First implemented in 1983 in a federal attempt to control escalating Medicare costs, DRGs were used to categorize patients for billing purposes. The introduction of DRGs marked the shift from a reimbursement system based on retrospective charges (after the delivery of care) to one based on "prospective" payment. This meant that hospitals would receive compensation for a patient's care based on the qualifying DRG instead of services provided. In other words, Medicare provides the same reimbursement for each diagnosis regardless of how complex the case or high-risk the patient.

Based on a patient's primary and secondary diagnoses, primary and secondary procedures, age, and length of stay, DRGs are used to determine compensation for all acute care, nonspecialty hospitals in the United States. With a uniform cost established for each category, each DRG is one of about 500 possible classifications in which patients with similar lengths of stay and resource use are grouped together. DRGs set the maximum amount that Medicare will pay for a patient's care, motivating hospitals and healthcare providers to keep costs down, as they profit only if their costs for providing a patient's care are less than the amount indicated by the DRG category.

Total DRG payment is derived from a calculation involving several factors: a national standard amount, a local wage index, and DRG weight, which weighs diagnoses against each other in terms of resource consumption. For example, a DRG with a weight of 3 would be expected to consume twice the resources of one at 1.5. The national standard amount is divided into a labor-related portion and a nonlabor portion, and a formula exists for annual inflation index adjustment.

A sample DRG payment for a patient admitted to a hospital in Madison, Wisconsin for a kidney transplant would be calculated as follows: Payment = [(labor amount x wage index) + nonlabor amount] x DRG relative weight. The national standardized amount for large urban areas for labor-related costs is $3,022.60 and $1,228.60 for nonlabor-related costs. The wage index for Madison is 1.0467, and the relative weight of the DRG for kidney transplants is 3.3000. Therefore, payment = [(3022.60 x 1.0467) + 1228.60] x 3.3 = $14,494.77. Table 19.1 shows an example of relative organ transplant DRG-based payments.

Only one DRG is assigned per patient admission, but certain circumstances or institutions may be eligible for rate adjustments or new DRGs. For instance, additional costs for managing complications during the initial hospitalization are expected to be covered by the DRG payment. If the patient is discharged and readmitted because of a medical problem such as rejection, the hospital is assigned a new DRG for the patient. Some transplant centers may be eligible for potential adjustments to DRG payment rates if they are part of a teaching hospital, where payments may consider factors such as graduate medical education (GME) costs and indirect medical education (IME) costs. Certain teaching institutions also may be eligible for disproportionate share (DSH) adjustments, which approximate the higher costs associated with treating indigent patients, the special need for translators, social services, higher security costs, and other related

19

Table 19.1. Approximate organ transplant DRG-based payments

DRG	DRG Title	Amount
103	Heart transplant	$87,327.73
148	Major small and large bowel procedure (used for intestinal transplants)	$14,667.92
302	Kidney transplant	$14,028.96
480	Liver transplant	$44,129.58
495	Lung transplant	$39.117.84
512	Simultaneous pancreas/kidney transplant	$24,917.56
513	Pancreas transplant	$26,897.76

costs that often occur in large teaching hospitals. Changes in factors such as IME following the Balanced Budget Act (BBA) in 1997 may result in changes in DRG reimbursement for various services.

Hospitals also may be reimbursed for unusually high discharge expenses. A hospital may receive an outlier payment if the cost of a discharge, calculated through a hospital-specific, cost-to-charge ratio, exceeds a certain threshold amount. In fiscal year 2003 the threshold was $33,450 above the base payment for the DRG, so if a discharge reaches that threshold, Medicare will pick up about 80 percent of the overage.

The Medicare Payment Advisory Commission (MedPAC) re-calibrates and re-classifies the DRG on a periodic basis based partly on cost reports that all hospitals file annually. MedPAC is an independent federal body made up of 17 members that meet publicly to discuss policy issues and formulate recommendations to the U.S. Congress on improving Medicare policies. Changes are published every year in the Federal Register. For fiscal year 2003, there were 510 DRGs.

MEDICARE PHYSICIAN FEE SCHEDULE

The Medicare physician fee schedule determines reimbursement to physicians and surgeons for services rendered in the care of a transplant recipient under Medicare Part B (as opposed to Part A, which goes to the hospital). Published annually by the U.S. Department of Health and Human Services (HHS), the fee schedule identifies a prespecified reimbursement rate for each service it identifies. Services are described by Current Procedural Terminology (CPT) codes, a uniform coding system for healthcare procedures developed by the American Medical Association (AMA) that is used for submitting claims.

The fee schedule is a resource-based relative value system (RBRVS). Payment for each service in the schedule is based on three factors:

1. A nationally uniform relative value for the specific service. This relative value is based on calculations for each service based on components of work (RVU_w), practice overhead (RVU_{pe}) and professional liability (RVU_L) and is referred to as a relative value unit, or RVU.

2. A geographically specific modifier that considers variation in different areas of the country. Each area of the country has its own geographic practice cost indices (GPCI) for each of the relative value factors of work, practice overhead, and professional liability.

3. A nationally uniform conversion factor that is updated annually. The conversion factor for 2003 is $36.7856. This rate is up slightly from 2002, $36.1992.

The fee schedule is calculated according to several variables: the procedure performed, the CPT code for the procedure, the RVU factor for the procedure, the GPCI modifiers for the geographic area, and the national conversion factor for the year.

For surgeons, most of the payment is based on the number of work RVUs because it reflects the physician's services. The annual fee schedule increase for physicians is based on the Medicare economic index. This index is limited by the sustainable growth rate (SGR), which HHS determines to estimate how much Medicare expenditures for physician services should grow each year.

The Medicare fee schedule also is modified with input from the Relative Value Update Committee (RUC), a body convened by the AMA. The RUC is comprised of 29 members. Twenty-three are appointed by major national medical specialty societies, including three rotating seats (two of which are reserved for an internal medicine subspecialty with the other open to any other specialty) whose membership rotates every two years. The RUC Chair, Practice Expense Advisory Committee Chair, Co-Chair of the Health Care Professionals Advisory Committee (an advisory committee representing non-MD/DO health professionals), and representatives of the AMA, American Osteopathic Association, and CPT Editorial Panel hold the remaining six seats.

The Advisory Committee to the RUC, which is made up of representatives from the medical specialty societies whose members provide the services being valued, develops relative value recommendations for new and revised codes. One physician representative is appointed from each of the 98 specialty societies seated in the AMA House of Delegates to serve on the committee, and members are responsible for presenting their societies' recommendations to the RUC. The Advisory Committee member for each specialty is supported by an internal specialty RVS committee that manages the process of gathering information about the new or revised code(s) by developing vignettes and selecting reference services for use in surveying physicians in their specialties about the work involved in the service; reviewing survey results; and developing relative value recommendations for presentation to the RUC.

One fundamental feature about the Medicare physician fee schedule is that any changes must be budget neutral—in other words, any increase in one budget area must be taken from elsewhere in the budget. Beginning in 1999, the Medicare system began making the practice expense component of the fee schedule resource based. Transplant surgeons fees were significantly reduced for 1999 to 2002 in the fee schedule amounts. Heart-lung transplants lost 31%, kidney transplants about 15%, and liver transplants about 9%. The RUC evaluates all CPT codes every five years and makes its recommendations to CMS, which in turn interprets these recommendations and makes final rules. These are published in the Federal Register and, after a period for public comment, are considered final and comprise the new fee schedule.

19

ORGAN ACQUISITION COST CENTERS (OACC)

Medicare separates payment for organ acquisition costs from both the payment the hospital receives on a DRG basis and the physician fee schedule payment to physicians for services rendered in transplant recipient care.

Organ acquisition cost centers are an accounting category created outside the transplant DRG for each transplanted organ. OACCs were designed to compensate the hospital for reasonable expenses of organ acquisition as well as both living donor and recipient evaluation and selection, as well as maintenance and reevaluation of recipient candidates on waiting lists until transplantation occurs. For certified transplant centers, Medicare pays for organ acquisition costs on a reasonable cost basis, one of the very few areas of hospital payment that is still cost reimbursed. Reimbursement is based on full cost, allowing the hospital indirect costs on all components that reflect overhead, with the presumption that overhead is allocated fairly.

Dealing with both live donor and cadaver organs, organ acquisition costs cover costs related to acquiring the organ and evaluating the recipient prior to transplant. Medical directors must establish separate cost centers for each of the different types of organs that Medicare reimburses. One major exception to the general rule of pretransplant costs is costs related to professional fees for physician services rendered to live donors during the admission for surgery. Otherwise, examples of appropriate charges against OACCs include:

- Tissue typing,
- Donor and recipient evaluation,
- Costs associated with procurement of organs such as general routine and special care services for the donor, and
- Operating room and other inpatient ancillary services applicable to the donor.

Acquisition costs are divided into direct and indirect costs. Direct costs are those related to the organ acquisition itself, while indirect costs are transplant center overhead costs, or basic facility costs that all hospitals need to operate. From a cost reporting standpoint, direct and indirect costs are accumulated in the OACC on Worksheet A. Direct organ acquisition costs cover a wide scope of services and other transplant center costs including:

- Salaries of the staff involved in organ acquisition (procurement coordinators, administrative and support staff, clerical staff, medical directors, social workers or financial coordinators who may be working with potential transplant recipients, etc.);
- Outpatient services related to pretransplant workup such as evaluation services, tissue typing, and other laboratory services that occur on an ongoing basis;
- United Network for Organ Sharing (UNOS) registration fees;
- Purchase of the cadaver organ; and
- Transportation and preservation services.

For living donation, organ acquisition costs also include outpatient pretransplant workup for the donor; costs relating to the operating room such as anesthesia and other types of ancillary services related to the surgical procedure; and postoperative services to the live donor for any complications from the donation.

19

Some physician services can be included as direct organ acquisition costs, but they must be kept entirely separate from other types of physician professional services that can be billed under Medicare Part B. Donor and recipient pretransplant evaluation services, physician services that may relate to tissue typing and related laboratory services, and professional surgeon fees for cadaver organ procurement excisions (kidney excision fee is currently limited to $1250 per donor, although extra-renal excision fees which have no RVU allotment are reimbursed on a market value (carrier-based) basis) are all covered as direct organ acquisition costs. However, physician fees associated with the operative transplant and post-transplant services are not considered acquisition costs because they are paid under the physician fee schedule as part of the global surgery fee. In contrast to cadaver organ procurement, live donor organ procurement for a Medicare transplant recipient is paid directly to the physician under Part B, rather than through the OACC. Live donor organ procurement is paid at 100 percent of the physician fee schedule (deductibles and coinsurance do not apply). Postoperative physician services for a live donor and physician services that are related to other medical conditions when a patient may have been admitted for another medical reason besides organ donation also get paid directly under Part B.

When physician services are paid as direct organ acquisition services, transplant centers can pay physicians for their services and report those costs as organ acquisition costs as long as they provide appropriate documentation outlining the services and how much they cost. Accounting records should identify the recipient of the services and the services performed, and they should confirm the recipient's status as a potential organ donor or transplant recipient. If physician compensation includes other services, the transplant center and physician need some type of documentation that can be audited to identify how much of the compensation is attributable to organ acquisition services. Unless a provider is able to identify how much relates to organ acquisition services, it will not be able to directly assign those costs to the OACC.

Indirect costs of organ acquisition include various transplant center overhead costs such as:

- Fringe benefits based on the respective salaries of those whose services have been included in the OACC;
- Space costs such as depreciation expense, plant operation cost, utility cost, maintenance and repairs, and general costs that are necessary to operate a hospital;
- Equipment depreciation expense and social services costs that were not solely related to organ acquisition;
- General administrative costs (referred to as administrative and general costs on the Medicare cost report);
- Some portion of the salary of the hospital's CEO, CFO, and other various administrative personnel; and
- Other costs needed to run the hospital such as telephones, data processing, and insurance.

19

A portion of these costs gets brought over to the OACC. The percentage Medicare pays of the indirect organ acquisition costs is determined by apportionment, the process of identifying Medicare's share of the hospital's overall costs. Apportionment is based on a ratio of Medicare usable organs to total usable organs. The numerator of the ratio is the number of Medicare recipient organs that are *actually transplanted*. That includes the number of organs that were obtained or excised at the certified transplant center and sold to organ procurement organizations (OPOs). Medicare allows organs that are sold to OPOs to be included in the numerator because centers that excise organs and sell them to an OPO usually do not know who is the final recipient of the organ. The denominator is the total number of organs transplanted plus the number of organs that were excised at the transplant center and sold to OPOs.

To calculate the actual amount Medicare will pay for organ acquisition costs, total organ acquisition costs are multiplied by the apportionment ratio, and the amount of revenue the transplant center obtained when it sold organs to an OPO is subtracted. Any revenues the transplant center may have received from payors primary to Medicare also must be subtracted. For example, a transplant recipient for whom Medicare is a secondary payor is counted as a Medicare organ in computing the Medicare percentage, but on the tail end of the process, the revenues that the primary payor paid to the transplant center for organ acquisition are subtracted. The process of determining total organ acquisition costs and determining the Medicare share of organ acquisition costs is done differently for each type of organ, with the total amount of direct and indirect organ acquisition costs accumulated in cost report Worksheet D-6.

The average kidney (organ) acquisition cost is calculated by dividing the total full cost of all costs incurred by the hospital in the 'acquisition process' for that organ (kidney) divided by the number of kidneys transplanted. The average cost per organ is not inversely proportional to the number of total usable organs because cost is driven more by the average waiting time for the organ and length of the waiting list. Since periodic laboratory testing services (e.g., panel reactive antibody level testing) and the interval medical evaluations make up a large component of organ acquisitions costs, costs are higher the longer the potential transplant recipients are on the waiting list.

TRANSPLANTATION MARKET OVERVIEW

The market for organ transplantation is quickly becoming a mature one, which carries certain implications for a program's pricing and cost structures. Ten years ago, when liver transplants or lung transplants were more rare, transplant centers could charge more because very few others could provide the same services. Today, however, with many more providers in the marketplace, both hospital and physician transplantation services have become commodities to a large extent.

When a product or service becomes a commodity, it becomes essentially interchangeable with other similar products as far as the market and prices are concerned. In the case of the transplant marketplace, the ability of a competitor to do the same transplant in the same way as another center becomes what is known as a substitute. Commodities and their substitutes are subject to price sensitivity, where demand is tied inversely to cost—as cost goes down, demand goes up.

While transplants are not completely price sensitive because of the limit on organ donors, the commoditization of transplants has caused a very important cross-price elasticity, which affects the maximum amount payors are willing to pay for a center's product or service. How much a payor values the product and how much a center *thinks* the payor values the product will directly affect cost and pricing strategies.

BUSINESS FUNDAMENTALS

While Medicare accounts for transplant costs in a very specific way, hospitals often use other methods to examine transplant programs' fiscal health. One common way is a responsibility-based accounting system that looks at fixed and variable costs as well as direct and indirect costs. In this system, fixed costs are business-unit related costs that are not related to volume. These include hospital allocation of costs for nonrevenue functions like clinic costs and indirect costs, finance costs such as interest expenses, and administration and compliance. Because fixed costs are unrelated to transplant volume, they remain the same whether a center performs one transplant a year or 100.

On the other hand, variable costs are those directly related to each patient, and they can be either direct or indirect. Direct costs here would include professional fees for transplant surgeons and other physicians; acquisition costs for the patient; ancillary services such as lab, radiology and esoteric; and supplies and prescription medications. Indirect costs in this model are business-unit related costs that are indirect to the patient. These include malpractice costs, teaching/research labs, marketing and business development, and business unit administration.

One other element that comes into play during reimbursement is uncollectible costs, or the costs that Medicare does not cover. Transplant centers accrue these costs as part of delivering service to Medicare patients, and they are becoming increasingly higher as Medicare shifts its burden to providers.

As a transplant program matures, the relationship between its fixed and variable costs evolves. In the earliest stages of growth, when a program still is doing a low volume of transplants and trying to add patients to its list, fixed costs constitute the bulk of total program costs. As a program grows and the volume of transplants becomes more moderate (100 to 250 people on program waiting lists), fixed and variable costs become more evenly split. And once a program matures and begins providing a high volume of transplants, variable costs begin to outweigh fixed costs. At this stage, variable costs also start rising disproportionately to volume, and it becomes increasingly important for physicians to take part in monitoring them.

Understanding what drives costs in each area of reimbursement is critical to maximizing efficiency and margins. The biggest drivers on the DRG side are large critical care pathway deviations and longer patient stays in the hospital. On the physician side, salary, practice expense, and liability insurance make up 90% of cost. And on the OACC side, the cost a program pays to the OPO for the

19

organ is notable, but this area also contains the most critical cost driver by far: waiting lists.

As suggested in the organ acquisition cost section above, long waiting lists can drive up variable costs more than any other factor. Long lists can be a double-edged sword for transplant programs: On one hand, a high volume of patients provides more income to the physician group. On the other, a program with excessively long lists of patients to evaluate and maintain suffers from a cumulative effect of waiting list time and cost compounded with price increases and inflation.

In a mature, high-volume transplant program, waiting list costs are akin to inventory costs, so the longer patients spend on the waiting list, the higher costs go up per patient. For example, if a program's waiting list today has 100 patients with an acquisition cost of $40,000 per patient, the program currently has a $4 million inventory cost. Assuming monthly monitoring for patients, annual reevaluations, and cost increases of 7% annually each year over a 60-month waiting list period for a cadaveric transplant, if that waiting list grows by 15% each year, that waiting list will have 200 patients by the end of five years with an acquisition cost of $58,000—or an inventory cost of $11.6 million.

Because excessively long waiting lists trap capital, the most effective way to manage costs is to manage the ratio between the number of people on the waiting list and the number of transplants. In some cases, living donors can be extremely helpful in reducing inventory. In many cases, though, there is little physicians can do to manage the waiting lists themselves, as they cannot very well remove patients from the list because inventory is too high. However, physicians and surgeons at least need to recognize the fiscal effects of long waiting lists and long wait times, as payors also look closely at the relationship between list and transplant growth.

To calculate acquisition costs and responsibility accounting thoroughly, a hospital must look at the rate of inventory turnover, or the number of people on the transplant waiting list, divided by the number of annual transplants. Financial executives also use this ratio to compare their programs against others in the marketplace. Because facilities with longer waiting lists are technically higher cost ones than those that turn their inventory over sooner, the average waiting list in terms of years is a key variable that should be factored into pricing strategies.

PRICING STRATEGIES

The vast majority of hospitals use cost-based pricing for transplant services, meaning they use their own transplant and acquisition costs to determine how much to bill payors. Such cost-based pricing models are focused internally on the institution, as hospital administrators review hospital costs and set price targets based on the cost of delivering the service and running the hospital, plus a minimal acceptable return on capital. The target price in this case usually is 102% to 105% of costs for all business units and products—regardless of how they compare to similar services in the marketplace—and this target price is used as a guide for achieving overall institutional profitability. This model also is consistent with financial reporting and Medicare accounting practices.

Some transplant programs may use an externally focused, market-based pricing strategy. In this case, the program provides a discounted or premium price compared to the market average based on brand strength, quality difference, product maturity or payor buying power. However, a market-based pricing strategy is a much more sophisticated and somewhat riskier model requiring solid market intelligence on factors such as competitors' services, competitors' pricing strategies, payor buying power, and market maturity.

Many third-party payors pay a global fee for the transplant episode that also is intended to cover organ acquisition costs. Since most institutions use a cost-based pricing approach, an effective pricing strategy should focus on vendors and stratify customers based on volume of business. Using this model, preferred vendors, or those that historically have provided the greatest share of cases (10% or more), would be billed 102% to 107% of costs. The next tier down would be vendors that provide a large amount of business (but still less than 10%) and who eventually could become preferred vendors. These vendors would be charged 105% to 120% of costs. Finally, small independent groups that contract for one or two transplants at a time would be charged 120% of costs, as these are fairly labor intensive, and efficiency on return is fairly low.

PROFIT MARGINS VS. CONTRIBUTION MARGINS

Regardless of the pricing strategy, net reimbursement for a transplant equals billed charges less uncollectible costs. Profit margin equals net reimbursement less the fixed, variable, and capital costs; and contribution margin equals any reimbursement above fixed costs. In other words—how much the program's net reimbursement revenues offset the cost of having that program in that space in the hospital. Targets for profit margins and contribution margins are examined annually and set by the culture of the organization.

According to the Healthcare Financial Management Association, profit margins are about 3% for academic hospitals nationwide. This is derived almost entirely from DRG reimbursement for growth and development and is meant to subsidize nonrecoverable costs. In the Medicare reimbursement model for transplant services, DRGs are the only place where the hospital can make a profit margin—and it can only do that if the physicians are efficient and provide services totaling less than the preset DRG payment. (OACC are reimbursed on a full cost basis without opportunity for profit margins.)

On the physician reimbursement side, physicians have little incremental ability to affect variable cost because charges are a reflection of Association of American Medical Colleges (AAMC) charges across the country rather than physicians' and surgeons' true costs, which are very difficult to calculate. Also, physicians' practices constitute a fixed cost structure because the number of surgeons and physicians required to perform the services required for the operation of the transplant center is fixed and, therefore, the cost is fixed (unless a new physician or surgeon is hired). In other words, there is no mechanism to recognize variable cost with each service or procedure.

19

While hospitals look most closely at profit margins when examining their business units, contribution margins can better define the significance of transplant programs to their institutions. In order to understand their program's fiscal health and the significance of the transplant program to the hospital, physicians and surgeons must be aware of their program's contribution margin in both dollars and as a percentage relative to the institution's fixed costs.

Because Medicare reimburses organ acquisition cost centers on a full cost basis (including direct costs) and not on a charge basis, there is *no* profit margin ever possible on OACCs. However, the contribution margin provided by the OACC constitutes a significant proportion of the total contribution margin of the transplant program. This impacts positively both the hospital and physicians because if OACCs provide a strong contribution margin, the transplant center is much more likely to apply resources to transplant programs.

CONTRACTING STRATEGIES

The importance of well-structured contracts cannot be overemphasized when it comes to maximizing reimbursement. Contracts with payors are usually episode oriented, defining billing around four key events for the patient:

- Pretransplant evaluation,
- Acquisition, or the period between being listed and before the transplant (includes the start of acquisition services and other associated requirements),
- A very defined transplant admission, and
- Post-transplant care.

Under a global contract, a commercial payor generally provides one lump sum to the hospital, which then divides payment between itself and transplant physicians and surgeons based on prearranged terms. Alternatively, a discount contract reimburses all transplant charges at a discount. Medicare, on the other hand, usually regards each stage of the transplant experience as a discrete episode (defined by DRG), reimburses acquisition costs on an annual "pass-through" to the hospital based on the Medicare cost report, and pays on a fee schedule for the professional fees.

Because regulated payors such as Medicare have very specific rules that cover reimbursement, contract negotiations are extremely limited in this area except insofar as transplant centers can manage the number of cases that come in through these channels. Commercial payors technically contract with hospitals on either a global or discount basis, and hospitals, in turn, vary in how they distribute payments to their physicians and surgeons. The hospital-physician arrangement may take the form of an employment relationship, where the physician is an employee of the transplant center, or a contractual arrangement where payment can be based on fee-for-service, an hourly rate, some type of global rate based on a percentage, or any other financial arrangement.

Unfortunately, payor compensation usually benefits either the hospital or physicians at the expense of the other. Hospitals historically have benefited from Medicare reimbursement, but physicians have not: Reimbursement versus recovery of the fixed and variable costs (charges) usually leaves physicians significantly below their actual cost to deliver services. Meanwhile, physicians tend to fare better under global contracts while hospitals do not. Some centers rely on internal transfers of funds to help physicians offset shortfalls from Medicare by shifting the physician component up in the global package, transferring an increased portion of the package to physicians and surgeons. Only one reimbursement formula, the discount from billed charges, aligns financial incentives for both hospitals and physicians and can benefit both equally. Under this structure, the transplant center and physicians recover total costs plus profit margin at a predetermined percent of charges.

One critical aspect of contracting strategies is that physicians and hospitals can mix and match these modalities. If a transplant center is in a global rate agreement with a payor, the entire transplant experience does not have to be on a global arrangement. Certain portions of a transplant patient's care can be provided on a percent discount from charges, and other portions can be provided as a global case rate. A typical arrangement may have the transplant experience from admission to discharge negotiated on a global case rate, while evaluation and post-op care are negotiated on a per diem or percent discount from charges.

In most cases, patients also have two insurers, further underscoring the need for thorough contract agreements with payors. For example, a patient may have a commercial payor as the primary insurance and Medicare as the secondary. In the case of kidney transplantation, an entitlement to those undergoing transplantation is that Medicare is the secondary payor. Therefore, there exists an opportunity for coordination of benefits (COB) that allows the hospital to obtain full reimbursement from Medicare for costs not covered by the primary payor in full. The system is full of examples of such opportunities, which eventually could reflect badly on the fiscal health of a transplant program that in fact is being undermined by an ineffective contract or accounting system.

MANAGING THE BOTTOM LINE

Complex transplant centers cannot remain fiscally sound without rigorous participation from physicians and surgeons. The focus should be on managing costs through efficiency, quality, and forward fiscal thinking.

On the DRG side, physicians play a very important role in managing the bottom line because they control the cost (cost drivers) of the services provided. Since payment is fixed for each DRG, more efficient care leads to a higher DRG profit. However, managing the bottom line requires a thorough knowledge and understanding of all components of cost and reimbursement. This effort requires a high level of collaboration between the medical directors of each transplant program and the hospital administrators responsible for the transplant program.

19

CONCLUSIONS

In today's environment of increasing health insurance premiums, growing ranks of uninsured, and decreasing Medicare reimbursement, transplant centers and physicians face growing pressure to be more cost effective and efficient in their processes. In order to be successful on these fronts and maximize reimbursement, physicians and surgeons absolutely must understand the cost drivers of organ transplantation, as well as the various reimbursement models.

Certain pricing and contracting strategies may provide competing financial incentives. Physicians and surgeons must bear in mind that they and their hospitals are partners. Although hospital representatives negotiate on physicians' behalf with payors, the direct involvement of the physicians and surgeons in the financial aspects of the program is essential to the fiscal health of the program.

Regulatory and Fiscal Relationships between Transplant Centers and Transplant Surgeons/Physicians

Frank P. Stuart, Michael M. Abecassis and Dixon B. Kaufman

INTRODUCTION

Organ transplantation is an unusually complex and expensive undertaking that depends on organ prostheses from either cadaveric or living donors, as well as long lists of potential recipients whose health fails steadily as they wait for a transplant. During the past thirty years Congress and the Health Care Financing Administration (HCFA) have written the rules that regulate organ transplantation in the United States. It is imperative that transplant centers and their medical directors understand the rules and accounting practices peculiar to transplantation if they are to be fully reimbursed for the cost of their services.

Chronic hemodialysis for end stage renal disease and kidney transplantation following World War II. Reimbursement by insurance was scarce until the mid-1960s. By 1972, approximately 3500 individuals in the United States survived because of regular hemodialysis treatments, yet insurance coverage was still haphazard. After a dramatic demonstration before a U.S. Senate committee in 1972, Congress amended the Social Security Act to include coverage for end stage renal disease as a Medicare entitlement for any worker who had paid into the Social Security system at least fourteen quarters of three-months each. The worker's spouse and all children under the age of 26 were also entitled to the benefit.[1] Medicare became the primary insurer for kidney transplantation for over 90% of the United States population.

Thus, Congress and the United States Department of Health and Human Services (HHS), rather than the commercial health insurance industry, wrote the rules and accounting practices for kidney transplantation. As transplantation of heart, lung, liver, pancreas and intestine evolved, the rules and accounting practices applicable to the kidney were extended to the other transplantable organs.[2] Even though Congress has moved Medicare coverage of dialysis and kidney transplantation into a secondary position as co-insurer for the first 30 months of care, HCFA rules and accounting practices take precedence.

REVIEW

Organ transplantation in the United States is probably the most intensely regulated of medical disciplines. Congress creates relevant law and publishes the text in the United States Code of Federal Regulations and the Federal Register. The

Organ Transplantation, 2nd edition, edited by Frank P. Stuart, Michael M. Abecassis and Dixon B. Kaufman. ©2003 Landes Bioscience.

20

HHS then develops and disseminates detailed rules for implementing transplant law via three government publications: Commerce Clearinghouse Medicare and Medicaid Guide, Medicare Intermediary Manual, and Medicare Hospital Manual. The HCFA contracts with intermediaries to manage its day to day business with hospitals, physicians and other service providers. The intermediaries, usually large insurance companies, exercise considerable discretion as they interpret the Medicare Intermediary Manual. For example, specific regulations may be interpreted differently by intermediaries in various parts of the country. Successive intermediaries that contract for the same locale may also differ in their interpretation of the Medicare Intermediary Manual.

In order for a hospital to bill Medicare for transplant services it must first apply for HCFA certification for each organ it plans to transplant. If approved, the hospital then becomes a certified transplant center (CTC) for those organs. Certified transplant centers must apply for membership in the United States Organ Procurement and Transplant Network, which has been administered for the Department of Health and Human Services under contract by the United Network for Organ Sharing (UNOS) since 1986. UNOS bylaws interpret Federal Code with respect to personnel, facilities and other resources required to operate a certified transplant center. The transplant center must also establish working relationships with the HCFA certified Organ Procurement Organization (OPO) in its area.[3] HCFA recognizes approximately 60 OPO's (organ banks) throughout the county; their service areas do not overlap.

Clearly the hospital "owns" and is responsible for administration and operation of transplant programs. It is the hospital that is certified by Medicare; the hospital staff and the affiliated transplant physicians and surgeons constitute the hospital's transplant team and program. As the owner of its transplant programs, the hospital incurs obligations, many of which are executed by its surgeons and physicians. Because most of the obligations are directly or indirectly related to organ acquisition, HCFA permits the hospital to compensate surgeons and physicians for their role and recover its compensation costs through charges against its acquisition cost center for each organ.[4]

Examples of certified centers' obligations are listed below. Surgeons and physicians who direct transplant programs play a major role in helping the hospital meet these obligations.[3]

- Participate in governance of UNOS and creation of organ allocation bylaws.
- Adhere to constantly evolving bylaws of UNOS, which control all aspects of cadaveric organ allocation through complex algorithms.
- Interface with UNOS national computerized waiting list via all HCFA certified OPOS.
- Participate in governance and donor organ related activities of the local OPO through organ specific committee structures to maximize organ procurement and ensure equitable sharing of organs. Provide teams of surgeons and surgical technicians to procure (harvest) cadaver organs.

- Ensure equal access to the public of organs (a scarce national resource) as required under Title VI of the Civil Rights Act of 1964 (no person shall be subjected to discrimination on the basis of race, color, or national origin under any program or activity that receives Federal financial assistance). Because organ transplant services are unique and concentrated in fewer than 300 hospitals nationwide, transplant centers must be proactive in educating the public and physicians about transplantation so that equal access to it is meaningful.
- Develop and maintain organ specific lists of potential recipients who will wait for available cadaver organs. Evaluation of potential recipients is an on-going labor intensive process that stretches from first contact through sequential reevaluation as the candidate's health and priority status change between listing and transplantation. Extensive data are maintained so that queries from UNOS, organ banks and the Inspector General's office, with respect to equal access, can be answered.
- Transplant centers must evaluate potential living donors who might provide an intact organ or part of an organ. Evaluation of living donors is at least as complex as it is for recipients. Moreover, several potential donors are usually evaluated for each candidate who actually qualifies and proceeds to donation.
- Provide complex long-term posttransplant outpatient care. Management of immunosuppression and the array of problems associated with organ transplantation require that the transplant center be the primary provider of outpatient care for the first posttransplant year and secondary provider thereafter. Visits are frequent during the first few months, interspersed with laboratory tests performed at other hospitals and faxed to the transplant center. Recipients must be able to reach outpatient nurses by telepage at anytime. UNOS requires detailed reports from the recipient's medical record and the transplant center's databases as long as either the transplanted organ or the recipient survives. The reports required by UNOS include the candidate registration report (listing), the recipient registration report at the time of transplant (which includes data about the donor, the operative procedure, and the entire inpatient stay), and posttransplant follow-up reports at 6 months, the 1st year, and yearly thereafter.
- Appoint a Medical Director to supervise each specific organ transplant program. The physician or surgeon is responsible for planning, organizing, conducting, and directing the transplant center.[5]

Of the six organs transplanted (heart, lung, kidney, pancreas, liver and intestine) only kidney transplantation is an entitlement under Medicare via the End Stage Renal Disease amendment of the Social Security Act in 1972.[1] As successful transplantation of the other organs evolved, Congress declined to amend the Social Security Act to specifically cover services for end stage heart, lung and liver disease or for diabetes and intestinal failure. Nevertheless, potential recipients of non-renal organs may become eligible for Medicare via two separate provisions of the Social Security Act for the aged and disabled. To qualify under these provi-

20

sions the candidate must be at least 65 years of age or fully disabled for at least 24 months. In addition, insulin dependent diabetes qualifies potential recipients for Medicare coverage of pancreas transplantation if it occurs at the same time as or after kidney transplantation in a recipient who was eligible for Medicare at the time of kidney transplantation.

When Medicare introduced prospective global reimbursement of hospitals for inpatient care by diagnosis related grouping (DRG), each DRG with its own dollar value, it instructed transplant centers to separate acquisition costs for both cadaveric and living donor organs from the cost of inpatient care. Organ acquisition cost centers (OACC's) were created outside the transplant DRG in each hospital for each transplanted organ. The organ cost centers were set up so as to compensate the hospital for reasonable expenses of organ acquisition as well as evaluation, selection, maintenance and reevaluation of recipient candidates on waiting lists until transplantation occurred.[6] Examples of appropriate charges against acquisition cost centers included in the Code of Federal Regulations are:

- tissue typing;
- donor and recipient evaluation;
- other costs associated with excising organs such as general routine and special care services for the donor;
- operating room and other inpatient ancillary services applicable to the donor;
- preservation and perfusion costs;
- charges for registration of recipient with a transplant registry;
- surgeon's fees for excising cadaver organs;
- transportation;
- costs of organs acquired from other providers or organ procurement organizations;
- hospital costs normally classified as outpatient cost applicable to organ excisions (services include donor and donee tissue typing, work-up and related services furnished prior to admission);
- costs of services applicable to organ excisions which are rendered by residents and interns not in approved teaching programs;
- all pre-admission physician services, such as laboratory, electro-encephalography, and surgeon fees for cadaver excisions, applicable to organ excisions including the costs of physician's services.

For kidney transplantation, the full cost of organ acquisition is approximately twice the full cost of the inpatient transplant stay. When Medicare is the payor, the hospital is compensated for the inpatient stay through the Part A DRG case rate. Organ acquisition is treated as a full cost "pass through". If a commercial payor is primary, both the inpatient charge and the standard acquisition charge are submitted to the carrier. Depending on the wording of the contract between the transplant center and the commercial payor, Medicare might become the secondary payor for any portion of the inpatient care or standard acquisition charge denied by the commercial payor (this assumes that the recipient is eligible for Medicare benefits). Through a process referred to as "coordination of benefits" HCFA intends that its allowable charges and allowable reimbursement will determine its payment to hospitals and physicians when Medicare is the secondary payor.

For pancreas and liver transplantation, the inpatient costs represent a larger fraction of total cost than is true for kidney transplantation; the inpatient liver transplant costs usually exceed the standard acquisition cost for a liver. Coordination of benefits with Medicare as a secondary payor is possible for the pancreas and liver as it is for kidney transplantation.

Many commercial health insurers "exclude" organ transplantation from their general policies; the employer client must then contract separately with a reinsurance company that specializes in transplant insurance. Unfortunately for transplant centers most transplant reinsurance networks sell global managed care contracts that fail to identify inpatient care and organ acquisition coverage as separate components of the benefit package in the same way that Medicare and HCFA accounting practices do. By disregarding HCFA accounting practices transplant insurers compromise the transplant center's option to turn to Medicare for reimbursement via coordination of benefits for the full cost of organ acquisition.

Surgeons and physicians bill through part B Medicare for their services to transplant recipients during the inpatient stay and are paid 80 percent of the fee allowed by Medicare. When Medicare is the secondary payor to commercial insurance that pays less than 80% of Medicare allowable, Medicare may be billed for the difference if that possibility is not precluded by terms of the contract between the physician and the commercial payor.

When the organ is removed from a living donor, surgeons and physicians bill the recipient's part B Medicare for their services to the donor and are paid at 100% of Medicare allowable fees. If a commercial payor is primary it should be contacted prior to transplantation to determine whether it will accept separate charges from physicians and surgeons for care of the living donor. Some commercial insurers require that physician's bills be submitted with the hospital's standard acquisition charge. If the recipients' commercial payor refuses altogether to pay for physician's services to the living donor, the transplant hospital becomes payor of last resort by charging those services to its organ acquisition cost center.

In most transplant centers those physicians and surgeons most directly involved in evaluating potential donors and recipients, maintaining ready tests, organ procurement and overall direction of transplant centers are largely unaware that compensation for many of their daily activities is appropriately charged to the hospital's organ acquisition cost centers. Transplant hospitals rarely solicit charges from transplant surgeons and physicians; it is incumbent on physicians to understand Medicare rules concerning charges against organ acquisition cost centers and to initiate the process of billing the hospital appropriately for their services to organ acquisitions.

20

Physicians and surgeons should be aware that organ acquisition cost centers have four distinct components:

1. Normal operating costs associated with program operations;
   a. space, phone, supplies, pagers, answering services, utilities, computers, maintenance, computers, pre-transplant patient records, storage;
   b. personnel costs of clerical and professional staff, financial and insurance counselors, social service, nurse coordinators. (salary and benefits such as travel reimbursement for relevant meetings, continuing education, seminars, memberships, dues and subscriptions);
   c. program direction and administration;
   d. UNOS recipient registration charges;
   e. medical center overhead;
   f. educational materials, presentations.
2. Medical consultation/evaluation and testing service costs associated with pre-transplant evaluation of both potential recipients and potential living donors:
   a. dental evaluation;
   b. psychological evaluation;
   c. multidisciplinary assessment conferences;
   d. tissue typing and other assessment of immunological activity, cross-matches;
   e. outpatient services related to living donor after donation.
3. Costs associated with maintaining the evaluated patient/potential recipient on the waiting list such as monitoring to determine whether he remains suitable for transplant:
   a. exchange of information with potential recipient's physicians;
   b. laboratory tests and x-rays;
   c. interval history and physical examination;
   d. specialty consultations.
4. Costs associated with acquiring organs for transplant (cadaver donor and living donor):
   a. charges by OPOs;
   b. educational materials concerning transplantation for use with potential recipients and living donors;
   c. preservation, perfusion and organ preparation laboratory.

The Federal Code permits compensating transplant centers for all reasonable expenses of organ acquisition.[6] Aggregate acquisition expenses are fully reimbursed as a pass through outside the DRG prospective payment system. Hospital administrators should be receptive to compensating well-articulated, reasonable costs of physicians and surgeons as allowed by law. In return, transplant physicians should work closely with the hospital's Chief Financial Officer and staff to keep the overall transplant program's full cost and reimbursement in balance.

Complex transplant centers cannot remain fiscally sound without rigorous participation by the physicians as listed below:

- Practice efficient exemplary care based on sound medical decisions.
- Develop critical care pathways for patients and efficient protocols and systems for pre-and post-transplant care.
- Determine full cost of critical care pathways and reassess their cost-effectiveness regularly.
- Maintain extensive outcome data and assist hospital in negotiating payor contracts that will cover full costs of care.

Analysis of full costs and expected reimbursement on a quarterly basis should track pre-transplant care, the inpatient transplant hospitalization, and post-operative outpatient care for each type of organ transplanted; data should also be analyzed by payor to identify those commercial insurers that reimburse an unreasonably low fraction of full costs.

Transplant centers in which the hospital and its physicians/surgeons are equally concerned about the other's fiscal integrity will thrive and maintain the privilege of providing high quality organ transplant services to the American public.

ACKNOWLEDGEMENT

This article is reprinted from Graft 2001; Vol. 4, No. 6:398-402, with permission from the authors.

REFERENCES

1. United States Code of Federal Regulations. Title 42, Chapter IV, Subchapter B: 42 C.F.R. 405
2. United States Code of Federal Regulations. Title 42, Chapter IV, Subchapter B: 42 C.F.R. 486.302
3. United States Code of Federal Regulations. Title 42, Chapter IV, Subchapter B: 42 C.F.R. 405.2171
4. United States Code of Federal Regulations. Title 42, Chapter IV, Subchapter B: 42 C.F.R. 412.100
5. United States Code of Federal Regulations. Title 42, Chapter IV, Subchapter B: 42 C.F.R. 405.2170
6. United States Code of Federal Regulations. Title 42, Chapter IV, Subchapter B: 42 C.F.R. 412.113

Pregnancy and Transplantation

Vincent T. Armenti, Michael J. Moritz, John S. Radomski, Gary A. Wilson,
William J. Gaughan, Lisa A. Coscia and John M. Davison

INTRODUCTION

The first reported post-transplant pregnancy occurred in March 1958 and was published several years later.[1] This female recipient had received a kidney transplant from her identical twin sister and delivered a healthy baby boy by cesarean section. Over the ensuing years, numerous case reports, single center reports, surveys, and registry publications have reported outcomes of pregnancies in female transplant recipients. Since even large individual centers have limited experience with post-transplant pregnancies, registry data and surveys provide important information for comparing outcomes among different recipient groups.

In 1991, the National Transplantation Pregnancy Registry (NTPR) was established at Thomas Jefferson University to study the safety of pregnancy outcomes and sequelae for both female transplant recipients and male transplant recipients who father pregnancies. To date we have received entries on 732 female recipients (1121 pregnancies) and on 855 pregnancies fathered by 603 male recipients. Here we review recent reports in the literature as well as current findings and background information from the registry. Because of the concerns of immunosuppressive agents on fetal development, a brief review of these agents will follow. Additionally, we will discuss obstetrical management issues for female organ recipients.

IMMUNOSUPPRESSION DURING PREGNANCY

The U. S. Food and Drug Administration (FDA) categorizes the potential fetal risks of drugs using the following classification system: A = controlled studies, no risk; B = no evidence of risk in humans; C = risks cannot be ruled out; D = positive evidence of risk; X = contraindicated. The commonly used immunosuppressive agents are listed in Table 21.1: none of these drugs is Category A, the corticosteroids and basiliximab are Category B, and most are Category C. In infants born to renal transplant recipients exposed to azathioprine (Category D), two early reports described the incidence of congenital anomalies as 9% and 6.4% respectively.[2,3] There was, however, no specific pattern noted among the kinds of anomalies that occurred. More recent reports have been more reassuring. The extensive European Dialysis and Transplant Association (EDTA) report on 490 pregnancies (500 babies) concluded that azathioprine and prednisone immunosuppression was not associated with more congenital malformations in the newborn of renal recipients than seen in the normal population.[4] Other issues that have been raised with respect to azathioprine have included: fatal neonatal anemia,

Organ Transplantation, 2nd edition, edited by Frank P. Stuart, Michael M. Abecassis and Dixon B. Kaufman. ©2003 Landes Bioscience.

Table 21.1. Pregnancy safety information for immunosuppressive drugs used in transplantation

	Pregnancy Category
corticosteroids (prednisone, methylprednisolone)	B
cyclosporine (Sandimmune®, Neoral®)	C
cyclosporine (SangCya™ Oral Solution)	C
tacrolimus, FK506 (Prograf™)	C
sirolimus, rapamycin (Rapamune®)	C
azathioprine (Imuran®)	D
mycophenolate mofetil (CellCept®)	C
antithymocyte globulin (ATGAM®, ATG)	C
antithymocyte globulin (Thymoglobulin®)	C
muromonab-CD3 (orthoclone OKT®3)	C
basiliximab (Simulect®)	B
daclizumab (Zenapax®)	C

21

thrombocytopenia and leukopenia, and risk of acquired chromosomal breaks.[2,5-8] One approach that has been utilized adjusted the azathioprine doses to maintain the maternal leukocyte count within the normal limits for pregnancy, which minimized neonatal effects.[8] NTPR data in azathioprine-based renal recipients similarly have not revealed a specific malformation pattern in the newborn.

Among the Category C agents is a range of safety profiles. For several of the agents including daclizumab, orthoclone OKT®3, Thymoglobulin® and ATGAM®, there are no available reproductive data. In reports to the NTPR, rejections during pregnancy have been treated with corticosteroids (prednisone, methylprednisolone) with a few cases using OKT®3. With both cyclosporine (CsA) and tacrolimus, animal reproductive studies have revealed fetotoxicity and fetal resorptions at higher than therapeutic doses. In contrast, animal studies with mycophenolate mofetil (MMF) have suggested teratogenesis at dosages below those causing maternal toxicity and at dosages potentially within the human therapeutic range, thus raising a greater level of concern for the potential for adverse effects on fetal development. NTPR data have not revealed malformations in 5 female renal recipients taking MMF during pregnancy. Animal studies with sirolimus have not revealed teratogenesis.

With combinations of the newer agents, it may be more difficult to identify a specific cause and effect. However, with lowered dosages of these multiple agents, there will be less exposure to each specific drug so theoretically, the potential for teratogenesis may be less. Potentiating effects among drugs as well as unknown interactions in multiple drug regimens may result in adverse fetal effects. These will most likely be discovered only by clinical outcome analyses of viable and nonviable pregnancies.

REPORTS OF PREGNANCY OUTCOMES IN FEMALE TRANSPLANT RECIPIENTS

Shown in Table 21.2 are results of a recent large single center report and recent surveys compared with selected NTPR data.

21

Table 21.2. Comparison of pregnancy outcome data reports in the literature and from the NTPR

	Recipients/ Pregnancies	Immunosuppression	Pre-Eclampsia	% Liveborn	Mean GA[1] (wks)	Mean BW2 (gms)	Newborn Complications	Neonatal Deaths	Graft Dysfunction/ Rejection During	Graft Loss
Renal Toma et al	189/194	CsA-based 52% Aza-based 38% Tacro-based 0.5% Not reported 10%	24%	82%	35.7	2360	3%[3]	1.4%	19%	13%
NTPR	115/154 146/238	CsA-based Aza-based 92% Steroid only 8%	25% 21%	69% 83%	35.6 36.2	2407 2684	22% 30%	0.9% 2.4%	15% 6%	8%[4] 4%[4]
Heart Branch et al[5]	35/47	CsA-based	20%	74%	37	2543	20%	0%	24%	26%
Liver Jain et al	21/27	Tacro-based	4%	100%[6]	36.6[7]	2638[7]	11%	7%	11%	10%
NTPR	15/21 14/18	Tacro-based Neoral-based	5% 24%	76% 72%	37.4 36.9	3069 2565	38% 23%	0% 0%	14% 6%	0% 6%[4]
Pancreas-Kidney Barrou et al	17/19	CsA-based	NR[8]	100%[6]	35	2150	11%	0%	0%	12%[9]
NTPR	18/23	CsA-based	25%	87%	34.8	2041	35%	0%	8.7%	11%[4,10]
Lung NTPR	6/6	CsA-based 5 Tacro-based 1	0%	50%	32.8	2202	67%	0%	33%	17%[4]

[1]GA-gestational age; [2]BW-birthweight; [3]Newborn malformations; [4]Graft loss within 2 years of delivery; [5]International survey and NTPR data; [6]Only liveborns reported; [7]Neonatal deaths excluded from analysis; [8]NR-not reported; [9]1 kidney and 1 pancreas loss in 2 different recipients; [10]1 kidney and 1 pancreas/kidney loss in 2 different recipients.

RENAL

Toma et al reported that the 57 babies born to mothers on azathioprine had a significantly higher mean birthweight (2567 ± 491.1 gms) than the 94 babies born to mothers on CsA (2252 ± 629.2 gms). There were no congenital anomalies in either group. In this survey of 143 transplant units, graft function deteriorated in 20% of patients after delivery, with graft loss in approximately 10% of the patients.[9] In an analysis of NTPR data on female renal recipients, significant differences between CsA and non-CsA (azathioprine, steroids) recipients were noted for birthweight (CsA group lower, p=0.003) and drug-treated hypertension (CsA group higher, p=0.0001). When the data were analyzed using a multivariate analysis of all renal recipients, the most significant predictors of lower birthweight in newborns were maternal hypertension (p=0.0073), serum creatinine (≥ 1.5 mg/dL, p=0.0439), and diabetes mellitus (p=0.05). There were no specific patterns of malformation noted in the offspring of either recipient group (CsA vs. non-CsA) with a lower incidence of complications in the newborn of CsA-recipients. Neonatal death rates were 0.9% in the CsA-based recipients and 2.4% in the non-CsA-based recipients.[10]

As there are more extensive data available regarding renal transplant recipients and pregnancy, there have been efforts to identify predictors of adverse maternal and fetal outcomes as well as to determine whether pregnancy has long-term effects on graft function. Two well-designed case studies showed no significant effect on graft function when a group of pregnant women was compared to nonpregnant controls.[11,12] A longer-term follow-up of one of these studies, however, did suggest that a minor deleterious effect might result.[13] A retrospective case control study from the EDTA suggested that pregnancy rarely if ever has a deleterious effect on graft function.[14] NTPR data have shown a minor increase in serum creatinine post-partum when compared to prepregnancy levels.

A report from the National Transplant Database Pregnancy Register from the U.K. has recently shown that pregnancy success, defined as a livebirth and survival after at least a 24 weeks gestation, was significantly related to the serum creatinine level measured within 3 months prior to conception (p=0.04).[15]

Variables affecting post-partum graft loss were analyzed in female recipients from NTPR data.[16] Forty recipients whose graft failed any time after pregnancy were compared to 81 randomly selected recipients who did not experience a graft loss; all were on CsA-based regimens. In an analysis using a Cox proportional hazards model, those recipients with a prepregnancy serum creatinine ≥ 2.5 mg/dL were approximately 3 times more likely to experience graft loss than recipients with a prepregnancy serum creatinine < 1.5 mg/dL. An increase in serum creatinine during pregnancy was also associated with an increased risk of graft loss. In a recent single center analysis of 33 pregnancies in 29 female renal recipients, all recipients with prepregnancy serum creatinine > 200 mmol/l had progression of renal impairment and required renal replacement therapy within 2 years of delivery.[17] The authors also suggested that no evidence of chronic rejection before pregnancy and proteinuria < 1 gm/day allowed for a good obstetric outcome. In another single center study, 0.3 gms per day of urine protein loss prior to pregnancy was felt to be a useful marker of subclinical chronic rejection and that these women should undergo renal biopsy to assess graft status prior to pregnancy.[18]

Although not frequent, both acute and chronic rejections during pregnancy have been reported to the NTPR (4% in CsA based renal recipients) suggesting that immunosuppressive doses and levels be monitored during pregnancy.

Pregnancy potentially alters drug distribution and it has been noted that cyclosporine levels drop during pregnancy.[19] Some authors have suggested adjusting CsA doses during pregnancy including dose increases.[20] One group advised no dose changes despite decreases in trough levels as they observed continued stable graft function during pregnancy with this approach.[21] In an analysis of registry data, we noted that recipients with graft dysfunction and/or rejection or graft loss related to pregnancy had lower mean CsA doses prepregnancy and during pregnancy.[22] In this group of CsA renal recipients, 37% of recipients with peripartum graft problems either decreased or discontinued CsA during pregnancy. Of interest is a recent report of a successful pregnancy outcome in a nonimmunosuppressed renal recipient. In this case the recipient had discontinued her immunosuppression 3 years prior to pregnancy with stable graft function.[23]

Successful pregnancies in renal recipients have also been reported with successive pregnancies, with long transplant to conception intervals, and with multiple birth outcomes.[14,24-26] With regard to multiple births, it has been suggested that fetal reduction of triplets to twins would increase the likelihood of success.[27]

Case reports have recently appeared in the literature of successful pregnancies in renal recipients maintained on tacrolimus.[28-30] One report described a twin pregnancy, where both newborns developed respiratory and cardiopulmonary difficulties, one of whom died as a result of a thrombotic cardiomyopathy (the only such case reported to the NTPR).[31]

Tables 21.3–21.5 show comparisons of pregnancy outcomes for female renal recipients reported to the NTPR, including two of the newer immunosuppressive agents.[32] Reports to the registry of dose management with these agents during pregnancy are summarized in Table 21.6.[22,32]

Liver, Liver-Kidney

Successful pregnancy outcomes have been described in female liver recipients on CsA-based and tacrolimus-based regimens.[33-37] In a recent report from the Pittsburgh group (Table 21.2), one newborn had unilateral polycystic renal disease, which was the only structural anomaly.[37] There were no neonatal deaths in the NTPR group of tacrolimus- treated liver recipients (Table 21.2). In a prior registry report of 58 female liver recipients (89 pregnancies, the majority CsA-based), poorer newborn outcomes were noted in 10 recipients with biopsy-proven rejection during pregnancy. Overall, there were 4 maternal deaths with recipients having recurrent hepatitis C with successive pregnancies suggesting a group at higher risk. No specific patterns of malformations were reported in the newborn.[38] Case reports of successful pregnancy outcomes in female liver-kidney recipients and in female liver recipients transplanted during pregnancy have appeared in the literature and have been reported to the NTPR.[37,39-44]

Table 21.3. NTPR: Female renal recipients

Maternal Factors	CsA	Neoral®	Tacro
Transplant to conception interval	3.1 yrs	4.7 yrs	2.4 yrs
Hypertension during pregnancy	62%	70%	48%
Diabetes during pregnancy	12%	9%	17%
Graft dysfunction during pregnancy	11%	0%	9%
Infection during pregnancy	22%	29%	32%
Rejection episode during pregnancy	4%	3%	17%
Pre-eclampsia	27%	23%	38%
Creatinine (mean) mg/dL			
before pregnancy	1.4	1.3	1.4
during pregnancy	1.4	1.4	1.9
after pregnancy	1.6	1.5	1.8
Graft loss within 2 yrs of delivery	8%	1.8%	10.5%

CsA (304 recipients, 456 pregnancies), Neoral® (56 recipients, 68 pregnancies),
Tacro (19 recipients, 23 pregnancies)
CsA—Sandimmune® brand cyclosporine, Neoral®—Neoral® brand cyclosporine,
tacro—Prograf™ (tacrolimus, FK506)

Table 21.4. NTPR: Outcomes of female renal recipients

	CsA	Neoral®	Tacro
Outcomes (n)[a]	(465)	(71)	(24)
Therapeutic abortions	8%	1.4%	0%
Spontaneous abortions	13%	17%	29%
Ectopic	0.7%	0%	0%
Stillborn	3%	1.4%	0%
Livebirths	75%	80%	71%

[a]includes twins, triplets
CsA—Sandimmune® brand cyclosporine, Neoral®—Neoral® brand cyclosporine, tacro—
Prograf™ (tacrolimus, FK506)

Table 21.5. NTPR: Newborn outcomes of female renal recipients

	CsA	Neoral®	Tacro
Livebirths (n)	(350)	(57)	(17)
Gestational age (mean)	35.9 wks	35.8 wks	33 wks
Birthweight (mean)	2485 gms	2449 gms	2151 gms
Premature (<37 wks)	52%	51%	63%
Low birthweight (<2500 gms)	46%	54%	63%
C-Section	51%	48%	44%
Newborn complications	40%	49%	53%
Neonatal deaths (within 30 days of birth)	1%	0%	6%

CsA—Sandimmune® brand cyclosporine, Neoral®—Neoral® brand cyclosporine, tacro—
Prograf™ (tacrolimus, FK506)

21

Table 21.6. NTPR: Immunosuppressive dose adjustments during pregnancy in female kidney recipients

	CsA (260 pregnancies)	Neoral® (59 pregnancies)	Tacro (23 pregnancies)
Increased	8.9%	44%	22%
Varied	1.9%	12%	9%
Decreased	5.4%	0%	4%
No change	82.3%	32%	52%
Switch	0%	12%	9%
Discontinued	1.5%	0%	4%

CsA—Sandimmune® brand cyclosporine, Neoral®—Neoral® brand cyclosporine, tacro—Prograf™ (tacrolimus, FK506)

PANCREAS-KIDNEY

The international survey conducted by Barrou and colleagues includes some patients that overlap with the NTPR report (Table 21.2).[45,46] Two congenital malformations were reported (bilateral cataract and double aortic arch) and one child developed Type I diabetes at age 3 years.[45] There were no structural malformations noted among the offspring of the pancreas-kidney recipients reported to the NTPR. Recipients maintained normoglycemia during pregnancy. Occasional graft losses have occurred in the peripartum period.

HEART, HEART-LUNG AND LUNG

Pregnancies post-heart transplant, with a few cases post-heart-lung transplant, have also been successful.[47-50] An international survey by Wagoner et al combined with NTPR data (Branch, Table 21.2) examined the risk of having more than one post-transplant pregnancy and did not find a significant difference in neonatal complications and/or maternal graft survival.[50] There were 7 deaths in the first pregnancy group, 3 due to noncompliance, and a single recipient death due to allograft vasculopathy (5.6 yrs post-partum) in the subsequent pregnancy group. No structural malformations were noted in the newborn.

Data are still limited among lung recipients but significant rejection and mortality rates have been noted, prompting concern among practitioners (Table 21.2).[51] However, there are a few successful reports to the registry so that it is not possible to make clear-cut recommendations to these recipients with regard to pregnancy safety.

OTHER MATERNAL ISSUES

Maternal infections and other medical complications and their effects on pregnancy have been extensively reviewed.[21,36,52-54] Of the infectious complications reported to the registry, most have involved the urinary tract. There are occasional reports to the NTPR and to the literature of CMV infection during pregnancy.[36,55,56] A recent single case report described the use of ganciclovir during pregnancy with no untoward effects.[56]

PREGNANCIES FATHERED BY MALE TRANSPLANT RECIPIENTS

A registry study of 204 male recipients who fathered 288 pregnancies revealed 290 evaluable outcomes with 2 neonatal deaths.[38,57] One was an infant with Potter's syndrome and another a premature smaller twin with respiratory distress from birth. Analysis of outcomes (n=26) in relation to the new immunosuppressive regimens including Neoral®, tacrolimus and MMF, revealed 1 child who required surgery for ureteral obstruction and another for hydrocele. Specifically, none of the offspring of fathers with MMF exposure were reported to have congenital malformations.[38]

OBSTETRIC MANAGEMENT ISSUES

Pregnancies in female recipients of all organ types must be monitored as high-risk cases and teamwork is essential.[53] Management requires attention to serial assessment of transplant function, hematology, blood pressure control, diagnosis and treatment of rejection, treatment of any infections and serial fetal surveillance. It is essential to carefully assess the woman's emotional attitude and the overall support she receives from her family.

Preterm delivery (before 37 weeks gestation) is common because of intervention for obstetric reasons and the tendency to premature labor. Unless there are specific problems, however, spontaneous onset of labor can be awaited. Vaginal delivery is the aim and cesarean section is only necessary for obstetric reasons. Vaginal delivery does not cause mechanical injury to a renal transplant, and neither does the graft obstruct the birth canal.

Augmentation of steroids is necessary to cover delivery and aseptic technique is advisable at all times. Any surgical procedures, however trivial, should be covered by prophylactic antibiotics. Fetal monitoring is advisable. Pain relief is conducted as for healthy women.

Whatever the organ transplant, after surgery, endocrine and sexual functions return rapidly. If according to a suitable set of guidelines, prepregnancy assessment is satisfactory, pregnancy can be advised. In most, a wait of 18 months to 2 years post-transplant has been recommended. By then, the recipient will have recovered from surgery, graft function will have stabilized and immunosuppression will be at maintenance levels. Couples who want a child should be encouraged to discuss all the implications, including the harsh realities of maternal survival prospects.

ACKNOWLEDGMENTS

The authors are indebted to transplant coordinators, physicians, and recipients nationwide who have provided their time and information to the registry. The NTPR is supported by grants from Novartis Pharmaceuticals Corporation, Fujisawa Healthcare, Inc., and Roche Laboratories Inc. The authors acknowledge Ann Nickolas for her assistance with the preparation of the manuscript.

This article is reprinted from Graft 2000; Vol. 3, No. 2:59-63, with permission from the authors.

REFERENCES

1. Murray JE, Reid DE, Harrison JH et al. Successful pregnancies after human renal transplantation. N Engl J Med 1963; 269:341-343.
2. Penn I, Makowski EL, Harris P. Parenthood following renal transplantation. Kidney Int 1980; 18:221-233.
3. Registration Committee of the European Dialysis and Transplant Association: Successful pregnancies in women tetreated by dialysis and kidney transplantation. Br J Obstet Gynaecol 1980; 87:839-845.
4. Rizzoni G, Ehrich JHH, Broyer M et al. Successful pregnancies in women on renal replacement therapy: Report from the EDTA Registry. Nephrol Dial Transplant 1992; 7:279-287.
5. Lower GD, Stevens LE, Najarian JS et al. Problems from immunosuppressives during pregnancy. Am J Obstet Gynecol 1971;111:1120-1121.
6. Price HV, Salaman JR, Laurence KM et al. Immunosuppressive drugs and the foetus. Transplantation 1976; 21:294-298.
7. Rudolph JE, Schweizer RT, Bartus SA. Pregnancy in renal transplant patients. Transplantation 1979; 27:26-29.
8. Davison JM, Dellagrammatikas H, Parkin JM. Maternal azathioprine therapy and depressed haemopoiesis in the babies of renal allograft patients. Br J Obstet Gynaecol 1985; 92:233-239.
9. Toma H, Kazunari T, Tokumoto T et al. Pregnancy in women receiving renal dialysis or transplantation in Japan: a nationwide survey. Nephrol Dial Transplant 1999; 14:1511-1516.
10. Armenti VT, Ahlswede KM, Ahlswede BA et al. National Transplantation Pregnancy Registry—Outcomes of 154 pregnancies in cyclosporine-treated female kidney transplant recipients. Transplantation 1994; 57:502-506.
11. First MR, Combs CA, Weiskittel P et al. Lack of effect of pregnancy on renal allograft survival or function. Transplantation 1995; 59:472-476.
12. Sturgiss SN, Davison JM. Effect of pregnancy on long-term function renal allografts. Am J Kidney Dis 1992; 19:167-172.
13. Sturgiss SN, Davison JM. Effect of pregnancy on the long-term function of renal allografts: An update. Am J Kidney Dis 1995; 26:54-56.
14. Ehrich JHH, Loirat C, Davison JM et al. Repeated successful pregnancies after kidney transplantation in 102 women (Report by the EDTA Registry). Nephrol Dial Transplant 1996; 11:1314-1317.
15. Briggs D, Davison JM, Redman C et al. National Transplant Database Pregnancy Register—First Report. United Kingdom Transplant Support Service authority (UKTSSA) Users' Bulletin 1999; 33:8-9.
16. Armenti VT, McGrory CH, Cater JS et al. Pregnancy outcomes in female renal transplant recipients. Transplant Proc 1998; 30:1732-1734.
17. Crowe AV, Rustom R, Gradden C et al. Pregnancy does not adversely affect renal transplant function. Q J Med 1999; 92:631-635.
18. Kozlowska-Boszko B, Lao M, Gaciong Z et al. Chronic rejection as a risk factor for deterioration of renal allograft function following pregnancy. Transplant Proc 1997; 29:1522-1523.
19. Thomas AG, Burrows L, Knight R et al. The effect of pregnancy on cyclosporine levels in renal allograft patients. Obstet & Gynecol 1997; 90:916-918.
20. Biesenbach G, Zazgornik J, Kaiser W et al. Cyclosporin requirement during pregnancy in renal transplant recipients. Nephrol Dial Transplant 1989; 4:667-669.
21. Bumgardner GL and Matas AJ. Transplantation and Pregnancy. Transplantation Reviews 1992; 6:139-162.

22. Armenti VT, Jarrell BE, Radomski JS et al. National Transplantation Pregnancy Registry (NTPR): Cyclosporine dosing and pregnancy outcome in female renal transplant recipients. Transplant Proc 1996; 28:2111-2112.

23. Josephson MA, Lindheimer MD, Hibbard JU. Pregnancy in a non-immuno-suppressed transplant recipient. Am J of Kidney Dis 1998; 32:661-663.

24. Gaughan WJ, Moritz MJ, Radomski JS et al. National Transplantation Pregnancy Registry: Report on outcomes in cyclosporine-treated female kidney transplant recipients with an interval from transplant to pregnancy of greater than five years. Am J Kidney Dis 1996; 28:266-269.

25. Jimenez E, Gonzalez-Caraballo Z, Morales-Otero L et al. Triplets born to a kidney transplant recipient. Transplantation 1995; 59:435-436.

26. Rahbar K and Forghani F. Pregnancy in renal transplant recipients: An Iranian experience with a report of triplet pregnancy. Transplant Proc 1997; 29:2775.

27. Furman B, Wiznitzer A, Hackmon R et al. Multiple pregnancies in women after renal transplantation case report that rises a management dilemma. European J Ob & Gyn & Reproductive Biology 1999; 84:107-110.

28. Midtvedt K, Hartmann A, Brekke IB et al. Successful pregnancies in a combined pancreas and renal allograft recipient and in a renal graft recipient on tacrolimus treatment. Nephrol Dial Transplant 1997; 12:2764-2765.

29. Resch B, Mache CJ, Windhager T et al. FK506 and successful pregnancy in a patient after renal transplantation. Transplant Proc 1998; 30:163-164.

30. Yoshimura N, Oka T, Fujiwara Y et al. A case report of pregnancy in a renal transplant recipient treated with FK506 (tacrolimus). Transplantation 1996; 61:1552-1553.

31. Vyas S, Kumar A, Piecuch S et al. Outcome of twin pregnancy in a renal transplant recipient treated with tacrolimus. Transplantation 1999; 67:490-492.

32. Armenti VT, Wilson GA, Radomski JS et al. Report from the National Transplantation Pregnancy Registry (NTPR): Outcomes of Pregnancy after Transplantation. In: Cecka JM and Terasaki PI, eds. Clinical Transplants 1999. 15. Los Angeles: UCLA Tissue Typing Laboratory 2000; 111-119.

33. Scantlebury V, Gordon R, Tzakis A et al. Childbearing after liver transplantation. Transplantation 1990; 49:317-321.

34. Radomski JS, Moritz MJ, Muñoz SJ et al. National Transplantation Pregnancy Registry: Analysis of pregnancy outcomes in female liver transplant recipients. Liver Transplantation and Surgery 1995; 1:281-284.

35. Rayes N, Neuhaus R, David M et al. Pregnancies following liver transplantation—how safe are they? A report of 19 cases under cyclosporine A and tacrolimus. Clin Transpl 1998; 12:396-400.

36. Armenti VT, Herrine SK and Moritz MJ. Reproductive function after liver transplantation. Clinics in Liver Disease 1997; 1:471-485.

37. Jain A, Venkataramanan R, Fung JJ et al. Pregnancy after liver transplantation under tacrolimus. Transplantation 1997; 64:559-565.

38. Armenti VT, Radomski JS, Moritz MJ et al. Chapter 8 Report from the National Transplantation Pregnancy Registry (NTPR): Outcomes of Pregnancy after Transplantation. In: Cecka JM and Terasaki PI, eds. Clinical Transplants 1997. 13. Los Angeles: UCLA Tissue Typing Laboratory, 1997:101-112.

39. Skannal DG, Dungy-Poythress LJ, Miodovnik M et al. Pregnancy in a combined liver and kidney recipient with type 1 primary hyperoxaluria. Obstet Gynecol 1995; 86:641-643.

40. Catnach SM, McCarthy M, Jauniaux E et al. Liver transplantation during pregnancy complicated by cytomegalovirus infection. Transplantation 1995; 60:510-511.

21

21

41. Fair J, Klein AS, Feng T et al. Intrapartum orthotopic liver transplantation with successful outcomes of pregnancy. Transplantation 1990; 50:534-535.

42. Finlay DE, Foshager MC, Longley DG et al. Ischemic injury to the fetus after maternal liver transplantation. American Institute of Ultrasound in Medicine 1994; 13:145-148.

43. Laifer SA, Darby MJ, Scantlebury VP et al. Pregnancy and liver transplantation. Obstet Gynecol 1990; 76:1083-1088.

44. Moreno EG, Garcia GI, Gomez SR et al. Fulminant hepatic failure during pregnancy successfully treated by orthotopic liver transplantation. Transplantation 1991; 52:923-926.

45. Barrou BM, Gruessner AC, Sutherland DER et al. Pregnancy after pancreas transplantation in the cyclosporine era. Transplantation 1998; 65:524-527.

46. McGrory CH, Groshek MA, Sollinger HW et al. Pregnancy outcomes in female pancreas-kidney recipients. Transplant Proc 1999; 31:652-653.

47. Scott JR, Wagoner LE, Olsen SL et al. Pregnancy in heart transplant recipients: Management and outcome. Obstet & Gynecol 1993; 82:324-327.

48. Troche V, Ville Y, and Fernandez H. Pregnancy after heart or heart-lung transplantation: a series of 10 pregnancies. Br J of Obstet & Gyn 1998; 105:454-458.

49. Wagoner LE, Taylor DO, Olsen SL et al. Immunosuppressive therapy, management, and outcome of heart transplant recipients during pregnancy. J Heart & Lung Transpl 1993; 12:993-1000.

50. Branch KR, Wagoner LE, McGrory CH et al. Risks of subsequent pregnancies on mother and newborn in female heart transplant recipients. J Heart and Lung Transpl 1998; 17:698-702.

51. Armenti VT, Gertner GS, Eisenberg JA et al. National Transplantation Pregnancy Registry: Outcomes of pregnancies in lung recipients. Transplant Proc 1998; 30:1528-1530.

52. Armenti VT, Moritz MJ and Davison JM. Medical management of the pregnant transplant recipient. Advances in Renal Replacement Therapy 1998; 5:14-23.

53. Davison JM. Pregnancy in renal allograft recipients: problems, prognosis and practicalities. In: Lindheimer MD, Davison JM, eds. Balliere's Clin Obstet Gynaecol. 8. London: Bailliere Tindall, 1994:501-527.

54. Hou S. Pregnancy in organ transplant recipients. Med Clin North Am 1989; 73:667-683.

55. Blau EB and Gross JR. Congenital cytomegalovirus infection after recurrent infection in a mother with a renal transplant. Pediatric Nephrology 1997; 11:361-362.

56. Pescovitz MD. Absence of teratogenicity of oral ganciclovir used during early pregnancy in a liver transplant recipient. Transplantation 1999; 67:758-759.

57. Ahlswede KM, Ahlswede BA, Jarrell BE et al. National Transplantation Pregnancy Registry: Outcomes of pregnancies fathered by male transplant recipients. Mattison DR and Olshan AF, eds. Male-Mediated Developmental Toxicity New York: Plenum Press, 1994:335-338.

Dominant Transplantation Tolerance

Luis Graca and Herman Waldmann

Abbreviations

AICD   Activation induced cell death
CsA   Cyclosporin-A
MHC   Major histocompatibility complex

E1

ABSTRACT

The ideal therapy for the prevention of graft rejection would be one given short-term to achieve life-long tolerance without incurring side effects nor diminishing immunocompetence to infectious agents. Recent advances in the understanding of peripheral transplantation tolerance suggest that this may eventually be possible. The demonstration of regulatory cells with the ability to tame alloreactive clones may provide the framework for this advance. This review focuses on the challenging prospects of dominant tolerance and some of its characteristics, namely linked suppression and infectious tolerance.

Transplantation tolerance can be achieved therapeutically through two distinct approaches: inactivation of alloreactive clones and the induction of regulatory circuits. Although the approaches might seem incompatible, we here argue that most tolerance induction strategies involve, to a certain degree, both inactivation of alloreactive cells and the amplification of regulatory cells.

Current immunosupressive regimens target the whole immune system. However, an elective ablation of only the alloreactive clones, if feasible, offers a way of preventing graft rejection while sparing host's immunocompetence. One possible approach to achieving this involves the transfer of a high dose of bone marrow cells from the donor to establish mixed hemopoietic chimerism or macro-chimerism.[1-3] This permits for in vitro monitoring of the tolerant state by sampling lymphocytes from the host and testing their reactivity against donor-type cells. Such "functional" assays may be impracticable, inconvenient and not always reliable. Furthermore, it might prove difficult to deplete all alloreactive T-cell clones, and any expansion of residual cells might give rise to delayed transplant rejection.

The complementary strategy aims to control alloreactive cells in a different way. It is based on the induction of a dominant tolerance state and its hallmark is the emergence of regulatory CD4+ T cells.[4] Unlike tolerance by deletion, here cells with the ability to react in vitro with donor type cells may still be demonstrated, but grafts are still accepted indefinitely. Furthermore, tolerance is very robust and resists the adoptive transfer of cells with the potential to mediate graft rejection – the reason why it is termed dominant.[5,6] The regulatory cells can even do more

---

Organ Transplantation, 2nd edition, edited by Frank P. Stuart, Michael M. Abecassis and Dixon B. Kaufman. ©2003 Landes Bioscience.

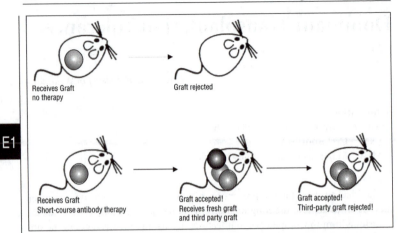

Figure E1.1. Demonstration of tolerance in antibody treated animals. Mice accept a second challenge with a graft of the same type, but readily reject third party grafts. Alloreactive cells, as demonstrated by proliferation assays, are present at any time point.

than just "suppress": if they are allowed to coexist with the naïve cells, they have the capacity to recruit new regulatory CD4+ T cells from that naïve pool. After this recruitment, the initial regulatory T cells can be removed experimentally and one observes that the new regulators can maintain tolerance themselves.[7] This process can be repeated experimentally for several cell transfers, and has therefore been named infectious tolerance.[8]

ACHIEVING DOMINANT TRANSPLANTATIONTOLERANCE

Short courses of therapeutic antibodies have been shown to lead to long-term acceptance of foreign grafts in several experimental systems (reviewed in ref. 4). The first examples of peripheral tolerance induced with monoclonal antibodies were reported in 1986.[9,10] In these experiments tolerance to foreign immunoglobulins was achieved after a short-term treatment with depleting anti-CD4 antibodies. It was soon demonstrated that depletion of CD4+ cells was not required for tolerance induction, as similar results were found using F(ab')2 fragments,[11-13] non-depleting isotypes[5] or non-depleting doses of synergistic pairs of anti-CD4 antibodies.[14] Treatment with anti-CD4 antibodies was also shown to lead to long-term acceptance of skin grafts differing by "multiple-minor" antigens[5] even in pre-sensitised recipients.[15] The same results were also demonstrated for heart grafts across MHC barriers[16,17] or concordant xenografts.[16]

Further demonstrations of transplantation tolerance were later reported with anti-LFA-1 antibodies, alone[13] or in combination with anti-ICAM-1[18] and also with anti-CD2 and anti-CD3 antibodies.[19]

More recently, co-stimulation blockade of CD28,[20] CD40L (CD154)[21] or both in combination[22] has been shown effective. These findings have recently been extended to non-human primates. In one study, long-term survival of renal allografts was achieved following blockade of CD40L alone.[23] Another group achieved pro-

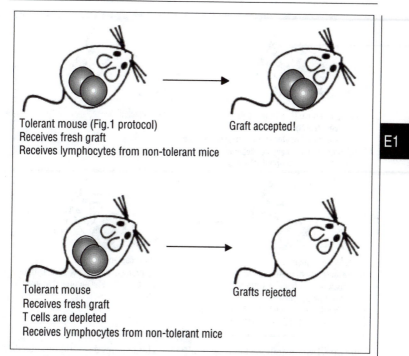

Figure E1.2. Demonstration of dominant tolerance. This requires the demonstration of tolerance being imposed on cells with the ability to reject a graft in the absence of regulatory cells.

longed islet allograft acceptance after a similar treatment.[24] Interestingly, the association of tacrolimus or steroids to the therapeutic regime abrogated tolerance.[23]

INFECTIOUS TOLERANCE

Models of transplantation tolerance induced with anti-CD4 or anti-CD40L antibodies showed that tolerant mice did not reject the grafts even after the adoptive transfer of lymphocytes from a non-tolerant animal.[5,6,25] It was also demonstrated that spleen cells from animals made tolerant to skin and heart grafts using anti-CD4 or anti-CD40L antibodies could regulate naïve T cells, and in so doing, rendering them regulatory in their own right.[7,26,25] Using transgenic mouse strains carrying specific cell surface markers in their lymphocytes, it was possible to selectively eliminate the host-type T cells from the tolerant animal.[7,25] If this cell-depletion was performed immediately after cell transfer, the tolerance state was broken and indicator grafts were readily rejected by the transferred non-tolerant lymphocytes.[7,25] If, however, the host cells were allowed to coexist with those adoptively transferred for 4-6 weeks, then tolerance was maintained even after the depletion of the host cells.[7,25] The remaining cells were nevertheless fully competent to reject an unrelated graft. Not only were they unable to reject a

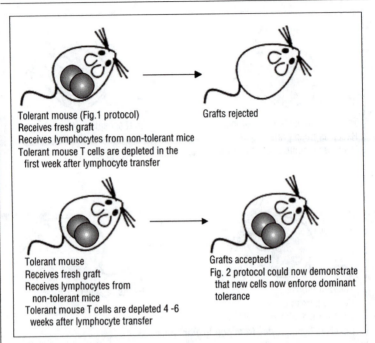

Tolerant mouse (Fig.1 protocol)
Receives fresh graft
Receives lymphocytes from non-tolerant mice
Tolerant mouse T cells are depleted in the
  first week after lymphocyte transfer

Grafts rejected

Tolerant mouse
Receives fresh graft
Receives lymphocytes from
non-tolerant mice
Tolerant mouse T cells are depleted 4 -6
weeks after lymphocyte transfer

Grafts accepted!
Fig. 2 protocol could now demonstrate
  that new cells now enforce dominant
  tolerance

Figure E1.3. Demonstration of infectious tolerance. This requires the demonstration that cells with the ability to reject are converted into the regulatory type after coexistence with cells from a tolerant animal.

graft from a similar donor, but they could now regulate another population of spleen cells from a non-tolerant animal in a similar transfer experiment.[7] This effect, named "infectious tolerance", provides compelling evidence for the existence of regulatory T cells: the regulatory cells from a tolerant animal can suppress the aggressive action of graft-reactive T-cells and induce members of that population to become regulatory as well.

A further important finding underlining the significance of infectious tolerance comes from the demonstration of a phenomenon named "linked suppression". In the original experiment[27] mice of type A made tolerant to type B skin grafts with non-depleting anti-CD4 and anti-CD8 treatment, readily rejected a third party graft of type C. However, if they were grafted with (BxC)F1 skin instead, rejection was delayed or absent while (AxC)F1 grafts were readily rejected. Furthermore, mice that accepted the (BxC)F1 skin grafts later accepted C type skin. The same phenomenon was recently demonstrated for anti-CD40L antibody induced tolerance.[28]

T-CELL REGULATION

Evidence for the existence of regulatory T cells does not come exclusively from studies of transplantation tolerance. Regulatory T cells have been found in several

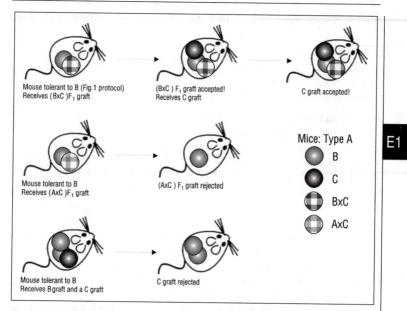

Figure E4.1. Demonstration of linked suppression. This requires the demonstration that tolerant animals accept grafts where a third party antigen is present in cells that also have the tolerated antigens (BxC), but reject third party grafts (C) if the tolerated antigen absent from the graft cells (even if a concomitant tolerated-type graft (B) is given). The animals that accept the grafts with the "linked" third party antigen (BxC) should accept later grafts of the third party (C).

autoimmunity models (reviewed in [29]). Even among the T cell population of normal individuals T cells with the capacity of causing autoimmune disease have been identified, as well as regulatory cells that prevent this pathological autoaggression.[30,31] It is therefore likely that, in addition to thymic tolerance, peripheral tolerance mechanisms operate to safeguard tolerance to extra-thymic antigens.

The phenotype of these regulatory cells, and their proposed mechanisms of action is not yet totally clear. Although it is possible to induce transplantation tolerance with mAbs in thymectomised mice,[14,28] there is evidence suggesting that regulatory cells in some autoimmunity models are a defined lineage originating in the thymus (reviewed by Seddon & Mason[32]). This lineage was shown to have some distinctive surface markers: they are included in the CD45RClo population of CD4+ cells in the rat,[30] or in the CD45RBlo in the mice.[33] It also seems that expression of the IL2 receptor a-chain (CD25) reflects the presence of a putative regulatory CD4+ cell that further subdivide the CD45RBlo population.[34,35] Given that CD25 seems to be a marker of suppressor cells it may seem paradoxical that an antibody targeting CD25 is licensed for use as immunosuppressive agent in clinical transplantation (reviewed in [36]). A theoretical risk for a therapy that targets CD25 expressing cells might be the loss of potential to induce tolerance to the graft, as well as a possible disruption of normal regulatory mechanisms that prevent autoimmunity.

Other markers, such as L-selectin[37] or CD38[38] have also been suggested as possible surface markers of regulatory cells. It is hoped that purification and cloning of these elusive regulatory cells will allow a better understanding of their biology.

TOLERANCE AND CELL DEATH

There are now many examples where evidence is found for alloreactive T-cell death in response to transplanted tissue without the need for purposeful chimerism. For example, two interesting recent papers demonstrate that tolerance induction with therapeutic anti-CD40L mAbs requires cell death.[39,40] In fact, blockade of activation induced cell death (AICD) either by using transgenic mice resistant to apoptosis,[40] or by using Cyclosporin-A (CsA)[39] resulted in graft rejection in animals subjected to antibody blockade of CD28 and CD40L.

In spite of the importance of AICD in anti-CD40L transplantation tolerance, regulatory cells also play a role in its maintenance. In fact, tolerance induction with therapeutic anti-CD40L results in linked-suppression[28] and in infectious tolerance.[25] Thus, regulatory CD4+ T cells emerge, following tolerance induction, and actively enforce a dominant tolerance state.

We can safely speculate that amplification of regulatory cells and induction of AICD are probably general mechanisms exploited in the different tolerance inducing strategies. It is likely, although not yet demonstrated, that anti-CD4 therapeutic mAbs also require some cell death for the induction of transplantation tolerance. Recently it has been shown that anti-CD4 tolerance is independent of the Fas (CD95) pathway.[41] Probably in all tolerance inducing strategies some cell clones will remain fully committed towards an aggressive phenotype and their physical (AICD) or functional (anergy) deletion is required if tolerance induction is to be successful.

Information is lacking on whether therapeutic protocols that aim at the deletion of alloreactive clones, such as the ones based on macrochimerism,[2] also support the emergence of regulatory cells. Such studies need to be performed.

HOW CAN THIS KNOWLEDGE TRANSLATE TO THE CLINIC?

Current immunosuppressive agents, although the best option available, are far from being ideal drugs. However, their known efficacy in preventing acute allograft rejection makes it ethically difficult to displace them in clinical trials of potential tolerogenic drugs. CsA is known to hinder tolerance induction with therapeutic mAbs.[39] Is it wise though to give transplanted patients an experimental therapeutic regime in the absence of CsA?

One reason why CsA exerts a tolerance-blocking effect is due to its capacity to interfere with AICD.[39] In fact, both CsA and sirolimus (FK506) are calcineurin inhibitors that block transcriptional activation of the IL-2 gene in response to antigen stimulation. As lymphocytes are prevented from being activated, AICD does not occur. In that respect the new immunosupressive drug rapamycin might be a good alternative. It does not interfere with activation and AICD. It rather functions by arresting the cell cycle, rendering lymphocytes insensitive to proliferative signals. Therefore, although CsA prevents tolerance induction with anti-CD40L antibodies, rapamycin does not affect tolerance in this system.[39] One can predict that also anti-CD4 tolerance induction might be achieved in spite of concomitant administration of rapamycin.

Another issue to bear in mind is the practical feasibility of any therapeutic strategy. Some of the experimental protocols might be too complex or involve potential side-effects too risky for widespread clinical use. The ideal agent should be easily administered and with low impact on the immune system as a whole.

A potentially simple approach that has not yet been exploited, though offering therapeutic potential, is linked suppression. Although very little is known of its mechanisms, it can mediate powerful immuregulatory effects: for example, after tolerance induction with non-depleting CD4 and CD8 mAbs to a minor mismatched skin, tolerant animals subsequently accept skin from donors that in addition to the tolerised minors have a major histocompatibility mismatch.[27] In practise one might be able to "tolerise" to a series of polymorphic allo-antigens in advance of a transplant to pre-expand regulatory cells. Following organ transplantation this first cohort of regulatory cells may facilitate spread of tolerance to clones reactive to the "linked" antigens. Thus inducing tolerance to the whole organ.

The administration of epitopes by oral,[42,43] nasal[44] and even intra-peritoneal[45] routes can lead to tolerance. It is also possible to modify the characteristics of the peptide, such as the affinity for the MHC, to modulate this effect.[46]

As we get to know some of the most important or dominant epitopes involved in graft rejection, we may be able to use them to induce transplantation tolerance. Furthermore, tolerance induced with a few dominant epitopes might then "spread" by linked suppression to other epitopes that are also present in the graft. It should therefore be possible to "build tolerance in stages": it is probably not necessary to tolerise to the whole set of major and minor antigens of the allograft, since tolerance to a few dominant ones will subsequently spread to the rest. Ultimately it might be possible to identify a group of dominant epitopes that could be used as a universal therapy to induce transplantation tolerance in any host-donor combination.

ACKNOWLEDGEMENTS

This work was funded by grants from the Medical Research Council (UK). LG is supported by a scholarship from the Gulbenkian Foundation and the Portuguese Foundation for Science and Technology.

This article is reprinted from *Graft* 2001; Vol. 4, No. 3:174-179, with permission from the authors.

REFERENCES

1. Ildstad ST, Sachs DH. Reconstitution with syngeneic plus allogeneic or xenogeneic bone marrow leads to specific acceptance of allografts or xenografts. Nature 1984; 307(5947):168-170.

2. Wekerle T, Kurtz J, Ito H et al. Allogeneic bone marrow transplantation with co-stimulatory blockade induces macrochimerism and tolerance without cytoreductive host treatment. Nat Med 2000; 6(4):464-469.

3. Wekerle T, Sykes M. Mixed chimerism as an approach for the induction of transplantation tolerance. Transplantation 1999; 68(4):459-467.

4. Waldmann H, Cobbold S. How do monoclonal antibodies induce tolerance? A role for infectious tolerance? Ann Rev Immunol 1998; 16:619-644.

5. Qin SX, Wise M, Cobbold SP et al. Induction of tolerance in peripheral T cells with monoclonal antibodies. Eur J Immunol 1990; 20(12):2737-2745.

6. Scully R, Qin S, Cobbold S et al. Mechanisms in CD4 antibody-mediated transplantation tolerance: kinetics of induction, antigen dependency and role of regulatory T cells. Eur J Immunol 1994; 24(10):2383-2392.

7. Qin S, Cobbold SP, Pope H et al. Infectious transplantation tolerance. Science 1993; 259(5097):974-977.

8. Cobbold S, Waldmann H. Infectious tolerance. Curr Opin Immunol 1998; 10(5):518-524.

9. Gutstein NL, Seaman WE, Scott JH et al. Induction of immune tolerance by administration of monoclonal antibody to L3T4. J Immunol 1986; 137(4):1127-1132.

10. Benjamin RJ, Waldmann H. Induction of tolerance by monoclonal antibody therapy. Nature 1986; 320(6061):449-451.

11. Carteron NL, Wofsy D, Seaman WE. Induction of immune tolerance during administration of monoclonal antibody to L3T4 does not depend on depletion of L3T4+ cells. J Immunol 1988; 140(3):713-716.

12. Carteron NL, Schimenti CL, Wofsy D. Treatment of murine lupus with F(ab')2 fragments of monoclonal antibody to L3T4. Suppression of autoimmunity does not depend on T helper cell depletion. J Immunol 1989; 142(5):1470-1475.

13. Benjamin RJ, Qin SX, Wise MP et al. Mechanisms of monoclonal antibody-facilitated tolerance induction: a possible role for the CD4 (L3T4) and CD11a (LFA-1) molecules in self-non-self discrimination. Eur J Immunol 1988; 18(7):1079-1088.

14. Qin S, Cobbold S, Tighe H et al. CD4 monoclonal antibody pairs for immunosuppression and tolerance induction. Eur J Immunol 1987; 17(8):1159-1165.

15. Marshall SE, Cobbold SP, Davies JD et al. Tolerance and suppression in a primed immune system. Transplantation 1996; 62(11):1614-1621.

16. Chen Z, Cobbold S, Metcalfe S et al. Tolerance in the mouse to major histocompatibility complex-mismatched heart allografts, and to rat heart xenografts, using monoclonal antibodies to CD4 and CD8. Eur J Immunol 1992; 22(3):805-810.

17. Onodera K, Lehmann M, Akalin E et al. Induction of infectious tolerance to MHC-incompatible cardiac allografts in CD4 monoclonal antibody-treated sensitized rat recipients. J Immunol 1996; 157(5):1944-1950.

18. Isobe M, Yagita H, Okumura K et al. Specific acceptance of cardiac allograft after treatment with antibodies to ICAM-1 and LFA-1. Science 1992; 255(5048):1125-1127.

19. Chavin KD, Qin L, Lin J et al. Combined anti-CD2 and anti-CD3 receptor monoclonal antibodies induce donor-specific tolerance in a cardiac transplant model. J Immunol 1993; 151(12):7249-7259.

20. Lenschow DJ, Zeng Y, Thistlethwaite JR et al. Long-term survival of xenogeneic pancreatic islet grafts induced by CTLA4lg [see comments]. Science 1992; 257(5071):789-792.

21. Parker DC, Greiner DL, Phillips NE et al. Survival of mouse pancreatic islet allografts in recipients treated with allogeneic small lymphocytes and antibody to CD40 ligand. Proc Natl Acad Sci U S A 1995; 92(21):9560-9564.

22. Larsen CP, Elwood ET, Alexander DZ et al. Long-term acceptance of skin and cardiac allografts after blocking CD40 and CD28 pathways. Nature 1996; 381(6581):434-438.

23. Kirk AD, Burkly LC, Batty DS et al. Treatment with humanized monoclonal antibody against CD154 prevents acute renal allograft rejection in nonhuman primates. Nat Med 1999; 5(6):686-693.

24. Kenyon NS, Chatzipetrou M, Masetti M et al. Long-term survival and function of intrahepatic islet allografts in rhesus monkeys treated with humanized anti-CD154. Proc Natl Acad Sci U S A 1999; 96(14):8132-8137.

25. Graca L, Honey K, Adams E, Cobbold SP, Waldmann H. Cutting edge: Anti-cd154 therapeutic antibodies induce infectious transplantation tolerance. J immunol 2000; 165:4783-4786.

26. Chen ZK, Cobbold SP, Waldmann H et al. Amplification of natural regulatory immune mechanisms for transplantation tolerance. Transplantation 1996; 62(9):1200-1206.

27. Davies JD, Leong LY, Mellor A et al. T cell suppression in transplantation tolerance through linked recognition. J Immunol 1996; 156(10):3602-3607.

28. Honey K, Cobbold SP, Waldmann H. CD40 ligand blockade induces CD4+ T cell tolerance and linked suppression. J Immunol 1999; 163(9):4805-4810.

29. Mason D, Powrie F. Control of immune pathology by regulatory T cells. Curr Opin Immunol 1998; 10(6):649-655.

30. Fowell D, Mason D. Evidence that the T cell repertoire of normal rats contains cells with the potential to cause diabetes. Characterization of the CD4+ T cell subset that inhibits this autoimmune potential. J Exp Med 1993; 177(3):627-636.

31. Powrie F, Mason D. OX-22high CD4+ T cells induce wasting disease with multiple organ pathology: prevention by the OX-22low subset. J Exp Med 1990; 172(6):1701-1708.

32. Seddon B, Mason D. The third function of the thymus. Immunol Today 2000; 21(2):95-99.

33. Powrie F, Carlino J, Leach MW et al. A critical role for transforming growth factor-beta but not interleukin 4 in the suppression of T helper type 1-mediated colitis by CD45RB(low) CD4+ T cells. J Exp Med 1996; 183(6):2669-2674.

34. Asano M, Toda M, Sakaguchi N et al. Autoimmune disease as a consequence of developmental abnormality of a T cell subpopulation. J Exp Med 1996; 184(2):387-396.

35. Suri-Payer E, Amar AZ, Thornton AM et al. CD4+CD25+ T cells inhibit both the induction and effector function of autoreactive T cells and represent a unique lineage of immunoregulatory cells. J Immunol 1998; 160(3):1212-1218.

36. Waldmann TA, O'Shea J. The use of antibodies against the IL-2 receptor in transplantation. Curr Opin Immunol 1998; 10(5):507-512.

37. Herbelin A, Gombert JM, Lepault F et al. Mature mainstream TCR alpha beta+CD4+ thymocytes expressing L-selectin mediate "active tolerance" in the nonobese diabetic mouse. J Immunol 1998; 161(5):2620-2628.

38. Read S, Mauze S, Asseman C et al. CD38+ CD45RB(low) CD4+ T cells: a population of T cells with immune regulatory activities in vitro. Eur J Immunol 1998; 28(11):3435-3447.

E1

39. Li Y, Li XC, Zheng XX et al. Blocking both signal 1 and signal 2 of T-cell activation prevents apoptosis of alloreactive T cells and induction of peripheral allograft tolerance. Nat Med 1999; 5(11):1298-1302.

40. Wells AD, Li XC, Li Y et al. Requirement for T-cell apoptosis in the induction of peripheral transplantation tolerance. Nat Med 1999; 5(11):1303-1307.

41. Honey K, Cobbold SP, Waldmann H. Dominant tolerance and linked suppression induced by therapeutic antibodies do not depend on Fas-FasL interactions. Transplantation 2000; 69(8):1-7.

42. Chen Y, Kuchroo VK, Inobe J et al. Regulatory T cell clones induced by oral tolerance: suppression of autoimmune encephalomyelitis. Science 1994; 265(5176):1237-1240.

43. Miller A, Lider O, Roberts AB et al. Suppressor T cells generated by oral tolerization to myelin basic protein suppress both in vitro and in vivo immune responses by the release of transforming growth factor beta after antigen-specific triggering. Proc Natl Acad Sci U S A 1992; 89(1):421-425.

44. Bai XF, Zhu J, Zhang GX et al. IL-10 suppresses experimental autoimmune neuritis and down-regulates TH1-type immune responses. Clin Immunol Immunopathol 1997; 83(2):117-126.

45. Liu GY, Wraith DC. Affinity for class II MHC determines the extent to which soluble peptides tolerize autoreactive T cells in naive and primed adult mice--implications for autoimmunity. Int Immunol 1995; 7(8):1255-1263.

46. Anderton S, Burkhart C, Metzler B et al. Mechanisms of central and peripheral T-cell tolerance: lessons from experimental models of multiple sclerosis. Immunol Rev 1999; 169:123-137.

Antigen Receptor Revision as a Mechanism of Peripheral T Cell Tolerance

Cristine J. Cooper and Pamela J. Fink

Abbreviations

TCR	T cell receptor
MHC	Major histocompatibility complex
RAG	Recombination activating gene
Tg	Transgenic
Vβ^{endo}	Endogenously-derived TCR Vβs
Mtv	Mammary tumor virus
GFP	Green fluorescent protein
GC	Germinal center

Tolerance induction among mature T cells in the lymphoid periphery operates through many mechanisms, including the induction of anergy and cell death. By one newly described pathway, CD4$^+$ T cells that encounter a tolerogen are either deleted or are driven to reexpress the proteins that mediate DNA recombination and to rearrange and express diverse novel antigen receptor genes encoding proteins that no longer recognize the tolerogen. T cells that have successfully completed such receptor revision are both functional and self tolerant.

The broad antigen receptor repertoire that results from receptor revision benefits the individual faced with decreasing CD4$^+$ T cell counts due to elimination of T cells recognizing a wide-spread self antigen that cannot be cleared. However, reexpression of the recombinase machinery in mature peripheral T cells offers the potential for illegitimate recombination and subsequent dysregulation of cellular functions. Why would such a risky venture be undertaken? Perhaps the downregulation of receptor expression that precedes revision decreases the basal level of signaling through the receptor, signaling that is critical for T cell survival. The cell may interpret this loss of signaling capacity as a developing thymocyte would, by generating alternate antigen receptors whose expression levels are conducive to cell survival. In this way, receptor revision may recapitulate thymocyte maturation.

INTRODUCTION

The immune system is charged with the dual tasks of defense against invading pathogens and preservation of self. For T cells, carrying out these simultaneous

Organ Transplantation, 2nd edition, edited by Frank P. Stuart, Michael M. Abecassis and Dixon B. Kaufman. ©2003 Landes Bioscience.

Summary

- Although intrathymic negative selection operates primarily to eliminate overtly self-reactive thymocytes, the induction of tolerance among mature peripheral T cells operates through many pathways.
- Through the process of TCR revision, mature peripheral CD4[+] T cells can be driven to reinitiate DNA rearrangement within the TCR loci and express diverse, newly generated, nonautoreactive TCRs.
- The decision to upregulate RAG expression and undergo TCR revision may be a byproduct of T cell maturation.
- The surprisingly broad TCR repertoire that results from TCR revision is an obvious benefit to the individual faced with decreasing CD4[+] T cell counts due to elimination of cells recognizing a self antigen it is unable to clear.
- Reexpression of the recombinase machinery in mature peripheral T cells offers the potential for aberrant juxtaposition of cellular oncogenes and lymphocyte-specific promoters.

E2

duties requires careful discrimination between self and non-self, a distinction whose borders are constantly reassessed throughout the lifetime of the cell. Of clear importance to T cell function is the nature of the T cell receptor for antigen (TCR), a heterodimeric cell surface molecule monoclonally expressed by each individual T cell. This TCR recognizes short peptide antigens bound to a groove within molecules encoded by the major histocompatibility complex (MHC) or to longer glycoproteins called superantigens, presented outside the peptide binding groove of the MHC.[1,2] One of the first challenges facing developing T cells within the thymus, the organ in which T cells mature, is the assembly of diverse TCRs through recombination of the separate gene elements that together encode this protein.[3] TCR-α and -b gene rearrangement occurs by a developmentally-regulated process mediated by the products of recombination activating gene (RAG)1 and RAG2.[4] The TCR repertoire is selected within the thymus for recognition of the multitude of peptide antigens presented by self MHC molecules and culled of overt self reactivity.[5] This latter intrathymic process, termed negative selection, requires that thymocytes be exposed to the relevant self antigens.[6] Although intrathymic expression of antigens once believed to be strictly tissue-specific has recently expanded the pool of known contributors to negative selection,[7] alternate forms of tolerance induction must exist to handle mature peripheral T cells recognizing age-dependent or tissue-specific antigens.

Although intrathymic negative selection operates primarily to eliminate overtly self-reactive thymocytes through programmed cell death, the induction of tolerance among mature peripheral T cells has been shown to operate through many pathways.[8] Autoreactive T cells may be prevented from encountering antigen in a context that could lead to cell activation.[9] On the other hand, these T cells may meet antigen and be rendered anergic (nonfunctional) to further stimulation through their TCRs.[10] Anergic T cells generally have a shortened lifespan and may appear phenotypically normal or may express reduced surface levels of TCR and/ or CD4 or CD8 coreceptor molecules. Self reactive cells may be directly eliminated without traversing an anergic state, or may be driven into terminal differen-

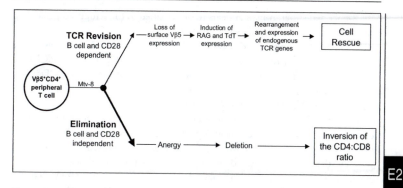

Figure E2.1. Schematic diagram of the alternate pathways to tolerance for mature CD4⁺ T cells in Vβ5 Tg Mtv-8⁺ mice. Vβ5⁺CD4⁺ T cells that encounter Mtv-8 in the lymphoid periphery either become anergic and die, or revise their TCRs, thereby eliminating Mtv-8 reactivity. T cells that undergo TCR revision express a diverse repertoire and contribute to the self-tolerant, functional T cell pool.

E2

tiation, thereby temporarily exhausting the supply of antigen-reactive cells.[11] The activity of autoaggressive T cells may also be suppressed by veto cells,[12] antigen-specific regulatory T cells,[13] or nutrient deprivation in specialized sites such as the maternal-fetal interface.[14] What follows is a review of a novel form of tolerance induction by which mature, peripheral CD4⁺ T cells reinitiate rearrangement of TCR loci to transform a self reactive TCR into one that is self tolerant. This tolerance mechanism, termed TCR revision, rescues self reactive lymphocytes and generates from them a diverse population of functional, self tolerant CD4⁺ T cells. However TCR revision is a risky form of tolerance induction, as it entails potential illegitimate DNA rearrangement events and generation of TCRs outside the selective thymic microenvironment.

A WEAK TOLEROGEN CAN DRIVE PERIPHERAL CD4⁺ T CELLS DOWN ALTERNATE PATHWAYS OF DEATH AND ANTIGEN RECEPTOR REVISION

The majority of T cells from young mice carrying a functionally rearranged TCR-β chain transgene express that gene, constituting a population of T cells expressing diverse TCR-α chains paired with a uniform, transgene-encoded TCR-β chain.[15] In Vβ5 transgenic (Tg) mice, transgene expression among CD8-T cells remains high at all ages. However, Vβ5 expression among CD4⁺ peripheral T cells decreases with age, and concomitantly, expression of TCR-β chains encoded by rearranged endogenous genes increases.[16] The lymphoid periphery of Vβ5 Tg mice is also characterized by a striking age-dependent inversion of the CD4:CD8 ratio. Both the inversion of the CD4:CD8 ratio (caused by the loss of CD4⁺ peripheral T cells) and the loss of transgene expression (caused by the appearance of cells expressing endogenous Vβs or Vβ^endo) are dependent on a superantigen encoded by a defective endogenous mammary tumor virus (Mtv)-8.[17] Expression of the

Mtv-8-encoded superantigen appears to be confined to the lymphoid periphery, and provides for an unusually weak interaction with Vβ5⁺ TCRs.[18]

The interaction between Vβ5⁺CD4⁺ T cells and MHC class II⁺Mtv-8⁺ cells drives the T cell partner down one of two tolerance pathways (Fig. E2.1). The CD4⁺ T cell can be rendered anergic and die, thereby effecting an inversion of the CD4:CD8 ratio, or it can be rescued by losing Mtv-8 reactivity upon extinction of Vβ5 surface expression. This latter pathway is called TCR revision because the loss of Vβ5 surface expression occurs hand-in-hand with the acquisition of endogenously-derived TCR-β chains.[19] T cells from Vβ5 Tg mice do not undergo TCR revision in the absence of either B cells or CD28 molecules,[20,21] and revision appears to be highly inefficient in lethally irradiated Mtv-8⁺ hosts whose hematopoietic systems have been reconstituted with bone marrow from Mtv-8⁺ donors (McMahan CJ, Fink PJ, unpublished observations). In contrast, the deletional pathway is fully operative in Vβ5 Tg mice lacking B cells or CD28 molecules, and after lethal irradiation and bone marrow reconstitution. Thus, encounter with the same weak tolerogen can drive CD4⁺ T cells down alternate pathways leading to cell deletion or cell rescue through TCR revision.

TCR REVISION RESULTS IN EXPRESSION OF A DIVERSE, SELF TOLERANT RECEPTOR REPERTOIRE

The TCR repertoire of Vβ^endo+Vβ5⁻CD4⁺ T cells from Mtv-8⁺ Vβ5 Tg mice is so diverse that it effectively recreates the nontransgenic TCR-β chain repertoire within each individual mouse.[21] This diversity is apparent even at the molecular level, of individual rearrangements of one particular Vβ gene element to one particular Jβ element. These newly generated TCRs can deliver proliferative signals upon antibody-mediated crosslinking, and are therefore fully functional. Vβ^endo+CD4⁺ T cells from Vβ5 Tg mice do not appear to be autoreactive, either in vivo or in vitro.[21]

CELLS UNDERGOING REVISION ARE ACUTELY ACTIVATED, TCR^LOW, AND RECOMBINATION COMPETENT

As Mtv-8⁺ Vβ5 Tg mice age, they accumulate Vβ5^low cells within the CD4⁺ T cell compartment, and it appears to be these cells that are undergoing TCR revision. Vβ5^low cells are CD44^high, CD62L^low, both markers consistent with an activated T cell phenotype (Table E2.1). However, unlike antigen-activated lymphocytes, Vβ5^low cells are Thy-1^low (our unpublished observations) and express both RAG1 and RAG2.[20] The presence of recombination intermediates within the TCRβ and β loci in these unusual cells indicates both that the RAG1 and RAG2 gene products are functional and that the TCR loci accessible to the recombinase.[20] To more easily focus only on T cells undergoing TCR revision, the Vβ5 transgene has been crossed onto a line of mice Tg for green fluorescent protein (GFP) under the control of the RAG2 promoter.[22] Cells from these mice glow green when RAG2 is expressed. GFP⁺CD4⁺Vβ5^low peripheral T cells from these Mtv-8⁺ RAG reporter mice are larger than their GFP-Vβ5^high counterparts and are CD45RB^high (our unpublished observations), both markers of acutely activated rather than memory

Table E2.1. *Relative size and surface phenotype of naïve or antigen-experienced CD4+ T cells*

	Type of Mature CD4+ T Cell			
Marker	Naïve	Acutely Activated	Revising	Memory
size	small	large	large	small
TCR	high	low	low	high
Thy1	high	high	low	high
CD45RB	high	high	high	low
CD62L	high	low	low	low
CD44	low	low	high	high
CD69	low	high	low	low
CD25	low	high	low	low

cells (Table E2.1). Although this small population of cells appears to be acutely activated, expression of the transient activation markers CD69 and CD25 does not appear to be significantly upregulated (our unpublished observations). Thus, T cell interaction with a weak tolerogen that leads to TCR revision appears to initiate some but not all of the events associated with full activation of T cells encountering a foreign conventional antigen.

Cells whose phenotype is consistent with their position as intermediates in a TCR revision process have also been reported in normal humans[23] and, in increased numbers, in patients with defective responses to DNA damage.[24] These CD4+ peripheral T lymphocytes are TCR[low], RAG-expressing cells that contain recombination intermediates at the TCR-β loci. It is unclear whether these cells are undergoing TCR revision, and if so, what triggers this response.

REVISION TARGETS MATURE PERIPHERAL T CELLS

Several observations indicate that TCR revision is a peripheral event targeting mature T cells. Vβ[endo+]Vβ5-CD4+ T cells, the products of TCR revision, appear with similar kinetics in both thymectomized and unmanipulated Mtv-8+ Vβ5 Tg mice.[25] Conversely, Vβ[endo+]Vβ5-CD4+ thymocytes cannot be detected in mice at any age, although the diversity of the expressed endogenous TCR repertoire and the minimal mouse-to-mouse variation in this diversity together suggest their generation is not a rare event.[21] Sequence analyses of revised TCR-β chain genes indicate they contain regions of nontemplated nucleotides that are atypical of those generated in the adult thymus.[21] Most definitively, GFP Tg Vβ5 Tg mice that have been thymectomized more than 4 weeks previously can generate GFP+Vβ[endo+]Vβ5-CD4+ T cells, and these cells are acutely activated and RAG-expressing (our unpublished observations). Together, these results indicate that through the process of TCR revision, mature peripheral CD4+ T cells can be driven to reinitiate recombination within the TCR loci and express newly generated, nonautoreactive TCRs.

TCR REVISION IS NOT LIMITED TO TCR TG MICE

While using TCR Tg mice has the obvious benefits of creating an artificial situation in which a uniformly expressed TCR is known to interact with a given self antigen and in which the TCR expression history is known for T cells that can be physically tracked, this artificiality brings with it a set of caveats. It is therefore important to stress that TCR revision is not limited to ectopic, multicopy transgene-encoded receptors. Work from the Kanagawa lab demonstrates that TCR revision in response to recognition of exogenous superantigen can occur within the normally configured TCR-β locus.[26] Furthermore, it has been shown recently that in Mtv-8+ Vβ5 nontransgenic GFP Tg RAG reporter mice thymectomized at least 4 weeks previously, a significantly greater proportion of Vβ5+ cells are GFP+ relative to Vβ5- or Vβ8+ cells (our unpublished observations). These findings indicate that Mtv-8-driven RAG-mediated TCR revision occurs even in TCR nontransgenic mice. The appearance of CD4+TCR^{low}RAG+ T cells in normal humans also suggests that TCR revision occurs in individuals carrying normal TCR loci. Thus, the notion that a weak tolerogen can initiate TCR revision appears to be generalizeable to unmanipulated individuals. However, it should not be inferred from these studies that TCR revision is a common response to tolerogen encounter in the lymphoid periphery. Outside of the Vβ5+CD4+ population of mature T cells, the frequency of GFP+ T cells in the RAG reporter mice is very low (less than 2%, our unpublished observations). TCR revision may be initiated within a narrow window defined by TCR/superantigen affinity, superantigen expression levels, or frequency of encounter between T cells and superantigen-expressing cells.

TCR REVISION MAY BE INDUCED BY ENCOUNTER WITH SUPERANTIGENS BUT NOT CONVENTIONAL ANTIGENS

In recently published experiments, Huang et al[26] initiated RAG expression and TCR-α chain revision in TCR Tg animals injected with spleen cells expressing a viral superantigen (Mtv-6) capable of interacting with the transgenic TCR.[26] Similar to the Vβ5/Mtv-8 system described above, this TCR revision resulted in the gradual appearance of CD4+ T cells no longer expressing the transgenic TCR, although in these mice, expression of the TCR-α chain, rather than the TCR-β chain, was gradually lost. Although this distinction is likely due to the configuration of the TCR loci in these engineered mice (the TCR-α transgene was incorporated within the endogenous TCR-α locus and subsequent rearrangement events would thereby physically eliminate the TCR-α and not the β transgene), it does emphasize that revision at both the TCR-α and -β loci is possible. It is still unclear whether both genetic regions are equally accessible to the recombinase. Interestingly, Kanagawa and coworkers failed to induce TCR revision in these same animals immunized not with Mtv-6+ cells, but with cells expressing cytochrome c, the foreign antigen recognized in the context of self MHC by the transgenic TCR-αβmolecules.[26] These striking results remain to be generalized by data from other conventional antigen/ superantigen systems, but they may point to distinct biological outcomes resulting from recognition of these two classes of antigens. It is unclear whether these distinctions result from differences in the type of antigen presenting cell, the affinity of the TCR/ligand interaction, the frequency of these interactions, or some qualitatively different signal transmitted by a TCR bound to a conventional peptide antigen presented in the groove of MHC[1] versus that same receptor bound to a superantigen presented outside the MHC antigen presenting groove.[2]

POTENTIAL RISKS AND BENEFITS OF TCR REVISION

As a tolerance mechanism, TCR revision appears to be a risky proposition. Reexpression of the recombinase machinery in mature peripheral T cells offers the potential for aberrant juxtaposition of cellular oncogenes and lymphocyte-specific promoters.[27,28] Such genome instability can result in dysregulated cellular functions and transformation. It is not clear yet whether TCR revision increases the risk of oncogenesis. However a relationship between the increased frequency of TCR^{low}RAG$^+$CD4$^+$ peripheral T cells in ataxia telangiectasia and Nijmegen breakage syndrome patients and their frequent lymphoma-specific chromosomal translocations has been suggested.[24] TCR revision may also serve to modulate T cell reactivity to superantigen-expressing bacterial or viral pathogens.[26] Loss of superantigen reactivity could influence the outcome of an infection with such an organism.[29-32] A further danger in TCR revision lies in the fact that by not eliminating the autoreactive cell outright, the individual exposes itself to the possibility of continued autoaggression. Although the endproduct of TCR revision is a population of cells that appears to be self tolerant,[20,21] it is unclear if the revision process itself is associated with stringent selection against overt self reactivity, as in the thymus during T cell maturation, or whether subsequent selection events in the lymphoid periphery are called into play to eliminate newly generated autoreactive T cells. Regardless of the means of selection, this secondary process is unlikely to be infallible. Why, then, would evolution select for such risk-taking behavior?

The surprisingly broad TCR repertoire that results from TCR revision is an obvious benefit to the individual faced with decreasing CD4$^+$ T cell counts due to elimination of cells recognizing a self antigen it is unable to clear. The benefits of receptor revision may therefore outweigh the risks, although the age-dependency of TCR revision makes it likely these benefits are enjoyed most commonly by mice 5-6 months of age, well past the age of sexual maturity. One solution to this conundrum may be that the decision to upregulate RAG expression and undergo TCR revision is a byproduct of T cell maturation. During thymocyte development, RAG expression is maintained and TCR loci remain accessible to the recombinase until the proper signals are delivered into the cell through a functional TCR.[33] In the absence of such a signal, TCR rearrangement continues.[34] One of the first phenotypic changes apparent in cells undergoing TCR revision is the partial loss of TCR expression at the cell surface.[20] In fact, following this TCR^{low} trait alone led to the isolation of RAG-expressing CD4$^+$ T cells from human donors.[23,24] Perhaps this downregulation, whether ligand-mediated or not, serves to decrease the basal level of signaling through the TCR, signaling that is thought to be key for T cell survival.[35,36] The cell may interpret this loss of signaling capacity

E2

in the same way a developing thymocyte would—by upregulation of RAG expression and generation of alternate TCR genes whose protein products will be tested for their signaling capacity. Thus, rather than being selected for directly, TCR revision may be a byproduct of the way in which the TCRs expressed by developing thymocytes are selected to meet the dual requirements for self tolerance and recognition of foreign peptides in the context of self MHC molecules. The added flexibility in immune recognition provided to the aging mouse by TCR revision could then be considered an unexpected bonus.

UNANSWERED QUESTIONS

E2

HOW DOES THE SAME WEAK TOLEROGEN DRIVE CD4+ T CELLS DOWN DISTINCT PATHWAYS LEADING TO CELL ELIMINATION, ON THE ONE HAND, AND CELL RESCUE THROUGH TCR REVISION ON THE OTHER?

No definitive experiments have yet shed light on this question, although the B cell requirement for TCR revision may be informative.[20] One viable hypothesis suggests that cell death requires a strong signal delivered to a T cell, perhaps by a dendritic cell, while the signal that initiates TCR revision would be delivered by a B cell, a less potent antigen presenting cell. The exact nature of the TCR-_ chain paired with the Tg TCR-_ chain may also modulate the strength of signal delivered by a particular superantigen.[37-39] In support of this argument, the TCR-_ chain repertoire of V_5 Tg mice has been shown to vary with age, becoming less diverse as V_5 expression decreases.[17]

HOW IS TCR REVISION TRIGGERED?

If further experiments substantiate the notion that superantigens but not conventional antigens can induce reactive T cells to undergo TCR revision, it becomes important to undertand how these cellular interactions differ. The phenotype of the cells actively engaged in revision suggests that one initial trigger may be an interaction that initiates partial, but not complete, cellular activation. All work to date does suggest that one important characteristic of cells undergoing revision is their TCRlow status,[20,23,24] although how this phenotype is achieved is still unclear.

WHERE DOES TCR REVISION TAKE PLACE?

The germinal center (GC) offers one potential site for TCR revision that is consistent with all available data. At this point, only CD4+ and not CD8+ T cells have been shown to undergo TCR revision.[20,23,26] CD4+ T cells can enter GCs, while CD8+ T cells are excluded.[40,41] TCR revision, but not cellular elimination, requires B cells[20] and CD28 molecules[21] both known to be required for efficient GC formation. GC T cells are activated and Thy-1low,[42] as are cells undergoing TCR revision (our unpublished observations). The GC provides a niche in which B cells undergo stringent selection on the basis of their expressed antigen receptors. It is conceivable that such a selective microenvironment could impose self tolerance on a population of CD4+ T cells expressing newly generated TCRs. Using the RAG reporter mice, it should now be possible to pinpoint the location of those T cells undergoing antigen receptor revision.

IS TCR REVISION ASSOCIATED WITH AN INCREASED RISK OF AUTOIMMUNITY, ILLEGITIMATE RECOMBINATION, OR SUSCEPTIBILITY TO PATHOGENS THOUGHT TO EXPRESS SUPERANTIGENS?

It is these three areas that are most likely to be impacted negatively by TCR revision. Will cells undergoing TCR revision be found at the site of tissue-specific autoimmunity? Can the expression of RAG genes in T cells within the lymphoid periphery lead to chromosomal translocations within the TCR loci? Will the loss of superantigen-reactivity in T cells increase an individual's susceptibility to pathogens expressing superantigens?

Clearly, many key questions remain to be answered as this newly discovered means of peripheral T cell tolerance induction is explored.

E2

DEFINITIONS

Superantigens: Virally or bacterially-encoded glycoproteins presented outside the groove of MHC class II molecules and recognized by the TCR primarily through its Vβ domain

Negative Selection: Elimination of overtly autoreactive T cells within the thymus as a means of inducing central tolerance (tolerance of immature lymphocytes within a generative organ)

Peripheral Tolerance: Tolerance induced by multiple mechanisms among mature lymphocytes in the lymphoid periphery

TCR REVISION

One mechanism of tolerance induction among mature peripheral CD4[+] T cells, in which an autoreactive TCR is replaced with a nonself-reactive TCR

ACKNOWLEDGEMENTS

The authors thank Dr. C.J. McMahan for her contributions to these studies. This work was supported by NIH grant AG13078 (P.J.F.) and the Juvenile Diabetes Research Foundation Chet Edmonson postdoctoral fellowship (C.J.C.).

This article is reprinted from *Graft* 2002; Vol. 5, No. 7:383-389, with permission from the authors.

REFERENCES

1. Hennecke J, Wiley DC. T cell receptor-MHC interactions up close. Cell 2001; 104:1-4.
2. Li H, Llera A, Malchiodi EL, Mariuzza RA. The structural basis of T cell activation by superantigens. Annu Rev Immunol 1999; 17:435-66.
3. Lewis SM. The mechanism of V(D)J joining: lessons from molecular, immunological, and comparitive analyses. Adv Immunol 1994; 56:27-150.
4. Oettinger MA, Schatz DG, Gorka C, Baltimore D. RAG-1 and RAG-2, adjacent genes that synergistically activate V(D)J recombination. Science 1990; 248:1517-23.
5. von Boehmer H. Positive selection of lymphocytes. Cell 1994; 76:219-28.
6. Sprent J, Kishimoto H. The thymus and central tolerance. Transplantation 2001; 72(8):S25-8.

7.   Derbinski J, Schulte A, Kyewski B, Klein L. Promiscuous gene expression in medullary thymic epithelial cells mirrors the peripheral self. Nat. Immunol. 2001; 2:1032-9.

8.   Stockinger B. T lymphocyte tolerance: From thymic deletion to peripheral control mechanisms. Adv Immunol 1999; 71:229-65.

9.   Ohashi PS, Ochen S, Buerki K, Pircher H, Ohashi CT, Odermatt B, et al. Ablation of "tolerance" and induction of diabetes by virus infection in viral antigen transgenic mice. Cell 1991; 65:305-17.

10. Schonrich G, Kalinke U, Momburg F, Malissen M, Schmitt-Verhulst AM, Malissen B, et al. Down-regulation of T cell receptors on self-reactive T cells as a novel mechanism for extrathymic tolerance induction. Cell 1991; 65:293-304.

11. Rocha B, Grandien A, Freitas AA. Anergy and exhaustion are independent mechanisms of peripheral T cell tolerance. J Exp Med 1995; 181:993-1003.

12. Fink PJ, Shimonkevitz RP, Bevan MJ. Veto cells. Annu Rev Immunol 1988; 6:115-37.

13. Sakaguchi S. Regulatory T cells: key controllers of immunologic self-tolerance. Cell 2000; 101:455-8.

14. Mellor AL, Sivakumar J, Chandler P, Smith K, Molina H, Mao D, et al. Prevention of T cell-driven complement activation and inflammation by tryptophan catabolism during pregnancy. Nat Immunol 2001; 2:64-8.

15. Uematsu Y, Ryser S, Dembi'c Z, Borgulya P, Krimpenfort P, Berns A, et al. In transgenic mice the introduced functional T cell receptor b gene prevents expression of endogenous b genes. Cell 1988; 52:831-41.

16. Fink PJ, Fang CA, Turk GL. The induction of peripheral tolerance by the chronic activation and deletion of CD4+Vb5+ cells. J Immunol 1994; 152:4270-81.

17. Blish CA, Gallay BJ, Turk GL, Kline KM, Wheat W, Fink PJ. Chronic modulation of the T cell receptor repertoire in the lymphoid periphery. J Immunol 1999; 162:3131-40.

18. Scherer MT, Ignatowicz L, Pullen A, Kappler J, Marrack P. The use of mammary tumor virus (Mtv)-negative and single-Mtv mice to evaluate the effects of endogenous viral superantigens on the T cell repertoire. J Exp Med 1995; 182:1493-504.

19. Fink PJ, McMahan CJ. Lymphocytes rearrange, edit and revise their antigen receptors to be useful yet safe. Immunol Today 2000; 21:561-6.

20. McMahan CJ, Fink PJ. RAG reexpression and DNA recombination at T cell receptor loci in peripheral CD4+ T cells. Immunity 1998; 9:637-47.

21. McMahan CJ, Fink PJ. Receptor revision in peripheral T cells creates a diverse Vb repertoire. J Immunol 2000; 165:6902-7.

22. Yu W, Nagaoka H, Jankovic M, Misulovin Z, Suh H, Rolink A, et al. Continued RAG expression in late stages of B cell development and no apparent re-induction after immunization. Nature 1999; 400:682-7.

23. Lantelme E, Palermo B, Granziero L, Mantovani S, Campanelli R, Monafo V, et al. Recombinase-activating gene expression and V(D)J recombination in CD4+CD3low mature T lymphocytes. J Immunol 2000; 164:3455-9.

24. Lantelme E, Mantovani S, Palermo B, Campanelli R, Granziero L, Monafo V, et al. Increased frequency of RAG-expressing, CD4+CD3low peripheral T lymphocytes in patients with defective responses to DNA damage. Eur J Immunol 2000;30:1520-5.

25. Fink PJ, Swan K, Turk G, Moore MW, Carbone FR. Both intrathymic and peripheral selection modulate the differential expression of Vb5 among CD4+ and CD8+ T cells. J Exp Med 1992; 176:1733-8.

26. Huang CY, Golub R, Wu GE, Kanagawa O. Superantigen-induced TCRa locus secondary rearrangement: role in tolerance induction. J Immunol 2002;168:3259-65.

27. Hiom K, Melek M, Gellert M. DNA transposition by the RAG1 and RAG2 proteins: a possible source of oncogenic translocations. Cell 1998; 94:463-70.

28. Rabbitts TH. Chromosomal translocations in human cancer. Nature 1994; 372:143-9.

29. Abe J, Takeda T, Watanabe Y, Nakao H, Kobayashi N, Leung DY, et al. Evidence for superantigen production by Yersinia pseudotuberculosis. J Immunol 1993; 151:4183-8.

30. Leite-de-Moraes MC, Coutinho A, Hontebeyrie-Joskowicz M, Minoprio P, Eisen H, Bandeira A. Skewed Vb TCR repertoire of CD8+ T cells in murine Trypanosoma cruzi infection. Int Immunol 1994; 6:387-92.

31. Tripp RA, Hamilton-Easton AM, Cardin RD, Nguyen P, Behm FG, Woodland DL, et al. Pathogenesis of an infectious mononucleosis-like disease induced by a murine gamma-herpesvirus: role for a viral superantigen? J Exp Med 1997;185:1641-50.

32. Dalwadi H, Wei B, Kronenberg M, Sutton CL, Braun J. The Crohn's disease-associated bacterial protein I2 is a novel enteric T cell superantigen. Immunity 2001; 15:149-58.

33. Turka LA, Schatz DG, Oettinger MA, Chun JJM, Gorka C, Lee K, et al. Thymocyte expression of RAG-1 and RAG-2: termination by T cell receptor cross-linking. Science 1991; 253:778-81.

34. McGargill MA, Derbinski JM, Hogquist KA. Receptor editing in developing T cells. Nat Immunol 2000; 1:336-41.

35. Goldrath AW, Bevan MJ. Selecting and maintaining a diverse T-cell repertoire. Nature 1999; 402:255-62.

36. Polic B, Kunkel D, Scheffold A, Rajewsky K. How ab T cells deal with induced TCR a ablation. Proceedings of the National Academy of Sciences USA 2001; 98:8744-9.

37. Vacchio MS, Kanagawa O, Tomonari K, Hodes RJ. Influence of T cell receptor Va expression on Mls^a superantigen-specific T cell responses. J Exp Med 1992; 175:1405-8.

38. Blackman MA, Woodland DL. Role of the T cell receptor a-chains in superantigen recognition. Immunologic Research 1996; 15:98-113.

39. Pullen AM, Bogatzki LY. Receptors on T cells escaping superantigen-mediated deletion lack special b-chain junctional region structural characteristics. J Immunol 1996; 156:1865-72.

40. Przylepa J, Himes C, Kelsoe G. Lymphocyte development and selection in germinal centers. Curr. Topics Microbiol Immunol 1998; 229:85-104.

41. Cyster JG, Ansel KM, Reif K, Ekland EH, Hyman PL, Tang HL, et al. Follicular stromal cells and lymphocyte homing to follicles. Immunol Rev 2000; 176:181-93.

42. Zheng B, Han S, Kelsoe G. T helper cells in murine germinal centers are antigen-specific emigrants that downregulate Thy-1. J Exp Med 1996; 184:1083-91.

E2

Essay 3: T Cell Autoreactivity by Design: A Theoretical Framework for Understanding Tolerance, Autoimmunity and Transplant Rejection

Peter S. Heeger

During development in the thymus, T lymphocytes initially undergo positive selection so as to be able to preferentially recognize peptides expressed in the context of self-MHC molecules.[1-4] Subsequently, the process of intrathymic negative selection results in deletion of T cell clones with "high affinity" for many self-antigens. The end result is that the mature T cell repertoire is capable of responding to an enormous variety of foreign antigens that it has not previously encountered.[1-4] Nonetheless, central deletion of self-reactive T cells is incomplete and many relatively low affinity autoreactive T cells "escape" into the periphery.[5] Standard paradigms in immunology view these escapees as an unwanted consequence of T cell development and as problematic to the host. In this view, the immune system must make use of a variety of peripheral tolerance mechanisms, including deletion, ignorance, anergy, suppression and end organ resistance, to control these potentially pathogenic T cells. Autoimmune disease results under rare circumstances when such tolerance mechanisms are overcome. While this paradigm can explain many experimental observations, it falls short of providing a comprehensive basis for our understanding of natural tolerance to self-antigens, and of experimentally induced tolerance, particularly in light of some recent observations regarding the development of autoreactive T cells following allograft transplantation. It is the goal of this commentary to provide an alternative framework within which one can incorporate the known experimental findings and potentially better account for them. It is hoped that the model will provoke thought and discussion.

Emerging results from multiple laboratories showing that autoreactive T cells can exhibit regulatory properties (reviewed in ref. 6) raise the possibility that one function of the T cell repertoire selection process is to seed the periphery with autoreactive T cells. Based on this postulate, one can then hypothesize that autoreactive T cells are present by design and play an active role in the maintenance of self-tolerance through dominant, interactive regulatory mechanisms (Table E3.1 and Fig. E3.1A). This concept stands in contradistinction to the standard view that autoreactive T cells represent an unwanted "side effect" of T cell

Organ Transplantation, 2nd edition, edited by Frank P. Stuart, Michael M. Abecassis and Dixon B. Kaufman. ©2003 Landes Bioscience.

Table E3.1. Key features of autoimmunity by design theory

- Autoreactive T cells are released from the thymus by design
- Autoreactivity is not equivalent to autoimmune disease
- Autoreactive T cells usually prevent inflammation through dominant regulatory mechanisms but can be corrupted into becoming pathogenic
- The lack of inflammation is an active process involving autoreactive regulatory T cells interacting with parenchymal tissues
- The inflammatory (or quiescent) states of individual microenvironments are regulated independently
- The phenotypic expression of inflammation at a given site is dependent on the presence or absence of proinflammatory versus tolerogenic signals expressed by cells of that organ, and the relative numbers of pathogenic versus regulatory T cells at that site.

---

ontogeny that must be controlled in order to prevent autoimmune disease. The "autoimmunity by design" model assumes that autoreactivity is not equivalent to autoimmune disease; the specificity of a T cell does not define its functional capabilities. Implicit in the model is the concept that a naïve T cell has an ability to differentiate into an effector cell with proinflammatory features (for example, an IFNγ-producing TH1 cell) or into one with a protective phenotype (for example, a TGF-β- or IL-10-producing Tr1 cell) depending on the specific environmental conditions encountered by the lymphocyte. In this view the process of T cell ontogeny has evolved such that a small number of positively selected, autoreactive T cells are released from the thymus, rather than escape from the thymic deletion process. The TCRs expressed by the autoreactive T cells are likely to have relatively low affinities for their ligands (as the T cells expressing the highest affinity TCRs are presumably deleted centrally). There are some data to suggest that a proportion of these autoreactive T cells can be preconditioned centrally to have a regulatory or suppressor phenotype after encountering self-antigens on thymic APCs.[6,7] However, as all self-antigens are not expressed in the thymus, some autoreactive T cells are likely to be released into the periphery as naïve precursors. These latter autoreactive T cells are hypothesized to home to secondary lymphoid tissues where they have the opportunity to interact with self-APCs expressing self-antigens. If and when the naïve autoreactive T cells encounter their antigenic ligand on a nonactivated (or immature) APC, they have the potential to differentiate into a regulatory or suppressor cell.

Activated, regulatory T cells (either deriving directly from the thymus or after priming in the periphery) would then circulate widely where they could re-encounter their antigenic ligands expressed on normal tissues. These interactive events are hypothesized to result in reciprocal down regulatory signals: the autoreactive T cells are hypothesized to encounter self-antigen in the absence of proinflammatory stimuli (i.e., no costimulation) and thereby maintain and reinforce their anergic/suppressive phenotype. The induced regulatory characteristics would prevent activation or effector function of small numbers of other potentially pathogenic T cells that infiltrate the organ, either through direct cell: cell contact or through bystander (possibly cytokine-mediated) effects. At the same

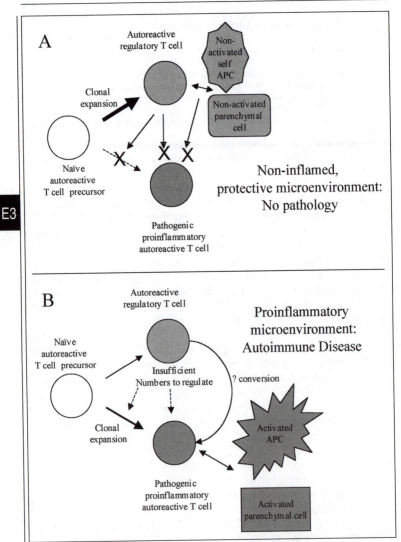

Fig. E3.1. Schematic depiction of autoimmunity by design in the absence (A) or the presence (B) of autoimmune disease. (A) Naïve autoreactive T cell precursors emerge from the thymus and differentiate into regulatory T cells. These regulatory T cells interact with quiescent or immature antigen-presenting cells (APCs) and normal parenchymal cells to create a protective microenvironment. This interactive process prevents activation and effector function of any potentially pathogenic autoreactive T cells. (B) Under unusual proinflammatory conditions pathogenic, autoreactive T cells expand in a proinflammatory microenvironment (comprised of activated APCs, chemoattractant signals, and the absence of protective signals) to an extent that cannot be controlled by regulatory cells and results in autoimmune disease.

time, signals from the regulatory T cells delivered to vascular cells, parenchymal cells, and/or bone marrow derived APCs of the organ would theoretically maintain its quiescent, tolerogenic state (possibly through upregulation of protective genes or inducing APCs to differentiate into a tolerogenic phenotype). The end result would be a self-perpetuating, protective microenvironment that is dependent on the interaction between the regulatory T cells and the induced protective state of the organ. It is hypothesized that the autoreactive regulatory T cells are required to induce the tolerant state, but may not be sufficient to maintain it; induced alterations in the peripheral organ tissue cells and/or APCs such that they have protective or tolerogenic properties would be necessary for maintenance. In contrast to standard paradigms, the hypothesis suggests that the lack of inflammation is an active process, involving an ever-evolving interaction between autoreactive (regulatory) T cells and the "normal" tissues of the peripheral organs. The immune system must overcome these active, protective processes in order to mediate local inflammation. Further implicit in this paradigm is the concept that individual microenvironments are regulated independently—how the immune system behaves in one location may not be identical to how it behaves at a different site in the organism (i.e., inflammation can be localized to a single site with the remainder of the host being unaffected).

Is the autoimmunity by design framework consistent with experimental observations made in tolerant animals? Certainly, normal hosts, presumably tolerant to self-antigens, have T cell repertoires containing autoreactive T cells, some of which are naïve and others of which seem to have regulatory properties.[6-8] Increasing evidence is accumulating to show that regulatory T cells are detectable in both mice and humans and that depletion of certain subpopulations of regulatory T cells (i.e., CD25[+] CD4[+]) can result in autoimmune pathology (although the specificity of these cells has not been defined in most studies) (reviewed in ref. 9). Studies from at least one transplant model show that normal, donor organs contain T cells with regulatory potential that have the capability of repopulating an immunodeficient recipient and can mediate tolerance.[10] In vivo expansion of autoreactive T cells can be elicited experimentally and can result in either pathogenic autoimmunity or protective tolerance, depending on the experimental protocol utilized.[11,12] Some induced regulatory populations of T cells can transfer tolerance to naïve animals.[13-15] There is also evidence that transfer of regulatory T cells alone may not be sufficient to mediate tolerance in naïve animals but that tolerogenic APCs may be required.[16,17] Finally, tolerance can be associated with end organ resistance, mediated by expression of protective gene products (i.e., heme oxyengase, indoleamine 2,3-dioxygenase, FasL) that, in turn, can be upregulated by interactions with primed T cells (some of which have been shown to be tolerogenic) or their secreted cytokines.[18-20] Thus in a broad sense, the autoreactivity by design hypothesis is consistent with many experimental observations regarding tolerance.

How would the conceptual framework of autoreactivity by design explain the development of pathologic autoimmune disease? Firstly, it is notable that the development of autoimmune disease is a relatively rare event and that it is difficult to induce autoimmune disease in animal models using even the most potent proinflammatory stimuli (for example complete Freund's adjuvant).[12] This observation in conjunction with the fact that autoreactive T cells are detectable in normal hosts, suggests that there are potent regulatory mechanisms controlling or preventing the development of pathogenic autoreactive T cells under normal conditions. The autoimmunity by design paradigm suggests that naïve or regulatory autoreactive T cell emigrants, designed to be protective, can be subverted into a pathogenic phenotype under unusual potent proinflammatory conditions (Fig. E3.1B). As an illustrative hypothetical example, a pulmonary viral infection can result in increased local expression of MHC and costimulatory molecules (among others) thereby altering the microenvironment of the lung (but not of other noninfected organs) from a permissive/tolerogenic phenotype to a proinflammatory phenotype. This would be entirely appropriate in order to cure the infection. Virus-specific T cells activated in the secondary lymphoid organs by APCs expressing processed viral determinants are then preferentially attracted to the infected organ where they re-encounter their antigenic ligands, leading to tissue destruction and, ultimately, control of the virus. It is possible that some viral antigens could exhibit cross reactive features to certain autoantigens such that priming of proinflammatory anti-viral T cells could inadvertently result in activation of cross reactive, pathogenic autoreactive T cells. The number of autoreactive T cells primed under these conditions would be dependent, in part, on the genetic composition of the individual (including the T cell repertoire and a number of other genes that determine responsiveness) and specific characteristics of the infectious agent. Circumstantial evidence for this type of cross reactivity has indeed been detected in a number of models of autoimmune disease.[21,22] An additional consequence of the local tissue destruction aimed at curing the viral infection would be the release, endocytosis, processing and presentation of peptides derived from a large number of self-proteins found in the normal cells (as well as foreign, virus-derived proteins). While the majority of these self-peptides would be innocuous, rare self-peptides expressed in the context self-MHC on activated APCs may act as cryptic antigens and elicit priming of naïve autoreactive T cell precursors into pathogenic autoreactive T cells. Such a scenario would be consistent with the well-established concept of epitope spreading, in which an initially focused immune response (in this case anti-viral) spreads to involve additional antigens (in this case self-antigens) presented to the immune system in the context of the proinflammatory microenvironment.[23] The local inflammatory phenotype of the infected organ would support and potentially accelerate attraction of these pathogenic autoreactive T cells and perpetuate the autoimmune reactivity. If sufficient numbers of these autoreactive T cells are activated and if the target organ expressing the self-antigen maintains the proinflammatory state (due to the ongoing viral infection), the proposed regulatory, autoreactive-T-cell-dependent mechanisms that maintain tolerance may be overwhelmed and the newly primed pathogenic autoreactive T cells could contribute to organ damage (and in fact could contribute to the cure of the infection). Resolution of the inciting infection should lead to a down regulation of the virus-specific immune repertoire such that only a few residual anti-virus, memory T cells remain in the host. It is hypothesized that normally, the number of autoreactive T cells activated during such infections is limited, and that this response also resolves, despite the fact that persistent self-antigens are always present (perhaps due to the persistence of the autoreactive regulatory repertoire already present in the host). Under extremely rare conditions, again determined by genetic predispositions of the host and various environmental factors, the autoreactive component of the proinflammatory immune repertoire may not fully resolve or may re-activate. In these latter instances, the target organ would not resolve back to the normal, quiescent, but actively protective state, with the end result being the development of self-perpetuating, organ-specific autoimmune disease.

It is intriguing to note that pathologic autoimmune reactions are generally organ-specific, and do not spread to involve other organs despite the fact that many normal tissues likely express some of the same autoantigens (although organ specific autoimmunity can be directed towards antigenic targets specifically found in a given organ and not another). The autoimmunity by design framework suggests a plausible explanation to account for this, based on the assumption that the various microenvironments of the host can differentially influence the autoimmune repertoire. The model would suggest that the noninvolved host tissues maintain a tolerant phenotype, consisting of infiltrating, autoreactive, regulatory T cells, a nonpathogenic microenvironment, and an absence of chemoattractant signals to attract pathogenic T cells. If small numbers of activated pathogenic T cells spill over into these tissues they would be controlled by the permissive microenvironment (just as outlined above for organs of the normal, noninfected host), thus preventing spread of the autoimmune disease to additional organs.

The development of pathogenic and protective autoreactivity within the conceptual context of autoimmunity by design can account for some recent observations in transplantation immunobiology as well. Emerging data, summarized in the accompanying articles by Benichou, Fedoseyeva and Wilkes (Editors of Graft will need to insert references here), provide convincing evidence that autoreactive T cells can contribute to destruction of a transplanted organ. Work by these investigators and by others showed that allograft transplantation primes pathogenic, recipient-MHC-restricted T cells specific for peptides derived from cardiac myosin (heart grafts), collagen V (lung transplants), heat shock proteins (skin grafts and heart grafts) and some unknown autoantigens.[11,24-30] The primed autoreactive T cells were not simply innocent bystanders, because 1) they could be isolated from allografts undergoing rejection, 2) immunization with these autoantigens prior to transplantation could accelerate allograft rejection and 3) induction of a pathogenic immune response to these autoantigens through experimental immunization could precipitate rejection of an isograft. Interestingly, the primed, autoreactive T cells capable of rejecting a transplanted isograft did not seem to cause injury to the native organs of the recipient.

The detection of autoreactive T cells following transplantation should be anticipated (Fig. E3.2). In addition to direct recognition of donor cells, recipient T

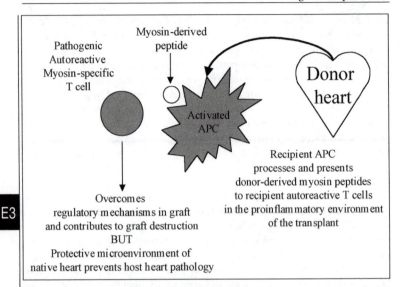

Fig. E3.2. Schematic depiction of the development of pathogenic autoreactive T cells following heart transplantation. See text for details.

cells recognize donor-derived antigenic determinants complexed to recipient MHC molecules expressed on recipient APCs.[31] This indirect pathway of allorecognition represents the usual method of immune recognition by T cells—the exogenous antigen is engulfed by the host's APCs, processed into peptide fragments, shunted through the MHC processing pathways and expressed on the antigen presenting cell surface. While many of the indirectly presented peptides derive from donor MHC molecules, any antigen found in donor cells (including so-called "minor" antigens and even nonpolymorphic antigens common to both donor and recipient) could theoretically be processed and presented by recipient APCs. The proinflammatory state of the transplanted organ (due, in part to surgical trauma and ischemia reperfusion injury), along with the enormous anti-allograft T cell immune response focused towards donor MHC molecules (direct pathway) could easily overcome the hypothesized tolerogenic state of the donor organ, permitting priming of autoreactive T cells, and facilitating/accelerating the migration to, and pathologic function of these autoreactive T cells in the transplant. The ability of such primed T cells to contribute to destruction of an allograft would also be anticipated, as the autoreactive T cells infiltrating the donor allograft could re-encounter self-antigens expressed on infiltrating self-APCs and mediate local tissue injury through release of pro-inflammatory cytokines, induction of delayed type hypersensitivity reactions and through initiation of other, secondary, macrophage-mediated effector mechanisms.

   The experimental data show that despite the development of pathogenic autoimmunity directed towards transplanted organs (including isografts), the autoreactive T cells do not cause injury to the native organs of the recipient (see accompanying articles). The endogenous host tissues seem to maintain a non-pathogenic microenvironment (theoretically due to the reciprocal interactions between regulatory T cells and the host tissue), and an absence of chemoattractant signals to attract pathogenic T cells. Although pathogenic autoreactive T cells are activated as a component of the alloimmune response, these cells seem to be preferentially attracted to the transplanted graft and do not accumulate in the native organ in large numbers. Small numbers of rogue, activated T cells that enter the normal tissue could theoretically be controlled by the permissive/tolerogenic microenvironment, preventing the development of diffuse autoimmune disease in the native tissues. If this hypothesis is true, then expression of inflammatory signals within an otherwise normal organ could precipitate organ-specific autoimmune disease. Indeed, studies in which TNFα was genetically over-expressed in islet cells confirmed that local production of this proinflammatory molecule could result in islet inflammation and diabetes.[32] Another potential test of the hypothesis would be to induce injury of the native heart (for example, by ischemia reperfusion via tying off a coronary vessel) at the time of heart allograft placement, with the premise being that the induced injury would result in attraction of primed, autoreactive (i.e., myosin-specific) T cells and thereby precipitate myocarditis of the native heart.

The autoimmunity by design hypothesis additionally provides a potential explanation for the intriguing observation that induction of tolerance to the organ-specific autoantigen prior to transplantation can delay or even prevent rejection of a subsequently placed allograft (Fig. E3.3). Experimental tolerization (for example, by administration of antigen in incomplete Freund's adjuvant) would expand the population of endogenous autoreactive regulatory T cells, creating a permissive microenvironment in the host through interaction with normal host tissues, and may be dependent on T cell mediated induction of "tolerant" APCs. Following transplant surgery, the expanded repertoire of tolerant APCs and regulatory T cells would inhibit priming or effector function of any pathogenic alloreactive T cells (functioning either in the secondary lymphoid organs and/or in the graft). In addition, this increased number of regulatory cells would partially restore the microenvironment of the inflamed graft towards a protective state, thereby raising the threshold number of pathogenic T cells required to mediate graft rejection. Prevention of graft rejection would be thus be dependent on the relative numbers of regulatory cells versus pathogenic T cells infiltrating the graft as well as the phenotype of the graft itself. In some cases (as outlined in the accompanying articles), tolerance induction may sufficiently raise the number of regulatory cells to fully prevent rejection of an allograft. In other situations the induced tolerance to autoantigens may expand the number of regulatory cells but not to a sufficient degree to prevent the eventual effects of a potent alloimmune response (and thus only delay, not prevent rejection). The hypothesis is again supported by recent results in the models of skin graft tolerance,[15] autoimmune diabetes[33] and experimental autoimmune encephalomyelitis[34] in which expanded populations of regulatory T cells (in some cases shown to be autoreactive) can localize to the target organ and can inhibit the development and effector function

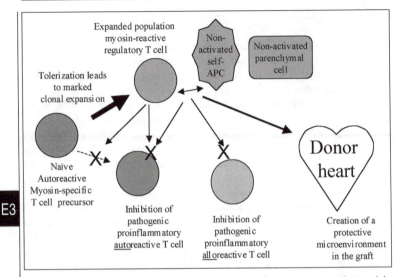

Fig. E3.3. Schematic representation of how tolerance induction to autoantigens might prevent or delay graft rejection. See text for details.

of pathogenic T cells, thereby preventing tissue injury. Overall, the experimental data suggest that one can harness the naturally developing, autoreactive regulatory cell repertoire, and that expansion of these T cells to a sufficient degree, can result in a regulatory immune repertoire capable of controlling or preventing the development of a pathogenic alloimmune response.

Autoimmunity by design as outlined, does not account for the presence of autoantibody-mediated processes, although analogous regulatory features could be envisioned (through controlling complement activation or signaling through inhibitory Fc receptors expressed on macrophages). Overall, while the idea of protective, autoimmunity by design is consistent with much of the published literature, there remain a number of unanswered questions (Table E3.2). The hypothesis would be bolstered by more experimental data clearly identifying the phenotypes, mechanisms of action, and origins of regulatory T cells, and of autoreactive, regulatory T cells in specific. Experiments designed to better test whether the numbers of pathogenic versus tolerogenic, autoreactive T cells affect the threshold for the expression of organ pathology/rejection, as well as experiments focusing on further isolation, characterization and mechanistic analysis of tolerogenic APCs are also needed.

The autoimmunity by design paradigm provides a conceptual framework through which to consider experimental results and requires a shift in thinking about why autoreactive T cells are present in a host. Instead of functioning as escaped prisoners that are dangerous to the community, this hypothesis suggests that autoreactive T cells act more like your friendly, local police department, constantly patrolling the neighborhood for signs of commotion, reinforcing the walls

---

Table E3.2. Questions raised by the autoimmunity by design theory

---

- Can naïve T cell precursors differentiate into multiple phenotypes of proinflammatory versus regulatory cells? If so, what influences the decision?
- Can one routinely isolate autoreactive regulatory T cells from normal organs?
- Do autoreactive, regulatory T cells proliferate and alter expression of cell surface markers upon activation?
- What signals attract regulatory cells to normal organs and do these signals differ from those that attract proinflammatory cells? Do regulatory T cells cross endothelial cell barriers and if so, how?
- Can autoreactive regulatory T cells be converted into pathogenic proinflammatory T cells or are they at an end-differentiated stage of development?
- Can one alter the expression of protective genes in parenchymal cells and affect the proinflammatory versus tolerant state of the organ?
- Do the relative numbers of regulatory versus proinflammatory autoreactive T cells contribute to the development of inflammation at a given site?
- What are the mechanisms employed by regulatory T cells? Bystander suppression? Prevention of precursor differentiation into proinflammatory effectors? Upregulation of protective genes on parenchymal cells?
- Can certain cells, other than T cells, transfer tolerance? If so, what are the phenotypic markers that define these cells, how are they induced, and how do they function?

E3

of protection against intruders and quenching any local disturbances. Under certain stimulatory conditions, additional members of the department can be recruited into active duty, affording a more potent protective police force. It is only under the most unusual combination of circumstances where such protective T cells are corrupted into criminal behavior that results in tissue destruction and true autoimmune disease. I hope that the readership will consider the merits (and the shortcomings) of this conceptual framework and I look forward to your thoughts and to your feedback.

Acknowledgements

I wish to especially thank Rob Fairchild, Anna Valujskikh and Charley Orosz for their helpful discussions, input and feedback.

This article is reprinted from Graft 2003; Vol. 6, No. 1:33-41, with permission from the authors.

References

1. Jameson SC, Hogquist KA, Bevan MJ. Positive selection of thymocytes. Annu Rev Immunol 13:93.
2. Sha WC, Nelson CA, Newberry RD et al. Positive and negative selection of an antigen receptor on T cells in transgenic mice. Nature 1988; 336:73.
3. Kisielow P, Teh HS, Bluthmann H et al. Positive selection of antigen-specific T cells in thymus by restricting MHC molecules. Nature 1988; 335:730.
4. von Boehmer H, Teh HS, P. Kisielow P. The thymus selects the useful, neglects the useless and destroys the harmful. Immunol Today 1989; 10:57.
5. Targoni OS, Lehmann PV. Endogenous myelin basic protein inactivates the high avidity T cell repertoire. J Exp Med 1998; 187:2055.

6.   Shevach EM. Regulatory T cells in autoimmmunity*. Annu Rev Immunol 2000; 18:423.

7.   Apostolou I, Sarukhan A, Klein L et al. Origin of regulatory T cells with known specificity for antigen. Nat Immunol 2002; 3:756.

8.   Burns JB, Bartholomew BD, Lobo ST. Isolation of CD45RO⁺, memory T cells recognizing proteolipid protein from neurologically normal subjects. Cell Immunol 2001; 212:44.

9.   Shevach EM. CD4⁺ CD25⁺ suppressor T cells: more questions than answers. Nat Rev Immunol 2002; 2:389.

10. Anderson CC, Matzinger P. Immunity or tolerance: opposite outcomes of microchimerism from skin grafts. Nat Med 2001; 7:80.

11. Fedoseyeva EV, Kishimoto K, Rolls HK et al. Modulation of tissue-specific immune response to cardiac myosin can prolong survival of allogeneic heart transplants. J Immunol 2002; 169:1168.

12. Heeger PS, Forsthuber T, Shive C et al. Revisiting tolerance induced by autoantigen in incomplete Freund's adjuvant. J Immunol 2000; 164:5771.

13. Cobbold S, Waldmann H. Infectious tolerance. Curr Opin Immunol 1998; 10:518.

14. Zelenika D, Adams E, Humm S et al. The role of CD4⁺ T-cell subsets in determining transplantation rejection or tolerance. Immunol Rev 2001; 182:164.

15. Graca L, Cobbold SP, Waldmann H. Identification of regulatory T cells in tolerated allografts. J Exp Med 2002; 195:1641.

16. Fairchild PJ, Waldmann H. Dendritic cells and prospects for transplantation tolerance. Curr Opin Immunol 2000; 12:528.

17. Jung S, Unutmaz D, Wong P et al. In Vivo Depletion of CD11c(+) Dendritic Cells Abrogates Priming of CD8(+) T Cells by Exogenous Cell-Associated Antigens. Immunity 2002; 17:211.

18. Lau HT, Yu M, Fontana A, Stoeckert CJ Jr. Prevention of islet allograft rejection with engineered myoblasts expressing FasL in mice. Science 1996; 273:109.

19. Munn DH, Zhou M, Attwood JT et al. Prevention of allogeneic fetal rejection by tryptophan catabolism. Science 1998; 281:1191.

20. Hancock WW, Buelow R, Sayegh MH et al. Antibody-induced transplant arteriosclerosis is prevented by graft expression of anti-oxidant and anti-apoptotic genes. Nat Med 1998; 4:1392.

21. Tian J, Lehmann PV, Kaufman DK. T cell cross-reactivity between coxsackievirus and glutamate decarboxylase is associated with a murine diabetes susceptibility allele. J Exp Med 1994; 180:1979.

22. Gross DM, Forsthuber T, Tary-Lehmann M et al. Identification of LFA-1 as a candidate autoantigen in treatment- resistant Lyme arthritis. Science 1998; 281:703.

23. Kaufman DL, Clare-Salzler M, Tian J et al. Spontaneous loss of T-cell tolerance to glutamic acid decarboxylase in murine insulin-dependent diabetes. Nature 1993; 366:69.

24. Rolls HK, Kishimoto K, Illigens BM et al. Detection of cardiac myosin-specific autoimmunity in a model of chronic heart allograft rejection. Transplant Proc 2001; 33:3821.

25. Fedoseyeva EV, Zhang F, Orr PL et al. De novo autoimmunity to cardiac myosin after heart transplantation and its contribution to the rejection process. J Immunol 1999; 162:6836.

26. Haque MA, Mizobuchi T, Yasufuku K et al. Evidence for immune responses to a self-antigen in lung transplantation: role of type V collagen-specific T cells in the pathogenesis of lung allograft rejection. J Immunol 2002; 169:1542.

E3

27. Yasufuku K, Heidler KM, Woods KA et al. Prevention of bronchiolitis obliterans in rat lung allografts by type V collagen-induced oral tolerance. Transplantation 2002; 73:500.

28. Yasufuku K, Heidler KM, O'Donnell PW et al. Oral tolerance induction by type V collagen downregulates lung allograft rejection. Am J Respir Cell Mol Biol 2001; 25:26.

29. Mares DC, Heidler KM, Smith GN et al. Type V collagen modulates alloantigen-induced pathology and immunology in the lung. Am J Respir Cell Mol Biol 2000; 23:62.

30. Valujskikh A, Fedoseyeva E, Benichou G et al. Development of autoimmunity after skin graft rejection via an indirect alloresponse. Transplantation 2002; 73:1130.

31. Gould DS, Auchincloss H Jr. Direct and indirect recognition: the role of MHC antigens in graft rejection. Immunol Today 1999; 20:77.

32. Green EA, Flavell RA. Tumor necrosis factor-alpha and the progression of diabetes in non-obese diabetic mice. Immunol Rev 1999; 169:11.

33. Green EA, Choi Y, Flavell RA. Pancreatic lymph node-derived CD4(+)CD25(+) Treg cells: highly potent regulators of diabetes that require TRANCE-RANK signals. Immunity 2002; 16:183.

34. Wildbaum G, Netzer N, Karin N. Tr1 cell-dependent active tolerance blunts the pathogenic effects of determinant spreading. J Clin Invest 2002; 110:701.

E3

Essay 4: Male Infertility in the Transplant Patient

Robert E. Brannigan and Robert Nadler

E4

INTRODUCTION

Advances in surgical technique and immunosuppressive therapy have led to increasing numbers of transplant procedures and extended postoperative survival among these patients. Not surprisingly, many of these individuals, especially younger transplant patients, ultimately become interested in having children. In the general population, approximately fifteen percent of couples are considered to be infertile. Of these, it is estimated that thirty percent are due to a male factor alone and twenty percent are due a combination of male and female factors. Patients who have undergone transplantation procedures often present with a unique set of circumstances and medical problems that can impair their reproductive health. In this article, we will focus on the appropriate evaluation of the infertile transplant patient, and we will discuss several different etiologies for infertility commonly seen in the transplant patient population. Finally, we will review the appropriate treatment of these individuals, keeping their primary pathology in mind.

SPECIAL ISSUES FOR THE TRANSPLANT PATIENT

Among patient groups, those receiving liver and cardiac transplants often tend to be older individuals. As such, fertility issues are typically not frequent health concerns. However, those patients with end stage renal disease (ESRD) are often young when afflicted, and their primary disease process itself has been shown to directly impair fertility. For this reason, many of the items discussed below apply specifically to the ESRD/uremic population. Other items, duly noted, apply to the larger transplantation population as a whole.

GONADOTROPIC FACTORS

The hypothalamic-pituitary-gonadal (HPG) axis is markedly disturbed by chronic renal failure, and this is manifest in diminished testosterone levels and impaired spermatogenesis. These patients often present with complaints of diminished libido, erectile dysfunction (ED), and infertility. Upon evaluation, they have low serum testosterone levels and elevated follicle stimulating hormone (FSH) and luteinizing hormone (LH) levels.[1] Interestingly, these patients usually retain a normal response to clomiphene citrate stimulation. Clomiphene citrate has antiestrogenic properties which result in a decrease in negative feedback by

Organ Transplantation, 2nd edition, edited by Frank P. Stuart, Michael M. Abecassis and Dixon B. Kaufman. ©2003 Landes Bioscience.

estrogens on the hypothalamus. This in turn leads to increased secretion of gonadotropin releasing hormone (GnRH) and a subsequent increase in LH and FSH secretion. A normal response to this test, as is generally seen in uremic patients, implies that the HPG axis is intact and that the impairments in testosterone synthesis and spermatogenesis are due to a direct effect of the uremia on the Leydig cells and germinal epithelium.[2] Usually this abnormality is not corrected by hemodialysis, and impairment often persists even after renal allograft transplantation.[3] This suggests the possibility of permanent testicular damage in some of these patients. The best marker of spermatogenic potential posttransplant may be the serum FSH. Persistently elevated FSH levels suggest irreversible testicular damage, and these patients may benefit from a three month trial of clomiphene citrate therapy. As explained above, this agent causes an increase in gonadotropin (FSH and LH) production with a subsequent rise in serum and intratesticular testosterone levels. In some patients, this may have a favorable effect on sperm production and semen quality. Patients should be placed on a minimum of three months of therapy, because this is approximately the length of time for the full cycle of spermatogenesis. A repeat semen analysis should be repeated at that time to determine the response to therapy. If the response to therapy is unfavorable, then the patient should be counseled regarding other options, including assisted reproductive techniques such as intrauterine insemination, in vitro fertilization (IVF), and intracytoplasmic sperm injection (ICSI). For all of these reproductive techniques, the patient's own sperm or else donor sperm may be used.

Although less common in the transplant population, patients with low FSH levels and an abnormal semen analysis may suffer from hypogonadotropic hypogonadism.[4] A trial of gonadotropin replacement therapy with human chorionic gonadotropin (HCG) may benefit the patient. It acts as an LH agonist to increase intratesticular testosterone and improve spermatogenesis in some patients. Human menopausal gonadotropin (HMG) has similar effects and is also a therapeutic option. Unfortunately, the impact of therapy on sperm production is not fully appreciated until three months after initiation of therapy. For these patients who do not respond adequately to the treatment, the options of assisted reproductive techniques including artificial intrauterine insemination, IVF, and ICSI remain available, often viable options.

Hyperprolactinemia

Elevated prolactin levels are sometimes discovered in the workup of the infertile male renal transplant patient. This may be found in association with uremia, a poorly functioning renal graft, or even occasionally a well-functioning graft. These individuals often have coinciding complaints of diminished libido and ED. The testosterone is often low, but testosterone supplementation does not correct the impaired libido nor the ED because elevated prolactin disrupts penile end organ responsiveness. Dopamine is normally produced by the hypothalamus and transported to the pituitary gland where it has an inhibitory effect on prolactin secretion. Bromocriptine and carbergoline are dopamine receptor agonists which have prolactin-lowering effects. These agents have also been shown to be helpful in restoring libido, erectile function, and spermatogenesis to normal.

ANATOMIC ABNORMALITIES

Transplant patients do not generally appear to be at increased risk for anatomic etiologies for infertility than the general population.[5] As such, the incidence of varicoceles is approximately forty-two percent and the incidence of obstruction within the reproductive tract (at the level of the epididymis, vas deferens, or ejaculatory duct) is approximately fifteen percent. These two common as well as other anatomic abnormalities (such as penile deformity, congenital absence of the vas deferens, etc.) should be kept in mind during the evaluation. It is important to consider, however, the possibility of iatrogenic injury to the spermatic cord and vas deferens, especially during the renal transplantation procedure. These structures are close to the renal allograft site and may be inadvertently injured during transplantation.

URINARY TRACT INFECTIONS/POSITIVE SEMEN CULTURES

Transplant patients are prone to opportunistic infections due to their immunosuppression medical regimen. As a result, they are at increased risk for genitourinary tract infections, including cystourethritis, prostatitis, and epididymitis. Each of these conditions can have a potentially detrimental impact on fertility status. These conditions may lead to a higher likelihood of pyospermia, or white blood cells (WBCs) within the semen. WBCs, when present in excessively high numbers, may release abnormally high levels of molecules called reactive oxygen species (ROS) into the semen.[6] These ROS have the ability to pass an extra electron, or "free radical", onto other molecules. This electron transfer can be severely detrimental to the fatty acids of the sperm membrane and thus impair sperm function, including sperm-oocyte binding. The treatment of this condition includes effectively eradicating the infection as well as supplementing the patient with antioxidant therapy, such as Vitamin E. (Dosages should be carefully determined and the patients closely monitored, especially in transplant patients with hepatic failure.)

RETROGRADE EJACULATION

This condition should be suspected in men with azoospermia/severe oligospermia and no or abnormally low ejaculate volume. The abnormality arises from an open bladder neck at the time of ejaculation which permits retrograde passage of semen into the bladder. The diagnosis is made by catheterizing the bladder after ejaculation and finding sperm in the pellet from the centrifuged urine sample. This condition is most often seen in men with autonomic neuropathy, such as diabetics and patients who have undergone extensive retroperitoneal surgery. The treatment of choice is sympathomimetic agents such as pseudephedrine, which increases bladder neck tone and thus facilitates antegrade ejaculation.

IMMUNOSUPPRESSIVE MEDICATIONS

One of the hallmarks and major advances in transplantation medicine has been the advent of effective immunosuppressive therapy. Although these agents are essential for the maintenance of allograft viability, many of them potentially contribute to male infertility. Below is a brief summary of the more commonly used agents and their known effects on fertility.[5]

• *Azathioprine*-No adverse effect on male infertility demonstrated, although occasional hematopoetic suppression was noted in offspring.

• *Prednisone*-Studies using a short course of high-dose exogenous steroids (prednisone 30 mg/day for 30 days) revealed the development of spermatogenic arrest, decreased sperm density, and impaired motility in men with normal baseline semen parameters. Baseline oligospermic men did not demonstrate a similar deterioration in their semen parameters. Fortunately, patients on a short course of low-dose exogenous steroids (prednisone 10 mg/day for 30 days) did not have changes in their semen analysis. Furthermore, testicular biopsy performed on these patients revealed normal spermatogenesis with preservation of Sertoli and Leydig cell microscopic architecture. Most organ transplant recipients receive maintenance doses of prednisone which are closer to the low-dose therapy regimen.

• *Cyclosporine A (CsA)*-This has perhaps been the most rigorously studied immunosuppressive agent in terms of its impact on male fertility. CsA at therapeutic levels often causes a decrease in serum testosterone, a decrease in intratesticular testosterone, and an increase in serum gonadotropin levels (LH and FSH). These changes are frequently associated with coinciding impairments in sperm density and motility. It is unclear whether the detrimental effects of CsA are mediated at the level of the HPG axis or else at the level of the testicle itself. One hypothesis, which is supported by some animal studies, suggests that CsA interferes with LH signal transduction at the Leydig cell level, with secondarily impaired testosterone production. Supplementation of these animals with exogenous HCG (an LH agonist) improved testosterone production, spermatogenesis, and sperm densities.

EVALUATION OF THE INFERTILE MALE TRANSPLANT PATIENT

HISTORY

A thorough history is the foundation of the workup of the infertile male. In addition to the standard history, the patient should be carefully questioned about any previous evaluation and treatment of infertility (i.e., previous semen analyses, use of exogenous hormones, prior IUI or IVF cycles with partner), intercourse technique (timing, frequency, and use of lubricants), and previous contraception. Information about the partner's reproductive health history and evaluation should be gathered. The patient should also be carefully questioned about the impact of his organ failure and transplant on his overall health. Additional issues that should be addressed in the course of history-taking include:[4]

Congenital Disorders/Anomalies

• *Cryptorchidism*-Congenitally undescended testes is prevalent in 1% of one year old boys. If left untreated, this condition can lead to progressive germ cell damage due to the increased temperature at the extrascrotal location. In extreme cases, the postpubertal untreated male may have complete loss of germ cells altogether.

• *Bladder exstrophy*-This is abnormal development of the bladder and bladder neck, often associated with severe ejaculatory disorder even after surgical correction.

• *Posterior urethral valves*-This condition was sometimes treated with Y-V plasty surgical procedure to facilitate bladder emptying. However, these patients as a result often experience poor bladder neck closure during ejaculation, leading to retrograde ejaculation and infertility.

• *Neurologic abnormalities*

Childhood Illnesses and Development

Testicular torsion-Occurs most commonly in adolescence, with an annual incidence of 1/4000 men under 25 years of age. The result is ischemic injury to the testicle with the degree of damage depending on the duration of torsion. There is some evidence that the torsed testicle, when left in situ, has a deleterious effect on the contralateral testicle over time. Evidence suggests that early detorsion may not only spare the testicle, but also prevent future subfertility and deterioration in sperm quality.

• *Testicular trauma*-Infertility is believed to be due to direct cell injury and also the formation of antisperm antibodies as a result of the disruption of the blood-testis barrier.

• *Infection*-Mumps orchitis is rare in prepubertal males, but it is the most common complication of the mumps infection in adults. It occurs in 5-37% of cases, with bilateral involvement in 16-65%. The virus directly attacks the testicular tissue, leading to massive cell destruction and inflammation. This inflammation, within the inelastic tunica albuginea, leads to swelling, pressure necrosis, and increased cell death. Testicular atrophy is seen in 40-70% of these men, and infertility in 30-87%.

• *Onset of puberty*

Medical History

• *Systemic illness*-Note the onset and duration of diseases such as diabetes mellitus, multiple sclerosis, etc. Information regarding previous and current therapy should be obtained.

Gonadotoxins

These are chemicals, drugs, or other agents that have a deleterious effect on spermatogenesis, sperm motility, or sperm morphology. The gonadotoxin effect can result from harmful input at the level of the HPG axis, the level of the testicle, or at the posttesticular level. The effects are often reversible if the offending agent is identified and removed. Some of the immunosuppressive medications, as previously mentioned, are potentially gonadotoxic. Common gonadotoxic agents

include:
- *Chemicals*-Certain pesticides (especially DBCP, or dibromochloropropane), some organic solvents, heavy metals, etc.
- *Drugs*-A partial listing includes some of the chemotherapeutic agents, cimetidine, sulfasalazine, nitrofurantoin, alcohol, marijuana, and androgenic steroids.
- *Thermal exposure*
- *Radiation*
- *Tobacco Use*

Family History
Information about the fertility status of first and second degree relatives may provide helpful insight into a patient's underlying pathology. Furthermore, familial diseases such as cystic fibrosis, androgen receptor deficiency, Noonan's syndrome, Kallman's syndrome, and myotonic dystrophy are associated with impaired fertility.

E4

PHYSICAL EXAMINATION
The physical examination of transplant patients presenting for infertility evaluation should be thorough, because any condition which impacts overall health can be detrimental to sperm production. A close inspection of the body should detect *signs of inadequate virilization or androgen deficiency*. These signs include eunuchoid body habitus, gynecomastia, and decreased body hair. The penis and scrotum should be carefully evaluated. *Curvature of the phallus or ectopic location of the urethral meatus* can result in abnormal deposition of the ejaculate within the vagina. Close attention should be paid to the scrotal exam, which should be done with the patient standing in a warm room. *Testicular size, consistency, and volume* (using an orchidometer) should be determined. The normal testicular length is > 4 cm, and normal testis volume is > 20 mL. The presence of the epididymis should be confirmed. *Cysts, induration, and other abnormalities* should be noted. Finally, the scrotum should be evaluated for the presence of *varicoceles*. Extremely large varicoceles are visible and have a characteristic "bag of worms" appearance. To examine for varicoceles, the patient should valsalva while the examiner palpates the spermatic cords. A dramatic impulse is usually felt in patients with clinically apparent varicoceles. This is due to transmission of the increased intraabdominal pressure to the veins of the pampiniform plexus; this pressure impulse is typically not felt in normal patients. A Doppler stethoscope, which can be easily used in the clinic, and scrotal ultrasonography (to evaluate for veins > 3 mm and reversal of venous blood flow) can be used to confirm clinically apparent varicoceles and to detect those which are subclinical.

LABORATORY EVALUATION
Laboratory testing is begun after the history and physical exam are completed. The tests ordered should be individually tailored to each patient. It is important to note that a semen analysis, although a vital component of the work-up, is not a "fertility test". The interpretation should take into account crucial factors, such as

how the specimen was collected and how it was analyzed. Semen collections should occur after an abstinence period of 2-3 days, and they should be presented for analysis within 1-1$^{1}/_{2}$ hours of ejaculation. The collection container should be wide-mouthed to ensure collection of the entire sample. The method of collection may be via masturbation, coitus interruptus, or condom collection (must be free of spermicidal agents). No one single semen analysis should be used to determine the patient's baseline. Usually 2-3 samples are collected and analyzed before a diagnosis and treatment plan are made. Semen parameters characteristically evaluated include: 1) volume, 2) density, 3) motility, and 4) forward progression. Normal values may vary slightly from lab to lab.

Several additional semen tests which are often utilized. An immumobead assay is done to detect the presence of antisperm antibodies, which can severely impair sperm motility.[4] These may respond to treatment with corticosteroids. An assay for the presence of reactive oxygen species is important, for abnormally high concentrations may damage the sperm plasma membrane and impair its function. Reactive oxygen species are treated with antioxidant therapy, such as Vitamin E. As previously discussed, the presence of genitourinary infections can be detrimental to sperm function. A microscopic examination of the semen alone is not sufficient to detect the presence of white blood cells (WBCs), because of the similarity between their appearance and the appearance of immature sperm (round cells). Therefore, a combination of semen culture as well as a monoclonal antibody test to specifically detect the presence of seminal WBCs is performed. Urethral swab cultures should be performed if there is a question of urethritis (chlamydia, ureaplasm, etc.). Antimicrobial therapy is tailored appropriately, based on culture results.

Another test which is often performed is the Kruger Strict Morphology assay.[6] It's purpose is to determine the percentage of normally shaped sperm within a sample, using a very rigid criteria. Normal values are >4% (in the Baylor Andrology Lab). Values less than 4% are associated with impaired sperm function and reduced potential for successful fertilization. A number of other tests are available to evaluate sperm function, however in this age of extensive use of in vitro fertilization (IVF) and intracytoplasmic sperm injection (ICSI), whereby a sperm can be directly microinjected into an egg, many of the functional sperm problems associated with male factor can be circumvented.

In addition to the semen analysis, hormonal testing is usually undertaken in the evaluation of the infertile male. Serum FSH should be measured prior to any intervention. Elevation of FSH above 2-3 times the normal value should be construed as an unfavorable sign associated with a probable primary testicular pathology. This may be the result of longstanding metabolic disturbances or gonadotoxic drug effects in the transplant patient. Finally, a prolactin level should be ordered to rule out the presence of abnormalities involving the HPG axis, as previously detailed.

TREATMENT OF THE INFERTILE MALE TRANSPLANT PATIENT

The presence of male factor infertility in a transplant patient can present a difficult clinical problem for the patient and treating physician. Some of the medications which are essential for maintaining allograft viability posttransplant are directly injurious to spermatogenesis. Furthermore, especially in patients with a longstanding history of renal failure and uremia, there may already be significant, permanent underlying damage to the testes. For these reasons, the appropriate treatment for the infertile couple depends on the patient's clinical problems, the age of the patient and his partner, and the desires of the couple. The specific appropriate treatments for the various causes of male infertility in the transplant patient have been outlined above. It is important to realize that although therapy targeted for these various disorders is quite often well tolerated and effective, some of these treatments do require several months time before the full clinical impact of treatment is realized. For those situations where this time interval is prohibitively long, such as in a couple with advanced maternal age (> 35 y.o.), it may be prudent to proceed directly to assisted reproductive techniques such as IVF and ICSI. Fortunately, with this technology, even those patients with only a few viable sperm within the ejaculate, epididymal fluid, or testicular tissue, are now able to achieve pregnancies.

REFERENCES

1. Holdsworth S, Atkins RC, de Kretser DM. The pituitary-testicular axis in men with chronic renal failure. New Eng J Med 1977; 296:(22)1245-1249.
2. Baumgarten SR, Lindsay GK, Wise GJ. Fertility problems in the renal transplant patient. J Urol 1977; 118:991-993.
3. Holdsworth SR, de Kretser DM, Atkins RC. A comparison of hemodialysis and transplantation in reversing the uremic disturbance of male reproductive function. Clin Nephrol 1978; 10(4):146-150.
4. Lipshultz LI, Howards SS. Infertility in the Male, Saint Louis, Mosby-Year Book Inc., 1997.
5. Killion D Rajfer J. The evaluation and management of male infertility following solid-organ transplantation. Seminars in Urology 1994; 12 (2):140-146.
6. Lipshultz LI, ed. Male Infertility. In: The Urologic Clinics of North America 1994; 21(3).

Essay 5: Spontaneous and Transplanted Malignancy

Israel Penn

Supported in part by a grant from the Veterans Administration

Problems with cancer may be encountered in transplant patients in three sets of circumstances: 1) Inadvertently grafted neoplasms transmitted with organs obtained from donors with malignancies, 2) Tumors arising spontaneously de novo after transplantation, and 3) Recurrence of cancers treated prior to transplantation. In this essay, one will briefly cover the first two topics. Most of the material presented is based on publications from the Cincinnati Transplant Tumor Registry (CTTR).[1-4]

INADVERTENTLY TRANSMITTED MALIGNANCIES

Immunosuppressive therapy may permit cancer cells inadvertently transmitted with donor organs to survive, multiply and metastasize.[1,2] Most transmitted neoplasms occurred in the pioneering era of transplantation, when the danger of grafting organs from donors with cancer into immunosuppressed recipients was not appreciated. Today inadvertent transplantation of organs containing tumor cells is a rare event. However, several cases of per annum do occur despite the most stringent selection criteria used by organ procurement teams. By November, 1997, the CTTR had received data on 279 patients who received organs from donors with malignancies.[1,2] The great majority received kidney allografts and the remainder had hepatic, cardiac, pancreatic, pulmonary or cardiopulmonary allografts. Eighty-three percent of donors were cadavers and, of these more than one third provided two or more organs to the recipients. Fifteen percent of donors were living related individuals who had been treated for malignancy within 10 years of donation, or were found to have cancer at the time of donation, or manifested evidence of one within 18 months of the procedure. Another 2% of donors were living unrelated individuals.

The cause of brain death was misdiagnosed in 27 cadaver donors (involving organs transplanted into 56 recipients). Cerebral metastases, particularly from choriocarcinomas, carcinomas of the bronchus, and malignant melanomas either masqueraded as primary brain tumors, or bled and mimicked hemorrhage from a cerebral aneurysm or arteriovenous malformation.

In the overall series 43% of allografts had evidence of cancer.[1,2] In nine instances these were small primary tumors of a renal allograft which were widely

Organ Transplantation, 2nd edition, edited by Frank P. Stuart, Michael M. Abecassis and Dixon B. Kaufman. ©2003 Landes Bioscience.

excised before transplantation of the kidney into the recipient. In the other cases the malignancies were discovered from days after transplantation (when several renal allografts were removed for various reasons) through various periods of time up to 63 months posttransplantation. In the case of the recipients who did not show evidence of cancer, we presume that the allografts were free of tumor or that transmitted neoplastic cells failed to survive after transplantation. The transmitted tumors usually were histologically identical to those in the original donors. In 39% of affected recipients the malignancies were confined to the allografts, in 56% there was invasion beyond the grafts, and in 56% there were distant metastases. The tumors that most frequently caused metastases were malignant melanomas, renal carcinomas, bronchial carcinomas and choriocarcinomas.

Sixty-six percent of recipients who developed metastases died of their cancers, many having not received any treatment. However, another 27% (all renal allograft recipients) had complete remissions and the remainder are currently undergoing treatment. Most remissions followed reduction of the tumor burden by transplant nephrectomy and cessation of immunosuppressive therapy or immuno-therapy. Presumably the recipients' depressed immune systems were able to recover and to reject the neoplasms. In several patients these measures were supplemented by chemotherapy, or radiotherapy, or immunotherapy. However, 22% of all patients with distant metastases died of their tumors, despite these treatments, suggesting that their immune systems were unable to recover, or were unable to handle the large residual tumor burden which led to a fatal outcome.[1,2]

While allograft removal and discontinuation of immunosuppression is an option in kidney transplant recipients who can be returned to dialysis, it can be utilized only to a limited extent in recipients of nonrenal organs.[1,2] Of 35 nonrenal allograft recipients 16 survived without evidence of malignancy, another 7 had cancer confined to the allograft, and another 12 died of metastatic tumor.

To avoid the inadvertent transmission of cancer, careful attention must be paid to the patient's history, such as past treatment of cancer or of menstrual irregularities following a pregnancy or abortion. Every effort must be made to exclude a metastasis as the cause of intracranial bleeding when the donor has no evidence of hypertension, and an intracranial aneurysm or arteriovenous malformation cannot be documented. One should be particularly careful with a female donor in the child bearing years, who has a history of menstrual irregularities, since a metastatic choriocarcinoma may be the underlying cause.[1,2] Measurement of beta-human chorionic gonadotropin (beta-HCG) levels is a major safeguard and, perhaps, is advisable in all female donors of childbearing age.[1,2]

With several exceptions, donors who have malignancies should not be used: low grade skin cancers such as basal cell carcinomas and many squamous cell carcinomas: carcinoma in situ of organs such as the uterine cervix; or primary brain tumors that rarely spread outside the central nervous system.[1,2] However, one must be certain that brain neoplasms arose there because, in some instances autopsy examinations performed after organ retrieval have shown that the apparent brain tumors were actually metastases from occult primary malignancies. We should also avoid using donors with brain cancers that were treated with radiotherapy, chemotherapy, ventriculoperitoneal or ventricyloatrial shunts, or extensive craniotomies, as they may open pathways for malignant spread.[1,2]

A much more difficult decision arises when a donor has a history of cancer treatment in the remote past. Most surgeons would accept a five year disease-free interval as evidence of "cure". However, it is well recognized that late metastases may occur from carcinomas of the breast, or colon, or from malignant melanomas. On rare occasions these may be present as micrometastases at the time of organ retrieval and a diseased organ could be transplanted. The transplant surgeon has to evaluate each donor on an individual basis, and weigh the small risk of transplanting cancer with organs from such a donor, against the chance of discarding potentially usable organs, at a time when there is a profound shortage of cadaver organs.

During organ retrieval surgeons should carefully examine all accessible intrathoracic and intra-abdominal organs for evidence of cancer.[1,2] This has occasionally yielded positive findings, particularly with primary renal carcinomas, so that a particular donor or organ was not used.

Theoretically every cadaver donor should have, an autopsy examination performed as expeditiously as possible and before any organs are transplanted. In actual practice permission for autopsy examination is seldom given, and if an autopsy is performed this is usually done after organs have been transplanted. Furthermore, the pathologists' need to fix the brain in preservative for a week or more often delays the results of an autopsy for several weeks. To complicate matters further, even when an autopsy is performed at the donor hospital, the results may not be made available to the various recipient teams. Therefore, an added onus falls on the procurement team to check with the donor hospital regarding any untoward autopsy findings.[1,2]

The danger of inadvertently transmitting malignancies from donors to recipients must be viewed perspective. Most reported cases occurred in the pioneering era of transplantation, when the risks not appreciated. Over 400,000 solid organ transplants have been performed, but only a handful of transmitted tumors have occurred. Nowadays, with careful selection of donors, inadvertent transplantation of cancer should be a rare event.[1,2]

SPONTANEOUS MALIGNANCIES

Overall there is a 3- to 4-fold increase compared with age matched controls in the general population.[3,4] Apart from skin tumors (mostly squamous cell carcinomas—SCCs), that show 4- to 21-fold increased incidence, cancers that are frequently encountered in the general population (carcinomas of the lung, breast, prostate, colon and invasive uterine cervical carcinomas) show no increase, but a variety of mostly uncommon neoplasms are encountered. Epidemiologic studies show increases of 28- to 49-fold of posttransplant lymphoproliferations and lymphomas, (frequently grouped together as posttransplant lymphoproliferative disease PTLD), 29-fold of lip carcinomas, 400- to 500-fold of Kaposi's sarcoma (KS) 100-fold of vulvar and anal carcinomas, 20- to 38-fold of hepatocellular carcinomas, 14- to l6-fold of in situ uterine cervical carcinomas, and small increases in sarcomas (excluding KS) and renal carcinomas.[3,4] The major part of this report is based on material collected by the CTTR up till January 1999, when it had received information on 11,017 organ allograft recipients who developed 11,729 types of de novo malignancy (Table E5.1).

The incidence of tumors increases with length of follow-up posttransplantation. An Australian study of 6596 patients shows that the percent probability of developing cancer following renal transplantation from cadaver donors 24 years postoperatively is 66% for skin neoplasms, 27% for nonskin cancers and 72% for any type of tumor.[5] These exceptional figures must be interpreted with caution as most cancers are skin malignancies (which are very common in Australia) and the number of 24-year survivors is small. Nevertheless, they emphasize the need to follow transplant patients indefinitely. Cancers occur a relatively short time posttransplantation with KS appearing at an average of 21 (median 13) months posttransplantation, PTLDs at an average of 34 (median 13) months, and vulvar and perineal carcinomas appearing at the longest time posttransplantation, at an average of 115 (median 114) months.[3,4] If all tumors are considered the average time of their appearance is 63 (median 47) months.

Malignancies that occur in organ allograft recipients frequently demonstrate more aggressive behavior than do similar cancers in the nontransplant population.[6]

The most common tumors affect the skin and lips and comprise 37% of all neoplasms in the CTTR.[3,4] They occur on sun exposed areas, mainly of the head

E5

Table E5.1. Most common spontaneous malignancies

TYPE OF TUMOR	NO. OF TUMORS[a]
Cancers of skin and lips	4435
Posttransplant lymphoproliferative disease	1962
Carcinomas of the lung	662
Kaposi's sarcoma	469
Carcinomas of uterus (cervix 357; body 68; unknown 5)	430
Carcinomas of the kidney	417
(host kidney 352; allograft kidney 43, both 1; unknown 21)	
Carcinomas of colon and rectum	394
Carcinomas of the breast	364
Carcinomas of the head and neck	326
(excluding thyroid, parathyroid and eye)	
Carcinomas of the vulva, perineum, penis, scrotum	286
Carcinomas of urinary bladder	254
Carcinomas of prostate gland	229
Metastatic carcinoma (primary site unknown)	225
Leukemias	204
Hepatobiliary carcinomas	188
Sarcomas (excluding Kaposi's sarcoma)	148

[a] There were 11,017 patients of whom 667 (6%) had two or more distinct tumor types involving different organ systems. Of these, 43 patients each had 3 separate types of cancer and 1 had 4.

and neck and upper extremities, especially in light-skinned individuals with blue eyes and blonde or red hair.[3,4,7] Exposure to sunshine is not the only etiologic factor. A surprisingly high incidence of SCCs is recorded from areas of low sunlight in some northern countries and may be related to malignant change in papillomavirus-induced warts, under the influence of immunosuppression, sunlight, HLA-antigens and other factors.[3,4] The incidence of cutaneous malignancies increases with length of follow-up after transplantation, as demonstrated in a Dutch study that showed a 10% incidence of nonmelanoma skin cancer at 10 years posttransplantation, that rose to 40% after 20 years.[3,4]

Skin tumors in transplant patients show several unusual features compared with similar lesions in the general population.[3-5,7] Whereas basal cell carcinomas (BCCs) outnumber SCCs in the general population by 5 to 1, in transplant recipients SCCs outnumber BCCs by 1.8 to 1. SCC is estimated to occur at a frequency between 40 and 250 times higher than in the general population, BCC ten times higher and malignant melanoma five times more commonly than expected. In the general population SCCs occur mostly individuals in their 60s and 70s, but the average age of transplant patients is 30 years younger. In addition, the frequency of multiple skin malignancies in the CTTR is remarkably high (43%) and, despite being a worldwide collection, is similar to that seen only in areas of copious sunlight. Some patients each have more than 100 skin tumors. In some patients there is an apparently widespread cutaneous abnormality with areas of unstable epithelium containing multifocal premalignant and malignant lesions.[7]

In the general population most lymph node metastases and deaths from skin malignancies are caused by melanomas. In contrast SCCs are much more aggressive in transplant patients than in the general population and account for the majority of lymph node metastases and deaths from skin cancer.[3,4] Thus, nearly 6% of patients with skin neoplasms in the CTTR have lymph node metastases. Of these 73% are from SCCs and only 17% from melanomas. Similarly 5% of patients die of skin cancer, with 60% of deaths being from SCC and only 30% from melanomas.[3,4]

Among the PTLDs, Hodgkin's disease and plasmacytoma/myeloma are much less common than in the general population[3,4] and most tumors represent a broad spectrum of lesions ranging from benign hyperplasias, such as infectious mononucleosis-like disorders at one end to frankly malignant lymphomas at the other extreme.[8,9]

The majority of PTLDs arise from B-lymphocytes but CTTR data indicate that 15% arise from T-lymphocytes, while rare cases are of null cell origin. In approximately 80-90% of PTLDs infection with Epstein-Barr virus (EBV) plays an important role.[8,9] Patients at particular risk for devoloping EBV-related PTLD are young children who are EBV negative pretransplantation but subsequently become seropositive.[3,4,8,9] Recipients of nonrenal organs are at higher risk for developing PTLD as they generally tend to be more heavily immunosuppressed than renal allograft recipients.[3,4]

PTLDs differ from lymphomas in the general population in several respects.[3,4] Whereas extra-nodal involvement occurs in from 24-48% of

E5

patients in the community at large, it is present in 70% of PTLDs. Surprisingly, one of the most common extranodal sites is the central nervous system, which is involved in 21% of cases. Another remarkable finding is the frequency of either macroscopic or microscopic allograft involvement, which occurs in 23% of patients with PTLD. In some patients the infiltrate is mistaken for rejection when allograft biopsies are studied microscopically. It is disappointing that 16% of patients with PTLD die without treatment, either because the diagnosis is missed, or is made too late to save them. Following treatment complete remissions are obtained in 38% of patients.

Kaposi's sarcoma (KS) is most common in transplant patients who are Arab, black, Italian, Jewish, Turkish or Greek.[1-3] It occurred in 1.6% of 820 Italian renal transplant recipients,[10] but was the most common neoplasm in renal transplant recipients in Saudi Arabia, comprising 76% of all malignancies.[11]

A clinician should suspect KS whenever a transplant patient, particularly one belonging to the ethnic groups described above, presents with reddish blue macules or plaques in the skin or oropharyngeal mucosa, or apparently infected granulomas that fail to heal.[3,4] If the diagnosis is confirmed, a thorough workup including CT scans of the chest and abdomen and upper and lower gastrointestinal endoscopy, is necessary to exclude any internal visceral involvement.

Nonvisceral KS occurs in 59% of patients and is confined to the skin, or oropharyngeal mucosa and 41 % have visceral disease, involving mainly the gastrointestinal tract, lungs, and lymph nodes, but other organs are also affected.[3,4] In patients with nonvisceral disease the lesions are confined to the skin in 98% and the mouth or oropharynx in 2%. Patients with visceral lesions have no skin involvement in 27%, but 3% have oral involvement which provides an accessible site for biopsy and diagnosis. The outlook of patients with nonvisceral disease is much more favorable than those with visceral disease, as 54% of the former group have complete remissions following treatment compared with only 30% in the latter.[3,4]

Most renal carcinomas in renal recipients arise in their own diseased kidneys although 10% occur in the allografts.[3,4] Unlike most other neoplasms, that arise as complications of immunosuppressive therapy, many renal carcinomas are related to the underlying kidney disease necessitating transplantation. One contributary factor is analgesic nephropathy in renal allograft recipients, which occurs in 8% of CTTR patients with carcinomas of their native kidneys. This disorder is known to cause carcinomas in various parts of the urinary tract. This is borne out in the CTTR series in which 59% of patients with analgesia-related renal carcinomas have similar neoplasms elsewhere in the urinary tract.[3,4] Another predisposing cause of cancers is acquired cystic disease of the native kidneys, which, in dialysis patients, is complicated by an increased incidence of renal carcinomas. The exact incidence of such carcinomas in renal transplant recipients is not known, but at least 17 patients in the CTTR have this disorder.[3,4]

A group of carcinomas arise in the vulva in females, the penis or scrotum in males, and in the perineum, perianal skin or anus in either sex.[3,4] Sometimes female patients have multifocal lesions that involve not only of the vulva and perineum but also the vagina and/or uterine cervix. Females outnumber males by 2.6:1 in

E5

contrast with most other posttransplant cancers where males outnumber females by more than 2:1, Thirty-eight percent of patients have in situ lesions. A worrying finding is that patients with invasive lesions are much younger (average age 42 years) than their counterparts in the general population, whose average age is usually between 50 and 80 years. Prior to the development of the malignancies 56% of transplant patients have condyloma acuminata ("genital warts"), suggesting that human papillomavirus plays an important role in the development of these tumors.

It is possible that many in situ carcinomas of the uterine cervix are missed because they are asymptomatic. Therefore all postadolescent female allograft recipients should have regular pelvic examinations and cervical smears to detect these lesions, and also carcinomas of the vulvar and anal areas.[3,4]

Most hepatobiliary tumors are hepatomas and a substantial number of patients have a preceding history of hepatitis B infection.[3,4] Increasing numbers of patients with a history of hepatitis C infection are now being encountered.

Most sarcomas (other than KS) involve the soft tissues or visceral organs whereas cartilage or bone involvement is uncommon. The major types in descending order are fibrous histiocytoma, leiomyosarcoma, fibrosarcoma, rhabdomyosarcoma, hemangiosarcoma, and mesothelioma.[3,4]

Posttransplant malignancies probably arise from a complex interplay of many factors.[3-5,7-9,12] Severely depressed immunity may impair the body's ability to eliminate malignant cells induced by various carcinogens.[3,4] Chronic antigenic stimulation by the foreign antigens of transplanted organs, by repeated infections, or transfusions of blood or blood products may overstimulate a partially depressed immune system and lead to PTLD.[3,4] Alternatively, defective feedback mechanisms may fail to control the extent of immune reactions and lead to unrestrained lymphoid proliferation and PTLD. Furthermore, once this loss of regulation occurs, the defensive ability of the immune system is weakened and other nonlymphoid malignancies may appear.[3,4]

Activation of oncogenic viruses probably plays an important role in the development of some tumors.[3,4,8,9] Epstein-Barr virus is strongly implicated in causing PTLD, some smooth muscle tumors, and, some cases of Hodgkin's disease; various strains of papillomavirus in causing carcinomas of the vulva, perineum, uterine cervix, and anus, but there is controversy concerning the role of these virus in causing skin cancers: hepatitis B or hepatitis C virus in causing hepatomas; and herpes virus type 8 (HHV-8) appears to play an important role in the development of KS.

Some immunosuppressive agents may directly damage DNA and cause cancers.[3,4] Immunosuppressive agents may enhance the effects of other carcinogens, such as sunlight in causing carcinomas of the skin, or papilloma virus in causing carcinomas of the uterine cervix or vulva.[3,4] Genetic factors may affect susceptibility to neoplasia by affecting carcinogen metabolism, level of interferon secretion, response to virus infections, or regulation of the immune response by the major histocompatibility system.[3,4]

HLA antigens play an important role in host defense against the development and spread of neoplasms, especially in virus-induced cancers. For example, in renal transplant recipients HLA-A1 1 may protect against skin tumors, whereas HLA-B27 and HLA-DR7 are associated with an increased risk of these malignancies.[12]

ACKNOWLEDGMENT
The author wishes to thank numerous colleagues, working in transplant centers throughout the world, who have generously contributed data concerning their patients to the Cincinnati Transplant Tumor Registry.

EDITORS' NOTE:
Dr. Penn died shortly before publication of the first edition of *Organ Transplantation*. This essay was reproduced without change from the first edition.

REFERENCES
1. Penn I. Transmission in transplanted organs (editorial). Transplant Int 1993, 6:1.
2. Penn I. Transmission of cancer from organ donors. Ann Transplant 1997; 2:7-12.
3. Penn I. Depressed immunity and the development of cancer. Cancer Detect Prev 1994; 18:241.
4. Penn I. Malignancy after immunosuppressive therapy: How can the risk be reduced? Clin Immunotherapy 1995; 4:207.
5. Sheil AGR, Disney APS, Mathew TH, Amiss N. De novo malignancy emerges as a major cause of morbidity and late failure in renal transplantation. Transplant Proc 199; 25:1383.
6. Barrett WL, First R, Aron BS, Penn I. Clinical course of malignancies in renal transplant recipients. Cancer 1993; 72:2186.
7. Sheil AGR. Skin cancer in renal transplant recipients. Transplant Science 1994; 4:42.
8. Nalesnik MA, Starzl TE. Epstein-Barr virus, infectious mononucleosis, and posttransplant lymphoproliferative . Transplant Science 1994; 4:61.
9. Hanto DW. Classification of Epstein-Barr virus-associated posttransplant lymphoproliferative diseases: Implications for understanding their pathogenesis and developing rational treatment strategies. Annual Rev Med 1995; 46:381.
10. Montagnino G, Bencini PL, Tarantino A, Caputo R, Ponticelli C. Clinical features and course of Kaposi's sarcoma in kidney transplant patients: Report of 13 cases. Amer J Nephrol 1994; 14:121.
11. Al-Sulaiman MH, Al-Khader AA. Kaposi's sarcoma in renal transplant recipients. Transplant Science 1994; 4:46.
12. Bouwes Bavinck JN. Epidemiological aspects of immunosuppression: Role of exposure to sunlight and human papillomavirus on the development of skin cancer. Hum Exp Toxicol 1995; 14:98.

E5

Retransplantation of Vital Organs

Susan M. Lerner, Paige Porrett, James F. Markmann,
and Ronald W. Busuttil

INTRODUCTION

The dramatic success of organ transplantation in the last 20 years has led to a growing imbalance in the number of patients awaiting transplantation and the number of organs available for that purpose. As a result, the prioritization of individual recipients for organ allocation, especially for vital organs such as the heart, lung and liver, has become the subject of heated debate. Nowhere is this issue more pressing than in the discussion of the appropriate allocation of vital organs to patients with a failed first graft.

The decision to retransplant a critically ill patient, who will almost surely die rapidly without such intervention, touches on many controversial aspects of modern medicine. Is it ethical to provide one individual with a second opportunity for a life-saving therapy when others will die awaiting their first? Is it appropriate to expend a precious health care resource on an individual whose outcome is known to be inferior and more costly than if used in a patient receiving first time transplant? The reduced efficacy of retransplantation, in conjunction with the increased cost relative to a first graft, is magnified as the financial constraints of health care provision continue to grow. How to balance these concerns with the obligation of physicians to provide maximal care to their patients is a complex and unresolved issue. Is it really appropriate to consider depriving a critically ill patient a proven therapy simply because their medical history includes prior similar resource utilization?

What is clear is that the scarcity of organs mandates difficult decisions regarding their rational use. This duty currently falls on UNOS (the United Network for Organ Sharing). This organization supervises the procurement and allocation of organs —compiling patient waiting lists and formulating list priority — ranking patients while attempting to balance fairness, efficacy, and medical urgency. The waiting list, for some organs such as kidney and pancreas, gives priority to time on the list and to better donor-recipient histocompatibility (a factor which has been shown to improve survival for these grafts). In contrast, for vital organs, for practical reasons, allocation is based predominantly on the time waited and medical urgency. Thus, even if one puts the issue of retransplantation aside, inherent in the system is the prioritization of those who are the most critically ill and the most likely to die soonest. As a result the system gives preference to individuals likely to benefit the most even though the average graft survival in these patients is less

Organ Transplantation, 2nd edition, edited by Frank P. Stuart, Michael M. Abecassis
and Dixon B. Kaufman. ©2003 Landes Bioscience.

than if the same graft were used to transplant a patient in better health.

Once on the list, primary transplant and retransplant candidates are treated identically. Practically, however, it should be noted that patients awaiting retransplantation generally represent a sicker group of patients. For example, only about 22% of patients awaiting primary liver transplantation are the highest UNOS status whereas more than 80% of patients awaiting retransplant are the highest status. Thus, on average, patients who undergo retransplantation wait less time on the list than patients waiting for their first transplant.

Current data shows that survival for heart, lung, and liver retransplantation is poorer than that demonstrated for primary transplants. If retransplantation were as likely to produce the same results as primary transplantation, would these questions then be unnecessary and would retransplant candidates then rightly deserve their equal place on the waiting list alongside primary transplant candidates? Although the adult population is more frequently studied, the pediatric population suffers from many of the same constraints. However, it is inherently more difficult to refuse a child repeat transplantations and most transplant teams have accepted the decreased survival statistics to continue the practice of pediatric retransplantation of vital organs. Some studies have shown that among candidates for retransplantation, there are easily identifiable parameters which, when followed, can accurately predict outcome, suggesting that safe modifications to our current system are possible. We review such attempts for heart, lung, and liver retransplantation.

E6

CARDIAC RETRANSPLANTATION

Cardiac transplantation has evolved as the best therapeutic option for those with end-stage heart disease.[1] However, despite improvements in selection processes, organ preservation, and immunosuppression, a percentage of patients still experience rejection of their graft necessitating a second transplant. Clinically, rejection is diagnosed and graded according to endomyocardial biopsy. In addition to acute and chronic rejection, some cardiac transplant patients are plagued by transplant coronary artery disease. This form of accelerated atherosclerosis differs from traditional atherosclerosis in that it is a concentric and diffuse hyperplastic process. The internal elastic lamina remains intact, calcification of the vessels is rare, and intramyocardial vessels are involved as well. While coronary angiography is useful for the diagnosis of progressive narrowing, coronary angioplasty is seldom effective as discrete stenoses are unusual. Graft failure is the inevitable result. In addition, hearts thus diseased often show hypertrophy and myocyte disarray and develop areas of scarring that can lead to electrical irritability and sudden death. Repeat orthotopic transplantation has become the best treatment for those with existing or impending graft failure, whether it be secondary to rejection or to the accelerated coronary artery disease it produces. (Table E6.1)

Patients accepted for retransplantation usually have terminal decompensation with severe functional disability and limited expected survival time.[2] Those with accelerated coronary artery disease are generally assessed to have impending coronary occlusion as demonstrated by coronary arteriography. Typically, these pa-

Table E6.1. Indications for the retransplantation of vital organs

Heart	Lung	Liver
Primary graft dysfunction	Early graft dysfunction	Primary nonfunction
Intractable acute rejection	Intractable airway healing	Rejection
Coronary graft diseases	Obliterative bronchiolitis	Technical Complications

tients are hemodynamically stable and retransplantation is usually performed in a semi-elective fashion. Candidates for retransplantation secondary to acute or chronic allograft rejection refractory to immunosuppressive therapy are usually hemodynamically unstable, often requiring intensive care nursing and inotropic support. The same absolute contraindications exist for retransplantation as for transplantation including: elevated pulmonary vascular resistance, active infection, and a positive donor-specific lymphocyte crossmatch. Relative contraindications include advanced age and psychosocial instability.

Since the first report of cardiac retransplantation in 1977, data has been collected by the registry of the International Society for Heart and Lung Transplantation (ISHLT). It indicates that approximately 4% of all reported transplants are retransplants. [3] While early data from the ISHLT registry demonstrated a much poorer survival rate among retransplanted patients than those receiving their first transplants (48 vs 78%, respectively), more recent data from the registry has shown improved outcomes. One year survival among retransplant patients as a whole has improved to 65%, while the one year survival rate among primary transplant recipients has remained at approximately 80%[4]. Some individual centers have even published retransplant survival rates that equal those of their primary transplant population when very stringent selection criteria are applied to the retransplant recipients.[5] Not only has retransplant survival improved, but the long-term survivors of cardiac retransplantation fare as well as those receiving a primary transplant from the standpoint of cardiac function and quality of life. (Table E6.2)

Table E6.2. Patient survival after transplantation and retransplantation

Organ Type		1 Year	2 Years	5 Years	10 Years
Cardiac	Primary	84%	75%	71%	23%
	Retransplant	65%	59%	55% (3 years)	——
Pulmonary	Primary	77%	65%	44%	16%
	Retransplant	47%	40%	33% (3 years)	——
Hepatic	Primary	83%	77%	74%	68%
	Retransplant	65%	49%	47%	44%

In light of data indicating that selection of the appropriate retransplant population may significantly impact retransplant outcome, various groups have attempted to identify prognostic indicators. The interval between the first and second grafts has consistently been shown to be of great importance. Patients receiving their second graft more than two years after their first graft have survival rates approaching those of primary transplants, while only 40-50% of those patients that are retransplanted within six months of their first graft live to one year. More recent analysis of the ISHLT data has also demonstrated that recipient age impacts retransplant survival as well, although prior data analysis had indicated that recipient age was not of prognostic importance. A recent multivariate analysis of the 514 cardiac retransplants performed from 1987-1998 listed in the registry revealed that older recipient age adversely affected survival both at one month and one year post-retransplantation. The ISHLT data also indicates that preoperative mechanical ventilation and lack of experience of the transplant center also adversely influence retransplant survival. Therefore, if one were to construct the "ideal" retransplant candidate from recent ISHLT registry data, it would be a 40 year old patient retransplanted for transplant-associated coronary artery disease more than two years after his inital graft. The results of cardiac retransplantation are similar in the pediatric population, with graft coronary artery disease being the primary factor limiting long-term survival. (Table E6.3)

PULMONARY RETRANSPLANTATION

E6

Although the results of primary lung transplantation have improved dramatically in the last decade, lung transplant recipients are still plagued by early graft

Table E6.3. Predictors of poor retransplant survival

Heart	Lung	Liver
Recipient Age	Nonambulatory Status (rel. risk = 1.93)	Elevated Serum Cr (rel. risk = 1.22-1.24)
< 6 months between Transplants	Preop Ventilator Support (rel. risk = .53)	Elevated Serum T bili (rel. risk = 1.03-1.4)
Mechanical Circulatory Support	< 2 years between Transplants (rel. risk = 1.96)	Recipient Age (rel. risk = 1.3-1.6)
	Low Center Volume (rel. risk = 2.04)	Preop Ventilator Support (rel. risk = 1.8)
		UNOS Status
		Total Number of Transplants
		Donor Age (rel. risk = 2.2)
		Organ Ischemia (rel. risk = 1.7)

dysfunction, intractable airway healing problems, and obliterative bronchiolitis — all conditions necessitating a second transplant. [6] Retransplantation accounts for 5-10% of all pulmonary transplants. Obliterative bronchiolitis, first described in the lung transplant population in 1984, is an inflammatory disorder of the small airways leading to obstruction and destruction of pulmonary bronchioles. It can occur after isolated single- and bilateral-lung transplantation and is found to some degree in up to two-thirds of all lung transplant recipients. The main risk factor for the development of this disease is recurrent, severe, and persistent acute lung rejection. It is not, however, associated with any other known recipient variables (such as, age, sex, or indication for primary transplantation.) The fibrosis resulting from obliterative bronchiolitis is irreversible, and there is no satisfactory treatment for the disease other than retransplantation. (Table E6.1)

The pulmonary retransplant registry was established in 1991 to help assess the outcome and survival after pulmonary transplantation. Analysis of the registry data indicates that a prior lung transplant increases the risk of 1-year mortality by more than three-fold. (Table E6.2) On multivariate analysis, survival was not significantly different according to the age, sex, original diagnosis, or the cytomegalovirus status of the recipient. Similarly, survival did not depend on the indication for retransplantation. The most significant predictors of survival were the preoperative ambulatory status of the recipient, as well as the preoperative ventilatory status. An ambulatory recipient was defined as a patient who was able to walk at least 50m. with or without assistance, immediately before retransplantation. In addition, there is an association between survival and the total center volume of retransplantation and between survival and the interval length between transplant procedures. There was no significant survival difference according to the type of retransplantation procedure that was performed (single versus bilateral lung transplants). Patients with and without an old retained contralateral graft have similar survival and pulmonary function.

Opportunistic infection is the leading cause of death after pulmonary retransplantation. [7] Recently, retransplant deaths due to infectious causes have declined, but the percentage of deaths caused by recurrent bronchiolitis obliterans has increased. The remainder of the deaths are due to acute failure of the second graft early after reoperation and airway complications. Death due to these complications has remained relatively constant. As with cardiac retransplant recipients, pulmonary retransplant recipients maintain good long-term functional status and the majority do not require supplemental oxygenation. Seventy-nine percent of all retransplant recipients are free of bronchiolitis obliterans at 1 year, 64% at 2 years, and 56% at 3 years. The prevalence of severe disease was 12% at 1 year, 15% at 2 years, and 32% at 3 years, comparable to primary lung transplantation data. Indeed, this shows that the pulmonary function of surviving retransplant recipients is preserved as well as that in recipients of first-time lung grafts.

This data suggests that lung retransplantation should be limited to ambulatory, non-ventilated patients and perhaps the procedure should only be performed in centers with extensive experience in pulmonary retransplantation. Survival after pulmonary retransplantation can be comparable to that after primary lung trans-

E6

plantation in experienced hands, as long as the patients are carefully selected. On the other hand, suboptimal patient selection can produce a very low probability of short-term survival. (Table E6.3)

LIVER RETRANSPLANTATION

Hepatic allograft failure continues to be a serious risk for the liver transplant recipient. [8] Because no effective method of extracorporeal support is currently available for these patients, hepatic retransplantation provides the only available option for patients in whom an existing graft has failed, and accounts for 10-20% of all liver transplants. The need for retransplantation can occur as the result of four principle causes — primary nonfunction, rejection, disease reoccurrence, or as a result of technical complications. These four diagnoses comprise almost 90% of liver retransplants. A diagnosis of primary non-function is a diagnosis of exclusion made if a graft never shows evidence of initial function and its dysfunction cannot be attributed to technical or other causes. By a more stringent definition, PNF is defined as graft dysfunction in the first week post transplant leading to patient death or retransplantation. In practice, this latter definition may be too limited as there is a spectrum in the degree of initial graft failure that may extend out to a month post transplant. These late initial failures (from 7-30 days) have been referred to by some as delayed non-function. The clinical syndrome accompanying initial non-function is also quite variable. In its most dramatic form, it is characterized by a constellation of findings in many ways analogous to fulminant hepatic failure; including a profound hepatic dysfunction accompanied by severe hypoglycemia, deep hepatic coma, renal failure, no bile output, marked coagulopathy, acidosis, and shock. Cerebral edema and elevated intracranial pressures are also possible. Initial graft failure can also present with a more indolent course, as a slow and inexorable loss of function with a rising bilirubin, a gradually worsening coagulopathy, and poor bile production. (Table E6.1)

If a graft fails secondarily, following initial evidence of good function, the cause usually represents either rejection, disease reoccurrence, or technical causes (such as vascular thrombosis and less frequently, complications of biliary reconstruction). Factors that support a decision to retransplant a patient for rejection include: 1) chronic rejection that is not likely to be reversible as shown by massively elevated liver function enzymes or a biopsy showing the disappearance of bile ductules, arteriolar thickening, and extensive periportal fibrosis; 2) the persistent elevation of serum bilirubin which is unresponsive to multiple courses of immunosuppression; or 3) poor liver function that deteriorates even further when immunosuppression is reduced to maintenance levels. Rejection is the most common reason for retransplantation in adults while technical failure is the predominant cause of retransplantation in the pediatric population. Primary non-function is the second most common reason for retransplantation in adults and the least common cause in children.

Disease reoccurrence, especially with regard to hepatitis C (the most common reason for primary transplantation) is a growing cause for considering retransplantation. Recent analysis of large data sets of liver recipients indicates

E6

that long-term survival following transplantation for HCV is compromised by the reoccurrence of the disease in the transplanted liver.[9] While intense efforts are underway to find medical therapies that treat viral replication post transplant, the decision whether to retransplant for this condition, which is likely to again reoccur in the second graft, is a subject of much debate. For the recipient dying of graft failure from recurrent HCV the benefit is obvious even if only a reduced survival compared with a first graft in a non-HCV patient is anticipated. When considered in terms of equitable and efficient use of a scarce resource the correct decision is less clear.

Operative complications can clearly impact the success of hepatic transplantation and retransplantation.[10] Liver retransplantation when done early in the postoperative period presents less of a technical challenge than the original operation owing to the relative simplicity of the recipient hepatectomy. However, when performed after a significant delay, these patients represent an arduous technical challenge. But, despite the greater technical complexity, patients transplanted late generally have shown better survival. An important exception is the finding that adults transplanted within the first week post transplant have survival rates equal to first grafts. Since those that succumb later often do so as the result of infection, the better survival in those retransplanted early may be a consequence of intervention before the patient become too ill and infected.

Several other technical considerations are important in hepatic retransplantation. Vascular grafts, usually of donor iliac artery and vein obtained at the time of liver procurement, have proven useful as an alternate method of arterial or portal venous reconstruction when there are difficulties with the standard anastomosis to the recipients own artery or portal vein. The use of an arterial graft is significantly more common in the retransplant setting (9.5% vs. 28.2 %). In addition, the biliary reconstruction can also present a challenge. If a duct-to-duct anastomosis is planned, all donor duct tissue must be removed while still ensuring that the recipient bile duct is viable and long enough to allow for a new anastomosis that is without tension. If any doubts exist, a choledochodochostomy should be abandoned in favor of an anastomosis to a Roux limb of jejunum. The need for enteric biliary drainage is required in approximately 18.8% of first time grafts compared with 47.7% in retransplants. If a jejunal limb is to be reused, the first site of duct anastomosis should be excised or closed primarily and a new site prepared

In addition to operative contributions, it is also possible that inferior patient survival after retransplantation may reflect an inherently sicker population. It is well documented that the outcome after a primary hepatic transplant correlates well with the patient's UNOS status and it is reasonable to assume that this might also apply to retransplantation. Moreover, sicker patients who are desperately awaiting retransplantation may be more apt to receive marginal grafts, thus further reducing their survival potential. In addition to being critically ill at the time they are undergoing major surgery, retransplanted patients are also highly immunosuppressed. In the case of intractable or chronic rejection, where immunosuppression is often increased, the further loss of immunocompetence and the infectious complications associated with this loss may aggravate the already precarious clinical situation. For this reason it is common to reduce the immunosuppression once the patient is re-listed so that the retransplant can be carried out under more optimal conditions.

We recently examined the survival of a large cohort of patients undergoing retransplantation at UCLA. We found that the survival of patients undergoing primary hepatic transplantation is 83% at 1 year, 74% at 5 years, and 68% at 10 years. In contrast, the survival of retransplanted patients at similar time points is significantly less, 62%, 47%, and 45% respectively. (Table E6.2) A number of variables were analyzed to determine what factors influenced outcome. We found that patient age, UNOS status, and total number of transplants were all negative prognostic factors in univariate analysis. In addition, patients retransplanted more than 30 days after their initial graft fared better than did those retransplanted between 8 and 30 days. The survival in patients transplanted within 1 week was nearly equivalent to that seen in the chronic group. This observation emphasizes the need for early recognition of patients who require early retransplantation. In contrast, pediatric retransplants performed within 1 week of the first graft fared worse than did pediatric patients retransplanted between 8 and 30 days.

To determine whether differences in illness severity, as reflected in UNOS status, could explain the poorer survival of second versus first transplants, a case-control analysis was performed comparing retransplant patients against primary transplant recipients. The transplant cases and controls were matched for age and UNOS status. Survival of retransplanted patients was still inferior to patients receiving their first graft even though the groups were objectively comparable in terms of their severity of illness.

A multivariate regression analysis was also performed on the UCLA cohort to determine independent risk factors predictive of poor patient survival. Donor cold ischemia time > 12 hours, preoperative mechanical ventilator requirement, age > 18 years, preoperative serum creatinine (Cr), and preoperative serum total bilirubin (T bili) were all independently predictive of a patient's bad outcome. Analogous findings have been noted in similar works done at the University of Pittsburgh. [11] They also identified donor age and donor gender as significant, as well as choice of primary immunosuppression. These findings have also been corroborated by the analysis of data collected at King's College Hospital in London. They saw their best results in those transplanted for chronic rejection, with an age cutoff of 39, and a better outcome occurring if the retransplanted patient was admitted from home. As in the other studies, those who were retransplanted late in the postoperative period and those with a lower serum bilirubin and creatinine fared better. These biochemical abnormalities are markers of disease severity and may well by themselves have specific effects on multiorgan function. For example, renal failure is known to be accompanied by deficiencies of cellular and humoral immunity, which predispose and aggravate the tendency to postoperative sepsis. Likewise, hyperbilirubinemia predisposes to endotoxemia, defects in cellular immunity, and Kupffer cell dysfunction. (Table E6.3)

E6

The incidence of death secondary to sepsis was also significantly higher in re-transplanted patients (60.7% of deaths in retransplants vs. 29% in primary recipients). Among those retransplanted patients in whom sepsis was the primary cause of death, there was a significantly increased incidence of fungal infection (16 of 34 patients in the retransplanted group vs. 1 of 9 patients in the primary group.) The total average hospital and ICU stay was significantly longer in both retransplanted patients as well as in pediatric patients (irrespective of whether retransplantation was performed during the same or different admission.) Both adult and pediatric retransplantation patients accrued higher total charges.

A mathematical model derived from the UCLA data can be applied to predict relative outcome based on characteristic donor and recipient variables.[12] This system can be employed to identify a sub-group of patients in whom the expected outcome is too poor to justify retransplantation. The regression equation that estimates 1-year survival in retransplanted patients is shown:

Estimated survival = $0.611 exp(R-1.6856)$

Where 0.611 is the mean 1 year survival for the patient group and R is the patient risk score calculated by: R=0.726 x ischemia + 0.561 x vent. + 0.0292 x tbil + 0.202 x Cr + 0.526 x age group. In this equation the three categorical variables are defined as follows: ischemia = 1 if >/= 12 hrs and 0 if < 12 hrs., vent.= 1 if pre-op ventilation is required and 0 if not, age group = 1 if adult and 0 if child. The two continuous variables T bil and Cr are represented as preoperative serum values in mg/dl. The mean overall risk score for the group is 1.6856. From the above equation it was determined that a risk score of > 2.3 corresponds to an expected 1 year survival of <40%. We have arbitrarily suggested that an expected survival of less than one-half of that expected from a primary transplant is an unacceptable use of a valuable organ.

Slightly cumbersome in nature, the above equation was used as the basis for a simplified five point scoring approach, in which all five variables were binary. Each covariate was assigned an equal weight of 1 point, (i.e., 1 point was awarded for adults, organ ischemia > 12h, pre-op ventilator requirement, tbili >/=13, and Cr>/ 1.6). Patients were then grouped into 1 of 6 risk classes based on the sum total of

E6

Table E6.4. Estimated survival of retransplanted adult liver patients categorized by UCLA risk class

Risk Class	Transplant Candidacy	Predicted Survival	Observed Survival
1	Acceptable	83%	88%
2	Acceptable	67-72%	60%
3	Acceptable	43-53%	55%
4	Unacceptable	20-27%	30%
5	Unacceptable	6%	———

Each variable is assigned an equal weight of 1 point (i.e., 1 point was awarded for adults, organ ischemia > 12h, preop ventilator requirement, T bili > /= 13, and Cr > 1.6). Patients are grouped into risk classes based on the sum total of points obtained.

points obtained. (Table E6.4) For example, for adults in whom only one of the four conditions was present, the predicted survival was 67-72% at 1 year. If all 5 conditions were present, estimated survival was 6% at 1 year. Using this model we have suggested that patients with a score of 4 or 5 (predicted survival < 40% at 1 year) be considered poor retransplant candidates.

Application of this model will theoretically result in improved overall survival following retransplantation and an increase in the efficiency of organ utilization. This would entail excluding from retransplantation a significant percentage of patients currently being transplanted. As with other organs, we do not recommend that these rules be applied to pediatric patients. In addition, in cases of rapid graft failure due to primary non-function or technical difficulties, a transplant surgeon would likely be more compelled to retransplant the patient despite all the evidence pointing to a poorer outcome.

In the year 2002, UNOS modified the liver allocation algorithm in an attempt to de-emphasize waiting time and promote allocation based on disease severity. To accomplish this, disease severity was estimated using a formula (MELD score) initially found to be predictive of 3 month survival in patients with end-stage liver disease who were undergoing TIPS procedure.[13] Further analysis found that it was also predictive of survival in patients awaiting liver transplantation. An analogous formula was developed for predicting survival of children awaiting transplant (PELD score). Although it might be assumed that an allocation schema more based on disease severity than waiting time would favor patients in need of retransplantation because of their severity of illness, this remains to be proven. Additional analysis of this unique subgroup of patients is required to determine whether their interests are appropriately served by the recent modification in liver allocation.

E6

CONCLUSION

It is generally agreed that the liberal retransplantation policies of the past can no longer be justified. The present challenge is to determine which patients should be offered retransplantation, and which should not. Some have proposed that the allocation system should direct all or most organs to primary transplant candidates based on the lower survival statistics of retransplant patients. This approach is clearly unfair as it fails to recognize the heterogeneity of retransplant patients in terms of expected outcome. Some non-physician bioethicists have suggested that allocation should not be determined on the basis of any special obligations that transplant teams might feel toward patients on whom they have already performed transplants — instead suggesting that they abandon those patients in their time of need.[14] They argue that health care workers cannot be expected to recognize when lifesaving methods should be curtailed, and that blinded by their role as patient advocates, the transplant team often makes poor decisions. Such a position denies the fact that such difficult decisions are faced throughout medicine and can only be made by the clinician at the bedside. Efforts to unduly restrict retransplantation would also negatively impact the field in general. For example, current efforts to expand the organ pool by the utilization of marginal donors would be impaired by the loss of the implicit safety net of retransplantation. A universal ban on retransplantation cannot be defended on ethical on practical grounds. The best approach at present is to work to improve the outcome of re-transplanted patients and to define the optimal situations in which retransplantation is appropriate. The use of statistical models to identify patients with unacceptably poor expected outcome has proven helpful – but must be applied with sound clinical judgment on a case-by-case basis.

REFERENCES

1. Michler R., McLaughlin M., Chen J., Geimen R., Schenkel F., Smith C., Barr M., Rose E. Clinical Experience with Cardiac Retransplantation. J Thorac Cardiovasc Surg 1993; 106: 622-31.

2. Dein J., Oyer p., Stinson E., Starnes V., Shumway N. Cardiac Retransplantation in the Cyclosporine Era. Ann Thorac Surg 1989; 48: 350-5.

3. The Registry of the International Society for Heart and Lung Transplantation.

4. Srivastava R., Keck B., Bennett L., Hosenpud JD. The Results of Cardiac Retransplantation: An Analysis of the Joint International Society for Heart and Lung Transplantation/United Network for Organ Sharing Thoracic Registry. Transplantation 2000; 70(4): 606-612.

5. Ranjit J., Chen J., et al. Long-term Survival After Cardiac Retransplantation: A Twenty-Year Single-Center Experience. The Journal of Thoracic and Cardiovascular Surgery 1999; 117 (3): 543-55.

6. Novick R., Stitt L., Al-Kattan K., Klepetko W., Schafers H., Duchatelle J., Khaghani A., Hardesty R., Patterson G., Yacoub M. Pulmonary Retransplantation: Predictors of Graft Function and Survival in 230 Patients. Ann Thorac Surg 1998; 65: 227-34.

7. Novick R., StittL., Schafers H., Andreassian B., Duchatelle J., Klepetko W., Hardesty R., Frost A. Pulmonary Retransplantation: Does the Indication for Operation In-fluence Postoperative Lung Function? J Thorac Cardiovasc Surg 1996; 112: 1504-14.

8. Shaw B., Gordon R., Iwatsuki S., Starzl T. Retransplantation of the Liver. Seminars in Liver Disease 1985; 5: 394-401.

9. Velidedeoglu E, Desai NM, Campos L, et al. The outcome of liver grafts procured from hepatitis C-positive donors. Transplantation 2002; 73: 582.

10. Markmann J., Markowitz J., Yersiz H., Morrisey M., Farmer D., Farmer D., Goss J., Ghobrial R., McDiarmid S., Stribling R., Martin P., Goldstein L., Seu P., Shackleton C., Busuttil R. Long-term Survival After Retransplantation of the Liver. Annals of Surgery 1997; 226: 408-20.

11. Doyle H., Morelli F., McMichael J., Doria C., Aldrighetti L., Starzl T., Marino I. Hepatic Retransplantation – An Analyis of Risk Factors Associated With Outcome. Transplantation 1996; 61: 1499-1505.

12. Markmann J., Gornbein J., Markowitz J., Levy M., Klintmalm G., Yersiz H., Morrisey M., Drazan K., Farmer D., Ghobrial M., Goss J., Seu P., Martin P., Goldstein L., Busuttil R. A Simple Model to Estimate Survival Following Retransplantation of the Liver. Transplantation 1999; 67: 422.

13. Malinchoc M, Kamath PS, Gordon FD, Peine CJ, Rank J, ter Borg PCJ. A model to predict poor survival in patients undergoing transjugular intrahepatic portosystemic shunts. Hepatology 2000; 31: 864

14. Ubel P., Arnold R., Caplan A. Rationing Failure – The Ethical Lessons of the Retransplantation of Scarce Vital Organs. JAMA 1993; 270: 2469-2474.

E6

Essay 7: Noncompliance with Immunosuppressive Regimens

Thomas E. Nevins and Arthur J. Matas

Supported by NIH Grant #13083

INTRODUCTION

As clinical transplantation and the care of transplant recipients have evolved, acute rejection is increasingly an unusual cause of graft loss. Instead, chronic rejection and death with function are currently responsible for the majority of losses. Presented elsewhere in this volume are data showing that the major risk factor for chronic rejection is an acute rejection episode.[1] Within the patient group experiencing acute rejection, those with multiple rejection episodes or whose first rejection episode occurs late posttransplant are at increased risk for developing chronic rejection and late graft loss. Importantly, noncompliance with the immunosuppressive regimen is associated with an increased risk for acute rejection (especially late acute rejection), chronic rejection, and graft loss.

The broadest definition of noncompliance is simply the failure of a patient to follow the advice of their healthcare providers. For the transplant patient compliance includes following complex advice, ranging from diet restrictions and weight goals, to regular attendance at laboratory and clinic visits. Conversely, whether behaviors are called compliance, adherence or persistency, each patient's active and effective participation in their own care is crucial to consistently successful outcomes.

Medication compliance focuses specifically on adhering to the dynamic multi-drug regimens routinely prescribed after transplantation. To better understand medication compliance in perspective, it is informative to consider the Physicians' Health Study.[2] This study recruited over 33,000 male US physicians to prospectively examine the effect of aspirin on cardiovascular events. Of interest, among the 33,200 physician volunteers indicating a definite desire to participate, only 22,000 (66%) regularly took a single daily pill during the study's 18 week run-in period.[3] Thus in spite of education, their obvious motivation and insight, faced with taking a chronic medication these physician volunteers fared no better than the majority of their own patients!

Significantly, post-transplant medication compliance is not a voluntary activity. Although the regimens are complex and the drugs themselves are of varying importance, overall medication noncompliance critically impacts transplant outcomes including acute rejection and graft loss.[4]

Organ Transplantation, 2nd edition, edited by Frank P. Stuart, Michael M. Abecassis and Dixon B. Kaufman. ©2003 Landes Bioscience.

RISK FACTORS FOR NONCOMPLIANCE

Frequently identified risk factors for noncompliance are shown in Table E7.1. Importantly, these risks may occur either early, late, or throughout the duration of followup. For example, a significant problem arising later posttransplant is the cost of the immunosuppressive medications. For almost all patients, medication is paid for (private insurance, Medicare, Medicaid) during the early post-transplant period. However, individual programs may have time limits or benefit maximums and if insurance coverage lapses, some recipients are unable to purchase necessary medications.

In a representative study, Frazier, et al., studied 241 kidney transplant recipients; about half reporting some degree of noncompliance.[5] Of this population, recipients who were younger, female, unmarried, retransplanted, and with lower incomes more frequently reported medication noncompliance (p<.05). Similarly followup compliance was worse for those recipients who were unmarried, had lower income, non-diabetic, or with a longer time since transplant. Patients reporting higher stress and more depression, who coped with stress by using avoidant coping strategies, or who believed that health outcomes are beyond their control were also less compliant with both medications and follow-up. Conversely, the excellent outcomes with spousal renal transplants highlight the importance of loving concern and social support for the patient.[6] Intuitively it seems very clear that significant medication noncompliance is unlikely to occur when a patient is eating breakfast and dinner with their donor.

Rudman, et al, noted that the best predictors of noncompliance were medication side effects, younger age, less education, and lack of comprehensive health insurance.[7] In addition, using a survey guided by protection motivation theory, they found that patients more likely to be noncompliant felt less susceptible to negative outcomes, less able to follow treatment regimens, had less faith in the treatment's efficacy, and were relatively younger. In spite of recognizing these high-risk groups, to date, no studies have demonstrated effective prevention or intervention strategies.

Table E7.1. Frequently identified risk factors for noncompliance

Demographic	- Younger age group (e.g., adolescent)
	- Lower socioeconomic/unemployment
	- Distance from transplant center
Social	- Cost
	- Lack of support (friends or family)
Transplant related	- Number of medications
	- Medication schedule
	- Medication side effects
	- Time from transplant
Psychological	- Depression
	- Behavior problems
Belief pattern	- Lack of belief that medication is related to outcome
	- Lack of belief in taking care of oneself

Although a lack of knowledge about the medications or the belief that they are ineffective is certainly associated with an increased risk of noncompliance, what is truly striking is that most noncompliant recipients seem to understand the importance of the immunosuppressive medication and the risks associated with stopping the drugs or missing doses. Especially for most noncompliant recipients, it seems unlikely that mere forgetfulness resulted in missed doses, instead there seem to be other powerful motivating forces that are as yet poorly understood. A number of studies have shown that the rate of noncompliance increases over time posttransplant. While this may be due to financial concerns, it may also be that recipients truly feel well and it is harder to take medicine when feeling well, or when taking drugs is the one thing that differentiates them from everyone else.

DEFINITION AND MEASUREMENT TECHNIQUES

It would be ideal to have a specific and standard definition of posttransplant noncompliance, yet no such definition exists. Authors may dichotomously group recipients as compliant or noncompliant using a variety of definitions. Others use a percentage of prescribed doses correctly taken, but as above these numbers are often crude estimates. More importantly, it is unknown if there is a "compliance threshold" associated with deteriorating graft function or loss. Is missing 20%, 10%, or even 5% of doses a problem? Is missing a certain percent of doses more of a problem early versus later posttransplant? Is missing successive doses ("drug holiday") more of a problem than intermittently missing single doses? No studies have clearly addressed any of these questions. Similarly, there is no consensus on which end points are most informative. Some studies have used acute rejection episodes as an endpoint, some have looked at the incidence of chronic rejection or graft loss, while still others have measured glomerular filtration rate.

Also a variety of techniques have been used to assess compliance, each has some disadvantages. Pill counts performed at clinic visits assume that the patient has taken the missing pills. But there are numerous reports of patients discarding (or hoarding) medication. Blood levels, especially when a drug has a short half-life (e.g., cyclosporine), can be manipulated unpredictably. Noncompliance has been noted in recipients with both low and high cyclosporine levels (either missing doses or taking an extra dose just before the clinic visit). Chart reviews and patient surveys routinely underestimate the incidence of noncompliance. A chart review can only document noncompliance if it has been previously recorded. Patient interviews or questionnaires are only as accurate as the information being provided. Patients simply can not remember and report forgotten medication doses! Furthermore, personal embarrassment and the desire to "please the physician" lead to substantial under-reporting of consciously missed medications. Precisely for these reasons the most credible patient reports are those documenting noncompliance, clearly an under-reporting bias. Even anonymous questionnaires are of limited use since if anonymous, the degree of compliance cannot be related to an outcome. In addition, while studies using these techniques may be important in elucidating risk factors, they provide data too late to be of any value in an intervention trial.

Electronic medication monitors have some unique advantages. These systems use microchip technology to register the date and time the medication container is opened. This provides a dynamic measure of medication dosing rather than a simple percentage. Still there remain several critical assumptions. The first is that even if the container is opened and closed at the appropriate time, a single medication dose was removed and taken. The second is that taking one medication at the correct time, is an appropriate surrogate for overall medication compliance. While both assumptions are intuitive, they remain unproven. However to date these medication monitors seem to provide the best estimates of medication compliance.[8]

ASSOCIATION OF NONCOMPLIANCE WITH OUTCOMES

Numerous studies indicate associations between noncompliant behaviors and rejection or graft loss. Only a few are summarized here, but an in depth review of the problem has recently been published.[9] In the first large-scale study demonstrating the importance of noncompliance in transplantation, Schweitzer, et al documented that based on chart review and report 18% of recipients transplanted between 1971 and 1984 were noncompliant.[10] Of these, 91% suffered graft rejection or died (vs. only 18% in compliant recipients). In a subsequent prospective study, the authors documented that 15% of patients transplanted between 1984 and 1987 were noncompliant; of these, 17% had graft rejection or died (vs. only 1.3% of compliant patients). Dunn, et al, noted that noncompliance was responsible for 27% of graft losses after the first year.[11]

Rudman, et al,[7] studied 374 adult transplant recipients, noting that noncompliance with clinical followup was associated with more frequent return to dialysis (p<0.01), number of rejection episodes (p<0.01), higher serum creatinine levels (p<0.01), and total number of transplants (p<0.01).

In an analysis critical to our interpretation of reports of long-term transplant follow-up, Gaston, et al, reviewed the records of 91 transplant recipients whose graft loss was characterized as being due to chronic rejection.[12] They found that 7 of the 91 experienced other events that played a significant role in graft loss (e.g., infection necessitating a decrease in immunosuppressive medications). Of the remaining recipients, 48 (58%) had noncompliance as a precipitating factor for development of chronic rejection. Gaston's data suggest that all long-term studies of transplant recipients should identify and quantify noncompliance associated with chronic rejection and graft loss.

DeGeest, et al, interviewed 150 adult kidney transplant recipients; of these, 22.3% had subclinical noncompliance.[13] This subgroup had significantly more late acute rejections (p=0.003) and decreased 5-year graft survival rates (p=0.03). Using electronic monitor technology, the same authors studied 100 heart transplant recipients more than 1 year post-transplant.[14] Using their monitor data, they subdivided the recipients into three groups: 84 excellent compliers, 7 minor subclinical noncompliers, and 9 moderate subclinical noncompliers. Late acute rejection (>1 year posttransplant) occurred in 1.2% of the excellent compliers, 14.3% of the minor subclinical noncompliers, and 22.2% of the moderate subclinical noncompliers (p<0.01). In addition, the authors correlated electronic monitor results with appointment noncompliance and drug holidays.[15] Appointment non-compliance was defined as missing a single clinic appointment in the previous year. A "drug holiday" was defined as not taking any cyclosporine over any 24-hour period in the previous three months. Of the 100 heart transplant recipients, 7% were appointment noncompliers; 43% took 1 drug holiday (vs. 5% of appointment compliers; p=0.01). In addition, 57% of appointment noncompliers experienced 1 late acute rejection (vs. 2% of appointment compliers; p=0.001).

At our institution, Nevins et al,[4] used the same technology to prospectively track 180 kidney transplant recipients for up to 4 years. A medication monitor was used on the azathioprine bottle, and compliance was defined as a single daily bottle opening. Most recipients had better than 90% compliance (median =97.2%). However, during each of the first six months posttransplant, an average of 43% of patients missed at least one dose of azathioprine. When medication compliance was expressed as quartiles, the risks of acute rejection (p=0.006) and graft loss (p=0.002) were highly predicted by worsening compliance. For a select patient subgroup whose compliance declined successively over the first three months; even after adjusting for gender, diabetes, donor source and early rejection, we found a 13-fold increase in the risk of acute rejection (p=0.0011) and a 4.3 fold increase in the risk for graft loss (p=0.03).[4] Conversely, patients in the best compliance quartile experienced no rejections or allograft losses over 3 years of followup!

PREVENTION/TREATMENT OF NONCOMPLIANCE

Today newer medication protocols have minimized the immunologic risk of rejection, however that very progress increasingly highlights the issue of patient noncompliance. At a practical level, the challenge for each physician and caregiver is to become more acutely aware of this clinical problem. At the same time we need to increase both our efforts to improve compliance as well as removing recognized barriers to compliance (Table E7.2).

Clearly, patient groups at higher risk for noncompliance can now be identified by demographic factors or by medication monitor technology. But at present there are no prospective randomized trials in transplant recipients to demonstrate effective strategies to improve compliance. While changing human behavior is difficult, data from the Diabetes Control and Complications Trial suggest that intensive interventions may be associated with overall improvement in compliance and better outcomes.[16]

Unfortunately, intervention studies are difficult and expensive. Still the obvious costs (human and financial) of graft failure and return to dialysis are even higher. Now is an appropriate time to begin such studies in transplant recipients, so every patient may fully benefit from the progress already made in transplantation.

E7

Table E7.2. Techniques to improve medication compliance

Enlist patient participation
- Ask specifically about missed doses (number & reasons)
- Ask about medication-related problems – listen to answers
- Ask specifically about medication side-effects
- Ask about drug costs/payments

Simplify regimen
- Prescribe more "forgiving" (longer acting) drugs
- Minimize number of daily doses (once-twice daily is best)
- Eliminate drugs not clearly needed at this time

Observe / Investigate
- Randomly measure blood levels of drug (note -?erratic levels)
- Consider noncompliance when clinical picture is unusual (lack of clinical response to drugs at usual doses)
- Followup on unusual prescription refill patterns
- Followup on unexplained, missed clinic or laboratory visits

REFERENCES

1. Matas AJ. Relationship between acute and chronic rejection (elsewhere in this volume).
2. Glynn RJ, Buring JE, Manson JE, LaMotte F, Hennekens CH. Adherence to aspirin in the prevention of myocardial infarction. Arch Intern Med 1994; 154:2649-2657.
3. Lang JM, Buring JE, Rosner B, Cook N, Hennekens CH. Estimating the effect of the run-in on the power of the physicians' health study. Statis in Med 1991; 10:1585-1593.
4. Nevins TE, Kruse L, Skeans MA, Thomas W. The natural history of azathioprine compliance after renal transplantation. Kidney Int 2001; 60:1565-1570.
5. Frazier P, Davis-Ali S, Dahl K. Correlates of noncompliance among renal transplant recipients. Clin Transplantation 1994; 8:550-557.
6. Terasaki PI, Cecka JM, Gjertson DW, Cho YW. Spousal and other living renal donor transplants. Clinical Transplants 1997; 269-284,.
7. Rudman LA, Gonzales MH, Borgida E. Mishandling the gift of life: Noncompliance in renal transplant patients. J Applied Social Psychology 1999; 29:835-852.
8. Boudes P. Drug compliance in therapeutic trials: A review. Controlled Clin Trials 1998; 19:251-268.
9. Laederach-Hofmann K and Bunzel B. Noncompliance in organ transplant recipients: A literature review. General Hospital Psychiatry 2000; 22:412-424.
10. Schweizer RT, Rovelli M, Palmeri D, Vossler E, Hull D, Bartus S. Noncompliance in organ transplant recipients. Transplantation 1990; 49:374-377.
11. Dunn J, Golden D, Van Buren CT, et al. Causes of graft loss beyond two years in the cyclosporine era. Transplantation 1990; 49(2):349-353.
12. Gaston R, Hudson S, Ward M, et al. Late renal allograft loss: noncompliance masquerading as chronic rejection. Transplant Proc 1999; 31(4A):21S-23S.
13. DeGeest S, Borgermans L, Gemoets H, et al. Incidence, determinants, and consequences of subclinical noncompliance with immunosuppressive therapy in renal transplant recipients. Transplantation 1995; 59(3):340-347.
14. DeGeest SD, Abraham I, Moons P, et al. Late acute rejection and subclinical noncompliance with cyclosporine therapy in heart transplant recipients. Journal of Heart and Lung Transplantation 1998; 17(9):854-863.

E7

15. DeGeest SD, Abraham II, Vanhaecke J. Clinical risk associated with appointment noncompliance in heart transplant recipients. Circulation 1996; 94(8, Suppl 1):I-180.

16. The Diabetes Control and Complications Trial Research Group. The effect of intensive treatment of diabetes on the development and progression of long-term complications in insulin-dependent diabetes mellitus. New Eng J Med. 1993; 329:977-986.

E7

Essay 8: Relationship between Acute and Chronic Rejection

Abhinav Humar and Arthur J. Matas

Supported by NIH DK13083

With the evolution of immunosuppressive protocols and the improvement in posttransplant care, acute rejection is now only rarely a cause of graft loss. Chronic rejection and death with function have become the predominant causes. Thus, current clinical research focuses on decreasing the incidence of both of these events.

Two major hypotheses attempt to explain pathogenesis of chronic rejection. The first hypothesis is that nonimmunologic factors are primary. Proponents feel that if a kidney with a limited nephron mass (relative to recipient size) is transplanted, hyperfiltration of the remaining nephrons occurs, and that this hyperfiltration results in further renal damage. The second hypothesis is that chronic rejection results mainly from immunologic injury. Proponents note that most of patients with chronic rejection have had a previous acute rejection episode.

These two hypotheses are not necessarily contradictory. There may be a final common pathway for renal injury. Rejection may act by limiting nephron mass, which results in hyperfiltration of the remaining nephrons. (Alternatively, rejection may be an important prognostic factor because acute and chronic rejection have associated underlying immunologic injury, because acute rejection continues to smolder after treatment, or because an acute rejection episode sets off a cytokine cascade that continues to have effects after treatment.)

Arguments for and against both hypotheses exist (reviewed in 1), but a major argument against nonimmunologic factors playing a predominant role is that chronic rejection occurs in extrarenal transplants (where hyperfiltration cannot be implicated) and the histologic picture of chronic rejection in extrarenal transplants is similar to that in renal transplants. We recently analyzed our kidney transplant population to study the relative role of immunologic and nonimmunologic factors. In our analysis, we excluded our recipients with graft loss due to death with function, technical failure, primary nonfunction, and recurrent disease—leaving a group of 1,987 recipients with potentially both immunologic and nonimmunologic graft loss.[2] For this group, 10-year graft survival was 72%. We then repeated the analysis, this time also excluding recipients who had had acute rejection episodes; presumably graft loss in the remaining group (n=1,128) would be primarily due to nonimmunologic causes. Importantly, overall 10-year graft

Organ Transplantation, 2nd edition, edited by Frank P. Stuart, Michael M. Abecassis and Dixon B. Kaufman. ©2003 Landes Bioscience.

survival for this group was 92%, suggesting that, in kidney transplantation, nonimmunologic factors are responsible for only a small percent of graft loss.

Numerous studies have demonstrated the association between acute rejection, chronic rejection, and graft loss.[1,2] This data may suggest that eliminating acute rejection could potentially eliminate chronic rejection and much late graft loss. Certainly, there has been a significant decrease in the incidence of chronic rejection, which has paralleled a decrease in the acute rejection rates. With our current immunosuppressive regimens, rates for acute rejection have fallen below 10%; yet, a similar substantial decrease has not been seen in chronic rejection rates. Of 701 recipients transplanted at our center between 1992-1996, the 6-month incidence of acute rejection was 31%, and the 5-year incidence of chronic rejection was 15%. Amongst 741 recipients transplanted more recently, from 1997 to 2001, 6-month acute rejection rates had decreased to 8% (p=0.001) but 5-year chronic rejection rates had remained stable at 14% (p=ns). By multivariate analysis, however, acute rejection remained the major risk factor for the development of chronic rejection in both or these time groups. A number of possible factors may explain these findings. It is possible that we have reached the limits of our current pharmacologic immunosuppression with regards to acute rejection rates and total elimination of rejection may be associated with unacceptable morbidity and mortality. The absence of a more significant drop in the chronic rejection rate suggests either that non-immunologic factors are having more of an impact or that the consequences of breaking through on modern-day immunosuppression (i.e. having an acute rejection episode) may be more significant than breaking through with lesser degrees of immunosuppression.

But, not all patients with acute rejection develop chronic rejection. In fact, most patients with a single acute rejection episode do not develop chronic rejection. Data from our center is shown in Table E8.1. For first transplant recipients who have a single early acute rejection episode, <10% develop chronic rejection; for those with late rejection or multiple rejection episodes, the rate of chronic rejection is higher.

What, then, are the risk factors for development of chronic rejection within the subgroup having an acute rejection episode? As shown in Table E8.1, a late rejection episode or multiple acute rejection episodes are significant risk factors. In addition, we and others have shown that the kidney biopsy at the time of the rejection episode is prognostic. In our series, 229 recipients with a single acute

Table E8.1. Percent of kidney transplant recipients with an acute rejection episode who do not develop chronic rejection

	% Chronic Rejection-Free	
	Living Donor	Cadaver Donor
1 Rejection Episode	88%	86%
<90 days	93%	92%
>90 days	73%	70%
>1 Rejection Episode	44%	46%

rejection episode were divided by their histopathologic assessment: mild, moderate, or severe tubular infiltrate with or without a vascular component.[3] Those with a mild or moderate infiltrate but without a vascular component (n=164) had a 17% incidence of biopsy-proven chronic rejection and a 78% five-year graft survival rate. In contrast, those with a severe infiltrate or with a vascular component (n=65) had a 27% incidence of biopsy-proven chronic rejection (p=0.02) and a 64% five-year graft survival rate (p=0.005).

Other measures of the severity of the rejection episode (e.g., changing glomerular filtration rate, steroid-sensitive vs. -resistant, and lack of return of creatinine to baseline) have also been associated with worse outcome.[1]

Thus, kidney transplant recipients with each of these characteristics could be targeted for changes in immunosuppressive therapy. Prospective studies are necessary to determine whether such changes will result in a decreased incidence of chronic rejection.

The role of noncompliance in recipients with multiple rejection episodes or late rejection needs to be considered. Noncompliance is discussed elsewhere in this volume. In our series of kidney transplant recipients, noncompliance was increased in recipients with late acute rejection episodes and in those with >1 acute rejection episodes (that is, the two groups at greatest risk for chronic rejection). Other investigators have noted this same finding. Thus, noncompliance should be considered in all recipients with rejection. Only if noncompliance is suspected can appropriate attempts at intervention take place.

While the above discussion on the relationship between acute and chronic rejection is concerned specifically with kidney transplant recipients, similar findings may or may not hold true for other organ transplant recipients. For example, chronic rejection after liver transplantation is relatively uncommon with modern day immunosuppression, and there is no good data to support a significant association between acute rejection episodes and eventual development of chronic rejection in these patients.

Pancreas transplant recipients, however, show a very strong association between episodes of acute rejection and eventual graft loss to chronic rejection [4]. Short-term results with pancreas transplants have improved dramatically in the last 10 years due to improvements in surgical techniques and immunosuppressive agents. As an increasing number of grafts continue to function beyond the first year posttransplant, chronic rejection is becoming the major cause of graft loss in these cases. Chronic rejection in pancreas transplant recipients is defined by a typical clinical course coupled with characteristic histologic findings. The clinical course is characterized by a gradual deterioration in pancreas graft function beginning at least 2 months posttransplant. The exocrine component is usually affected first (manifested by falling urine amylase levels in bladder-drained grafts), followed by the endocrine component (manifested by episodes of hyperglycemia and the need for insulin therapy). Histologically, the process is characterized by arteriopathy with concentric narrowing of the small vessels and parenchymal fibrosis with atrophy of acini.

In an analysis of 914 cadaver pancreas transplants performed at our center between 1999 and 2002, technical failure was found to be the most common cause of graft loss, accounting for 120 (13.1%) failed grafts. The second most common cause was chronic rejection, accounting for 80 (8.8%) of the failed grafts. The incidence of graft loss to chronic rejection was highest after isolated pancreas transplants (vs. simultaneous with a kidney). By multivariate analysis, the most significant risk factors for graft loss to chronic rejection were a previous episode of acute rejection (RR=13.75, $p < 0.0001$), an isolated (vs. simultaneous) transplant (RR=3.66, p=0.0004).

In summary, for the majority of transplant recipients, acute rejection and chronic rejection seem to be closely linked. Acute rejection episodes in these recipients represent the major risk factor for the eventual development of chronic rejection. As rates for acute rejection continue to decline, it will be interesting to see what impact this will have on the process of chronic rejection.

REFERENCES

1. Matas AJ. Risk factors for chronic rejection—a clinical perspective. Transplant Immunol 1998; 6:1-11.
2. Humar A, Hassoun A, Kandaswamy R, et al. The questionable importance of nonimmunologic risk factors in long-term graft survival. JASN 1998; 9:7-8A.
3. Chavers BM, Mauer M, Gillingham KF, et al. Histology of acute rejection (AR) impacts renal allograft survival (GS) in patients (pts) with a single rejection episode (SRE). J Am Soc Nephrol 1995; 6:1076.
4. Drachenberg CB, Papadimitriou JC, Klassen DK, Weir MR, Cangro CB, Fink JC, Bartlett ST. Chronic pancreas allograft rejection: morphologic evidence of progression in needle biopsies and proposal of a grading scheme. Transplant Proc. 1999;31(1-2):614.

E8

Essay 9: Elective and Emergency Surgery in the Stable Transplant Recipient

Amy L. Friedman, Giacomo P. Basadonna and Marc I. Lorber

INTRODUCTION

Although stable transplant recipients enjoy adequate allograft function, and an immunologic steady (or at least meta-stable) state, development of new medical problems and/or the requirement for new interventions to remedy those problems, threaten their precarious homeostasis. Resolution of acute medical challenges without disturbing the immunologic "status quo" is best accomplished under the direction of physicians with specific expertise managing immunosuppressed transplant recipients. Surgical intervention (elective or emergent) should be undertaken, whenever possible by individuals experienced in the care and management of these challenging patients. Beyond application of the general principles of good patient care, it is important to evaluate new problems within the context of the transplant recipient's immunosuppressed state, to anticipate factors which might impact transplant organ function, to ensure the appropriate use of antibiotics, and to maintain proper immunosuppression. This discussion will review various surgical illnesses commonly faced by transplant recipients, and specific management points will be considered.

GENERAL CONSIDERATIONS

The overall management of surgical disease after transplantation generally follows the fundamental principles appropriate for all surgical illnesses. However, among important considerations the effects of immunosuppressant medications that can result, at times, in serious underestimation of disease severity must be emphasized. Accordingly, a high level of suspicion is always important, and occasionally it becomes necessary to proceed with operative exploration to establish a firm diagnosis.

Additionally, when a disease management algorithm might reasonably include a conservative versus a bold option, we generally favor the less risky approach. This offers the likelihood of more frequent success, recognizing that immunosuppressed transplant recipients enjoy diminished reserve when compared to average patients. For example, a transplant patient with an intestinal perforation or necrosis is usually better managed with segmental resection and creation of an "ostomy," when

Organ Transplantation, 2nd edition, edited by Frank P. Stuart, Michael M. Abecassis and Dixon B. Kaufman. ©2003 Landes Bioscience.

E9

primary anastomosis might be considered in an otherwise healthy patient. Experience has taught that the tolerable margin of error among immunosuppressed patients is reduced, therefore management must be directed appropriately. Consequently, it has become our strong recommendation that serious diseases, particularly those requiring invasive procedures of a significant magnitude, should be managed at a transplant center with the full participation of the transplant team.

PREOPERATIVE PREPARATION

When an operation is planned in advance, preparation of the transplant recipient does not differ substantially from other patients. Maintenance medications, especially the immunosuppressants, antihypertensives and cardiac drugs should be administered with a minimal volume of water. Diabetics should reduce the dose of long and intermediate acting insulin, usually by one half. If hypoglycemia occurs or the start of the operation is delayed, intravenous dextrose must be initiated. When bowel preparation is necessary, preoperative overnight hospitalization to include intravenous fluids is important to avoid dehydration and the potential for intra- operative hypotension.

Antibiotic prophylaxis is important for invasive procedures transplant recipients might undergo, and therapy should be tailored according to the expected risk of contamination. Penicillin or ampicillin are appropriate for dental procedures, whereas broader spectrum coverage to include gram positive and gram negative bacteria is warranted for intra- abdominal procedures. When the anticipated period of bacteremia is brief and self-limited (e.g., following dental cleaning or laparoscopic cholecystectomy), we recommend initiation of antibiotic coverage just prior to the procedure, continuing for 24 hours afterwards. When ongoing microbial seeding is anticipated, a prosthetic material is implanted, or residual foci of infection or necrotic material might remain, the antibiotic course should be extended (Table E9.1).

INTRAOPERATIVE CONSIDERATIONS

The choice of suture materials and method of incisional closure are influenced by the effects of immunosuppression and underlying systemic illness on wound healing. Virtually all transplant recipients will experience delayed development of adequate tensile strength, therefore whenever feasible we recommend non-absorbable suture materials. When an absorbable suture is selected, monofilament, synthetic absorbable materials are preferable, because they maintain higher tensile strength over a longer duration. Also because of concern regarding delayed wound healing, skin sutures (or staples) should remain in place for 2-3 postoperative weeks.

MANAGEMENT OF IMMUNOSUPPRESSION

MAINTENANCE

Particularly since the introduction of tacrolimus (TCL) and the microemulsion cyclosporine (CY) preparation, Neoral®, efforts to maintain oral immuno-suppression administration whenever possible have become the usual practice. However, a practical challenge to successful administration results when diseases

E9

Table E9.1. Guidelines for antibiotic prophylaxis in stable transplant recipients

None	Percutaneous biopsy
	Incision & drainage of superficial abscess (no cellulitis)
	Cystoscopy
	Endoscopy (without extensive tissue manipulation)
One preoperative dose	Percutaneous nephrostogram
	ERCP
	Placement of temporary foreign body (e.g., Hickman cath)
One preoperative dose	Indwelling vascular prosthesis (e.g., prosthetic graft)
plus 1-5 postoperative days	Joint replacement
	Prosthetic abdominal wall mesh
	Dental surgery

---

requiring surgical intervention are encountered. Small bowel obstruction, or paralytic ileus are important considerations in this regard.

Although both cyclosporine and tacrolimus are available for intravenous use, the cremophore vehicle used to solubilize these agents has been associated with important toxicity. Therefore, even when clinical situations suggest gastrointestinal absorption may be compromised, including those causing increased motility (e.g., diarrhea), decreased absorption (e.g., small bowel obstruction) or during external biliary drainage, efforts to persist with oral administration are usually exhausted before proceeding with parenteral dosing. It is frequently possible to achieve adequate circulating drug levels when Cy or Tcl is administered through a nasogastric tube (the tube is clamped for 30-60 minutes). Cyclosporine or tacrolimus levels are monitored daily when concerns regarding absorption are apparent. Intravenous Cy or Tcl is usually reserved for situations when poor absorption has been documented with low circulating drug concentrations.

Azathioprine is also available as an intravenous preparation; it is administered at an equivalent (mg for mg) dose.

Mycophenolate mofetil is not clinically available in a parenteral form. Accordingly, this drug is usually transiently discontinued when oral administration is not feasible.

CORTICOSTEROIDS

Corticosteroids are converted to an appropriate dosage form using methylprednisone (5 mg for each 1 mg of hydrocortisone) or hydrocortisone (4 mg for each 1 mg of prednisone) when intravenous administration is required.

Because most posttransplant immunosuppression includes administration of chronic corticosteroids, when transplant recipients develop emergent problems requiring surgical intervention (e.g., perforated ulcer, motor vehicle accident. etc.) or even for a planned operation (e.g., herniorraphy), the physiologic response to stress must be considered. Accordingly, it is important to consider corticosteroid management as an integral component of the care algorithm. Although the penalty associated with temporary administration of "stress" corticosteroids is relatively small, available evidence suggests this is not usually necessary. Transplant recipients receiving ≥ 10 mg/day, prednisone rarely require additional steroids to accommodate acute stress.[1] Occasionally, individuals receiving lower doses require additional corticosteroid dosing. We generally treat patients who experience highly stressful circumstances (e.g., hypotension, septic shock, coronary bypass surgery) with hydrocortisone, 100 mg, every 8 hours during maximal stress with a rapid taper to maintenance when the stressful situation subsides (usually 3-5 days). It is also important to recognize that the diagnosis of "stress" is highly subjective; the clinician must have familiarity with the signs and symptoms of adrenal insufficiency (Table E9.2). Whenever a patient presents a clinical setting with consistent clinical signs or symptoms, such as unexplained hypotension or hyperkalemia, adrenal insufficiency should be prominently considered in the differential diagnosis. Rapid resolution after an initial intravenous dose of hydrocortisone, 100 mg, can be considered diagnostic in such a setting.

WITHHOLDING IMMUNOSUPPRESSION

In life threatening circumstances it is occasionally important to temporarily discontinue immunosuppression, hoping to permit resolution of an acute medical problem. For example, profound sepsis failing to respond to usual measures, when posing an immediate threat to survival, may justify cessation of transplant related immunosuppressants. When such a decision is made, the clinical course must be closely observed. Immunosuppression should be resumed concomitant with early clinical recovery to avoid acute allograft rejection. Decision making in situations such as this becomes quite difficult, because specific therapeutic guidelines are remarkably elusive. Rather, highly subjective clinical judgement is extremely important, particularly when considering reintroduction of immunosuppression.

Similar clinical judgement is also required during administration of chemotherapy in treatment of malignancy. We have discontinued immunosuppression for prolonged periods on several occasions with successful control of the underlying process and without compromise to allograft function. Again, timing becomes critical when considering the reintroduction of immunosuppression.

E9

HEAD AND NECK SURGERY

Tracheostomy is usually performed when patients require prolonged ventilator support. These patients are always seriously ill, and they are usually faced with other comorbid processes as well. Specific changes in immunosuppression are usually not required as a consequence of the operation itself. However, the expectation of poor

---

Table E9.2. Signs and symptoms of adrenal insufficiency

Refractory hypotension
Hyperkalemia
Hyponatremia
General malaise
Severe fatigue

wound healing, and the risk of infection represent important technical challenges. It is therefore advisable to keep the operative incision as small as is practical, and meticulous wound care is similarly important. Recognition of complicating underlying systemic illness in the context of the required immunosuppressive medications represent important considerations when estimating relative operative risk. When faced with particularly high risk patients, we have occasionally used the percutaneous tracheostomy technique with success.

Additionally, head and neck malignancies have been particularly challenging after transplantation. These are usually squamous cell cancers, and an aggressive approach including early biopsy to establish a specific tissue diagnosis is recommended. When the initial procedure can be performed under local anesthesia, no specific preparation is required. However, when general anesthesia is necessary for adequate examination (e.g., endoscopy) and biopsy, or when a mass threatens airway maintenance, the usual dose of oral immunosuppressive medications should be given preoperatively to avoid missed doses. Additionally when a more complex procedure is required, antibiotic prophylaxis covering oropharyngeal flora is recommended.

When the diagnosis of malignancy is established, a reduction in the immunosuppressive regimen should be considered in developing the treatment plan. A specific discussion with the patient, covering practical and theoretical issues concerning ongoing immunosuppressive management should allow ample opportunity for informed decision making. A well informed kidney or pancreas recipient might elect to discontinue immunosuppression, accepting the risk of rejection or even graft loss, to maximize the anti-neoplastic strategies. Alternatively, s/he may elect to continue maintenance immunosuppression, except perhaps for a brief time during chemo- or radiation therapy. Although anecdotal, our experience has suggested that most patients elect reduced immunosuppression, but only occasional individuals have concluded that permanent discontinuation of immunosuppression is desirable. If the patient and/or family are unable to participate knowledgeably, our preference is to use a reduced level of immunosuppression. We have concluded that the negative impact of acute graft loss detracts sufficiently from the quality of life that most patients are most comfortable with this approach. Additionally, avoiding acute rejection due to rapid reduction or cessation of immunosuppression seems logical when developing a therapeutic plan for patients with malignant disease. The stakes are even higher for orthotopic allograft recipients (heart, lung, liver) where certain death will result if the transplanted organ undergoes acute rejection and fails. Here too, we have usually been successful using lower doses of Cy or Tcl plus low dose prednisone.

Major head and neck resections also have important therapeutic implications for transplant recipients. Here, the route of immunosuppressant administration becomes an important consideration, because oral intake will be disrupted. Whenever possible, we advocate intra-operative placement of a nasogastric or nasoesophageal tube to allow early enteral medications. When early oral alimentation is impractical, placement of a gastrostomy or jejunostomy should be considered. Only when the gastrointestinal tract cannot be used do we routinely advocate parenteral immunosuppression.

DENTAL PROCEDURES

The importance of daily oral hygiene, and regular dental prophylaxis cannot be over emphasized. Patients must maintain good daily oral hygiene, and they should have regular interval cleaning and examinations. Importantly, when problems are identified they must be promptly addressed. It is mandatory to use antibiotic prophylaxis for cleaning, as well as more invasive dental procedures. Specific recommendations should follow established guidelines for antibiotic prophylaxis in cardiac disease.

An important side effect of cyclosporine is development of gingival hyperplasia.[2] This problem is well controlled with careful dosing, avoidance of medications that exacerbate gingival hyperplasia, and regular dental prophylaxis. However, this problem can become sufficiently severe to warrant extensive gingival resection. Cyclosporine dose reduction and when possible reduction or elimination of calcium channel blocking agents has been helpful. It has rarely been necessary to discontinue cyclosporine, usually with conversion to tacrolimus based immunosuppression for specific management of gingival hyperplasia.

OPHTHALMOLOGY

Most eye surgery is performed using local anesthesia. Accordingly, alteration of the standard posttransplant regimen is only rarely required. Cataracts occur more frequently and progress more rapidly in transplant patients, presumably due to chronic corticosteroid therapy. Laser photocoagulation, vitrectomy and retinal reattachment procedures are common among diabetic recipients; these procedures should not be postponed in the stable patient.

PULMONARY PROCEDURES

Lung or mediastinal procedures are sometimes required to manage infectious or neoplastic disease after transplantation. Again, we favor aggressive pursuit of definitive diagnosis when an abnormality is identified on x-ray or when apparently routine infections fail to resolve after appropriate initial treatment. Bronchoscopy, open lung biopsy, and less commonly formal pulmonary resection can usually be accomplished without disruption of oral immunosuppression. Antibiotic coverage should include oropharyngeal flora, plus common pulmonary pathogens.

CARDIAC SURGERY

The high incidence of coronary disease among transplant recipients has become increasingly more important as patients have enjoyed longer allograft survival. Some have speculated that this may reflect hyperlipidemia observed in association with immunosuppression. Certainly, the natural history of underlying diseases such as diabetes mellitus and hypertension are also important.

Because the clinical signs of vascular insufficiency are frequently masked by drug induced suppression of local and systemic inflammation, as well as by autonomic neuropathy, especially when considering diabetics, it is important to

E9

maintain a high index of suspicion. Too frequently, otherwise stable transplant recipients present with advanced disease and require urgent intervention.

Cardiac catheterization should be undertaken after thoughtful consideration, but when indicated it is important to proceed. Renal function must be carefully monitored in these patients, particularly because most receive potentially nephrotoxic agents (cyclosporine or tacrolimus). When transplant recipients undergo cardiac catheterization, adequate intravascular hydration must be maintained and nephrotoxic contrast volume should be minimized. With appropriate attention to these details, catheterization can be accomplished safely.

When cardiac operation is necessary, transplant recipients present several management challenges. Cannulation for cardiopulmonary bypass must assure adequate perfusion of the transplanted organ, and catheter trauma to the relevant anastomoses must be avoided. Similarly, the choice of a prosthetic material for valvular replacement might require special consideration. For example, reoperation may be particularly undesirable in these patients because of relatively poor healing, and the challenges of immunosuppressant management. Therefore, an artificial, instead of a porcine valve might be selected.

Transient postoperative deterioration of renal function is relatively common after cardiopulmonary bypass. Accordingly, it is important to maintain intravascular volume during the early postoperative period. Management may require continuing right heart monitoring for a longer postoperative period to provide objective parameters for judicious fluid replacement. Again, we recommend enteral dosing of maintenance immunosuppressive drugs whenever feasible. Additionally, it is important to consider potentially important drug interactions when prescribing medications for patients with cardiac disease.

GASTROINTESTINAL SURGERY

Abdominal operations can require considerable planning in transplant recipients. Potential allograft injury from operative trauma, or by compromise of its vascular supply must be avoided. Again, when possible enteral immunosuppression should be maintained. Hepatobiliary operations, pancreatic procedures, and splenectomy do not usually require perioperative change to intravenous immunosuppression. However, gastric resection or repair, and other intestinal procedures usually require a period of intravenous cyclosporine or tacrolimus.

Acute appendicitis can provide a challenging diagnostic dilemma after transplantation. This diagnosis should be prominently considered when evaluating patients with abdominal pain and tenderness. Diverticulitis and/or perforation of the sigmoid colon is also relatively common. Again, immunosuppressed patients may present with relatively advanced disease requiring emergent operation. When colon resection is required, fecal diversion using an end colostomy provides the most conservative, and therefore the most desirable approach. We usually wait at least 6-8 weeks before considering elective stoma closure in this population.

We have previously observed that biliary calculus disease is associated with cyclosporine administration, possibly related etiologically to cyclosporine induced cholestasis.[3] Therefore, even after cholecystectomy, de novo biliary stone formation should be strongly considered when transplant recipients develop right upper quadrant pain or unexplained jaundice. Cholecystectomy (open or laparoscopic) is usually well tolerated, and this procedure should be promptly undertaken when it is indicated. External biliary drainage, when necessary, rarely requires alteration of the immunosuppressive regimen when cyclosporine microemulsion or tacrolimus are used.

GYNECOLOGIC PROCEDURES

Similar to the general population, routine gynecologic care is an important part of health care maintenance for stable female transplant recipients. Some problems encountered are similar, but some are distinct from the general population. Mucocutaneous herpes infections from herpes simplex virus type 1 or 2 and condylomata acuminata from human papilloma virus are relatively common. Occasionally, extensive disease may warrant systemic antiviral therapy. Surgical resection is occasionally necessary, and this can be undertaken without jeopardy to patient or graft survival.

The incidence of cervical dysplasia and/or neoplasia is increased among immunosuppressed women, and some authors have suggested an etiologic relationship between the effects of immunosuppression and growth of the human papilloma virus.[4] As with other histopathology, an abnormal Papanicolaou smear should be managed aggressively. We recommend a cervical cone biopsy for definitive diagnosis. When indicated hysterectomy can be safely undertaken, but it is necessary to consider the anatomic variations caused by the kidney and/or pancreas transplantation. Often, the intra-operative participation of an experienced transplant surgeon can be helpful in this regard.

Ovarian pathology should also be pursued aggressively. Although laparoscopic management is made more challenging by the presence of adhesions from prior abdominal surgery, and in the case of renal and/or pancreas transplantation by the mass effect of the allograft, this approach to management of pelvic pathology is usually successful.

Pregnancy after organ transplantation is a complex topic, and a detailed discussion is beyond the scope of this essay. However, increasingly transplant recipients have delivered healthy babies. An experienced, high risk obstetrician and an experienced transplant physician should, collaboratively manage pregnancy in the transplant setting. Cyclosporine or tacrolimus dose adjustment is usually required to maintain target levels, and caesarean delivery has commonly been necessary. Importantly, this has been accomplished without jeopardizing the baby, the mother, or the allograft. We have not seen unusual complications of vaginal deliveries.

UROLOGIC PROCEDURES

Urologic procedures are occasionally necessary; they are well tolerated by the stable transplant recipient. Malignancy in one or both native kidneys can become particularly challenging, since hematuria, a cardinal sign, is often absent in small atrophic kidneys. When a suspicious lesion is identified (often as an incidental finding on a CT or MRI), we favor nephrectomy to establish the diagnosis. The procedure is usually well tolerated; in contrast a biopsy can be difficult to perform, and this often fails to provide a definitive diagnosis.

Cystoscopy is recommended when persistent hematuria is documented without clear etiology, as well as in the setting of recurrent infection. Removal of retained suture material is occasionally necessary; this can be safely accomplished via the cystoscope. Additionally, endoscopic evaluation of the urethra and stricture dilation is often comfortably performed in the outpatient setting.

Prostatic resections, usually for benign hypertrophy, have become more common as the transplant recipient population has aged. We favor aggressive application of this procedure to avoid irreversible obstructive damage to the renal transplant.

VASCULAR SURGERY

Vascular insufficiency is also common among transplant recipients, particularly those with diabetes and/or longstanding hyperlipidemia. When faced with serious peripheral arterial disease, it is sometimes important to favor amputation over distal bypass, particularly among fragile patients with multiple comorbidities. Alternatively, the likelihood of successful ambulation with a prosthesis may be reduced in some patients; in these settings aggressive attempts at limb salvage becomes substantially more important.

When revascularization requires manipulation and/or transient occlusion of distal vessels without ischemia to the allograft, no specific alterations to the usual treatment protocol are necessary. On the other hand, proximal occlusion of arterial inflow to an allograft may result in serious morbidity. The guiding principles include maintenance of transplant perfusion whenever possible. Mannitol is recommended to induce brisk diuresis, as well as for its protective properties as an oxygen free radical scavenger.

When managing aneurysmal disease, we favor aggressive early intervention using established principles of size, growth, etc. When transplant recipients require aortic replacement, outcomes have been excellent. However, the importance of meticulous attention to detail during the perioperative, operative and postoperative period, including optimal maintenance of allograft perfusion cannot be stressed too vigorously.

PLASTIC SURGICAL PROCEDURES

Major reconstructive procedures are occasionally necessary to repair abdominal wall defects resulting from peri-transplant wound complications. Since both diminished wound healing and posttransplant weight gain usually stabilize within the first posttransplant year, we favor reserving these interventions until that time. When a significant hernia is repaired, increased intra-abdominal pressure may occasionally compromise allograft function. Accordingly, intra-operative participation of a transplant surgeon is often helpful.

Wide debridement of infectious or malignant lesions can require creative reconstruction with skin and muscle flaps. Collaborative care including the plastic and transplant surgeons will afford optimal patient outcomes.

Demand for cosmetic surgical procedures has been relatively low among the transplant recipient population. While the impact of appearance on quality of life should not be underestimated, the benefits of such elective interventions must be carefully weighed against the risks of infection and poor wound healing in the transplant setting.

PERSPECTIVE

As greater numbers of people enjoy long-term success after organ transplantation, the need to properly manage nontransplant surgical disease has become increasingly apparent. Similar to the experience with transplantation itself, important insight into optimal management strategies have evolved over the past ten to fifteen years. The underlying principles of surgery remain applicable, and specific approaches are frequently identical to those for the general population. However, there are also important differences in management of some specific problems, and there are always important management peculiarities that result from the immunosuppressed state. Immunosuppressed transplant recipients live in a precarious homeostatic balance with unquestionably less reserve margin when compared to the general population. However, careful attention to detail and collaborative care including appropriate surgical specialists and the transplant team, have been demonstrated to afford excellent outcomes for this challenging group of surgical patients.

REFERENCES

1. Bromberg JS, Baliga P, Cofer. Stress steroids are not required for patients receiving a renal allograft and undergoing operation. J Amer Coll Surg 1995; 180 (5):532-6.
2. Dodd DA. Rapid resolution of gingival hyperplasia after switching from cyclosporine A to tacrolimus [letter]. J Heart & Lung Transpl 1997; 16(5):579.
3. Lorber MI, Basadonna GP, Friedman AL. Pathophysiology and treatment of gallstone disease in organ transplant recipients. J Neph 1996; 9(5):225-231.
4. Bornstein J. Rahat MA. Abramovici H. Etiology of cervical cancer: Current concepts. Obstetr Gynecol Surv 1995; 50(2):146-54.

E9

Essay 10: Dental Issues before and after Organ Transplantation

Peter Hurst

Despite the dramatic reduction in dental disease over the past three decades, resulting principally from fluoridation of public water supplies and increased public education, the mouth remains the most commonly infected site in the body. Almost all adults have some subacute, asymptomatic source of oral infection, arising most frequently as a result of chronic periodontal disease. Added to this is the fact that less than half the population chooses to receive regular maintenance care and instead seek treatment only episodically when pain or other acute symptoms force them to see a dentist. This means that a vast amount of untreated, often asymptomatic dental pathology exists in the population at large. For patients in robust health even well established, chronic oral infections can be tolerated with few, minor systemic symptoms. This, however, is not true of organ transplant patients who following successful immunosuppression can quickly and easily be overwhelmed by a fulminating oro-facial infection of odontogenic origin.

To reduce the possibility of serious oral problems arising during and after organ transplantation, it is prudent that all candidates for whom transplantation of any organ is proposed, should be first screened and, if necessary, treated by a dentist who is familiar with the specific problems facing this group of patients. Ideally, the best person to care for transplant patients is a hospital based dentist who works closely with the transplant team, and who understands fully the oral complications which can arise following organ transplantation and immuno-suppression. This individual can best ensure that any necessary oral care is carried out in a safe, timely and efficient manner. Once the transplant patient has been stabilized, there is no reason why the patient cannot return to the care of a dental general practitioner in the community, provided he/she understands their special status and needs.

The dentist's role in the care of organ transplant patients can be broadly divided into two phases, the pretransplant evaluation and treatment and post-transplant evaluation, maintenance and treatment.

PRETRANSPLANT EVALUATION

The initial oral screening and treatment planning of all prospective organ transplant patients should aim to identify and document any existing hard or soft tissue pathology. A complete and comprehensive clinical examination of all the teeth and oral structures, including full mouth periodontal probing should be carried

E10

Organ Transplantation, 2nd edition, edited by Frank P. Stuart, Michael M. Abecassis and Dixon B. Kaufman. ©2003 Landes Bioscience.

out. A thorough head and neck examination is also necessary to detect any problems associated with the surrounding structures. A full series of intra-oral x-rays and, a panoramic survey is also essential.

The list below outlines the most frequently identified problems that require treatment:

1. Decayed, broken-down teeth should be evaluated as to their restorability.
2. Nonvital teeth and those with periapical pathology should be assessed as to their potential for successful endodontic treatment.
3. Periodontal disease with associated pocketing and bone loss should be evaluated as to the possibility of successful long-term treatment. It should be borne in mind that periodontal disease can be notoriously difficult to control in an otherwise healthy patient, let alone in an immunosuppressed transplant patient.
4. Impacted and partially erupted teeth with associated soft tissue infection should be considered for extraction.
5. Traumatic ulceration arising from ill-fitting dentures or other appliances should be evaluated and the appliances adjusted accordingly.
6. Dental implants with evidence of infection should be removed.
7. Oral ulcerations and other oral mucous membrane disease should be evaluated, biopsied as necessary and treated appropriately.
8. Nonodontogenic, radiolucent lesions of the jaws should be identified, evaluated and, if necessary, treated.

In treatment planning for this group of patients, the goal is not simply to identify existing pathology, but rather to assess all areas for any potential future problems. The treatment planning process, therefore, should be more, rather than less, aggressive. In this way, it is anticipated that the oral cavity will remain stable for many years without further invasive treatment being necessary.

In general, patients who have shown little interest in their dental health in the past should not be treatment planned for heroic reconstructions, but rather should have problematic teeth extracted and dentures made. The same is true for those patients with established, progressive, periodontal disease where future treatment following immunosuppression is likely to have poor and unpredictable results.

E10

Patients with healthy dentitions should be told of the need for extra vigilance in the care of their mouths after transplantation. All patients should be placed on a frequent recall protocol which includes regular, professional cleanings, fluoride treatments, antibacterial rinses, and frequent, complete and comprehensive examinations b a dentist to detect any new changes in their oral status.

PRETRANSPLANT TREATMENT
Ideally, before transplantation, all active dental disease should be treated and acute and chronic infection eradicated. Impacted, periodontally involved and unrestorable teeth should be extracted. Nonvital teeth and those with periapical pathology should be endodontically treated or if a predictably successful outcome cannot be guaranteed, then extracted. Meticulous oral hygiene regimes should be established, taught and constantly reinforced for all patients.

While the pretransplant goals for the oral care of any organ transplant patient are essentially the same, namely to eradicate all oral infection and eliminate other significant oral pathology, the degree to which this can be accomplished in each different type of organ transplantation situation may vary considerably.

Hopefully, patients scheduled to undergo organ transplantation are stable enough medically to undergo any necessary dental treatment, including multiple extractions. This, however, is not always true, particularly in the case of patients who are suffering from end-stage renal, cardiac or hepatic disease. The frail and unstable status of many of these very sick patients often makes them poor candidates for even the most basic and straightforward dental procedure. Patients suffering from end-stage renal disease, for example, who are being maintained on dialysis, pose significant medical management difficulties that can seriously interfere with their dental treatment. The complexity of their fluid and electrolyte imbalance problems, their coagulopathies, their susceptibility to infection, their persistent anemia, their inability to metabolize and excrete certain medications and the possibility of septic emboli developing in their arterio-venous shunts, puts them at significantly increase risk during dental treatment. In patients scheduled for transplantation of other organs, their complicating medical problems may be quite different, but are no less important. When patients are seriously ill, it is still prudent to identify ahead of time any oral disease present even if definitive treatment has to be delayed until medical stabilization of the patient is achieved after transplantation.

In deciding which patients are fit enough to receive dental care prior to transplantation, the dentist and physician should consult together about the proposed treatment and its risks vs. benefits to the patient. Issues to be discussed include; the ability of the patient to tolerate the planned procedures, the need for additional pretreatment screening tests if surgical treatment is indicated; e.g., bleeding time, prothrombin time, partial thromboplastin time and platelet count. If significant abnormalities are found in the bleeding or coagulation profiles, the need to use antifibrinolytic agents, fresh frozen plasma or platelet replacement should be discussed.

The physician and dentist should also decide if antibiotic prophylaxis is needed prior to carrying out any invasive dental procedures. There are no hard and fast rules about the use of pretreatment antibiotics in this group of patients. The decision should be made taking into account the extent and virulence of any existing oral infection being treated and the degree to which the patient is vulnerable and susceptible to that infection spreading locally or systemically as a result of the planned dental manipulation. The final decision should be made on a case by case basis, depending upon the particular circumstances. With regard to patients with end-stage liver or kidney disease, drugs metabolized by the affected organ should be limited, so the consultation should include advice from the physician about any modification in the choice and dosage of medications which may be used by the dentist during treatment.

For those patients undergoing invasive dental treatment prior to organ transplantation they should be managed under the very strictest infection control conditions. The surgical techniques should be meticulous, with great attention being paid to atraumatic tissue handling and precise wound closure. Restorative treatments should be carried out quickly and efficiently with emphasis only on those procedures that have a predictably successful outcome.

The use of general anesthesia and sedation is often contraindicated because of the underlying medical instability of these patients. It becomes important therefore that the dentist should use stress reducing patient management techniques to help gently ease the patient through this potentially high stress experience.

POSTTRANSPLANTATION TREATMENT

Immediate Posttransplant Phase

In the 4-6 months immediately following organ transplantation, dental treatment should only be carried out if an acute and serious oral emergency arises. During this time, the possibility of organ rejection is high and the adjustment of immunosuppressive drugs to optimal levels takes utmost priority. In this phase of treatment, the application of vigorous home oral hygiene measures by the patient should be monitored and encouraged.

In the unfortunate event of a patient who did not receive definitive treatment preoperatively developing an acute oro-facial odontogenic infection during this period then aggressive management is needed. Culture and sensitivity of the organism, surgical drainage and appropriate antibiotics will be needed to control the infection.

Stable Posttransplantation Phase

Once patients have stabilized following organ transplantation, they should be followed up closely for future dental problems and other changes arising as a result of their immunosuppressed status. Gingival hyperplasia, a particularly florid, hemorrhagic type of gingival overgrowth, can occur as a result of cyclosporine administration. This gingival overgrowth can be particularly troublesome and on occasion can cover the teeth almost completely obscuring them from view. A similar gingival overgrowth can occur with nifedipine often used to control hypertension arising from cyclosporine use.

E10

The application of meticulous oral hygiene measures with brushing, flossing and regular professional cleaning, is the best method of preventing or reducing the incidence of this condition. The problem is however, that once the tissue overgrows, creating deep pseudopockets around the teeth, it becomes almost impossible to gain control of situation because the bacterial dental plaque that colonizes the pockets cannot be fully eliminated. It is often necessary therefore to surgically remove the excess tissue to recreate a normal gingival anatomy. It is for this reason that these patients must be motivated from an early stage to keep their mouths clean and free from plaque.

In evaluating the stable post-transplant patient for dental care it is always important to assess how "stable" in fact they are. At one end of the spectrum, patients who are maintained on only minimal immunosuppression pose fewer problems that those, for example, who are on high doses of immunosuppressants, antihypertensives and anticoagulants. Each patient should be evaluated and treated on a case-by-case basis following consultation between the physician and dentist. It is the unpredictability of the degree of stability that will be achieved after transplantation that makes it imperative to be firm and resolute in the presurgical treatment planning of these patients. The dental management of a fragile, unstable, heavily medicated transplant patient with multiple, complex and serious oral problem can be very difficult.

PREDISPOSITION TO INFECTION

Because patients who are immunosuppressed with cyclosporine, azthioprine and prednisone are at increased risk of infection generally, they probably should receive prophylactic antibiotics prior to any significantly invasive dental treatment. For those patients, however, with immaculate oral hygiene and healthy mouths, the need for antibiotic coverage for basic, minimally invasive procedures is less clear. There is no documented evidence that such patients are at an increased risk of infection, following dental manipulation. The decision to use prophylactic antibiotics should be made depending upon each patient's individual circumstances.

ADRENAL SUPPRESSION

During the stable posttransplantation phase, the possible effects of suppression of adrenal function should be considered in patient's undergoing dental treatment. For many patients even routine dental procedures are significantly stress provoking. When more major and complex surgical procedures are anticipated, the additional and unavoidable stress placed on the patient should be evaluated, when necessary a steroid boost may be indicated to see the patient through the stressful period. In cases of patients taking more than 5 mgs of prednisone daily, they should at least double their dose on the day of the scheduled procedure. In some cases, the steroid boost may need to increase even greater and continued longer, particularly if post-operative pain is expected. It is important in managing patients during this stage of their transplant treatment to use stress reduction techniques as part of their overall management plan.

E10

MEDICATION USE

Depending upon the level of function of the transplanted organ in liver and kidney transplants, it may be necessary for the dentist to select different drugs or modify dosages of drugs in order to avoid the production of toxic levels.

OVERSUPPRESSION

In the mouth, evidence of overimmunosuppression may be manifest by the development of fungal or viral infections of the mucous membranes. Depending on the severity of the condition, topical or systemic treatment with ketoconazole or acyclovir is indicated. Delayed healing of oral wounds and slow healing of simple traumatic oral ulcers may also indicate that the patient is being over-immunosuppressed. Additionally, the increase incidence of carcinoma of the lip and lymphoma in patients receiving too much immunosuppression should be kept in mind by the dentist who is seeing posttransplant patients on a regular basis.

BLEEDING PROBLEMS

When oral surgical procedures are planned for transplant patients who are therapeutically anticoagulated, the INR should be adjusted to < 3.0. Even at this level, bleeding may be difficult to control and sockets may need to be packed, pressure splints applied and antifibrinolytic agents used.

CONCLUSION

Although organ transplantation is now being carried out with increasing success and increasing frequency, the procedure still carries with it the possibility of many significant long- and short-term complications which can adversely affect the patient's outcome. Infection is still the leading cause of death in these patients and the ubiquitous nature of oral infection should always be borne in mind when considering transplantation of any organ.

Dentistry has an important role to play in the care of these patients by ensuring that the potentially serious complications arising from oral disease are kept to a minimum, or better still, are eliminated completely.

E10

Essay 11: The Living Organ Donor: Laparoscopic Donor Nephrectomy

Joseph R. Leventhal

There is a steadily increasing disparity between cadaver donor organ supply and demand. UNOS registry data indicate the number of patients on waiting lists for renal transplants has almost tripled from 13943 to 53044 between 1988 and 2001, while the number of kidney transplants being performed using cadaver donors increased by only 14%, from 7208 to 8203, during the same period.[1] As one would expect, the waiting time for cadaver organs has also increased; between 1992 and 1999 the mean waiting period for a kidney increased from 624 to 1144 days. Consequently, living donors have assumed increasing importance in renal transplantation. Living donors accounted for 5949 (42%) of the 11458 kidney alone transplants reported to UNOS in 2001, more than three times the 1809 reported in 1988.[2]

There are several advantages to living donor renal transplantation. The use of living donors is associated with improved patient and graft survival. UNOS registry reports for 1995-96 live donor one-year graft and patient survival rates of 91.2% and 97.2% respectively, compared with one-year cadaveric donor graft and patient survival rates of 80.6% and 93.3%. At three years graft and patient survival rates for living donor transplants are 83.9% and 94.3% respectively, while cadaveric organs have fallen to 69% and 87.4%.[3] Improved graft survival has been observed for recipients of both living-related and living-unrelated (i.e., so-called emotionally related) allografts. Living donor transplantation helps avoid the prolonged waiting times for a cadaveric organ, and offers the ability to plan such a transplant in advance. Other advantages of live donor renal transplantation include a decreased incidence of delayed graft function and a shorter recipient hospitalization. Furthermore, the elective nature of the live donor procedure allows for optimization of the recipient's medical condition before surgery.

Live renal donation has been performed since the 1950s. Longitudinal studies of patients undergoing unilateral nephrectomy have not shown them to have an increased incidence of renal failure or diseases attributable to having donated a kidney.[4-6] Currently, live renal donation is most commonly performed via a retroperitoneal flank incision. The operation is safe, with reported mortality rates of 0.03-0.06%.[7-9] However, this extraperitoneal flank approach is not without minor morbidity. Wound complications including infection and hernia occur in 9% of patients.[10] Pneumothorax requiring pleural space drainage may occur. Chronic

Organ Transplantation, 2nd edition, edited by Frank P. Stuart, Michael M. Abecassis and Dixon B. Kaufman. ©2003 Landes Bioscience.

E11

incisional pain and so-called wound diastasis has been reported in up to 25% of patients.[11,12] Patients undergoing a large flank incision have a duration of hospitalization averaging 4-5 days.[9,13] Adequate pain control often requires the use of epidural analgesics and the prolonged use of parenteral narcotics. There is a delayed return to normal activities for as long as 6-8 weeks after surgery. Finally, potential donors commonly express concerns regarding the cosmesis of the large flank incision.

The limitations of the extraperitoneal approach to donor nephrectomy, combined with advances in techniques of laparoscopic solid organ surgery, have provided the impetus for development of a minimally invasive approach to live renal donation. Potential benefits of a laparoscopic donor procedure include less postoperative pain, shorter hospitalization, less incisional morbidity, more rapid return to normal activity, and improved cosmesis. Moreover, the potential advantages of a minimally invasive operation could lead to increased acceptance of the donor operation and expansion of the pool of potential kidney donors.

EVALUATION OF THE POTENTIAL LIVING DONOR

The primary goal of the living donor evaluation is to ensure the safety and well-being of the donor (Table E11.1). Suitable living donor candidates include children, parents, and siblings of the patient with renal failure. Zero-haplotype matched donors should not be a contraindication to proceeding with a live donor transplant. Furthermore, persons with close emotional ties to the transplant recipient, including adopted siblings, spouses, and best friends, may be considered. It has been established that transplant outcomes in these aforementioned groups are superior to that occurring with cadaveric donors.[14]

It is appropriate to discuss the advantages and disadvantages of living kidney donation with potential transplant recipients during their evaluation at a transplant center. However, patients should not be pressured into approaching family members or friends regarding living donation if they are uncomfortable

---

Table E11.1. Living donor evaluation (adapted from sources refs. 18,19)

Donor and recipient blood and HLA typing
Donor and recipient crossmatch
Donor general history and physical examination
Electrolytes, BUN, creatinine repeated on two occasions
Liver function tests, coagulation profile
CBC
VDRL/RPR, HbsAg, HCV antibody, EBV titer, CMV titer, HIV
Glucose tolerance test if patient is diabetic
Urinalysis and urine culture, pregnancy test
24-hour urine collection for protein and creatinine clearance (on two occassions)
Chest x-ray, EKG, cardiac echo/noninvasive stress test for donors greater than 50 years of age/if indicated.
Pap smear /mammogram for female donors
Psychiatric evaluation
IVP and arteriogram vs. CT scan angio

E11

doing so. Furthermore, potential donors should not be coerced into the evaluation process. Frequently, motivated individuals who are considering live renal donation will accompany the transplant recipient to his or her initial pretransplant evaluations. Ideally, the issue of living donation should be brought to the potential donor by the patient, and not the patient's nephrologist or members of the transplant team. However, both the transplant team and nephrologist should be willing and able to facilitate discussion of donation if assistance by the patient is requested.

The initial donor evaluation should attempt to screen out those individuals who are not suitable candidates (Table E11.2). Living donors typically range in age between 18 and 65 years of age; juveniles are rarely considered for donation, while the elderly usually have an unacceptably higher risk of general anesthesia and surgery. Blood typing and crossmatch are cost-effective screening tests commonly used early in the donor evaluation. Individuals who are blood –group incompatible, or who demonstrate a positive crossmatch with the recipient, are ruled out without further expensive testing. A complete history and physical examination can then proceed with an appropriate blood group compatible and crossmatch negative candidate.

It is imperative that potential donors be excluded on medical grounds when it is believed that there may be a risk of unrecognized kidney disease, or there exist medical conditions which increase the risk of both short and long term morbidity and mortality from the donor procedure. Table E11.2 lists criteria frequently cited for exclusion of potentially compatible donors. These criteria are not absolute; when findings are borderline, it is always appropriate to exclude the candidate in the name of donor safety. Relative contraindications peculiar to laparoscopic kidney donation include prior left colonic or splenic surgery, retroperitoneal inflammatory processes (i.e., diverticulitis), and obesity. Precise definition of

---

Table E11.2.   Potential contraindications for living kidney donors (adapted from refs. 18,19)

ABO incompatibility
Positive crossmatch
Age less than 18 or over 65
Malignancy
Infection
Hypertension (> 140/90 mm Hg)
Diabetes
Proteinuria (> 150 mg/24 hrs)
Renal disease or reduced renal function (creatinine clearance < 80 ml/min)
Microscopic hematuria
Urologic abnormalities in donor kidneys
Nephrolithiasis
Increased medical risk of surgery
Inability to give informed consent
Psychiatric contraindications

renal anatomy with intravenous pyelogram plus conventional angiography, or high-resolution CT scan angiography is the final step in the donor evaluation and should be performed after medical clearance of the transplant recipient. Published reports of laparoscopic donation support the limited use of the left kidney to maximize the length of renal vein, and to avoid posttransplant vascular.[15] The presence of multiple renal arteries is not a contraindication to the use of the left kidney. If for any reason the left kidney is not suitable to donate, then open right nephrectomy may be performed.

There is often concern about the long-term effects of donation in relatives of patients with hereditary kidney disease, most notably polycystic kidney disease and hereditary nephritis. In families with polycystic kidney disease, a negative ultrasound or CT scan in individuals greater than 30 years of age safely rules out the disease and permits donation. When a family history of hereditary nephritis or Alport's syndrome is documented, it is usually inherited in an X-linked pattern. Donation is best limited to male relatives without hematuria or other Alport's –associated abnormalities or to those female relatives without hematuria. Females relatives with hematuria or other associated abnormalites are not suitable donors. Both donors and recipients in families with hereditary nephritis need to be advised of the risk of disease recurrence and subsequent graft failure.

It is essential that the voluntary nature of the kidney donation be preserved during the entire evaluation process. A psychological assessment of the donor early on may prove valuable in this regard. Health care workers should protect individuals who decide not to donate by offering to tell the patient's family and friends that the donor is not suitable on medical grounds.

CHOOSING THE BEST DONOR

The best donor is usually a blood relative. If there are multiple family members who are suitable donor candidates, it is logical to proceed with evaluation of the relative who is best MHC-matched to the transplant recipient. The best match possible is an identical twin, followed by a MHC-identical (two haplotype match) sibling, then a single haplotype matched parent or sibling. However, non-immunologic considerations, such as donor geographic location, age, occupational risk, and familial responsibilities need to be considered. If a suitable blood relative is not found, then emotionally-related donors can be considered. However, it is imperative to ensure the altruistic motivation of the donor in these circumstances.

THE LAPAROSCOPIC DONOR PROCEDURE

Laparoscopic live donor left nephrectomy is performed under general anesthesia with the patient placed in the right decubitus position. The operating table is flexed at a point midway between the patient's iliac crest and ribcage, and a kidney rest is elevated in order to maximize exposure during the procedure. Positioning of the patient and draping is carried out to allow—if necessary—for open conversion to an extended subcostal, or standard flank approach for completion of the procedure. Orogastric suction, foley catheter bladder drainage, prophylactic antibiotics and antithrombotic sequential leg compression devices are routinely used. The patient receives a bowel prep with magnesium citrate the night before surgery to help decompress the colon. The operating surgeon stands on the patient's right, with the camera operator caudad. An assistant and scrub nurse stand on the patient's left. Two television monitors are placed at the head of the operating room table. Standard laparoscopic instrumentation, along with a 30° laparoscope and ultrasonic scalpel are used. A pneumoperitoneum of no more than 15 mm Hg is created by Verres needle insertion in the left subcostal location. After creation of the pneumoperitoneum, the laparoscope is introduced into the abdomen using a 10 mm Visiport™. Two additional 12 mm operating ports are placed in the left subcostal location, as well as a 5 mm port in the left posterior axillary line. Port placement will vary slightly from patient to patient depending upon patient girth and the length of the torso. The operation is conducted as follows: mobilization of the left colon and spleen, dissection of the renal vein, dissection of the renal artery, dissection of the adrenal gland off the upper pole of the kidney, dissection of the ureter, mobilization of the kidney, creation of an extraction incision, systemic anticoagulation, division of the ureter, renal artery, and renal vein, and renal extraction. During the dissection, adequate urine output is maintained through vigorous intravenous hydration. Osmotic diuresis is instituted following volume loading, and at the beginning of the vascular dissection, with 12.5 grams of mannitol and 10-20 mg of furosemide. Renal artery vasospasm can be minimized through the use of topical papaverine. Once the kidney is completely free except for its vascular and ureteral attachments, a 6-7 cm extraction incision located either around the umbilicus or in the left lower quadrant is made without violation of the peritoneum. The patient is then anticoagulated with 5000 units of I.V. heparin sodium. The distal ureter is clipped and divided. Division of the renal artery, followed by the renal vein, is performed with a linear vascular laparoscopic stapler. The peritoneum at the extraction incision is opened, and the kidney is delivered through this wound by the surgeon's hand into an iced saline solution. The staple lines on the allograft are removed, and the kidney flushed with Collins solution. Heparin is reversed with protamine sulfate while the extraction incision is closed. Pneumoperitoneum is re-established and inspection of the operative field is performed. Once hemostasis has been deemed adequate, ports are removed under direct visualization, the abdomen desufflated, and the incisions closed.

RESULTS OF LAPAROSCOPIC DONOR NEPHRECTOMY

Published reports of laparoscopic nephrectomy indicate it is a safe procedure with excellent outcomes.[16,17] There have been no procedure-related mortalities. Perioperative morbidity from this procedure has compared favorably with that seen with the open approach. The incidence of delayed graft function, as well as short-term patient and graft survival, is comparable to that achieved with open donor nephrectomy. Open conversion does occur and is related to both variations in renal vascular anatomy and donor size/obesity. Laparoscopic nephrectomy has been associated with statistically significant reductions in hospital stay, use of intravenous analgesics, time to resumption of diet, and time to return of normal daily activities. In addition, an increase in the rate of live renal donation has been reported by centers using the laparoscopic approach.

At our center, we have performed 450 laparoscopic donor nephrectomies from 10/97 to 3/03, with an open conversion rate of 2% (9/450). Open conversion was associated with donor obesity and the presence of abnormal vascular anatomy. All patients made complete recoveries. Patients undergoing laparoscopic nephrectomy experienced less postoperative pain, earlier resumption of diet, and shorter hospital stays. There has been only one recipient of an LDN kidney who experienced delayed graft function with ATN requiring dialysis. There have been no graft losses due to CDN technique. There have been no short of long-term allograft urologic complications in our series.

SUMMARY

Laparoscopic donor nephrectomy is technically demanding but nonetheless feasible, and can be performed with morbidity and mortality rates comparable to those of open nephrectomy, with marked improvements in patient recovery after the laparoscopic approach. The procedure is best performed only by surgeons with well developed laparoscopic skills. Laparoscopic donor nephrectomy may improve willingness to donate kidneys and expand the potential pool of organ donors.

REFERENCES

1. U.S. Scientific Registry of Transplant Recipients and the Organ Procurement and Transplantation Network-Transplant Data, 1988-1998. UNOS web site.
2. Ibid.
3. US Renal Data System, USRDS 1994 Annual Report, The National Institutes of Health, National Institute of Diabetes and Digestive and Kidney Diseases, Bethesda MD, 1994.
4. Velosa JA, Offord KP, Schroeder DR. Effect of age, sex, and glomerular filtration rate on renal function outcome of living kidney donors. Transplantation 1995; 60: 1618-1621.
5. Ohishi A, Suzuki H, Nakamoto H et al. Status of patients who underwent uninephrectomy in adulthood more than 20 years ago. Am J Kidney Dis 1995;26:889-97.
6. Narkun-Burgess DM, Nolan CR, Norman JE et al. Forty-five year follow-up after uninephrectomy. Kidney Int 1993;43:1110-1115.
7. Najarian JS, Chavers BM, McHugh LE et al. 20 years or more follow-up of living kidney donors. Lancet 1992; 340:807-810.
8. Bay WH, Herbert LA. The living donor in kidney transplantation. Ann Intern Med 1987;106:719-727.
9. Johnson EM, Remucal MJ, Gillingham KJ et al. Complications and risks of living donor nephrectomy. Transplantation 1997;64:1124-1128.
10. Cecka JM. Living donor transplants. In: Cecka JM, Terasaki PI, eds. Clinical Transplants 1995. Los Angeles:UCLA Tissue Typing Laboratory; 1995:363-377.
11. Blohme I, Fehrman I, Norden G. Living donor nephrectomy. Complication rates in 490 cases. Scand J Urol Nephrol 1992; 26:149-153.
12. Dunn JF, Richie RE, MacDonell RC et al. Living related kidney donors. A 14 year experience. Ann Surg. 1986; 203:637-642.

E11

13. Shaffer D, Sahyoun AI, Madras PN et al. Two-hundred-one consecutive living donor nephrectomies. Arch Surg 1998; 133:426-431.

14. Kaufman DB, Matas AJ, Arrazola L et al. Transplantation of kidneys from zero-haplotype matched living donors and from distantly related and unrelated donors in the cyclosporine era. Trans Proc 1993 25:1530.

15. Ratner LE, Kavoussi LR, Chavin KD et al. Laparoscopic live donor nephrectomy: Technical considerations and allograft vascular length. Transplantation 1998; 65:1657-1658.

16. Flowers JL, Jacobs SJ, Cho E et al. Comparison of open and laparoscopic live donor nephrectomy. Ann Surg 1997; 226:483-490.

17. Ratner LE, Kavoussi LR, Sroka M et al. Laparoscopic assisted live donor nephrectomy—A comparision with the open approach. Transplantation 1997 63:229-233.

18. Kasiske BL. The evaluation of prospective renal transplant recipients and living donors. In: Rao VK ed. Renal Transplantation Surg Clinic North Am 1998; 78(1):27-39.

19. Rosenthal JT, Danovitch GM. Live-related and cadaveric kidney donation. In: Danovitch GM ed. Handbook of Kidney Transplantation 2nd edition 1996, Little, Brown, and Co. 95-108.

E11

Essay 12: Xenotransplantation

Jonathan P. Fryer, Joseph R. Leventhal

INTRODUCTION

As the list of patients needing transplants grows, while the pool of available organs stagnates the search for new sources of donor organs becomes increasingly relevant. Efforts to enhance the yield from potential human donors have included campaigns to increase public and physician awareness of organ donation, greater emphasis on living donors, and relaxation of acceptance criteria for cadaver donors. Despite these efforts, an allogeneic solution seems unlikely. As a result, xenotransplantation, the transplantation of tissues and organs between species, is undergoing serious consideration.

In xenotransplantation, species combinations have typically been designated as either concordant or discordant, depending on whether or not hyperacute rejection (HAR) results. But over the past two decades, study of the pathobiology of xenograft rejection has demonstrated this classification system to be greatly oversimplified; it is not necessarily based on the pathogenesis of the rejection process observed in different species combinations. This review avoids broad categorization, and instead discusses xenografting in terms of the phylogenetic disparity of specific species pairs and the underlying mechanisms of xenograft loss. In addition, the processes which lead to graft loss even if HAR is prevented (i.e., beyond HAR) are discussed.

TRANSPLANTATION BETWEEN CLOSELY RELATED PRIMATE SPECIES

Clinical application of xenotransplantation has been limited almost exclusively to the use of nonhuman primates as organ donors. Indeed, encouraging results have been obtained with the transplantation of primate kidneys, hearts and livers into humans.

The first attempt at primate-to-human xenotransplantation was by Unger, who, in 1910, performed a chimpanzee-to-human renal transplant. The graft clotted and the recipient died. Over 50 years later, primate-to-human renal xeno-transplantation was again attempted, with markedly improved results. In 1964, Reemtsma and colleagues performed a series of chimpanzee-to-human renal transplants, achieving nine-month graft survival in one patient. Baboon-to-human renal transplants were attempted by Starzl, who noted poorer graft survival-despite an immunosuppressive protocol similar to the one used by Reemtsma. Although chimpanzee xenografts fared better than baboon xenografts, the rejection process was characterized in both by the infiltration of recipient cellular elements. Rejection of primate renal xenografts appeared to resemble an aggressive form of clinical

Organ Transplantation, 2nd edition, edited by Frank P. Stuart, Michael M. Abecassis and Dixon B. Kaufman. ©2003 Landes Bioscience.

allograft rejection. Clinical attempts at renal xenotransplantation were soon abandoned in favour of the use of cadaver and living related human donor organs. A limited number of unsuccessful nonhuman primate-to-human heart and liver transplants were also performed during the 1960s and 1970s.

In the 1980s, the introduction of cyclosporine (CsA) helped to revolutionize clinical solid organ transplantation. With this powerful new immunosuppressive agent, many felt the species barrier between humans and apes could be overcome. In 1984, Bailey et al performed a baboon-to-human heart transplant into a newborn infant (Baby Fae) using an ABO incompatible donor. Baby Fae's clinical course was characterized by progressive cardiac failure, beginning three days posttransplant and resulting in death 20 days posttransplant. Histological evaluation of the xenograft demonstrated humoral rejection. It is important to note that Baby Fae was blood type O-rare in baboons-and that an ABO-compatible donor was not available. This donor-recipient blood group incompatibility complicates analysis of the cause of xenograft loss. Her course did suggest that CsA alone may not be sufficient to bridge interspecies barriers, and that future strategies would probably require intervention against antibody.

The 1980s saw the advent of intensive scientific investigation in xenotransplantation using small and large animal models. Hamster organs placed into untreated rats are rejected over several days; this rodent model has been used as a small animal correlate of transplantation between closely related primate species. Using the hamster-to-rat species combination, several investigators reported extension of heart and liver xenograft survival with a variety of immunosuppressive regimens. In general, these studies indicate that protocols combining both antihumoral and anti-T-cell immunosuppression are most effective. Importantly, extended hamster xenograft survival appears to require prolonged suppression of rat xenoreactive antibody synthesis (a phenomenon likely to require agents with B-cell and T-cell immunosuppressive action) or prevention of the deleterious consequences of antidonor antibody binding to the graft (i.e., complement activation), or both.

Results from these rodent studies have been applied to nonhuman primate models with some success. Using different nonhuman primate combinations, prolonged xenograft survival of hearts and kidneys has been achieved with protocols directed at both cellular and humoral immune responses. These studies indicate that different primate species combinations vary in the aggressiveness of the immune response generated against the donor and that the intensity of the humoral response is a significant determinant of graft survival.

Successful extension of liver xenograft survival in the hamster-to-rat combination with FK506 and cyclophosphamide served as the basis for two clinical attempts at baboon-to-human liver transplantation by Starzl et al. Patients received an aggressive immunosuppressive regimen, including FK506, cyclophosphamide, steroids, and perioperative prostaglandin E1. Although extended graft survival was achieved in both cases, the patients died of infectious and neurological complications. Both liver xenografts displayed evidence of antibody-mediated

E12

injury posttransplant and vascular and sinusoidal deposition of antibody was also seen on later biopsies. At autopsy the baboon livers showed extensive steatosis and biliary tract changes, which resembled the pathological lesions observed in ABO-incompatible liver transplants. Both livers failed to show evidence of severe cellular rejection. Thus, antibody-dependent mechanisms of graft rejection appear to play a significant role in the baboon-to-human species combination.

Despite the promising results with nonhuman primate-to-human xeno-transplantation, species such as the chimpanzee and baboon are unlikely to be widely used as donors. First, chimpanzees and other anthropoid apes, the species which are most closely related to humans, are endangered. The few that are made available for biomedical research are used in AIDS or hepatitis research. Old world monkeys (baboons, cynomologous monkeys or macaques, and vervet monkeys) still exist in the wild in fairly large numbers so are now most common in xenotransplant research. However, a growing segment of society is outwardly opposed to the use of primates in medical research. It is unclear whether an increase in the use of old world monkeys-an increase sufficient to solve the current organ shortage-would be acceptable to society.

Secondly, there is a theoretical risk that nonhuman primate viruses will be transmitted when their organs are transplanted into humans. Some of these viruses, though harmless to nonhuman primates, can be deadly to humans. The risk involved is unclear, but the development of virus-free primate colonies necessary to address these concerns would be an arduous task.

Thirdly, the limited size range of primate species such as the baboon may pose problems for appropriate donor-recipient size matching for certain organs, such as the heart and lungs.

In light of the many real and potential obstacles to use of nonhuman primate organs, focus has shifted to the use of more phylogenetically disparate species as donors.

TRANSPLANTATION BETWEEN PIGS AND HUMANS

Although there may always be objections to the transplantation of animal organs into humans, it may be more acceptable to obtain organs from a species that is already being bred and killed for human use. Pigs are easy and inexpensive to breed and have many physiological similarities to humans. Furthermore, they attain adult weights comparable to humans, and so can provide large enough organs for any age of recipient. Unfortunately, immediately vascularized porcine organs are rapidly rejected by human and nonhuman primate recipients in a matter of minutes to hours. This process of (HAR) hyperacute rejection is mediated by naturally occurring, complement-fixing xenoreactive antibodies that bind to the pig vascular endothelium. HAR has become a major focus of xeno-transplantation research.

Clinical experience with organ transplantation between pigs and humans has been very limited. In 1968, in separate instances occurring on the same day, pig hearts were incorporated into the circulation of two patients who could not be weaned from cardiopulmonary bypass support. In both cases, the grafts rejected within minutes. There have been numerous attempts to support patients with fulminant hepatic failure using ex vivo perfusion of liver xenografts. In most of these cases, pig livers were used. Routinely, antibodies to pig serum proteins developed and titres of antipig antibodies increased. These livers were functional, and demonstrated the ability to clear lactate and ammonia and to synthesize donor proteins for many hours. Porcine livers appear to be less susceptible to antibody-mediated graft damage than other porcine organs perfused by human blood, suggesting and inherent resistance to antibody-mediated graft damage. This is similar to observations made in clinical ABO-incompatible liver transplants.

Makowka et al transplanted a pig liver into a young woman with fulminant hepatic failure, with the intent of using it as a 'bridge' until a human liver became available. Pretransplant, the recipient's blood was perfused ex vivo through the pig's kidneys to absorb antipig antibodies. Posttransplant, antipig antibodies rapidly rebounded, perhaps induced by the administration of unabsorbed blood products to the patient. Immunopathological evidence of graft rejection occurred in about three hours, and the patient died before successful bridging to an allograft could be achieved. Pathological analysis of the xenograft revealed diffuse vascular thrombosis, haemorrhagic necrosis, and extensive infiltration by neutrophils. Vascular deposits of IgG, IgM, and complement components were found. These results corroborate laboratory studies demonstrating the importance of antibody, complement, and elements of acute inflammatory reaction during HAR.

THE PROBLEM OF HYPERACUTE REJECTION

Before porcine xenografting can become a reality, the first hurdle to overcome is HAR. Evidence now strongly suggests that this immunological barrier to xenotransplantation between disparate species combinations derives from one or more of three factors: (1) the specific interaction of recipient xenoreactive antibodies with antigens present on the endothelium of the donor organ, followed by activation of the complement cascade, (2) the direct activation of the recipient complement system by donor endothelial cells, (3) the relative failure of complement inhibitory proteins in the donor organ to impede activation of the recipient complement system. This improved understanding of the pathogenesis of HAR has provided a scientific foundation on which to devise and evaluate therapeutic approaches to extend porcine xenograft survival.

PREFORMED NATURAL XENOANTIBODY (PNXAb)

A critical role for preformed natural xenoantibody (XNA) in the pathogenesis of HAR follows from several lines of evidence: 1) the rejection process occurs extremely rapidly, indicating a role for circulating factors, rather than a cell-mediated response, 2) the immunopathology of rejected xenografts in disparate species combinations reveals antibody deposition on donor endothelium, 3) removal of PNXAb is frequently associated with prolonged xenograft survival, and 4) infusion of PNXAb leads to rejection of organs previously engrafted in modified recipients.

Recent studies have provided important information on the characteristics and porcine targets of PNXAb in humans and nonhuman primates. HAR results from activation of complement initiated by binding of xenoreactive antibodies to carbohydrate antigens on donor endothelium. The presence of anti-gal specific antibodies in certain primates has been linked to their lack of expression of the enzyme (α-galactosyl-transferase) needed to produce nonreducing terminal α-gal sugar residues. In vitro studies suggest that membrane proteins and lipids bearing α-gal structures are widely expressed on porcine cells, and that absorption of antibodies with anti-gal specificity prevents the in vitro lysis of porcine cells. Cooper et al have describes antibodies eluted from porcine organs perfused with human plasma to bind carbohydrates with terminal α-gal residues. Furthermore, in vivo administration of oligosaccharides capable of binding anti-gal antibodies has been shown to reduce antipig PNXAb titres in baboons and result in extended porcine xenograft survival.

COMPLEMENT

In the pig-to-human combination, HAR occurs within minutes to hours of engraftment. It is initiated when PNXAb binding to antigens on the endothelial cell activate the classical pathway of the complement cascade. Activation of the classical pathway in turn allows for subsequent recruitment of the complement alternative pathway. As a result, several biologically active complement fragments complex (membrane attack complexes, MAC). These products of complement activation may play significant roles in vascular injury by inducing an increase in vascular permeability (C3a, C5a)-contributing to development of a procoagulant state (MAC), promoting cell adhesion (C3bi), or inducing direct endothelial cell damage (MAC).

Studies in primates support a process wherein complement-fixing PNXAb of the IgM isotype are thought to initiate HAR of porcine organs.

The severity of hyperacute rejection results in part from failure of membrane-associated regulators of complement in the donor organ to inhibit the activation of the recipient complement system. These regulators of complement activation (RCA) are known to be inhibitory only for homologous complement for complement of a closely related species and-not for complement of distantly related species. Thus, pig RCA fail to inhibit activation of human complement. Incorporation of recipient RCA into the vascular endothelium of the xenogeneic donor organ could allow for marked inhibition of recipient complement activation, with an associated impact on HAR. In vitro studies where genes for human RCA have been introduced into xenogeneic cells have demonstrated the feasibility and cytoprotective effects of this approach. Using transgenic technology, expression of human RCA in porcine organs has been achieved, with significant prevention of HAR.

E12

PREVENTION OF HYPERACUTE REJECTION

Three basic strategies to prevent HAR are: (1) reduction or elimination of PNXAb; (2) prevention of complement activation; (3) alteration or elimination of endothelial cell xenoantigens.

REDUCTION OR ELIMINATION OF PREFORMED NATURAL XENOREACTIVE ANTIBODY

The problem of PNXAb has been approached through various selective and nonselective techniques. Plasmapheresis, a nonselective technique for antibody removal, has been employed in xenotransplantation with some success. It has proven useful for antibody removal in human ABO-in-compatible kidney transplants and for renal transplants into highly sensitized recipients. In pig-to-primate xenotransplants, Alexandre et al. Used plasmapheresis and immunosuppression in splenectomized baboons to extend porcine renal xenograft survival, with one organ surviving 22 days. Recently, we have described the ability of plasmapheresis and immunosuppression to extend cardiac xenograft survival from one hour in untreated controls to greater than two days. When we combined plasmapheresis and immunosuppression with complement depletion, xenograft survival was extended to 17.5 days. Despite these encouraging results with plasmapheresis, it is hampered by the concomitant reduction of other plasma proteins-such as clotting factors and complement system proteins-that may be essential for the recipient. The effect of plasmapheresis on the coagulation system is likely to limit its application in the peritransplant period, since bleeding complications may result. Replacement of clotting factors with fresh frozen plasma would be hazardous, since it contains high levels of xenoreactive antibodies. Another, more selective approach to remove xenoreactive antibodies is the use of immunosorbent columns. They avoid the consequences of wholesale plasma protein removal seen with plasmapheresis. Protein A and protein G effectively remove of antibodies of the IgG and IgM isotype, and have been used successfully in a variety of clinical situations. We have recently shown that columns containing polyclonal antihuman IgG or IgM antibodies conjugated to Sepharose are extremely effective for removing antipig antibody from human plasma. However, two potential disadvantages of column technologies are their nonspecific removal of all antibodies and the potential side effects of column-dependent complement activation.

The optimal approach to selective xenoreactive antibody removal may be an extracorporeal source of target xenoantigens. The crudest application of this approach is the perfusion of a donor organ, most commonly the spleen, kidney, or liver. This approach, though effective, has several disadvantages. Organ perfusion results in the sequestration of blood volume and in the activation of the complement and coagulation cascades as the perfused organ is rapidly rejected. In addition, organ perfusion is cumbersome to perform, requiring sacrifice of a donor animal to obtain target organs. A more elegant approach to selective xenoreactive antibody removal would be to use purified or synthesized target antigens. Specific columns using human ABO blood group antigens have proved successful in removing anti-A and anti-B antibodies. Similarly columns loaded with a-gal have proven effective in recovery α-gal specific XNA and extending xenograft survival. An attractive solution to the problem of xenoreactive antibodies is to selectively prevent antibody synthesis. To date, attempts to achieve selective B cell tolerance against α-gal antigens has proven unsuccessful.

PREVENTION OF COMPLEMENT ACTIVATION

FLUID PHASE INHIBITION

Manipulation of the normal pathway of complement activation has proven to be extremely effective for overcoming HAR. The most effective agent to date has been cobra venom factor (CVF). Cobra venom contains, in addition to several toxins and enzymes, a C3b-like molecule, CVF, that is resistant to human and rat C3b inactivating factors. This C3b analogue combines with components of the alternative pathway of complement to form a highly stable enzyme complex that causes massive consumption of C3, factor B, and members of the MAC. The result is exhaustion of the complement cascade. Removal of phospholipase contaminants in cobra venom has greatly reduced its toxicity in rodents. To date, toxicity has not been a significant factor in primates. CVF has extended xenograft survival in several rodent models, and has prolonged pig-to-baboon cardiac xenograft survival. Clinical use may be limited by the potential toxicity of CVF as well as its immunogenicity. Repeated use may induce anti-CVF antibodies. A variety of complement-inhibiting agents have been evaluated in small and large animal models. Agents such as soluble complement receptor (sCR1) have been shown to significantly prolong porcine xenograft survival in primates.

HOST-SPECIFIC REGULATORS

Several biologically important regulators of complement activation (RCA) are located on the membranes of most cells that are in contact with blood or other body fluids containing complement system proteins. Decay accelerating factor (DAF or CD55), membrane inhibitor of reactive lysis (CD59 or protectin), homologous restriction factor (HRF), and membrane cofactor protein (CD46) interfere with the complement reaction only at the membrane sites where they are located. DAF inhibits formation of the classical and alternative pathway C3 and C5 convertases; if such convertases have been formed it promotes the dissociation of these enzymatic complexes. CD59 and HRF interfere with formation of the MAC at the C8 and C9 binding steps. CD46 acts as a cofactor for the inactivation of C3b and C4b by factor I. DAF, CD59, and HRF are uniquely anchored to the cell membrane by a phosphatidylinositol tail, which affords these molecules great mobility in the plane of the cell membrane and allows purified forms of RCA to be inserted into cell membranes. Membrane-associated RCA display the phenomenon of homologous restriction, i.e., they are inhibitory to homologous complement but have little inhibitory capacity over xenogeneic complement. This phenomenon has led several investigators to explore whether the introduction of human RCA molecules into xenogeneic cells would afford protection against human complement-mediated damage. Dalmasso et al were the first to show that purified human DAF protected these cells from the cytotoxic effects of human complement. They obtained similar results with purified human CD59. However, the protection afforded by inserted human RCA molecules is relatively short-lived, as they undergo rapid turnover.

E12

Long-term protection from human complement-dependent damage could be achieved through genetic expression of human RCA molecules in xenogeneic cells. Toward this end, experiments have been conducted with xenogeneic cell types that were transfected with cDNA for various human RCA molecules, resulting in expression of these proteins on cell membranes and protection from human complement. These findings served as the basis for several groups to develop animals transgenic for human RCA molecules. Cells and organs from transgenic mice expressing different human RCA are protected from the effects of xenoactive antibody and complement. Similarly, transgenic swine expressing different human RCA show resistance to hyperacute rejection in both ex vivo and in vivo models. Although transgenic porcine organs have been shownrepeatly to avoid HAR in nonhuman primates, these grafts are lost through a process of acute vascular xenograft rejection.

ACUTE VASCULAR XENOGRAFT REJECTION (AVXR)

These complement inhibitory strategies have not entirely solved the xenograft rejection problem, as they have unmasked an acute and irreversible vascular rejection, which causes graft loss days to weeks posttransplant despite effective complement inhibition. This process, known as AVXR or delayed xenograft rejection (DXR), is currently the most significant immunological barrier to successful xenotransplantation. AVXR is typically associated with: endothelial antibody deposition; infiltration of macrophages and natural killer (NK) cells; evidence of endothelial activation and localized hypercoagulability; loss of endothelial integrity, and microvascular thrombosis. While AVXR is not the result of complement mediated lysis (CML), anti-porcine xenoantibodies still appear to play a central role.

AVXR VERSUS ACUTE VASCULAR ALLOGRAFT REJECTION (AVAR)

The fundamental role of the xenoantibody-αGal interaction in pig-to-primate AVXR is best supported by the fact that an almost identical interaction occurring in ABO incompatible allotransplantation is undeniably the initiating event in a rejection process which closely resembles AVXR. Human grafts transplanted into ABO incompatible recipients undergo an acute vascular allograft rejection (AVAR) which resembles the AVXR seen when hCRP transgenic pig xenografts are transplanted into primates. In both situations, the donor's complement regulatory mechanisms are functional and therefore MAC formation is inhibited. Strategies that have been successful in prolonging ABO incompatible allograft survival have also been successful in prolonging xenograft survival. However, the cross-species regulatory incompatibilities implicated in AVXR, clearly do not play a role in AVAR. The two primary ABO blood group antigens, A and B, are terminal endothelial carbohydrates similar to αGal. Since these antibody-ABO antigen interactions independently initiate AVAR, the almost identical xenoantibody-αGal interaction which occurs in pig-to-primate xenotransplantation is likely sufficient to initiate AVXR, without contribution from other mechanisms. Therefore, while other mechanisms may contribute to AVXR, the xenoantibody-αGal interaction likely plays a central role in its initiation.

E12

INDUCED XENOANTIBODIES IN AVXR

Although HAR has been averted by temporarily inhibiting xenoantibody-xenoantigen interactions, AVXR still occurs in association with an increase in anti-PAEC xenoantibody titers. Some prolongation of xenograft survival has been achieved by combining complement and/or antibody depletion with anti-B cell immunosuppression to suppress this induced response. Strategies targeting T cells or specific T-cell based costimulatory pathways, which have been very successful in preventing allograft rejection, fail to effectively inhibit the xenoantibody response or to prevent AVXR. The findings of these studies suggest that the induced xenoantibody response may be partially T-cell independent.

ROLE OF IgG IN AVXR

While IgM is often the more significant preformed primate xenoantibody, IgG becomes more prominent following exposure to porcine antigens. While IgM titers rise and fall, IgG titers undergo a more significant and sustained increase, likely representing an isotype switch. Exposure of humans to vascularized porcine organs also results in a rise and fall in IgM, and a sustained IgG rise. Since their functional avidities for αGal are similar, an increasing IgG:IgM ratio will also result in fewer antigenic sites bound by IgM. This results in a relative decrease in complement activating potential since IgGs are less efficient than IgM at fixing complement. Infusion of human IgGs into primate recipients of porcine cardiac xenografts has been shown to prevent HAR, presumably by successful competition with IgM for endothelial binding sites although other mechanisms may also contribute.

ANTIBODY-DEPENDENT CELL-MEDIATED CYTOTOXICITY (ADCC)

In lieu of CML, xenoantibodies may facilitate endothelial cell changes by ADCC. NK cells, triggered by binding to the Fc region of anti-endothelial xenoantibodies, can disrupt them through perforin/granzyme-related mechanisms, or activate them by releasing TNFα. Furthermore, activated NK cells release interferon-γ (IFNγ) which can activate nearby monocytes and macrophages upregulating their expression of adhesion molecules and increasing their local accumulation. Through Fc- and C1q-receptors, monocytes and macrophages can also bind to anti-endothelial xenoantibodies. Subsequent activation leads to release of TNFα, NO, and reactive oxidant intermediates which can activate or injure endothelial cells; and proinflammatory cytokines, procoagulant factors, and complement components which may overwhelm local regulatory mechanisms.

E12

FC-FC RECEPTOR (FCR) INTERACTIONS WITH PERIPHERAL BLOOD LEUKOCYTES (PBLS)

The Fc region of some IgG subclasses, especially IgG$_1$ and IgG$_3$, bind to Fc receptors of circulating human PBLs (hPBLs). IgM does not play a significant role in these Fc-FcR interactions. Three types of Fc receptors (FcγRI [CD64], FcγRIIa [CD32], FcγRIII [CD16]) are found on hPBLs. NK cells express FcγRIIIs, monocytes express FcγRI and FcγRII and macrophages express all 3 types. These Fc-FcR interactions can lead to ADCC and likely contribute to AVXR.

EARLY CLASSICAL PATHWAY COMPLEMENT PROTEINS

While CML is blocked using complement inhibitory strategies, xenoantibodies continue to deposit on the graft endothelium, binding C1q and activating other uninhibited classical pathway complement proteins prior to C3. C1q binding facilitates cellular adhesion and ADCC, while C2b and C4a activation have proinflammatory effects. We hypothesize that the consequence of leaving this segment of the complement cascade functional is the enhancement of interactions with PBLs which lead to the development of proinflammatory and procoagulant events that contribute to AVXR. Monocytes, macrophages, B cells, neutrophils, platelets, and endothelial cells have C1q receptors (C1qR) which can facilitate their binding to endothelial cells following xenoantibody and C1q deposition.

ENDOTHELIAL CHANGES IN AVXR

In all models studied, AVXR is initiated at the level of the endothelium. Two broad classifications of endothelial changes may predispose to AVXR: (a) endothelial disruption; and (b) endothelial activation.

ENDOTHELIAL DISRUPTION

Endothelial disruption leads to a loss of vascular integrity allowing circulating blood cells and clotting factors, usually isolated from surrounding thrombogenic tissues, to initiate procoagulant and proinflammatory events. With electron microscopy (EM), endothelial cell changes can be detected prior to their disruption in rejecting xenografts. In vitro morphologic changes have also been described using light microscopy. While destruction of cell membrane integrity appears to be the primary mechanism for endothelial disruption with CML and ADCC, apoptosis of PAECs has also been demonstrated in vitro following exposure to xenoantibodies.

ENDOTHELIAL ACTIVATION

While endothelial disruption is a terminal event, endothelial activation is an adaptive response that is potentially reversible. Endothelial activation occurs in response to many immunologic and nonimmunologic stimuli. Many of the changes associated with endothelial activation are thought to be dependent on NFκB. Endothelial changes associated with activation include: (i) upregulation of surface expression of E-selectin, VCAM, class I and II MHC antigens; (ii) loss of surface proteins that are central in inhibiting coagulation such as thrombomodulin (TM, and heparan sulfate (HS); and (iii) release of cytokines and factors which are proinflammatory such as IL-, IL-1α and IL-8; release of factors which are procoagulant such as tissue factor (TF), plasminogen activator inhibitor type 1 (PAI-1), platelet activating factor (PAF); and (iii) and induction of enzymes such as inducible nitrous oxide synthetase (iNOS), or release of factors such as and Endothelin-1 which influence vasoactivity.

ANTIBODY-INDEPENDENT EFFECTIVE CELL INTERACTIONS WITH PAECS

NK cells have been shown to bind directly to the α-gal antigen and subsequently activate or disrupt PAECs. Thus direct interactions between NK cells and PAECs may play a role in endothelial activation and AVXR. While this interaction may be important in initiating AVXR it is enhanced by endothelial-bound xenoantibodies. The role of NK cells in pig-to-primate AVXR may be a transient one since they have been a rare finding in rejected porcine xenografts. NK mediated events are also not unique to AVXR since they are also seen in allogeneic models where they are not thought to play a major role. Direct interactions between monocytes/macrophages and PAECs may also contribute to endothelial changes, although these interactions are also enhanced with anti-endothelial xenoantibodies. Furthermore, in all of these studies, the monocyte/macrophage isolation techniques used likely resulted in their activation, and therefore the results may not reflect de novo interactions between monocyte/macrophages and PAECs and thus are unlikely to represent initiating events in AVXR.

THE ROLE OF T-CELL MEDIATED RESPONSES

Human recipient T cells are activated by cells in allogafts through direct engagement of the T-cell receptor by MHC antigens expressed by donor professional antigen presenting cells (direct activation), or by graft-derived peptides expressed on autologous antigen presenting cells in the context of self MHC antigens (indirect activation). Early in vitro studies of discordant xenogeneic cellular interactions in mice suggested a strong dependence upon the indirect pathway of activation. Among the reasons proposed for failure of murine T cells to respond directly to stimulation by xenogeneic cells were species specificity of CD4-class II and CD8-class I interactions, lower precursor cell frequencies, incompatibilities in costimulatory receptor-ligand interactions, and defects in cytokine effects across species barriers. These in vitro studies helped formulate the hypothesis that cell mediated xenograft rejection might be weaker than allograft responses, and thus potentially easier to control. In addition, in vivo examination of disparate skin xenograft rejection suggested a critical role for the CD4+ T-cell population in development of an effective xenogeneic immune response.

Studies of xenogeneic mixed leukocyte cultures using the human anti-pig comination clearly indicate that porcine cells are able to directly elicit proliferation and IL-2 production by human T cells. Studies using highly purified T cells devoid of human APC indicate that direct activation of human responders occurs through recognition of MHC antigen on porcine APC. Furthermore, experiments in which using purified subpopulations of human T cells have been evaluated show that CD4+ T cells proliferate, express IL-2 receptors, and produce IL-2 following stimulation with porcine APC. Blocking experiments with unprimed CD4+ T cells or CD4+ swine-specific T-cell clones have shown these cells to be SLA class II-restricted, indicating the human TCR can directly engage swine MHC antigens, and that the CD4-class II MHC interaction is functional in the pig-to-human species combination. Limiting dilution analysis has indicated the precursor frequency of human xenoresponding cells to be comparable to those found in allogeneic responses. Indeed, most studies have observed direct responses by CD4+ T cells in vitro that are at least as strong or stronger than allogeneic responses measured under the same conditions. These findings may reflect the high degree of homology between pig and human class II antigens that has been noted previously at the DNA level.

The induction of human anti-pig CD8+ T cell responses in vitro has been documented in several studies. Whereas intact proliferation by purified CD8+ T cells directed against class I xenoantigens has been described, others have observed the proliferation of CD8+ T cells following direct recognition of porcine APC to be dependent on the presence of IL-2 or help from CD+ T cells. This finding would represent a departure from the response of CD8+ T cells under allogeneic conditions. Overall, however, the proliferative responses by CD8+ T cells are less robust when compared to those observed by CD4+ T-cell populations. This serves to underscore the distinct role of CD4+ T cells in the xenogeneic cellular immune response.

The so-called 'two-signal' model of T-cell activation requires appropriate TCR-MHC binding be accompanied by an effective costimulatory signal(s) generated through MHC unrestricted receptor-ligand interactions(s). Examples of second signal receptor-ligand pairs include the interaction between CD28 on T cells and the B7 family of molecules expressed on APCs, as well as CD2-LFA-3 and LFA-1-ICAM interactions. Available data from in vitro studies suggest that several important costimulatory molecule interactions are functionally intact in the pig-to-human species combination. In vitro studies have also shown a vigorous indirect activation of human T cells by autologous APC expressing processed xenoantigens. Indeed, some have suggested that stimulation of human lymphocytes by porcine cells may be independent of MHC expression. An indirect xenogeneic response in excess of that observed for allorecognition may be linked to the greater number of antigenic differences between man and swine, and raises the possibility that the xenogeneic response might be difficult to control using currently available therapies.

The effector phase of allorecognition is characterized by the generation of CD8+ CTLs specific for donor MHC antigen. In vitro studies of cytotoxicity in the human anti-pig response have shown that pig-specific cytotoxic cells can be generated, and that such CTL are MHC restricted. However, the in vitro cytotoxic xenogeneic response is also characterized by a profound nonMHC specific cytotoxic component, which has been linked to NK cells. The cytotoxic effects of NK cells against porcine targets has been shown to be mediated through antibody-dependent and independent mechanisms. Of great interest are recent data indicating that the epitopes recognized NK cells on xenogeneic cells are at least in part the same terminal alpha galactose residues that serve as the target of xenoreactive natural antibodies. It is likely that the marked cytotoxic effects of human NK cells upon porcine cell targets is linked to the failure by NK cell inhibitory receptors (KIRs) to properly interact with pig class I molecules. Transfection of pig cells leading to expression of appropriatge human MHC has recently been reported to afford significant protection against human NK cell mediated lysis.

In vivo analysis of the human anti-pig cellular immune response has been hampered by the lack of suitable models. A limited number of studies using non-human primates as recipients of neovascularized islet xenografts have been reported. Cellular rejection of these porcine islets appears to be a T-cell dependent process, also characterized by striking numbers of eosinophils and macrophages in cellular infiltrates. Other investigators have turned to the use of immunodeficient rodents reconstituted with human cells to study the human anti-pig response. Although these models indicate that hu-scid or hu-RAG-1 deficient mice can mount immune responses against porcine islets or skin xenografts, described defects in the T-cell repertoire of reconstituted immunodeficient mice raise questions regarding the applicability and relevance of observations in these models.

THE PROBLEM OF INFECTION: XENOZOONOSES

Although progress has been made in solving the immunologic problems of pig-to-human xenografting, concerns have been increasingly raised regarding the infectious disease potential of xenotransplantation. The use of organs of nonhuman origin would greatly expand the spectrum of potential infectious diseases presently encountered in clinical transplantation to include those pathogens derived from other animal species. Not suprisingly, concerns have been raised over the potential for the introduction of zoonotic diseases not only into the transplant recipient, but into the general human population. Although known infectious pathogens of a donor species can be easily monitored and very likely eliminated through controlled breeding of pathogen-free herds, there may exist novel, latent pathogens capable of inducing symptomatic disease in the xenograft recipient. Retroviruses represent such agents.

Many mammalian species possess retroviruses integrated into their cellular DNA, so-called endogenous retroviruses. These retroviruses are capable of replication and transmission to cells of a different species, and have the potential to cause disease if so transmitted. Endogenous retroviruses have been linked to development of cancers. Recently, pig endogenous retrovirus (PERV) has been shown to infect human cells. This observation raises serious questions regarding the short and long-term risks of xenotransplantation.

Available data suggest the PERV possesses zoonotic potential for human cells in vitro. Replication-competent PERV has been demonstrated in a variety of pig cells and tissues, including peripheral blood mononuclear cells, endothelial cells, islets, and hepatocytes. In addition, different subtypes of PERV, with possible differences in tissue tropisms, have been identified.

There are many unanswered questions regarding PERV. While in vitro infection of human cells by PERV has been demonstrated, an exhaustive review of humans with acute and long-term exposure to porcine tissues, including porcine extracorporeal hepatic assist devices, has failed to demonstrate PERV infection of humans. Although these data are encouraging, they are potentially misleading: the risk of infection in a heavily immunosuppressed recipient of an immediately vascularized porcine organ, with months to years of viral exposure, has not yet been addressed. Retroviral transmission from swine to human is likely to be dependent upon several factors, including degree of retroviral activation (i.e., viral load), as well as recipient immune responses. Unfortunately, many conditions associated with retroviral activation are present in the transplant recipient; these include exogenous immunosuppression, graft rejection, and viral confection. It remains to be determined to what extent PERV viral load is effected by any of these variables.

The sensitivity of mammalian C-type retroviruses such as PERV to inactivation by human serum has recently been shown to depend upon the expression of alpha gal antigen. In so far as dismantling of the host αGal response has been critical to overcoming hyperacute xenograft rejection, it stands to reason that successful xenotransplantation will depend upon deliberate creation of a milieu most favorable to PERV transmission. The exact contribution of anti αGal antibody and complement to neutralization of PERV has not been completely explored.

Retroviral infections often cause severe immunosuppression in many species, opening the door to opportunistic infections leading to fatal disorders). Severity of retrovirus induced immunosuppression has been shown to depend upon viral load, an association clearly observed with HIV and AIDS. The effect of PERV infection upon human immune responses has not been investigated, nor has the immunomodulatory effect(s) of PERV infection been correlated with viral load. Clearly, the issue of PERV will need to be addressed, along with other potential zoonotic agents, as we move toward clinical pig-to-human transplantation.

CONCLUSION

The ultimate success of xenotransplantation will depend upon our ability to safely protect the xenograft from the damaging consequences of AVR/DXR, as well as T-cell dependent immune responses. Although remarkable prolongation of porcine xenograft survival has been achieved in nonhuman primates, subjects invariably succumb to infectious complications from the protocols used to prevent rejection. Safe prolongation of xenograft survival may depend on modulation of the graft-host interaction such that the xenograft is able to survive despite the presence of ongoing immune responses, a process termed "accomodation". Recent data from rodent studies suggest that accomodation of xenografts may be linked to the expression of so-called "protective genes" by endothelial cells, such as A20 and bcl-2, which prevent upregulation of proinflammatory genes and avert the onset of apoptosis. In addition, enduring xenograft survival has been linked to the induction of a helper type 2 (Th2) cytokine immune response within the graft, while rejected grafts display a Th1 type response. Avoidance of AVR/DXR may therefore require the development of transgenic swine donors not only resistant to HAR by virtue of their possession of human complement regulatory proteins, but also "protected" through enhanced expression of certain genes.

Xenoreactive antibodies have generally been viewed in the context of their importance for induction of HAR. However, it is apparent from available in vitro and in vivo evidence that their pathogenetic effects are potentially much greater. Although some have suggested that AVR/DXR can occur in the absence of xenoreactive antibody, the binding of antibody to rejecting porcine xenografts has been a reproducible observation in preclinical primate studies. Xenoreactive antibodies may potentiate, even help initiate the process of endothelial cell activation, and can potentiate the cytotoxic effects of graft infiltrating effector cells. Further examination of the B-cell compartment in xenotransplantation is clearly warranted.

The strength of the immune response to xenografts has encouraged efforts to induce tolerance, with the hope of avoiding the deleterious consequences of excessive recipient immunosuppression. Approaches have included attempts to induce mixed hematopoietic chimerism through the use of donor bone marrow derived cells. More recently, transplantation of fetal pig thymus and liver tissue to thymectomized, irradiated, T- and NK cell depleted mice has been shown to induce long standing donor specific tolerance across a discordant species barrier. Less convincing results for tolerance induction have been achieved in nonhuman primates, where problems related to xenoractive antibodies and potential incompatibility of growth factors have been identified. Modifications, including the use of strategies to absorb pre-existing xenoractive antibodies, B-cell depletion of autologous bone marrow, transfection of autologous bone marrow cells to express xenogeneic MHC antigen, and the use of donor-specific growth factors to promote bone marrow chimerism may prove useful in achieving tolerance in the pig to human species combination. Alternatively, the use of newly developed reagents for blockade of costimulation, such as CTLAIg and CD40 ligand may be sufficient to achieve a state of peripheral T-cell tolerance or anergy without the need for aggressive ablative regimens. It remains to be seen if effective control of the xenogeneic cellular immune responses can be achieved in a safe and clinically applicable manner.

SELECTED READINGS

1. Xenotransplantation; The Transplantation of Organ and Tissues between species 2nd Edition Cooper DKC, Kemp E, Platt JL, White DJG (eds) Springer Verlag 1997.
2. Bach FH, Hancock WW, Ferran C. Protective genes expressed in endothelial cells: A regulatory response to injury. Immunol Today 1997 Oct; 18(10):483-6.
3. Auchincloss H, Sachs DM. Xenogeneric transplantation. Ann Rev Immunol 1998; 16:433-70.
4. Lambights D, Sachs DM, Cooper DK. Discordant organ xenotransplantation in primates: World experience and current status. [Review] Transplantation 1998 Sep; 66(5):547-61.
5. Platt JL, Lin SS, McCregor CG. Acute vascular rejection. Xentransplantation 1998 Aug; 5 (3):169-75.
6. Patience KC, Takeuchi Y, Weiss RA. Zoonoses in xenotransplantation. Curr Opin Immunol 1998 Oct; 10 (5):539-42.
7. Dorling A, Lechler RI. T-cell-mediated xenograft rejection: Specific tolerance is probably required for long term xenograft survival. Xenotransplantation 5 (4):234-245.

E12

Appendix I

Alemtuzumab

Brand Name	Campath®
Company	Ilex/Berlex
Class	• Humanized rodent monoclonal antibody against the CD-52 expressed on both B and T lymphocytes
Mechanism of Action	• Antibody and cell mediated lysis of B and T lymphocytes
Indication	• Primary: Treatment of chronic B cell lymphocytic leukemia
Adverse Reactions	• Fever • Chills shortly after infusion
Formulation	• 30 mg vials
Dosage	• up to 30 mg daily

Acknowledgment:

The drug tables in Appendix I were published in Transplantation Drug Manual. Pirsch J, Simmons W, Sollinger H, eds. ©2003 Landes Bioscience.

Antithymocyte Globulin (Equine)

Brand Name	ATGAM®
Company	Pharmacia & Upjohn, Inc.
Class	• Immunosuppressant gamma globulin, primarily monomeric IgG, from hyperimmune serum of horses immunized with human thymic lymphocytes
Mechanism of Action	• Antibodies of multiple specificities interact with lymphocyte surface antigens, depleting numbers of circulating, thymus-dependent lymphocytes and interfering with cell-mediated and humoral immune responses
Indications	• Management of renal allograft rejection • Adjunct to other immunosuppressive therapy to delay the onset of the first rejection episode
Contraindication	• Hypersensitivity to ATGAM or any other equine gamma globulin preparation
Warnings	• Should be administered in facilities equipped and staffed with adequate laboratory and supportive medical resources • Immunosuppressive activity may vary from lot to lot • Potential for the transmission of infectious agents • Treatment should be discontinued if the following occur: • Symptoms of anaphylaxis • Thrombocytopenia • Leukopenia
Special Precautions	• Risk of infection, leukopenia, and thrombocytopenia • Safety and effectiveness demonstrated only in patients who received concomitant immunosuppression • Pregnancy Category C
Adverse Reactions	• Fever (1 patient in 3) • Chills (1 patient in 7) • Leukopenia (1 patient in 7) • Dermatologic reactions (1 patient in 8) • Thrombocytopenia (1 patient in 9) *Reported in >1%, but <5% of Patients* • Arthralgia • Chest and/or back pain • Clotted A/V fistula • Nausea and/or vomiting • Night sweats • Pain at infusion site • Peripheral thrombophlebitis • Stomatitis
Drug Interaction	• Dextrose Injection, USP
Formulation	• 5 mL ampule containing 50 mg/mL
Dosage	*Delaying Onset of Allograft Rejection* • Fixed dose of 15 mg/kg/d IV for 14 days, then every other day for 14 days for a total of 21 doses in 28 days • First dose should be administered within 24 hours before or after transplantation

Continued

Antithymocyte Globulin (Equine) (cont'd)

	Treatment of Rejection
	• 10 mg/kg/d IV for 8-14 days, then every other day up to 21 doses
	Dose should be infused at least over 4 hours, through a 0.2-1 micron filter

Editors' Notes:

ATGAM® and Thymoglobulin® are the only two polyclonal antilymphocyte preparations which are currently available. ATGAM® is an immunoglobule against lymphocytes prepared in horses; Thymoglobulin® is prepared in rabbits. ATGAM® causes less leukopenia than Thymoglobulin®. Both agents have been used to prevent acute rejection after transplantation and to treat acute rejection episodes.

Antithymocyte Globulin (Rabbit)

Brand Name	Thymoglobulin®
Company	SangStat
Class	• Immunosuppressant gamma globulin, obtained by immunization of rabbits with human thymocytes
Mechanism of Action	• Antibodies of multiple specificities interact with lymphocyte surface antigens, depleting numbers of circulating T-lymphocytes and modulating T-lymphocyte activation, homing and cytotoxic processes
Indication	• Treatment of acute renal allograft rejection in combination with other immunosuppressants
Contraindication	• History of anaphylaxis or allergy to Thymoglobulin or rabbit proteins, or acute viral illness
Warnings	• Should be administered in facilities equipped and staffed with adequate laboratory and supportive medical resources • Anaphylaxis has been reported • Thrombocytopenia or neutropenia may result but are reversible with dose reduction or discontinuance
Special Precautions	• Risk of infections, leukopenia, thrombocytopenia, lymphoma, post-transplant lymphoproliferative disease or other malignancies • Pregnancy Category C
Adverse Reactions	• Fever • Chills • Leukopenia • Pain/ abdominal pain • Headache • Thrombocytopenia • Dyspnea • Malaise • Dermatologic reactions
Drug Interaction	• None reported
Formulation	• 25 mg vial of lyophilized powder
Dosage	• 1.5 mg/kg/d for 7 to 14 days IV, diluted and infused through a 0.22 micron filter into a high flow vein • First dose infused over a minimum of 6 hours, subsequent doses over a minimum of 4 hours

Editors' Notes:

Antithymocyte globulin (rabbit) is approved for the reversal of acute rejection. A double-blind, randomized trial of thymoglobulin vs ATGAM® was conducted in 163 renal recipients with rejection. Thymoglobulin had a higher reversal rate than ATGAM® (88% vs 76%, p=0.027). Transplantation 66:29-37, July 15, 1998.

A recent study of high-risk renal transplants compared Thymoglobulin® induction with Simulect. The overall risk of rejection, delayed graft function and graft loss as statistically less frequent with Thymoglobulin®. (Brennan DC. A prospective, randomized, multi-center study of thymoglobulin compared to simulect for induction of immunosuppression: preliminary results. ©2002 American Transplant Congress.)

Azathioprine

Brand Name	Imuran®
Company	Glaxo Wellcome Inc.
Class	• Immunosuppressive antimetabolite (6-mercaptopurine)
Mechanism of Action	• Interferes with DNA and RNA synthesis, thereby inhibiting differentiation and proliferation of both T and B lymphocytes
Indication	• Adjunct for the prevention of rejection in renal hom transplantation
Clinical Experience	• 35-55% 5-year patient survival in over 16,000 transplantations
Contraindication	• Hypersensitivity to azathioprine
Warnings	• Increased risk of neoplasia • Severe myelosuppression • Serious infection • Pregnancy Category D
Special Precautions	• Gastrointestinal hypersensitivity reaction • Periodic blood counts may be needed
Adverse Reactions	• Leukopenia (>50%) • Infection (20%) • Nausea and vomiting (NA) • Neoplasia (3.3%) • Hepatotoxicity (NA)
Drug Interactions	• Allopurinol • Agents affecting myelopoiesis • Angiotensin converting enzyme inhibitors
Formulations	• 50 mg scored tablets • 20 mL vial containing 100 mg azathioprine
Dosage	• 3 mg/kg/d to 5 mg/kg/d single dose given at time of transplantation • 1 mg/kg/d to 3 mg/kg/d for maintenance • Dose usually adjusted depending on WBCs • Lower doses should be considered in presence of renal dysfunction

Editors' Notes:

Combined administration of azathioprine and allopurinol may result in severe pancytopenia. The dose of Azathioprine should be reduced by two-thirds when given with allopurinol. Azathioprine rarely causes liver dysfunction, frequently manifested by an isolated rise in ALT and bilirubin. With the introduction of mycophenolate mofetil, azathioprine may be relegated to a second-line antimetabolite for the prevention of graft rejection.

Basiliximab

Brand Name	Simulect®
Company	Novartis
Class	• Immunosuppressive chimeric monoclonal antibody, specifically binds to and blocks the interleukin-2 receptor alpha chain on the surface of activated T- lymphocytes
Mechanism of Action	• Acts as an IL-2 receptor antagonist by binding with high affinity to the alpha chain of the IL-2 receptor complex and inhibits IL-2 binding • Competitively inhibits IL-2 mediated activation of lymphocytes
Indications	• Prophylaxis of acute renal allograft rejection when used as part of an immunosuppressive regimen that includes steroids and cyclosporine
Contraindications	• Hypersensitivity to basiliximab or any component of the formulation
Warnings	• Should be administered in facilities equipped and staffed with adequate laboratory and supportive medical resources • Administration of proteins may cause possible anaphylactoid reactions (none reported) • Immunosuppressive therapies increase risk for lympho-proliferative disorders and opportunistic infections (incidence in basiliximab-treated patients is similar to placebo)
Special Precautions	• Long-term effect and re-administration after initial course has not been studied • Pregnancy Category B
Adverse Reactions	• Similar to placebo-treated patients
Drug Interactions	• None reported
Formulation	• 20 mg vial of lyophilized powder
Dosage	• 20 mg within 2 hours of transplantation surgery and repeated 4 days after transplantation • Children 2 to 15 years is 12 mg/m^2, to a maximum of 20 mg/dose

Editors' Notes:

Basiliximab was approved after a study of 380 cadaver transplant recipients on cyclosporine and prednisone was completed. Basiliximab recipients had a rejection incidence of 29.8% compared to placebo (44%, p=0.01). Steroid-resistant rejection was also lower. Lancet 350(9086):1193-1998, October 25, 1997.

The clinical experience with Basiliximab had confirmed the results of this study with excellent efficacy and low toxicity.

A recent study of high-risk renal transplants compared Thymoglobulin® induction with Simulect. The overall risk of rejection, delayed graft function and graft loss was statistically less frequent with Thymoglobulin®. (Brennan DC. A prospective, randomized, multi-center study of thymoglobulin compared to simulect for induction of immunosuppression: preliminary results. ©2002 American Transplant Congress.)

Cyclosporine-A (Non-Modified)

Brand Name	Sandimmune®	Cyclosporine
Company	Novartis[1]	Apotex[2]
Class	• Immunosuppressant produced as a metabolite by the fungus species *Beauvaria nivea* Gams[1] *Tolypocladium inflatum* Gams[2]	
Mechanism of Action	• Preferential inhibition of T lymphocytes • Suppresses activation of T lymphocytes by inhibiting production and release of lymphokines, specifically interleukin-2	
Indications	• Prophylaxis of graft rejection in kidney, liver, and heart allogeneic transplantation • Treatment of chronic rejection previously treated with other immunosuppressants	
Contraindication	• Hypersensitivity to cyclosporine or polyoxyethylated castor oil	
Warnings	• Nephrotoxicity • Hepatotoxicity • Increased susceptibility to infection and lymphoma • Erratic absorption of soft gelatin capsules and oral solution necessitates repeated monitoring of cyclosporine blood levels • Anaphylactic reactions with IV formulation	
Special Precautions	• Hypertension may occur and require therapy with antihypertensives (potassium-sparing diuretics should not be used) • Repeated laboratory monitoring is required • Pregnancy Category C	
Adverse Reactions	• Renal dysfunction (25% - renal, 38% - cardiac, 37% - hepatic) • Tremor (21% - renal, 31% - cardiac, 55% - hepatic) • Hirsutism (21% - renal, 28% - cardiac, 45% - hepatic) • Hypertension (13% - renal, 53% - cardiac, 27% - hepatic) • Gum hyperplasia (9% - renal, 5% - cardiac, 16% - hepatic)	
Drug Interactions	*Drugs with Synergistic Nephrotoxicity* • Gentamicin • Amphotericin B • Tobramycin • Ketoconazole • Vancomycin • Melphalan • Cimetidine • TMP-SMX • Ranitidine • Azapropazon • Diclofenac • Naproxen • Sulindac *Drugs that Increase Cyclosporine Levels* • Diltiazem • Ketoconazole • Nicardipine • Erythromycin • Verapamil • Itraconazole • Danazol • Bromocriptine • Fluconazole • Methylprednisolone • Metoclopramide • Drugs which inhibit cytochrome P450 3A4 *Drugs that Decrease Cyclosporine Levels* • Rifampin • Phenobarbitol • Carbamazepine • Phenytoin • Drugs which induce cytochrome P450 3A4 *Reduced Clearance with Cyclosporine* • Prednisone • Lovastatin	

Continued

Cyclosporine-A (Non-Modified) (cont'd)

Formulations	• Digoxin • Grapefruit juice can increase cyclosporine concentrations *Other* • Vaccinations—live vaccinations should be avoided • 25 mg soft gelatin capsules • 50 mL bottle containing 100 mg/mL for oral solution • 5 mL vial containing 50 mg/mL
Dosage	*Gelatin Capsules & Oral Solution* • 15 mg/kg single dose given 4 to 12 hours prior to transplantation • Single daily dose is continued postoperatively for 1 to 2 weeks and then tapered by 5% each week until maintenance dose of 5 mg/kg/d to 10 mg/kg/d is reached *IV Infusion* • 5 mg/kg to 6 mg/kg single dose given 4 to 12 hours prior to transplantation • Single daily IV dose continued postoperatively until patient can tolerate oral formulations

Editors' Notes:

Sandimmune® is the original formulation of cyclosporine used in clinical transplantation. Erratic absorption of the formulation has limited its use in transplant recipients. Many long-term transplant recipients continue on Sandimmune®. The formulation has been supplanted by the micro-emulsion formula of cyclosporine.

Cyclosporine Capsules and Oral Solution (Modified)

Brand Name	Neoral®	Gengraf	Cyclosporine	Cyclosporine
Company	Novartis[1]	Abbott Laboratories/SangStat[2]	Eon[3]	Sidmak[4]
Class	• Immunosuppressant produced as a metabolite by the fungus species *Beauvaria nivea* Gams[1]; *Aphanocladium album*[2]; *Cordyceps militaris*[3,4]			
Mechanism of Action	• Preferential inhibition of T lymphocytes • Suppresses activation of T lymphocytes by inhibiting production and release of lymphokines, specifically interleukin-2			
Indications	• Prophylaxis of graft rejection in kidney, liver, and heart allogeneic transplantation • Rheumatoid arthritis • Psoriasis			
Contraindication	• Hypersensitivity to cyclosporine or any of the ingredients of the formulation			
Warnings	• Nephrotoxicity • Hepatotoxicity • Increased susceptibility to infection and lymphoma • Reports of convulsions in pediatric and adult patients, especially when used in conjunction with methylprednisolone • Modified cyclosporine is not bioequivalent to Sandimmune			
Special Precautions	• Any change in cyclosporine formulation should be made cautiously and under the advisement of a physician—Patients should be given detailed dosage instructions • Hypertension may occur and require therapy with antihypertensives (potassium-sparing diuretics should not be used) • Repeated laboratory monitoring is required • Pregnancy Category C			
Adverse Reactions	• Renal dysfunction • Hyperkalemia • Tremor • Hyperurecemia • Hirsutism • Encephalopathy • Hypertension • Gum hyperplasia			
Drug Interactions	*Drugs with Synergistic Nephrotoxicity* • Gentamicin • Amphotericin B • Naproxen • Tobramycin • Ketoconazole • Sulindac • Vancomycin • Melphalan • Colchicine • TMP-SMX • Tacrolimus • Non-Steroidal • Cimetidine • Diclofenac • Anti- • Ranitidine • Azapropazon inflammatory agents *Drugs that Increase Cyclosporine Levels* • Diltiazem • Clarithromycin			

Continued

Cyclosporine Capsules and Oral Solution (Modified) (cont'd)

Drug Interactions (cont'd)	• Nicardipine • Verapamil • Ketoconazole • Fluconazole • Itraconazole • Erythromycin • Quinupristin/Dalfopristin • Amiodarone	• Allopurinol • Danazol • Bromocriptine • Methylprednisolone • Metoclopramide • Drugs which inhibit cytochrome P450 3A4 • Colchicine • HIV-protease inhibitors
	Drug/Dietary Supplements that Decrease Cyclosporine Levels	
	• Rifampin • Nafcillin • Phenytoin • Phenobarbitol • St. John's Wort	• Carbamazepine • Octreotide • Ticlopidine • Drugs which induce cytochrome P450 3A4 • Orlistat
	Reduced Clearance with Cyclosporine	
	• Prednisone • Digoxin • Grapefruit and grapefruit juice can increase cyclosporine concentrations	• Lovastatin
	Other • Vaccinations—live vaccinations should be avoided	
Formulations	• Neoral 25 mg and 100 mg soft gelatin capsules in 30-count blister packages • Oral solution 100 mg/ml in bottle containing 50 ml • Gengraf 25 mg and 100 mg capsules in 30-count unit dose packages • Eon cyclosporine 25 mg and 100 mg capsules in 30-count unit dose blisters • Sidmak cyclosporine 25 mg and 100 mg capsules in 30-count unit dose blisters	

Dosage	**NOTE: Neoral and Sandimmune are not bioequivalent and should not be used interchangeably. Neoral has increased bioavailability and this should be taken into consideration when making dosing decisions.**
	• Daily dose should be given as two divided doses on a consistent schedule
	
	Newly Transplanted Patients
	• Initial dose of Neoral should be the same as a Sandimmune dose. Suggested initial doses include:
	  – 9 ± 3 mg/kg/day for kidney transplant patients
	  – 8 ± 4 mg/kg/day for liver transplant patients
	  – 7 ± 3 mg/kg/day for heart transplant patients
	• The dose is then subsequently adjusted to achieve a predefined cyclosporine blood concentration	

Continued

Cyclosporine Capsules and Oral Solution (Modified) (cont'd)

	Conversion from Sandimmune to Neoral • Neoral should be started with the same daily dose as was previously used with Sandimmune (1:1 dose conversion) • The Neoral dose should then be adjusted to achieve preconversion cyclosporine blood trough concentrations • Until cyclosporine blood trough concentrations reach preconversion levels, monitoring should be undertaken every 4 to 7 days

Editors' Notes:

Cyclosporine has been used in clinical transplantation for nearly 20 years. However, it is clear that trough monitoring of CSA is the best way to monitor for adequate immunosuppression. Recently, C2 monitoring is being advocated as better than trough monitoring for better efficacy and preventing toxicity. The CSA blood level at 2 hours (C2) is a more consistent and reliable measure of the absorption of cyclosporine. (Cyclosporine microemulsion (Neoral) absorption profiling and sparse-sample predictors during the first 3 months after renal transplantation. ©2002 Am J Transplant)

Daclizumab

Brand Name	Zenapax®
Company	Roche Laboratories
Class	• Immunosuppressive humanized monoclonal antibody, specifically binds to and blocks the interleukin-2 receptor alpha chain on the surface of activated T- lymphocytes
Mechanism of Action	• Acts as an IL-2 receptor antagonist by binding with high affinity to the alpha chain of the IL-2 receptor complex and inhibits IL-2 binding • Competitively inhibits IL-2 mediated activation of lymphocytes
Indications	• Prophylaxis of acute renal allograft rejection when used as part of an immunosuppressive regimen that includes steroids and cyclosporine
Contraindication	• Hypersensitivity to daclizumab or any component of the formulation
Warnings	• Should be administered in facilities equipped and staffed with adequate laboratory and supportive medical resources • Administration of proteins may cause possible anaphylactoid reactions (none reported) • Immunosuppressive therapies increase risk for lympho-proliferative disorders and opportunistic infections (incidence in daclizumab-treated patients is similar to placebo)
Special Precautions	• Long-term effect and re-administration after initial course has not been studied • Pregnancy category C
Adverse Reactions	• Similar to placebo-treated patients
Drug Interactions	• None reported
Formulation	• 5 ml vial containing 25 mg
Dosage	• 1 mg/kg/dose for 5 doses, the first dose within 24 hours of transplantation, then at intervals of 14 days for four doses. Dilute with 50 ml normal saline over 15 minutes.

Editors' Notes:

Daclizumab was approved after a study of 260 cadaver recipients on azathioprine, cyclosporine and prednisone was completed. Daclizumab recipients had a rejection incidence of 22% compared to placebo (35%, p<0.03). New Engl J Med 338(3):161-165, 1998.

Recent data from a kidney pancreas induction study suggests that 2 doses of Daclizumab (2 mg/kg) at day 0 and day 14 is equivalent to 5 doses of 1 mg/kg every 14 days. (Stratta AJ, Alloway RR, Hodge E et al. A multicenter, open-label, comparative trial of two Daclizumab dosing strategies vs. no antibody induction in combination with tacrolimus, mycophenolate mofetil, and steroids for the prevention of acute rejection in simultaneous kidney-pancreas transplant recipients: interim analysis. Clin Transplant ©2002; 16(1):60-8.)

Methylprednisolone

Brand Name	Solu-medrol®
Company	Pharmacia and Upjohn, Inc.
Class	• Anti-inflammatory steroid
Mechanism of Action	• Causes emigration of circulating T cells from intravascular tissue compartment to lymphoid tissue • Inhibits production of T-cell lymphokines that are needed to amplify macrophage and lymphocyte response
Applications in Transplantation	• Immunosuppressive adjunct for the prevention and treatment of solid organ rejection • Attenuation of cytokine release syndrome in patients treated with muromonab-CD3
Contraindications	• Hypersensitivity to methylprednisolone and/or its components • Systemic fungal infections • Prematurity in infants
Warnings	• May produce posterior subcapsular cataracts, glaucoma, and may enhance the establishment of secondary ocular infections due to fungi or viruses • Increased calcium excretion • Vaccination should not be undertaken during therapy • Cardiac arrhythmias, circulatory collapse, and/or cardiac arrest have occurred after rapid administration • Adequate human reproduction studies have not been undertaken
Special Precautions	• Use with caution in patients with hypothyroidism, cirrhosis, ocular herpes simplex, hypertension, congestive heart failure, and ulcerative colitis • Psychologic derangements may occur while on therapy
Adverse Reactions	• Sodium and fluid retention, congestive heart failure in susceptible patients, potassium loss, hypokalemic alkalosis, hypertension • Muscle weakness, steroid myopathy, loss of muscle mass, severe arthralgia, vertebral compression fractures, aseptic necrosis of humeral and femoral heads, pathologic fracture of long bones, osteoporosis • Peptic ulcer with possible perforation and hemorrhage, pancreatitis, abdominal distension, ulcerative esophagitis • Impaired wound healing, thin fragile skin, petechiae and ecchymoses, facial erythema, increased sweating, suppressed reaction to skin tests • Pseudotumor cerebri, convulsions, vertigo, headache • Cushingoid state, suppression of growth in children, secondary adrenocortical and pituitary unresponsiveness, menstrual irregularities, decreased carbohydrate tolerance, manifestations of latent diabetes mellitus, increased requirements for insulin or oral hypoglycemic agent • Posterior subcapsular cataracts, increased intraocular pressure, glaucoma, exophthalmos

Continued

Methylprednisolone (cont'd)

Drug Interactions	• Barbiturates	• Phenytoin	• Rifampin
	• Salicylates	• Vaccines	• Toxoids
Formulations	• 40 mg single dose vial • 125 mg single dose vial • 500 mg vial • 500 mg vial with diluent • 1 g vial • 1 g vial with diluent		
Dosage	Induction • 250 mg to 1000 mg at time of transplantation and for next 2 to 3 doses *Taper* • Start at 2 mg/kg/d, taper to a range of 0.15 mg/kg/d to 0.2 mg/kg/d after one year *Attenuation of Cytokine Release Syndrome* • 8 mg/kg given 1 to 4 hours prior to first injection of muromonab-CD3		

Editors' Notes:

Multiple different regimens of methylprednisolone have been utilized to treat rejection. The initial higher doses are the most important in controlling rejection. It is unclear whether prolonged tapering of steroids following rejection is of value.

Muromonab-CD3

Brand Name	ORTHOCLONE OKT®3
Company	Ortho Biotech Inc.
Class	• Immunosuppressive monoclonal antibody with singular specificity to CD3 antigen of human T cells
Mechanism of Action	• Blocks the function of CD3 molecule in the membrane of human T cells, which has been associated in vitro with the antigen recognition structure of human T cells that is essential for signal transduction
Indications	• Treatment of acute renal allograft rejection as soon as it is diagnosed • Treatment of steroid-resistant acute cardiac allograft rejection • Treatment of steroid-resistant acute hepatic allograft rejection
Contraindications	• Hypersensitivity to muromonab-CD3 and/or any product of murine origin • Antimurine antibody titers $\geq$1:1000 • Uncompensated heart failure or fluid overload • History of seizures • Determined to be or suspected of being pregnant or breast-feeding
Warnings	• Cytokine release syndrome, ranging from a mild, self-limited, flu-like syndrome to a less-frequently reported severe, life-threatening shock-like reaction, has been associated with first few doses • Can be attenuated by premedicating with methylprednisolone, 8 mg/kg, given 1 to 4 hours prior to first injection • Anaphylactic reactions • Neuropsychiatric events, including seizures, encephalopathy, cerebral edema, aseptic meningitis, and headache • Risk of infection and neoplasia • Patients should be managed in a facility equipped and staffed for cardiopulmonary resuscitation
Special Precautions	• Clear chest X ray and weight restriction of $\leq$3% above patient's minimum weight during the week prior to injection • If patient's temperature exceeds 37.8°C, it should be lowered with antipyretics before each injection • Periodic monitoring to ensure muromonab-CD3 levels >800 ng/mL and CD3$^+$ cell levels <25 cell/mm^3 • Potentially serious signs and symptoms with immediate onset are likely due to hypersensitivity and therapy should be discontinued • Antiseizure precautions should be undertaken • Patient should be monitored for signs of infection and/or lymphoproliferative disorders • Antimurine antibody titers should be monitored after therapy with muromonab-CD3 • As with other immunosuppressives, arterial or venous thromboses of allografts and other vascular beds have been reported • Pregnancy Category C

Continued

Muromonab-CD3 (cont'd)

Adverse Reactions	• Pyrexia • Tachycardia • Chills • Rigor • Dyspnea • Hypertension • Nausea and vomiting • Infection with herpes simplex • Chest pain virus, cytomegalovirus, • Diarrhea *Staphylococcus epidermidis,* • Tremor *Pneumocystis carinii, Legionella,* • Wheezing *Cryptococcus, Serratia,* and • Headache gram-negative bacteria • Hypersensitivity reactions • Posttransplant lympho- proliferative disorders
Drug Interactions	• Indomethacin alone and in conjunction with muromonab-CD3 has been associated with CNS effects • Corticosteroids alone and in conjunction with muromonab-CD3 have been associated with psychosis and infection • Azathioprine alone and in conjunction with muromonab-CD3 has been associated with infection and malignancies • Cyclosporine-A alone and in conjunction with muromonab-CD3 has been associated with seizures, encephalopathy, infection, malignancies, and thrombotic events
Formulation	• 5 mL ampule containing 5 mg
Dosage	• 5 mg/d IV for 10 to 14 days

Editors' Notes:

Muromonab-CD3 is used infrequently to treat or prevent rejection in the present era. It has significant toxicity due to the cytokine defense syndrome (CDS).

Evidence suggests that giving 250 mg of methylprednisolone 6 hours prior and 1 hour prior to administration of muromonab-CD3 can decrease the incidence of cytokine release syndrome (CRS). Pentoxifylline has not been shown to attenuate CRS. Indomethacin, however, can be effective.

If CD3 levels remain high during therapy with muromonab-CD3, the dose may be increased or one may switch to an alternate antilymphocyte preparation.

Adjunctive prophylaxis with ganciclovir in CMV-positive patients or in recipients of CMV-positive organs may be effective in reducing risk of CMV infection.

Mycophenolate Mofetil

Brand Name	CellCept®
Company	Roche Laboratories
Class	• Immunosuppressive antimetabolite
Mechanism of Action	• Selectively inhibits inosine monophosphate dehydrogenase in the de novo pathway of purine synthesis, producing potent cytostatic effects on T and B lymphocytes
Indication	• Prophylaxis of graft rejection in patients receiving renal and cardiac allogeneic transplantation
Contraindication	• Hypersensitivity to mycophenolate mofetil, mycophenolic acid or any components of the drug
Warnings	• Increased susceptibility to infection and lymphoma • Adverse effects on fetal development have been observed in pregnant rats and rabbits—mycophenolate mofetil should not be used in pregnant women and contraception should be used during therapy • Neutropenia has been observed
Special Precaution	• Gastrointestinal hemorrhage may occur • Patients with renal impairment have shown higher MPA and MPAG AUCs than normal volunteers • Should not be used in conjunction with azathioprine • Repeated laboratory monitoring is required • Pregnancy Category C
Adverse Reactions	• Diarrhea • Sepsis • Leukopenia • Vomiting
Drug Interactions	• Acyclovir • Antacids with magnesium and aluminum hydroxides • Cholestyramine • Drugs that alter gastrointestinal flora may interact with mycophenolate mofetil by disrupting enterohepatic recirculation • Probenecid
Formulation	• 250 mg capsules supplied in bottles of 100 and 500 • 500 mg tablets in bottles of 100 and 500 • CellCept intravenous: 20 ml, sterile vial containing 500 mg mycophenolate mofetil
Dosage	• 1.0 g twice a day used in combination with corticosteroids and cyclosporine • Initial dose should be given within 72 hours following transplantation

Editors' Notes:

Mycophenolate mofetil reduces the risk of first acute rejection by 50%. Toxicity is minor, but includes bone marrow suppression and gastrointestinal complaints. A higher incidence of CMV disease compared to azathioprine control was observed in the clinical trials. The efficacy of the long-term use of mycophenolate mofetil has not been established.

The metabolism of mycophenolate is altered by coadministration with cyclosporine. Mycophenolic acid (MPA) levels are lower when mycophenolate mofetil is compared with cyclosporine. Lower doses should be considered in recipients receiving tacrolimus or steroids alone without a calcineurin inhibitor.

Prednisone

Brand Name	Deltasone®
Company	Pharmacia and Upjohn, Inc.
Class	• Anti-inflammatory steroid
Mechanism of Action	• Causes emigration of circulating T cells from intravascular tissue compartment to lymphoid tissue • Inhibits production of T cell lymphokines that are needed to amplify macrophage and lymphocyte response
Application in Transplantation	• Immunosuppressive adjunct for the prevention and treatment of solid organ rejection
Contraindications	• Hypersensitivity to prednisone and/or its components • Systemic fungal infection
Warnings	• May produce posterior subcapsular cataracts, glaucoma, and may enhance the establishment of secondary ocular infections due to fungi or viruses • Increased calcium excretion • Vaccination should not be undertaken during therapy • Adequate human reproduction studies have not been undertaken
Special Precautions	• Use with caution in patients with hypothyroidism, cirrhosis, ocular herpes simplex, hypertension, congestive heart failure, and ulcerative colitis • Psychologic derangements may occur while on therapy
Adverse Reactions	• Sodium and fluid retention, congestive heart failure in susceptible patients, potassium loss, hypokalemic alkalosis, hypertension • Muscle weakness, steroid myopathy, loss of muscle mass, vertebral compression fractures, aseptic necrosis of humeral and femoral heads, pathologic fracture of long bones, osteoporosis • Peptic ulcer with possible perforation and hemorrhage, pancreatitis, abdominal distension, ulcerative esophagitis • Impaired wound healing, thin fragile skin, petechiae and ecchymoses, facial erythema, increased sweating, suppressed reaction to skin tests • Pseudotumor cerebri, convulsions, vertigo, headache • Cushingoid state, suppression of growth in children, secondary adrenocortical and pituitary unresponsiveness, menstrual irregularities, decreased carbohydrate tolerance, manifestations of latent diabetes mellitus, increased requirements for insulin or oral hypoglycemic agent • Posterior subcapsular cataracts, increased intraocular pressure, glaucoma, exophthalmos
Drug Interactions	• Barbiturates • Phenytoin • Rifampin • Salicylates • Vaccines • Toxoids
Dosage Strengths	• 2.5 mg tablets • 5 mg tablets • 10 mg tablets • 20 mg tablets • 50 mg tablets

Continued ...

Prednisone (cont'd)

Dosage	*Maintenance—Adults* • 0.1 mg/kg/d to 2 mg/kg/d usually given once daily *Maintenance—Pediatric* • 0.25 mg/kg/d to 2 mg/kg/d or 25 mg/m² to 60 mg/m² usually given daily or on alternate days

Editors' Notes:

Steroid avoidance or early steroid withdrawal is under active investigation in many transplant centers. The short-term results are acceptable; long-term efficacy is yet to be established.

Rituximab

Brand Name	Rituxan®
Company	IDEC Pharmaceuticals
Class	• Chimeric/murine monoclonal antibody directed against CD20 expressed on B lymphocytes both normal and malignant
Mechanism of Action	• Antibody and cell mediated lysis of B lymphocytes
Indications	• Primary indications: treatment of CD20 positive B-cell non Hodgkin's Lymphoma
Adverse Reactions	• Fever • Chills shortly after infusion
Drug Interactions	• None
Formulation	• 100 mg and 500 mg vials
Dosage	• 375 mg/m^2 I.V.

Sirolimus

Brand Name	Rapamune®
Company	Wyeth-Ayerst
Class	• Kinase inhibitor macrocyclic lactone antibiotic with immunosuppressant properties
Mechanism of Action	• Binds to an immunophilin protein to form a complex which inhibits the activation of the mammalian Target of Rapamycin (mTOR)ulatory kinase. This inhibits T lymphocyte activation and proliferation by IL-2, IL-4, and IL-5.
Indications	• The prophylaxis of organ rejection in patients receiving renal transplants.
Contraindications	• Hypersensitivity to sirolimus or its derivatives or any component of the drug product.
Warnings	• Increased risk of infection and lymphomas.
Special Precautions	• Increased serum cholesterol and triglycerides. • Lymphocele • Impaired renal function in combination with cyclosporine • Pregnancy Category C
Adverse Reactions	• Hypertension • Rash • Acne • Anemia • Arthralgia • Diarrhea • Hypokalemia • Thrombocytopenia • Leukopenia • Fever
Drug Interactions	• Sirolimus is known to be a substrate for both cytochrome CYP3A4 and P-glycoprotein. *Drugs that Increase Sirolimus Levels* • Cyclosporine (amount affected by coadministration schedule and formulation). • Diltiazem • Ketoconazole • Rifampin *Drugs that may Increase Sirolimus Levels Include:* • Calcium channel blockers: nicardipine, verapamil • Antifungal agents: clotrimazole, fluconazole, itraconazole • Macrolide antibiotics: clarithromycin, erythromycin, troleandomycin • Gastrointestinal prokinetic agents: cisapride, metoclopramide • Other drugs: bromocriptine, cimetidine, danazol, HIV-protease inhibitors (e.g., ritonavir, indinavir) • Grapefruit and grapefruit juice may increase sirolimus concentrations. *Drugs that may Decrease Sirolimus Levels Include:* • Anticonvulsants: carbamazepine, phenobarbital, phenytoin • Antibiotics: rifabutin, rifapentine • Herbal preparations: St. John's Wort (hypericum perforatum) could result in reduced sirolimus levels.

Continued

Sirolimus (cont'd)

Formulations	• Oral solution in a concentration of 1 mg/ml in:
	    2 oz (60 ml fill) amber glass bottles
	    5 oz (150 ml fill) amber glass bottles
	• Cartons containing 30 unit-of-use laminated aluminum pouches of 1 ml, 2 ml and 5 ml
	• 1 mg tablets supplied in a bottle of 100 tablets
	• 1 mg tablets supplied in a carton of 100 tablets (10 blister cards of 10 tablets each)

Dosage	• De novo transplant recipients, a loading dose of sirolimus of 3 times the maintenance dose should be given. A daily maintenance dose of 2 mg is recommended for use in renal transplant patients, with a loading dose of 6mg. • The initial dose in patients >12 years old, who weigh less than 40 kg the loading dose should be 3 mg/m^2 followed by 1 mg/m^2/day.

Editors' Notes:

To minimize the variability of blood concentrations, sirolimus should be taken consistently in relation to time of administration with or without cyclosporine and/or food.

Sirolimus solution - For simultaneous administration, the mean C_{max} and AUC of sirolimus were increased by 116% and 230%, respectively, relative to administration of sirolimus alone. However, when given 4 hours after Neoral® Soft Gelatin Capsules (cyclosporine capsules [MODI-FIED]) administration, sirolimus C_{max} and AUC were increased by 37% and 80%, respectively, comparedto sirolimus alone. Cyclosporine clearance was reduced only after multiple-dose administration over 6 months.

Sirolimus tablets - For simultaneous administration, mean C_{max} and AUC were increased by 512% and 148%, respectively, relative to administration of sirolimus alone. However, when given 4 hours after cyclosporine administration, sirolimus C_{max} and AUC were both increased by only 33% compared with administration of sirolimus alone.

Sirolimus is an excellent immunosuppressant without apparent nephrotoxicity. The major problems with clinical use are delayed wound healing in some recipients and hyperlipidemia.

Tacrolimus (FK-506)

Brand Name	Prograf®
Company	Fujisawa
Class	• Macrolide antibiotic with immunosuppressant properties
Mechanism of Action	• Binds to a T-cell binding protein and prevents synthesis of interleukin-2 and other lymphokines essential to T-lymphocyte function
Indications	• Prophylaxis of graft rejection in liver and kidney allogeneic transplantation • It is recommended that tacrolimus be used concomitantly with adrenal corticosteroids
Contraindications	• Hypersensitivity to tacrolimus • Hypersensitivity to HCO-60 (polyoxyl 60 hydrogenated castor oil) with IV formulation
Warnings	• Increased incidence of posttransplant diabetes mellitus and insulin use at 24 months in kidney transplant recipients • Neurotoxicity • Nephrotoxicity • Hyperkalemia • Increased risk of infection and lymphomas • Patients should be monitored closely for at least the first 30 minutes of therapy for signs of anaphylactic reactions
Precautions	• Hypertension is a common occurrence with tacrolimus and may require treatment with antihypertensive agents. Since tacrolimus may cause hyperkalemia, potassium-sparing diuretics should be avoided • Hyperglycemia may occur and require treatment • Lower doses should be used in patients with renal insufficiencies • Patients with hepatic impairment may have a higher risk of developing renal insufficiency • Patients should be informed of the need for regular laboratory monitoring • Pregnancy Category C
Adverse Events	• Tremor • Parathesia • Headache • Hypomagnesemia • Diarrhea • Hypertension • Nausea • Renal dysfunction
Drug Interactions	*Drugs with Synergistic Nephrotoxicity* • Gentamicin • Amphotericin B • Tobramycin • Ketoconazole • Vancomycin • Melphalan • TMP-SMX • Diclofenac • Cimetidine • Azapropazon • Ranitidine

Continued

Appendix I

Tacrolimus (FK-506) (cont'd)

Drug Interactions (cont'd)	*Drugs that Increase Tacrolimus Levels* • Diltiazem • Clarithromycin • Nicardipine • Cimetidine • Verapamil • Danazol • Ketoconazole • Bromocriptine • Fluconazole • Methylprednisolone • Itraconazole • Metoclopramide • Erythromycin • Cyclosporine • Nifedipine • Clotrimazole • Ketoconazole • Troleandomycin • Ethinyl estradiol • Omeprazole protease inhibitors • Nefazodone • Drugs that inhibit cytochrome P450 3A4 • Grapefruit and grapefruit juice can increase tacrolimus levels. *Drugs that Decrease Tacrolimus Levels* • Rifampin • Carbamazepine • Rifabutin • Phenobarbitol • Phenytoin • St. John's Wort • Drugs that induce cytochrome P450 3A4 *Other* • Vaccinations—live vaccinations should be avoided
Formulations	• 1 mg capsules supplied in 100 count bottles • 5 mg capsules supplied in 100 count bottles • 1 mL ampules containing the equivalent of 5 mg of anhydrous tacrolimus per mL—supplied in boxes of 10 ampules
Dosage and Administration	*IV Infusion* • 0.03 to 0.05 mg/kg/d to 0.10 mg/kg/d as a continuous infusion • Patients should be converted to oral therapy as soon as can be tolerated *Capsules* • Liver: 0.10 to 0.15 mg/kg/d • Kidney: 0.2 mg/kg/d • First dose should be given 8 to 12 hours after discontinuing IV infusion

Editors' Notes:

Tacrolimus appears to be a more effective drug than cyclosporine for liver transplantation. Rescue for refractory liver and kidney rejection has also been shown to be an important use for the drug. A double-blind study comparing Sandimmune and FK506 in cadaveric renal transplantation demonstrated superior efficacy of FK506 for rejection prophylaxis. A higher incidence of PTDM was noted, but was reversible in 50% of patients at 2 years. The 5-year data from this study was recently published. Treatment failure was significantly less frequent in tacrolimus-treated recipients (43.8% vs. 56.8%; p=0.008). With cross-over due to rejection counted as graft failure, graft survival was significantly better with tacrolimus (63.8% vs. 53.8%; p=0.014). Hypertension and hyperlipidemia were less with tacrolimus. Nearly 50% of patients who required insulin with tacrolimus were able to discontinue insulin. (Vincenti F, Jensik SC, Filo RS et al. A long-term comparison of tacrolimus (FK-506) and cyclosporine in kidney transplantation: evidence for improved allograft survival at five years. ©2002 Transplantation. Lippincott, Williams and Wilkins)

Avoid taking with meals, antacids and divalent cation supplements.

A

ABO 1, 29, 99, 114, 118, 120, 133,
246, 248, 268, 283, 285, 301, 311,
322, 323, 569, 574-576, 578, 580
Accelerated rejection 2, 27, 29
Acute liver failure 205-209, 211, 213,
217, 224, 243
Acute rejection 1, 2, 25, 28, 29, 32, 55,
57, 59, 61, 62, 118, 119, 132-135,
137, 139, 142, 143, 145, 146, 153,
169, 171, 173, 175, 176, 238, 273,
294, 297, 298, 301-303, 327-329,
339, 340, 382, 383, 426, 437, 438,
531, 539, 541-544, 546-549, 554
Acyclovir 252, 296, 297, 318, 392,
393, 401, 418, 419, 420-422, 445
Alanine aminotransferase (see ALT)
ALF (see Acute liver failure)
Alports syndrome 137, 443
ALT 95, 96, 206
Amphotericin B 247, 393, 297, 401,
412-416, 432
Anti-GBM disease 137, 443
Antigen processing 7, 8
Antithymocyte globulin 56, 61, 139,
143, 273, 297, 298, 302, 315, 468
Aspartate aminotransferase (see AST)
Aspergillus 145, 188, 297, 303, 401,
412, 413
AST 95, 96, 206
ATGAM® 44, 56, 139, 252, 257, 298,
302, 304, 468, 469
Avascular necrosis of the hip 446
Azathioprine 44, 48, 51, 55, 58, 59,
61, 62, 139-142, 174, 175, 189,
198, 232, 272-274, 298, 314, 315,
339, 351, 431, 443, 468, 469, 471,
515, 543, 544, 552

B

B cell antigen receptor 9
BAL 209
Biliary leaks 222, 224, 228, 229, 233,
234, 342
BK virus (BKV) 136, 422

Bladder-drained pancreas transplantation 171
Bleeding 111, 128, 129, 132, 171, 173,
195, 219, 220, 224, 225, 228, 241,
252, 257, 270, 350, 521, 562, 565,
578
BOS 297, 303-305
Brain death 46, 90, 93, 94, 98, 105,
121, 123, 206, 267, 283, 285, 349,
520
Bronchial anastomosis 289, 291, 292,
298, 299, 350
Bronchiolitis obliterans syndrome
(see BOS)

C

CAD (coronary artery disease) 32,
111-113, 116, 117, 123, 157-159,
161, 182, 187, 188, 261, 268, 276,
277, 279, 283, 316, 318, 356, 370,
374, 375, 441, 432, 439, 529, 531
Calcineurin inhibitors 57, 61, 115,
132-135, 139, 140, 141, 143, 174,
175, 177, 185, 232, 233, 236, 238,
239, 409, 429, 442, 443, 484
Candida spp. 411, 412
Cardiac allograft vasculopathy (see
CAV)
Cardiomyopathy 112, 262, 309-311,
318, 365, 366, 374, 376, 472
Cardiopulmonary bypass 265, 269,
270, 287, 289, 295, 314, 347, 350,
351, 365, 367, 369-373, 556, 576
Cataracts 143, 239, 321, 433, 555
CAV 276-279
CD2 15, 21, 23, 480, 584
CD25 57, 59, 60, 483, 484, 492, 493,
503, 510, 511
CD28 13-15, 23, 480, 484, 584
CD4 3, 4, 6, 8, 10, 11, 13, 15, 18, 19,
21, 22, 27, 29, 31, 479-485,
489-498, 503, 510, 511, 583, 584
CD58 23
CD8 3-6, 8, 10, 11, 13, 15, 18-20, 22,
27, 29, 31, 482, 485, 490-492,
496, 498, 499, 510, 583, 584

Index

Vademecum
More Handbooks in this Series

Five ways to order:

- at our website www.landesbioscience.com
- by email: orders@landesbioscience.com
- by fax: 512.863.0081
- by mail: Landes Bioscience, 810 S. Church Street, Georgetown TX 78626
- by phone: 1.800.736.9948

Chemotherapy Regimens and Cancer Care
Alan D. Langerak, Luke P. Driesbach
Video-Assisted Thoracic Surgery
Todd L. Demmy
Cardiothoracic Surgery 2ed
Fritz J. Baumgartner
Hemostasis and Thrombosis
Thomas G. DeLoughery
Perioperative Care in Cardiac Anesthesia and Surgery
Davy C.H. Cheng, Tirone E. David
Clinical Transfusion Medicine
Joseph Sweeney, Yvonne Rizk
Vascular Disease 2ed
Darwin Eton
Bone Marrow Transplantation
Richard K. Burt, Scott T. Lothian, H. Joachim Deeg, George W. Santos
Organ Transplantation
Frank P. Stuart, Michael M. Abecassis, Dixon B. Kaufman
Transplantation Drug Manual, 3ed
John Pirsch, William Simmons, Hans Sollinger
Organ Procurement and Preservation
Goran B. Klintmalm, Marlon F. Levy
Aortic Surgery
Jeffrey Ballard
Surgical Oncology
David N. Krag
Cardiac Pacing
Charles J. Love
Breast Diseases
Patrick I. Borgen, Arnold D.K. Hill
Microsurgery
Konstantinos Malizos
Tropical Neurology
U.K. Misra, J. Kalita, R.A. Shakir
Musculoskeletal Procedures: Therapeutic and Diagnostic
Jacqueline Hodge

Access for Dialysis: Surgical and Radiologic Procedures, 2ed
Ingemar J.A. Davidson
Dermatopathology
Ramon Sanchez, Sharon Raimer
Extracorporeal Life Support
Dan M. Meyer
Endosurgery for Cancer
Steve Eubanks, Ricardo Cohen, Riad Younes, Fredrick Brody
Kidney-Pancreas Transplantation 2ed
Ingemar J.A. Davidson
Radioguided Surgery
Eric D. Whitman, Douglas Reintgen
Clinical Immunology
Reginald Gorczynski, Jacqueline Stanley
Mechanical Circulatory Support
Wayne E. Richenbacher
Tropical Dermatology
Roberto Arenas, Roberto Estrada
Pediatric Surgery
Robert Arensman, Daniel Bambini, P. Stephen Almond
Endocrine Surgery
Richard A. Prinz, Edgar D. Staren
Trauma Management
D. Demetriades, Juan A. Asensio
Surgical Arithmetic
Lawrence Rosenberg, Lawrence Joseph, Alan Barkin
Burn Care
Steven Wolf, David Herndon
Hepatobiliary Surgery
Leslie H. Blumgart, Ronald Chamberlain
Liver Transplantation
Michael R. Lucey, James Neuberger
Pediatric Laparoscopy
Thom E Lobe
Vital Signs and Resuscitation
Joseph P. Stewart

Titles Coming Soon:

The Biology and Practice of Current Nutritional Support, 2ed
Rifat Latifi, Stanley Dudrick
Tropical Surgery
Glenn W. Geelhoed
Heart and Lung Transplantation
Arthur J. Crumbley III

Ultrasound for Surgeons
Heidi L. Frankel
Urologic Oncology
Daniel A. Nactsheim
GI Endoscopy
Jacques Van Dam, Richard Wong
Hand Surgery
Roger Daley